T0135351

Advances in Intelligent Systems and Computing

Volume 1036

The series "Advances in Intelligent Systems and Computing" contains publications on theory, applications, and design methods of Intelligent Systems and Intelligent Computing. Virtually all disciplines such as engineering, natural sciences, computer and information science, ICT, economics, business, e-commerce, environment, healthcare, life science are covered. The list of topics spans all the areas of modern intelligent systems and computing such as: computational intelligence, soft computing including neural networks, fuzzy systems, evolutionary computing and the fusion of these paradigms, social intelligence, ambient intelligence, computational neuroscience, artificial life, virtual worlds and society, cognitive science and systems, Perception and Vision, DNA and immune based systems, self-organizing and adaptive systems, e-Learning and teaching, human-centered and human-centric computing, recommender systems, intelligent control, robotics and mechatronics including human-machine teaming, knowledge-based paradigms, learning paradigms, machine ethics, intelligent data analysis, knowledge management, intelligent agents, intelligent decision making and support, intelligent network security, trust management, interactive entertainment, Web intelligence and multimedia.

The publications within "Advances in Intelligent Systems and Computing" are primarily proceedings of important conferences, symposia and congresses. They cover significant recent developments in the field, both of a foundational and applicable character. An important characteristic feature of the series is the short publication time and world-wide distribution. This permits a rapid and broad dissemination of research results.

** Indexing: The books of this series are submitted to ISI Proceedings, EI-Compendex, DBLP, SCOPUS, Google Scholar and Springerlink **

More information about this series at http://www.springer.com/series/11156

Leonard Barolli · Hiroaki Nishino ·
Tomoya Enokido · Makoto Takizawa
Editors

Advances in Networked-based Information Systems

The 22nd International Conference
on Network-Based Information Systems
(NBiS-2019)

 Springer

Editors
Leonard Barolli
Department of Information
and Communication Engineering,
Faculty of Information Engineering
Fukuoka Institute of Technology
Fukuoka, Japan

Hiroaki Nishino
Division of Computer Science
and Intelligent Systems,
Faculty of Science and Technology
Oita University
Oita, Japan

Tomoya Enokido
Faculty of Business Administration
Rissho University
Tokyo, Japan

Makoto Takizawa
Department of Advanced Sciences,
Faculty of Science and Engineering
Hosei University
Tokyo, Japan

ISSN 2194-5357   ISSN 2194-5365   (electronic)
Advances in Intelligent Systems and Computing
ISBN 978-3-030-29028-3   ISBN 978-3-030-29029-0   (eBook)
https://doi.org/10.1007/978-3-030-29029-0

This Springer imprint is published by the registered company Springer Nature Switzerland AG
The registered company address is: Gewerbestrasse 11, 6330 Cham, Switzerland

Welcome Message from NBiS
Steering Committee Co-chairs

Welcome to the 22nd International Conference on Network-Based Information Systems (NBiS-2019), which will be held at Oita University, Japan from September 5 to September 7, 2019.

The main objective of NBiS is to bring together scientists, engineers, and researchers from both network systems and information systems with the aim of encouraging the exchange of ideas, opinions, and experiences between these two communities.

NBiS started as a workshop and was held for 12 years together with DEXA International Conference as one of the oldest among DEXA Workshops. The workshop was very successful and in 2009 edition the NBiS was held at IUPUI, Indianapolis, the USA as an independent International Conference supported by many international volunteers. In following years, the NBiSs was held in Takayama, Gifu, Japan (2010), Tirana, Albania (2011) Melbourne, Australia (2012), Gwangju, Korea (2013), Salerno, Italy (2014), Taipei, Taiwan (2015), Ostrava, Czech Republic (2016), Toronto, Canada (2017), and Bratislava, Slovakia (2018).

In this edition of NBiS, many papers were submitted from all over the world. They were carefully reviewed and only high-quality papers will be presented during conference days. In conjunction with NBiS-2019 are held also eight International Workshops where specific and hot topics are deeply discussed.

Many volunteer people have kindly helped us to prepare and organize NBiS-2019. First of all, we would like to thank General Co-Chairs, Program Co-Chairs, and Workshops Co-Chairs for their great efforts to make NBiS-2019 a very successful event. We would like to thank all NBiS-2019 Organizing Committee Members, Program Committee Members, and other volunteers for their great help and support. We have special thanks also to Finance Chair and Web Administrator Co-Chairs. Finally, we thank the Local Organization Team at Oita University, Japan, for their good arrangements.

We do hope that you will enjoy the conference and have a good time in Oita, Japan.

NBiS Steering Committee Co-chairs

Leonard Barolli   Fukuoka Institute of Technology (FIT), Japan
Makoto Takizawa   Hosei University, Japan

Welcome Message from NBiS-2019 General Co-chairs

We would like to welcome you to the 22nd International Conference on Network-Based Information Systems (NBiS-2019), which will be held at Oita University, Japan, from September 5 to September 7, 2019.

It is our honor to chair this prestigious conference, as one of the important conferences in the field. Extensive international participation, coupled with rigorous peer reviews, has made this an exceptional technical conference. The technical program and workshops add important dimensions to this event. We hope that you will enjoy each and every component of this event and benefit from interactions with other attendees.

Since its inception, NBiS has attempted to bring together people interested in information and networking, in areas that range from the theoretical aspects to the practical design of new network systems, distributed systems, multimedia systems, Internet/Web technologies, mobile computing, intelligent computing, pervasive/ubiquitous networks, dependable systems, semantic services, Grid, P2P, and scalable computing. For NBiS-2019, we have continued these efforts as novel networking concepts emerge and new applications flourish. NBiS-2019 consists of the main conference and eight workshops. We received 120 papers, and out of them, 30 were accepted (25% acceptance ratio), which will be presented during the conference days.

The organization of an international conference requires the support and help of many people. A lot of people have helped and worked hard for a successful NBiS-2019 technical program and conference proceedings. First, we would like to thank all the authors for submitting their papers. We are indebted to Track Co-Chairs, Program Committee members, and reviewers who carried out the most difficult work of carefully evaluating the submitted papers.

We would like to give our special thanks to Prof. Leonard Barolli and Prof. Makoto Takizawa the chairs of the Steering Committee for giving us the opportunity to hold this conference and for their guidance on organizing the conference. We would like to thank Program Co-Chairs and Workshops Co-Chairs for their excellent work. We would like to express our great appreciation to our keynote speakers for accepting our invitation as keynote speakers of NBiS-2019.

We would like to thank the Local Arrangements Chairs and Local Organization Team at Oita University, Japan, for making good local arrangements for the conference.

We hope that you have an enjoyable and productive time during this conference and have a great time in Oita, Japan.

NBiS-2019 General Co-chairs

Hiroaki Nishino   Oita University, Japan
Kin Fun Li   University of Victoria, Canada
Chuan-Yu Chang   National Yunlin University of Science and Technology, Taiwan

Welcome Message from NBiS-2019 Program Committee Co-chairs

Welcome to the 22nd International Conference on Network-Based Information Systems (NBiS-2019), which will be held at Oita University, Japan, from September 5 to September 7, 2019.

The purpose of NBiS conference is to bring together developers and researchers to share ideas and research work in the emerging areas of network and information systems.

The contributions included in the proceedings of NBiS-2019 cover all aspects of theory, design, and application of computer networks and information systems. There are many topics of information networking such as cloud computing, wireless sensor networks, ad hoc networks, peer-to-peer systems, grid computing, social networking, multimedia systems and applications, security, distributed and parallel systems, and mobile computing.

In this edition, 120 submissions were received from all over the world. Each submitted paper was peer-reviewed by Program Committee members and external reviewers who are experts in the subject area of the paper. Then, the Program Committee accepted 30 papers (25% acceptance ratio). In the conference, program are also included the distinguished keynote addresses.

The organization of an International Conference requires the support and help of many people. First, we would like to thank all authors for submitting their papers. We would like to thank all Track Chairs and Program Committee Members, who carried out the most difficult work of carefully evaluating the submitted papers. We also would like to thank NBiS-2019 Workshops Chairs for organizing excellent workshops with NBiS-2019 conference.

We would like to give special thanks to Prof. Leonard Barolli and Prof. Makoto Takizawa, the Chairs of the Steering Committee of NBiS for their strong encouragement, guidance, insights, and for their effective coordination of conference organization. We would like to greatly thank General Co-Chairs for their great support and invaluable suggestions to make the conference a very successful event.

We hope you will enjoy the conference and have a great time in Oita, Japan.

NBiS-2019 Program Committee Co-chairs

Tomoya Enokido	Rissho University, Japan
Marek R. Ogiela	AGH University of Science and Technology, Poland
Isaac Woungang	Ryerson University, Canada

Welcome Message from NBiS-2019 Workshops Co-chairs

Welcome to the NBiS-2019 Workshops to be held in conjunction with the 22nd International Conference on Network-Based Information Systems (NBiS-2019) at Oita University, Japan, from September 5 to September 7, 2019.

The goal of NBiS-2019 workshops is to provide a forum for international researchers and practitioners to exchange and share their new ideas, research results, and ongoing work on leading-edge topics in the different fields of network-based information systems and their applications. Some of the accepted workshops deal with topics that open up perspectives beyond the ordinary, thus enriching the topics usually addressed by the NBiS-2019 conference.

For this edition, the following workshops will be held with NBiS-2019.

1. The 14th International Workshop on Network-based Virtual Reality and Tele-existence (INVITE-2019)
2. The 13th International Workshop on Advanced Distributed and Parallel Network Applications (ADPNA-2019)
3. The 10th International Workshop on Heterogeneous Networking Environments and Technologies (HETNET-2019)
4. The 10th International Workshop on Intelligent Sensors and Smart Environments (ISSE-2019)
5. The 10th International Workshop on Trustworthy Computing and Security (TwCSec-2019)
6. The 9th International Workshop on Information Networking and Wireless Communications (INWC-2019)
7. The 8th International Workshop on Advances in Data Engineering and Mobile Computing (DEMoC-2019)
8. The 8th International Workshop on Web Services and Social Media (WSSM-2019)

We would like to thank the research community for their great response to NBiS-2019 workshops. The excellent technical program of the workshops was the result of professional work from workshop chairs, workshop program committees, reviewers, and authors.

We would like to give our special thanks to Prof. Leonard Barolli and Prof. Makoto Takizawa, the Steering Committee Chairs of NBiS conference for their strong encouragement and guidance to organize the NBiS-2019 workshops. We would like to thank NBiS-2019 General Co-Chairs and Program Co-Chairs for their advices to make the possible organization of NBiS-2019 workshops.

We would like to express special thanks to NBiS-2019 Web Administration Chairs for their timely unlimited support for managing the conference systems. We also would like to thank NBiS-2019 Local Arrangement Co-Chairs and Local Organization Team at Oita University, Japan, for making the local arrangement for the conference and workshops.

We wish all of you an entertaining and rewarding experience in all workshops and NBiS-2019 conference.

NBiS-2019 Workshops Co-chairs

Shunsuke Okamoto   Seikei University, Japan
Shih-Hao Chang   Tamkang University, Taiwan
Hiroaki Kikuchi   Meiji University, Japan

NBiS-2019 Organizing Committee

Honorary Chair

Seigo Kitano   Oita University, Japan

General Co-chairs

Hiroaki Nishino   Oita University, Japan
Kin Fun Li   University of Victoria, Canada
Chuan-Yu Chang   National Yunlin University of Science
          and Technology, Taiwan

PC Co-chairs

Tomoya Enokido   Rissho University, Japan
Marek R. Ogiela   AGH University of Science and Technology,
          Poland
Isaac Woungang   Ryerson University, Canada

Workshop Co-chairs

Shunsuke Okamoto   Seikei University, Japan
Shih-Hao Chang   Tamkang University, Taiwan
Hiroaki Kikuchi   Meiji University, Japan

Award Co-chairs

Minoru Uehara   Toyo University, Japan
David Taniar   Monash University, Australia
Fumiaki Sato   Toho University, Japan

Publicity Co-chairs

Akio Koyama	Yamagata University, Japan
Markus Aleksy	ABB AG, Germany
Farookh Hussain	University Technology of Sydney, Australia

International Liaison Co-chairs

Wenny Rahayu	La Trobe University, Australia
Lidia Ogiela	Pedagogical University of Cracow, Poland
Arjan Durresi	IUPUI, USA
Manuel Moreira da Silva	CEOS. PP, Politécnico do Porto, Portugal

Local Arrangement Co-chairs

Ken'ichi Furuya	Oita University, Japan
Makoto Nakashima	Oita University, Japan

Finance Chair

Makoto Ikeda	Fukuoka Institute of Technology, Japan

Web Administrator Co-chairs

Miralda Cuka	Fukuoka Institute of Technology, Japan
Kevin Bylykbashi	Fukuoka Institute of Technology, Japan
Donald Elmazi	Fukuoka Institute of Technology, Japan

Steering Committee

Leonard Barolli	Fukuoka Institute of Technology, Japan
Makoto Takizawa	Hosei University, Japan

Track Areas and PC Members

Track 1: Mobile and Wireless Networks

Track Co-chairs

Tetsuya Shigeyasu	Prefectural University of Hiroshima, Japan
Vamsi Krishna Paruchuri	University of Central Arkansas, USA
Evjola Spaho	Polytechnic University of Tirana, Albania

PC Members

Nobuyoshi Sato	Iwate Prefectural University, Japan
Kanunori Ueda	Kochi University of Technology, Japan
Masaaki Yamanaka	Japan Coast Guard Academy, Japan

Takuya Yoshihiro	Wakayama University, Japan
Tomoya Kawakami	Nara Institute of Science and Technology, Japan
Masaaki Noro	Fujitsu Laboratory, Japan
Admir Barolli	Aleksander Moisiu University of Durresi, Albania
Makoto Ikeda	Fukuoka Institute of Technology, Japan
Keita Matsuo	Fukuoka Institute of Technology, Japan
Elis Kulla	Okayama University of Science, Japan
Noriki Uchida	Fukuoka Institute of Technology, Japan
Arjan Durresi	IUPUI, USA
Sriram Chellappan	University of South Florida (USF), USA

Track 2: Internet of Things and Big Data

Track Co-chairs

Nik Bessis	Edge Hill University, UK
Chun-Wei Tsai	National Ilan University, Taiwan
Patrick Hung	University of Ontario Institute of Technology, Canada

PC Members

Ella Perreira	Edge Hill University, UK
Sergio Toral	University of Seville, Spain
Stelios Sotiriadis	Birkbeck, University of London, UK
Eleana Asimakopoulou	Hellenic National Defence College, Greece
Xiaolong Xu	University of Posts and Telecommunications, China
Kevin Curran	Ulster University, UK
Kamen Kanev	Shizuoka University, Japan
Shih-Chia Huang	National Taipei University of Technology, Taiwan
Jorge Roa	UTN Santa Fe, Argentina
Alvaro Joffre Uribe	Universidad Militar Nueva Granada, Colombia
Marcelo Fantinato	University of Sao Paulo, Brazil
Marco Zennaro	Wireless and T/ICT4D Laboratory, Italy
Priyanka Rawat	University of Avignon, France
Francesco Piccialli	University of Naples Federico II, Italy
Chi-Yuan Chen	National Ilan University, Taiwan

Track 3: Cloud, Grid, and Service-Oriented Computing

Track Co-chairs

Ciprian Dobre	Polytechnic University of Bucharest, Romania
Olivier Terzo	Istituto Superiore Mario Boella (ISMB), Italy
Muhammad Younas	Oxford Brookes University, UK

PC Members

Zia Rehman	COMSATS University Islamabad, Pakistan
Walayat Hussain	University of Technology Sydney, Australia
Farookh Hussain	University of Technology Sydney, Australia
Adil Hammadi	Sultan Qaboos University, Oman
Rui Pais	University of Stavanger, Norway
Raymond Hansen	Purdue University, USA
Antorweep Chakravorty	University of Stavanger, Norway
Rui Esteves	National Oilwell Varco, Norway
Constandinos X. Mavromoustakis	University of Nicosia, Cyprus
Ioan Salomie	Technical University of Cluj-Napoca, Romania, Romania
George Mastorakis	Technological Educational Institute of Crete, Greece
Sergio L. Toral Marín	University of Seville, Spain
Marc Frincu	West University of Timisoara, Romania
Alexandru Costan	IRISA/INSA Rennes, France
Xiaomin Zhu	National University of Defense Technology, China
Radu Tudoran	Huawei, Munich, Germany
Mauro Migliardi	University of Padua, Italy
Harold Castro	Universidad de Los Andes, Colombia
Andrea Tosatto	Open-Xchange, Germany
Rodrigo Calheiros	Western Sydney University, Australia

Track 4: Multimedia and Web Applications

Track Co-chairs

Takahiro Uchiya	Nagoya Institute of Technology, Japan
Tomoyuki Ishida	Fukuoka Institute of Technology, Japan
Nobuo Funabiki	Okayama University, Japan

PC Members

Shigeru Fujita	Chiba Institute of Technology, Japan
Yuka Kato	Tokyo Woman's Christian University, Japan
Yoshiaki Kasahara	Kyushu University, Japan
Rihito Yaegashi	Kagawa University, Japan
Kazunori Ueda	Kochi University of Technology, Japan
Ryota Nishimura	Keio University, Japan
Shohei Kato	Nagoya Institute of Technology, Japan
Shinsuke Kajioka	Nagoya Institute of Technology, Japan
Atsuko Muto	Nagoya Institute of Technology, Japan
Kaoru Sugita	Fukuoka Institute of Technology, Japan

Noriyasu Yamamoto	Fukuoka Institute of Technology, Japan
Hiroaki Nishino	Oita University, Japan

Track 5: Ubiquitous and Pervasive Computing

Track Co-chairs
Chi-Yi Lin	Tamkang University, Taiwan
Joan Arnedo Moreno	Open University of Catalonia, Spain
Elis Kulla	Okayama University of Science, Japan

PC Members
Jichiang Tsai	National Chung Hsing University, Taiwan
Chang Hong Lin	National Taiwan University of Science and Technology, Taiwan
Meng-Shiuan Pan	Tamkang University, Taiwan
Chien-Fu Cheng	Tamkang University, Taiwan
Ang Chen	University of Pennsylvania, USA
Santi Caballe	Open University of Catalonia, Spain
Evjola Spaho	Polytechnic University of Tirana, Albania
Makoto Ikeda	Fukuoka Institute of Technology, Japan

Track 6: Network Security and Privacy

Track Co-chairs
Takamichi Saito	Meiji University, Japan
Sriram Chellappan	University of South Florida, USA
Feilong Tang	Shanghai Jiao Tong University, China

PC Members
Satomi Saito	Fujitsu Laboratories, Japan
Kazumasa Omote	University of Tsukuba, Japan
Koji Chida	NTT, Japan
Hiroki Hada	NTT Security (Japan) KK, Japan
Hirofumi Nakakouji	Hitachi, Ltd., Japan
Na Ruan	Shanghai Jiaotong University, China
Chunhua Su	Osaka University, China
Kazumasa Omote	University of Tsukuba, Japan
Toshihiro Yamauchi	Okayama University, Japan
Masakazu Soshi	Hiroshima City University, Japan
Bagus Santoso	The University of Electro-Communications, Japan
Laiping Zhao	Tianjin University, China
Jingyu Hua	Nanjing University, China
Xiaobo Zhou	Tianjin University, China

Yuan Zhan   Nanjing University, China
Yizhi Ren   Hangzhou Dianzi University, China
Arjan Durresi   IUPUI, USA
Vamsi Krishna Paruchuri   University of Central Arkansas, USA

Track 7: Database, Data Mining, and Semantic Computing

Track Co-chairs
Wendy K. Osborn   University of Lethbridge, Canada
Eric Pardade   La Trobe University, Australia
Akimitsu Kanzaki   Shimane University, Japan

PC Members
Asm Kayes   La Trobe University, Australia
Ronaldo dos Santos Mello   Universidade Federal de Santa Catarina, Brazil
Saqib Ali   Sultan Qaboos University, Oman
Hong Quang Nguyen   Ho Chi Minh City International University,
                                    Vietnam
Irena Holubova   Charles University Prague, Czech Republic
Prakash Veeraraghavan   La Trobe University, Australia
Carson Leung   University of Manitoba, Canada
Marwan Hassani   Aachen University, Germany
Tomoki Yoshihisa   Osaka University, Japan
Tomoya Kawakami   NAIST, Japan
Atsushi Takeda   Tohoku Gakuin University, Japan
Yoshiaki Terashima   Soka University, Japan
Yuuichi Teranishi   NICT, Japan
Jackie Rice   University of Lethbridge, Canada
Yllias Chali   University of Lethbridge, Canada
John Zhang   University of Lethbridge, Canada

Track 8: Network Protocols and Applications

Track Co-chairs
Irfan Awan   University of Bradford, UK
Sanjay Kuamr Dhurandher   NSIT, University of Delhi, India
Hsing-Chung Chen   Asia University, Taiwan

PC Members
Samia Loucif   ALHOSN University, UAE
Abdelhamid Mammeri   Ottawa University, Ontario, Canada
Jun He   University of New Brunswick, Canada
Peyman Kabiri   University of Science and Technology, Iran
Chen Chen   University of Texas, USA

Ahmed Abdelgawad	Central Michigan University, USA
Wael Elmedany	University of Bahrain, Bahrain
Behrouz Maham	School of Electrical and Computer Engineering, Iran
Rubem Pereira	Liverpool John Moores University, UK
Carlos Juiz	University of the Balearic Islands, Spain
Faheem Ahmed	Thompson Rivers University, Canada
Paulo Gil	FCT-UNL, Portugal
Michael Mcguire	University of Victoria, Canada
Steven Guan	Xi'an Jiaotong-Liverpool University, China
Gregorio Romero	Universidad Politecnica de Madrid, Spain
Amita Malik	Deenbandhu Chhotu Ram University of Science and Technology, India
Mayank Dave	NIT, Kurukshetra, India
Vinesh Kumar	University of Delhi, India
R. K. Pateriya	MANIT, Bhopal, India
Himanshu Aggarwal	Punjabi University, India
Neng-Yih Shih	Asia University, Taiwan
Yeong-Chin Chen	Asia University, Taiwan
Hsi-Chin Hsin	National United University, Taiwan
Ming-Shiang Huang	Asia University, Taiwan
Chia-Cheng Liu	Asia University, Taiwan
Chia-Hsin Cheng	National Formosa University, Yunlin County, Taiwan
Tzu-Liang Kung	Asia University, Taiwan
Gene Shen	Asia University, Taiwan
Jim-Min Lin	Feng Chia University, Taiwan
Chia-Cheng Liu	Asia University, Taiwan
Yen-Ching Chang	Chung Shan Medical University, Taiwan
Shu-Hong Lee	Chienkuo Technology University, Taiwan
Ho-Lung Hung	Chienkuo Technology University, Taiwan
Gwo-Ruey Lee	Lung-Yuan Research Park, Taiwan
Li-Shan Ma	Chienkuo Technology University, Taiwan
Chung-Wen Hung	National Yunlin University of Science and Technology University, Taiwan
Yung-Chen Chou	Asia University, Taiwan
Chen-Hung Chuang	Asia University, Taiwan
Jing-Doo Wang	Asia University, Taiwan
Jui-Chi Chen	Asia University, Taiwan
Young-Long Chen	National Taichung University of Science and Technology, Taiwan

Track 9: Intelligent and Cognitive Computing

Track Co-chairs

Lidia Ogiela	Pedagogical University of Cracow, Poland
Farookh Hussain	University of Technology Sydney, Australia
Hae-Duck Joshua Jeong	Korean Bible University, Korea

PC Members

Yiyu Yao	University of Regina, Canada
Daqi Dong	University of Memphis, USA
Jan Platoš	VŠB Technical University of Ostrava, Czech Republic
Pavel Krömer	VŠB Technical University of Ostrava, Czech Republic
Urszula Ogiela	AGH University of Science and Technology, Poland
Jana Nowaková	VŠB Technical University of Ostrava, Czech Republic
Hoon Ko	Chosun University, Korea.
Chang Choi	Chosun University, Korea
Gangman Yi	Gangneung-Wonju National University, Korea
Wooseok Hyun	Korean Bible University, Korea
Hsing-Chung Jack Chen	Asia University, Taiwan
Jong-Suk Ruth Lee	KISTI, Korea
Hyun Jung Lee	Yonsei University, Korea
Ji-Young Lim	Korean Bible University, Korea
Omar Hussain	UNSW Canberra, Australia
Saqib Ali	Sultan Qaboos University, Oman
Morteza Saberi	UNSW Canberra, Australia
Sazia Parvin	UNSW Canberra, Australia
Walayat Hussain	University of Technology Sydney, Australia

Track 10: Parallel and Distributed Computing

Track Co-chairs

Kin Fun Li	University of Victoria, Canada
Bhed Bista	Iwate Prefectural University, Japan
Giovanni Cozzolino	University of Naples Federico II, Italy

PC Members

Deepali Arora	University of Victoria, Canada
Kosuke Takano	Kanagawa Institute of Technology, Japan
Masahiro Ito	Toshiba Lab, Japan
Watheq ElKharashi	Ain Shams University, Egypt
Martine Wedlake	IBM, USA

Jiahong Wang	Iwate Prefectural University, Japan
Shigetomo Kimura	University of Tsukuba, Japan
Chotipat Pornavalai	King Mongkut's Institute of Technology Ladkrabang, Thailand
Danda B. Rawat	Howard University, USA
Gongjun Yan	University of Southern Indiana, USA
Shu-Shaw Wang	Dallas Baptist University, USA
Naonobu Okazaki	Miyazaki University, Japan
Yoshiaki Terashima	Soka University, Japan
Atsushi Takeda	Tohoku Gakuin University, Japan
Tomoki Yoshihisa	Osaka University, Japan
Akira Kanaoka	Toho University, Japan
Flora Amato	University of Naples Federico II, Italy
Vincenzo Moscato	University of Naples Federico II, Italy
Walter Balzano	University of Naples Federico II, Italy
Francesco Moscato	University of Campania Luigi Vanvitelli, Italy
Francesco Mercaldo	Consiglio Nazionale delle Ricerche (CNR), Italy
Alessandra Amato	University of Naples Federico II, Italy
Francesco Piccialli	University of Naples Federico II, Italy

NBiS-2019 Reviewers

Ali Khan Zahoor
Barolli Admir
Barolli Leonard
Bista Bhed
Caballé Santi
Chang Chuan-Yu
Chellappan Sriram
Chen Hsing-Chung
Chen Xiaofeng
Cui Baojiang
Di Martino Beniamino
Durresi Arjan
Enokido Tomoya
Ficco Massimo
Fun Li Kin
Funabiki Nobuo
Gotoh Yusuke
Hussain Farookh
Hussain Omar
Javaid Nadeem

Jeong Joshua
Ikeda Makoto
Ishida Tomoyuki
Izu Tetsuya
Kikuchi Hiroaki
Kohana Masaki
Koyama Akio
Kulla Elis
Lee Kyungroul
Matsuo Keita
Na Ruan
Nishigaki Masakatsu
Nishino Hiroaki
Ogiela Lidia
Ogiela Marek
Okada Yoshihiro
Omote Kazumasa
Palmieri Francesco
Paruchuri Vamsi Krishna
Rahayu Wenny

Rawat Danda
Shibata Yoshitaka
Saito Takamichi
Sato Fumiaki
Spaho Evjola
Sugita Kaoru
Takizawa Makoto
Tang Feilong
Taniar David
Terzo Olivier
Uchida Noriki
Uchiya Takahiro
Uehara Minoru
Venticinque Salvatore
Wang Xu An
Woungang Isaac
Xhafa Fatos
Yamaguchi Toshihiro
Yim Kangbin
Younas Muhammad

Welcome Message from INVITE-2019 Workshop Organizers

Welcome to the 14th International Workshop on Network-based Virtual Reality and Tele-existence (INVITE-2019), which will be held in conjunction with the 22nd International Conference on Network-Based Information Systems (NBiS-2019) from September 5 to September 7, 2019, at Oita University, Japan.

The INVITE international workshop serves as a forum for the exchange of information and ideas regarding network-based Virtual Reality (VR) and tele-existence, 2D and 3D computer graphics, computer animation, multimedia, object-oriented approach, Web technology, and e-Learning. All participants of this workshop can share ideas and research work in the emerging areas of network-based VR and tele-existence.

Many people have kindly helped us to prepare and organize the INVITE workshop. First of all, we would like to thank the NBiS-2019 organization committee for their support, guidance, and help for making the workshop. We would like to express our special thanks to all of the INVITE program committee members and reviewers for organizing the workshop and reviewing the submitted papers, respectively.

We hope you will enjoy the workshop and have a great time in Oita, Japan.

INVITE-2019 Workshop Organizers

Yoshitaka Shibata   Iwate Prefectural University, Japan
Tomoyuki Ishida   Fukuoka Institute of Technology, Japan

INVITE-2019 Organizing Committee

Workshop Co-chairs

Yoshitaka Shibata   Iwate Prefectural University, Japan
Tomoyuki Ishida   Fukuoka Institute of Technology, Japan

Program Committee Members

Koji Koyamada   Kyoto University, Japan
Yasuo Ebara   Kyoto University, Japan
Hideo Miyaji   Tokyo City University, Japan
Tetsuro Ogi   Keio University, Japan
Akihiro Miyakawa   Nanao City, Japan
Kaoru Sugita   Fukuoka Institute of Technology, Japan
Noriki Uchida   Fukuoka Institute of Technology, Japan

Welcome Message from ADPNA-2019 Workshop Co-chairs

Welcome to the 13th International Workshop on Advanced Distributed and Parallel Network Applications (ADPNA-2019), which will be held in conjunction with the 22nd International Conference on Network-Based Information Systems (NBiS-2019) from September 5 to September 7, 2019, at Oita University, Japan.

The purpose of this workshop is to bring together developers and researchers to share ideas and research work in the emerging areas of distributed and parallel network applications.

The papers included in ADPNA-2019 cover aspects of theory, design, and application of distributed and parallel systems and applications. ADPNA-2019 contains high-quality research papers. Each submitted paper was peer-reviewed by Program Committee members who are experts in the subject area of the paper.

For organizing a workshop, the support and help of many people are needed. First, we would like to thank all of the authors for submitting their papers. We also appreciate the support from Program Committee members and reviewers who carried out the most difficult work of carefully evaluating the submitted papers.

We would like to thank NBiS-2019 Co-Chairs, Program Co-Chair, and Workshops Co-Chairs, for their support and strong encouragement and guidance in organizing the workshop. We would like to express special thanks to NBiS-2019 Web Administrator Chairs for their timely unlimited support for managing the conference system.

We also would like to thank NBiS-2019 local arrangement members for making the local arrangement for the conference and workshops.

We hope you will enjoy ADPNA-2019 and NBiS-2019 International Conference and have a great time in Oita, Japan.

ADPNA-2019 Workshop Organizers

Makoto Takizawa   Hosei University, Japan
Leonard Barolli   Fukuoka Institute of Technology (FIT), Japan

ADPNA-2019 Workshop PC Chair

Tomoya Enokido   Rissho University, Japan

ADPNA-2019 Organizing Committee

Workshop Organizers

Makoto Takizawa   Hosei University, Japan
Leonard Barolli   Fukuoka Institute of Technology (FIT), Japan

Workshop PC Chair

Tomoya Enokido   Rissho University, Japan

Program Committee

Makoto Ikeda   Fukuoka Institute of Technology, Japan
Elis Kulla   Okayama University of Science, Japan
Akio Koyama   Yamagata University, Japan
Admir Barolli   Aleksander Moisiu University of Durres, Albania
Wenny Rahaya   La Trobe University, Australia
Evjola Spaho   Polytechnic University of Tirana, Albania
Fumiaki Sato   Toho University, Japan
Minoru Uehara   Toyo University, Japan
Fatos Xhafa   Technical University of Catalonia, Spain
David Taniar   Monash University, Australia
Keita Matsuo   Fukuoka Institute of Technology, Japan
Tomoyuki Ishida   Fukuoka Institute of Technology, Japan
Kaoru Sugita   Fukuoka Institute of Technology, Japan

Welcome Message from HETNET-2019 Workshop Organizers

Welcome to the 10th International Workshop on Heterogeneous Networking Environments and Technologies (HETNET-2019), which will be held in conjunction with the 22nd International Conference on Network-Based Information Systems (NBiS-2019) from September 5 to September 7, 2019, at Oita University, Japan.

Due to advances in networking technologies, and especially, in wireless and mobile communication technologies, and devices with different networking interfaces, the networking systems are moving toward the integration of wired and wireless resources leading to Heterogeneous Networking Environments and Technologies (HETNET). Such environments comprise and integrate wireless LANs, multiple MANETs, Cellular Networks, Wireless Sensor Networks, Wireless Mesh Networks, and so on. Besides, other types of networking systems such as Grid and P2P systems are going wireless and mobile adding yet another dimension to the HETNET.

The integration of various networking paradigms into one networking system is requiring new understandings, algorithms and insights, frameworks, middleware and architectures for the effective design, integration, and deployment aiming at achieving secure, programmable, robust, transparent, and ubiquitous HETNETs.

The aim of this workshop is to present innovative researches, methods, and techniques related to HETNETs and their applications.

Many people contributed to the success of HETNET-2019. First, we would like to thank the organizing committee of NBiS-2019 International Conference for giving us the opportunity to organize the workshop. We would like to thank all authors for submitting their research works and for their participation. We are looking forward to meet them again in the forthcoming editions of the workshop.

We hope you will enjoy the workshop and have a great time in Oita, Japan.

HETNET-2019 Workshop Chair

Leonard Barolli   Fukuoka Institute of Technology (FIT), Japan

HETNET-2019 Workshop PC Chair

Jugappong Nawichai   Chiang Mai University, Thailand

HETNET-2019 Organizing Committee

Workshop Chair

Leonard Barolli   Fukuoka Institute of Technology, Japan

Workshop PC Chair

Jugappong Nawichai   Chiang Mai University, Thailand

Program Committee Members

Markus Aleksy	ABB, Germany
Paul Chung	Loughborough University, UK
Ciprian Dobre	University Politehnica of Bucharest, Romania
Makoto Ikeda	Fukuoka Institute of Technology, Japan
Axel Korthaus	Swinburne University of Technology, Australia
Kin Fun Li	University of Victoria, Canada
John Mashford	CSIRO, Melbourne, Australia
Hiroaki Nishino	University of Oita, Japan
Wenny Rahayu	La Trobe University, Australia
Makoto Takizawa	Hosei University, Japan
David Taniar	Monash University, Australia
Runtong Zhang	Beijing Jiaotong University, China
Pruet Boonma	Chiang Mai University, Thailand
Paskorn Champrasert	Chiang Mai University, Thailand
Yuthapong Somchit	Chiang Mai University, Thailand

Welcome Message from ISSE-2019 Workshop Co-chairs

To automatically provide human-centric services, it is desired to build smart environments for such purposes. The background system needs to first detect and collect needed information in the environment via various sensors like video cameras, RFIDs, infrared sensors, motion sensors, and pressure sensors. The information is then processed for understanding the context and intention of the people in the environment. Finally, the services are provided in a timely manner. Sensing technology, embedded systems, wireless communications, computer vision, and intelligent methods are the core to build integrated smart environments. Previous ISSE's were held in Fukuoka, Japan (2010), Barcelona, Spain (2011), Melbourne, Australia (2012), Gwangju, Korea (2013), Salerno, Italy (2014), Taipei, Taiwan (2015), Ostrava, Czech Republic (2016), Toronto, Canada (2017), and Bratislava, Slovakia (2018).

The 10th International Workshop on Intelligent Sensors and Smart Environments (ISSE-2019) will provide a platform for researchers to meet and exchange their thoughts. ISSE-2019 will be held in conjunction with the 22nd International Conference on Network-Based Information Systems (NBiS-2019) at Oita University, Japan, from September 5 to September 7, 2019.

Many people contributed to the CFP and paper review of ISSE-2019. We wish to thank the program committee members for their great effort. We also would like to express our gratitude to Steering Committee Co-Chairs, Program Co-Chairs, and Workshops Co-Chairs of NBiS-2019. Last but not least, we would like to thank and congratulate all the contributing authors for their support and excellent work.

ISSE-2019 Workshop Co-chairs

Chuan-Yu Chang   National Yunlin University of Science
                        and Technology, Taiwan
Leonard Barolli   Fukuoka Institute of Technology (FIT), Japan

ISSE-2019 Organizing Committee

Workshop Co-chairs

Chuan-Yu Chang	National Yunlin University of Science and Technology, Taiwan
Leonard Barolli	Fukuoka Institute of Technology (FIT), Japan

Program Committee Co-chairs

Takahiro Uchiya	Nagoya Institute of Technology, Japan
Chuan-Yu Chang	National Yunlin University of Science and Technology, Taiwan

Program Committee Members

Keita Matsuo	Fukuoka Institute of Technology (FIT), Japan
Evjola Spaho	Polytechnic University of Tirana, Albania
Markus Aleksy	ABB AG Corporate Research Center, Germany
Fatos Xhafa	Technical University of Catalonia, Spain
Makoto Ikeda	Fukuoka Institute of Technology (FIT), Japan
Chien-Cheng Lee	Yuan Ze University, Taiwan
Jose Bravo	Castilla-La Mancha University, Spain
Ching-Lung Chang	National Yunlin University of Science and Technology, Taiwan
Chu-Song Chen	Academia Sinica, Taiwan
Hsu-Yung Cheng	National Central University, Taiwan
Ding-An Chiang	Tamkang University, Taiwan
Wu-Chih Hu	National Penghu University, Taiwan
Lei Jing	University of Aizu, Japan
Wei-Ru Lai	Yuan Ze University, Taiwan
Wen-Ping Lai	Yuan Ze University, Taiwan
Jiann-Shu Lee	National University of Tainan, Taiwan
Kin Fun Li	University of Victoria, Canada
Chi-Yi Lin	Tamkang University, Taiwan
Jiming Liu	Hong Kong Baptist University, Hong Kong
Wei Lu	Keene University, USA
Mihai Sima	University of Victoria, Canada
Wen-Fong Wang	National Yunlin University of Science and Technology, Taiwan
Zhiwen Yu	Northwestern Polytechnical University, China

Welcome Message from TwCSec-2019 Workshop Co-chairs

It is our great pleasure to welcome you to the 10th International Workshop on Trustworthy Computing and Security (TwCSec-2019), which will be held in conjunction with the 22nd International Conference on Network-Based Information Systems (NBiS-2019) at Oita University, Japan, from September 5 to September 7, 2019.

This international workshop is a forum for sharing ideas and research work in the emerging areas of trustworthy computing, security, and privacy. Computers reside at the heart of systems on which people now rely, both in critical national infrastructures and in their homes, cars, and offices. Today, many of these systems are far too vulnerable to cyber attacks that can inhibit their operation, corrupt valuable data, or expose private information. Future systems will include sensors and computers everywhere, exacerbating the attainment of security and privacy. Current security practices largely address current and known threats, but there is a need for research to consider future threats.

We encouraged contributions describing innovative work on security of next-generation operating systems (OS), secure and resilient network protocols, theoretical foundations and mechanisms for privacy, security, trust, human–computer interfaces for security functions, key distribution/management, intrusion detection and response, secure protocol configuration and deployment, and improved ability to certify and analyze system security properties, and integrating hardware and software for security.

Many people contributed to the success of TwCSec-2019. First, we would like to thank the organizing committee of NBiS-2019 International Conference for giving us the opportunity to organize the workshop. We would like to thank all authors of the workshop for submitting their research works and for their participation. We are looking forward to meet them again in the forthcoming editions of the workshop.

Finally, we would like to thank the Local Arrangement Chairs for the local arrangement of the workshop.

We hope you will enjoy the workshop and have a great time in Oita, Japan.

TwCSec-2019 Workshop Co-chairs

Leonard Barolli   Fukuoka Institute of Technology (FIT), Japan
Arjan Durresi   Indiana University–Purdue University
                              Indianapolis (IUPUI), USA
Hiroaki Kikuchi   Meiji University, Japan

TwCSec-2019 Organizing Committee

Workshop Co-chair

Leonard Barolli	Fukuoka Institute of Technology (FIT), Japan
Arjan Durresi	Indiana University–Purdue University Indianapolis (IUPUI), USA
Hiroaki Kikuchi	Meiji University, Japan

Advisory Co-chairs

Makoto Takizawa	Hosei University, Japan
Raj Jain	Washington University in St. Louis, USA

Program Committee Members

Sriram Chellappan	University of South Florida, USA
Koji Chida	NTT, Japan
Qijun Gu	Texas State University–San Marcos, USA
Tesuya Izu	Fujitsu Ltd., Japan
Youki Kadobayashi	Nara Institute of Science and Technology, Japan
Akio Koyama	Yamagata University, Japan
Michiharu Kudo	IBM Japan, Japan
Sanjay Kumar Madria	Missouri University of Science and Technology, USA
Masakatsu Morii	Kobe University, Japan
Masakatsu Nishigaki	Shizuoka University, Japan
Vamsi Paruchuri	University of Central Arkansas, USA
Hiroshi Shigeno	Keio University, Japan
Yuji Suga	Internet Initiative Japan Inc., Japan
Keisuke Takemori	KDDI Co., Japan
Ryuya Uda	Tokyo University of Technology, Japan
Xukai Zou	Indiana University Purdue–University Indianapolis (IUPUI), USA
Wenye Wang	North Carolina State University, USA
Hiroshi Yoshiura	University of Electro-Communications, Japan
Farookh Hussain	University of Technology Sydney, Australia

Welcome Message from INWC-2019 Workshop Organizers

It is our great pleasure to welcome you to the 9th International Workshop on Information Networking and Wireless Communications (INWC-2019), which will be held in conjunction with the 22nd International Conference on Network-Based Information Systems (NBiS-2019) at Oita University, Japan, from September 5 to September 7, 2019.

This international workshop is a forum for sharing ideas and research work in the emerging areas of information networking and wireless communications. Nowadays, information networks are developing very rapidly and evolving into heterogeneous networks. Thus, we have an increasing number of applications and devices. To optimize the communication in these heterogeneous systems, we need to propose and evaluate more complex schemes, algorithms, and protocols for wired and wireless networks.

Wireless communications are characterized by high bit error rates and burst errors, which arise due to interference fading, shadowing, terminal mobility, and so on. Since the traditional design of the algorithms, methods and protocols of the wired Internet did not take wireless networks into account, the performance over wireless networks is largely degraded. Especially, multi-hop communication aggravates the problem of wireless communication even further. To solve these problems, there has been increased interest to propose and design new algorithms and methodologies for wireless communication.

The aim of this workshop is to present the innovative researches, methods, and numerical analysis for wireless communication and wireless networks.

Many people contributed to the success of INWC-2019. First, we would like to thank the organizing committee of NBiS-2019 International Conference for giving us the opportunity to organize the workshop. We would like to thank all the authors of the workshop for submitting their research works and for their participation. We are looking forward to meet them again in the forthcoming editions of the workshop.

Finally, we would like to thank the Local Arrangement Chairs for the local arrangement of the workshop.

We hope you will enjoy the workshop and have a great time in Oita, Japan.

INWC-2019 Workshop Co-chairs

Leonard Barolli	Fukuoka Institute of Technology, Japan
Hiroshi Maeda	Fukuoka Institute of Technology, Japan

INWC-2019 Workshop PC Chair

Makoto Ikeda	Fukuoka Institute of Technology, Japan

INWC-2019 Organizing Committee
Workshop Co-chairs

Leonard Barolli	Fukuoka Institute of Technology, Japan
Hiroshi Maeda	Fukuoka Institute of Technology, Japan

Workshop PC Co-chair

Makoto Ikeda	Fukuoka Institute of Technology, Japan

Program Committee Members

Arjan Durresi	Indiana University–Purdue University Indianapolis (IUPUI), USA
Shinichi Ichitsubo	Kyushu Institute of Technology, Japan
Elis Kulla	Okayama University of Science, Japan
Zhi Qi Meng	Fukuoka University, Japan
Irfan Awan	Bradford University, UK
Tsuyoshi Matsuoka	Kyushu Sangyo University, Japan
Fatos Xhafa	Technical University of Catalonia, Spain
Kiyotaka Fujisaki	Fukuoka Institute of Technology, Japan
Noriki Uchida	Fukuoka Institute of Technology, Japan
Keita Matsuo	Fukuoka Institute of Technology, Japan
Shinji Sakamoto	Seikei University, Japan

Welcome Message from DEMoC-2019 Workshop Chair

Welcome to the 8th International Workshop on Advances in Data Engineering and Mobile Computing (DEMoC-2019) which will be held in conjunction with the 22nd International Conference on Network-Based Information Systems (NBiS-2019) at Oita University, Japan, from September 5 to September 7, 2019.

This international workshop is to bring together practitioners and researchers from both academia and industry in order to have a forum for discussion and technical presentations on the current researches and future research directions related to these hot research areas: data engineering and mobile computing. We encouraged contributions describing innovative work on DEMoC-2019 workshop.

Many people contributed to the success of DEMoC-2019. First, we would like to thank the organizing committee of NBiS-2019 International Conference for giving us the opportunity to organize the workshop. We would like to thank all the authors of the workshop for submitting their research works and for their participation. We are looking forward to meet them again in the forthcoming editions of the workshop.

Finally, we would like to thank the Local Arrangement Chairs for the local arrangement of the workshop.

We hope you will enjoy the workshop and have a great time in Oita, Japan.

DEMoC-2019 Workshop Chair

Yusuke Gotoh   Okayama University, Japan

DEMoC-2019 Organizing Committee

Workshop Organizer

Yusuke Gotoh   Okayama University, Japan

Program Committee

Toshiyuki Amagasa	Tsukuba University, Japan
Akihito Hiromori	Osaka University, Japan
Akiyo Nadamoto	Konan University, Japan
Kenji Ohira	Nara Institute of Science and Technology, Japan
Shusuke Okamoto	Seikei University, Japan
Wenny Rahayu	La Trobe University, Australia
David Taniar	Monash University, Australia
Hideo Taniguchi	Okayama University, Japan
Tsutomu Terada	Kobe University, Japan
Tomoki Yoshihisa	Osaka University, Japan

Welcome Message from WSSM-2019 Workshop Co-chairs

Welcome to Bratislava and the 8th International Workshop on Web Services and Social Media (WSSM-2019), which is held in conjunction with the 22nd International Conference on Network-Based Information Systems (NBiS-2019) at Oita University, Japan from September 5 to September 7, 2019.

Web systems with IoT technology produce big data and they are becoming increasingly important in our society. The AI engine enables us to build novel web services for home and business use. In the next-generation social network services, not only consumer but also IoT will generate contents, which will drive our activity on the Internet.

The International Workshop on Web Services and Social Media (WSSM) encompasses a wide range of topics in the research, design and implementation of Web-related things, especially, social activity on the Internet. The 1st edition of the workshop (WSSM-2012) was held in Melbourne, Australia, the 2nd edition in Gwangju, Korea, the 3rd edition in Salerno, Italy, the 4th edition in Taipei, Taiwan, the 5th edition in Ostrava, Czech Republic, the 6th edition in Toronto, Canada, and the 7th edition in Bratislava, Slovakia. This time, we are honored to hold the 8th edition of the workshop.

This workshop provides an international forum for researchers and participants to share and exchange their experiences, discuss challenges, and present original ideas in all aspect of Web Services and Social Media.

Many people contributed to the success of WSSM-2019. First, we would like to thank the organizing committee of NBiS-2019 International Conference for giving us the opportunity to organize the workshop. We would like to thank our program committee members and all authors for submitting their research works and for their participation.

Finally, we would like to thank the Local Arrangement Chairs of NBiS-2019 conference. We hope you will enjoy WSSM-2019 workshop and NBiS-2019 International Conference and have a great time in Oita, Japan.

WSSM-2019 Workshop Co-chairs

Shusuke Okamoto   Seikei University, Japan
Masaki Kohana   Chuo University, Japan

WSSM-2019 Organizing Committee

Workshop Co-chairs

Shusuke Okamoto   Seikei University, Japan
Masaki Kohana   Chuo University, Japan

Program Chair

Masayuki Ihara   NTT Service Evolution Laboratories, Japan

Program Committee Members

Fatos Xhafa   Technical University of Catalonia, Spain
Hiroki Sakaji   The University of Tokyo, Japan
Jun Iio   Chuo University, Japan
Kin Fun Li   University of Victoria, Canada
Leonard Barolli   Fukuoka Institute of Technology, Japan
Makoto Ikeda   Fukuoka Institute of Technology, Japan
Makoto Takizawa   Hosei University, Japan
Masaru Kamada   Ibaraki University, Japan
Masaru Miyazaki   NHK, Japan
Tatsuhiro Yonekura   Ibaraki University, Japan
Tomoya Enokido   Rissho University, Japan
Yoshihiro Kawano   Tokyo University of Information Sciences, Japan
Yusuke Gotoh   Okayama University, Japan

NBiS-2019 Keynote Talks

3D Graphics Applications for Education and Visualization

Yoshihiro Okada

Kyushu University, Kyushu, Japan

Abstract. In this talk, I will introduce the research activities about 3D graphics applications. We have developed environments for 3D graphics applications for many years. We have proposed a new system for 3D graphics applications called IntelligentBox. I will present the IntelligentBox and several applications, especially for education and visualization. There are many education applications such as collaborative dental training system, Tai Chi-based physical therapy game, and so on. While as visualization applications, we can mention room layout system, Time-tunnel (a visual analytics tool for multi-dimensional data), Treecube (a visualization tool for browsing 3D multimedia data), and so on. I will introduce the development activities of our laboratory for e-learning materials using 3D graphics and VR/AR, such as web-based interactive educational materials for Japanese history and IoT security, and games for medical education.

Secure Resilient Edge Cloud Designed Network

Tarek Saadawi

The City University of New York, City College, New York, USA

Abstract. IoT systems have put forth new requirements in all aspects of their existence: a diverse QoS requirement, resiliency of computing and connectivity, and the scalability to support massive number of end devices in a plethora of envisioned applications. The trustworthy IoT/cyber-physical system (CPS) networking for smart and connected communities will be realized by distributed secure resilient Edge Cloud (EC). This distributed EC system will be a network of geographically distributed EC nodes, brokering between end devices and Backend Cloud (BC) servers. In this talk, I will present three main topics in the secure resilient cloud designed network: (1) resource management in mobile cloud computing; (2) information management in dynamic distributed databases; and (3) biological-inspired intrusion detection system (IDS). A focus in the presentation will be on the biological-inspired IDS.

Container-Leveraged Service Realization Challenges for Cloud-Native Computing

JongWon Kim

Gwangju Institute of Science & Technology (GIST), Gwangju, Korea

Abstract. Cloud-native computing, employing container-based microservices architecture, is accelerating its adoption for agile and scalable service deployment over worldwide multi-cloud infrastructure. In order to transparently enable diversified inter-connections for container-based cloud-native computing, by leveraging SDN/NFV technology, we need to tie distributed IoT things through multi-site edge clouds to hyper-scale core clouds. Thus, in this talk, I first attempt to relate the open-source-driven development for CNI (Container Networking Interface) and CSI (Container Storage Interface) to the required container-enabled cloud-native computing/storage with end-to-end (i.e., IoT–SDN/NFV–Cloud) inter-connections. Then, selected container-leveraged service realization challenges such as multi-tenant/multi-cluster Kubernetes orchestration, pvc (physical+virtual+containerized) harmonization, kernel-friendly accelerated and secured networking, and network-aware service meshes will be briefly discussed.

Contents

The 22nd International Conference on Network-Based Information Systems (NBiS-2019)

Evaluation of an OI (Operation Interruption) Protocol to Prevent Illegal Information Flow in the IoT 3
Shigenari Nakamura, Tomoya Enokido, and Makoto Takizawa

Evaluation of Data and Subprocess Transmission Strategies in the Tree-Based Fog Computing Model 15
Ryuji Oma, Shigenari Nakamura, Dilawaer Duolikun, Tomoya Enokido, and Makoto Takizawa

Performance Evaluation of WMNs by WMN-PSOHC System Considering Random Inertia Weight and Linearly Decreasing Vmax Replacement Methods 27
Shinji Sakamoto, Seiji Ohara, Leonard Barolli, and Shusuke Okamoto

Implementation of a Fuzzy-Based Simulation System and a Testbed for Improving Driving Conditions in VANETs Considering Drivers's Vital Signs 37
Kevin Bylykbashi, Ermioni Qafzezi, Makoto Ikeda, Keita Matsuo, and Leonard Barolli

Multi-cloud System for Content Sharing Using RAID-Like Fragmentation 49
Takumi Murakami and Shinji Sugawara

A VR System for Alleviating a Fear of Heights Based on Vital Sensing and Placebo Effect 62
Iku Kitanosono, Toshiyuki Haramaki, Tsuneo Kagawa, and Hiroaki Nishino

Exploring Bit Arrays for Join Processing in Spatial Data Streams 73
Wendy Osborn

2D Color Image Enhancement Based on Conditional Generative
Adversarial Network and Interpolation   86
Yen-Ju Li, Chun-Hsiang Chang, Chitra Meghala Yelamandala,
and Yu-Cheng Fan

Position Follow-up Control for Hand Delivery of Object
Between Moving Robot Arms of Remote Robot Systems
with Force Feedback   96
Qin Qian, Yutaka Ishibashi, Pingguo Huang, and Yuichiro Tateiwa

A SOA Based SLA Negotiation and Formulation Architecture
for Personalized Service Delivery in SDN.   108
Shuraia Khan and Farookh Khadeer Hussain

An Energy-Efficient Process Replication Algorithm
with Multi-threads Allocation.   120
Tomoya Enokido, Dilawaer Duolikun, and Makoto Takizawa

Application of Perceptual Features for User Authentication
in Distributed Systems   132
Marek R. Ogiela and Lidia Ogiela

User Oriented Protocols for Data Sharing and
Services Management.   137
Lidia Ogiela, Makoto Takizawa, and Urszula Ogiela

Development of Borehole Imaging Method with Using
Visual-SLAM   142
Tsuneo Kagawa

A Mechanism of Window Switching Prediction Based on User
Operation History to Facilitate Multitasking   154
Keizo Sato, Shotaro Imada, Shinya Mazume, and Makoto Nakashima

Opportunistic Communication by Pedestrians with Roadside Units
as Message Caches   167
Tomoyuki Sueda and Naohiro Hayashibara

VLSI Implementation of K-Best MIMO Detector with Cost-Effective
Pre-screening and Fast Sorting Design   178
Jheng-Jhan He and Chih-Peng Fan

Meal Information Recognition Based on Smart Tableware Using
Multiple Instance Learning   189
Liyang Zhang, Kohei Kaiya, Hiroyuki Suzuki, and Akio Koyama

From Ivory Tower to Democratization and Industrialization:
A Landscape View of Real-World Adaptation
of Artificial Intelligence   200
Toshihiko Yamakami

Data Exportation Framework for IoT Simulation Based Devices 212
Yahya Al-Hadhrami, Nasser Al-Hadhrami, and Farookh Khadeer Hussain

**Earthquake and Tsunami Workflow Leveraging the Modern
HPC/Cloud Environment in the LEXIS Project** 223
Thierry Goubier, Andrea Ajmar, Carmine D'Amico, Paul Dubrulle,
Susanna Grita, Stephane Louise, Jan Martinovič, Tomáš Martinovič,
Natalja Rakowsky, Paolo Savio, Danijel Schorlemmer, Alberto Scionti,
and Olivier Terzo

**Alternative Paths Reordering Using Probabilistic
Time-Dependent Routing** 235
Martin Golasowski, Jakub Beránek, Martin Šurkovský, Lukáš Rapant,
Daniela Szturcová, Jan Martinovič, and Kateřina Slaninová

**A Proposal of Recommendation Function for Solving Element
Fill-in-Blank Problem in Java Programming Learning
Assistant System** 247
Nobuo Funabiki, Shinpei Matsumoto, Su Sandy Wint,
Minoru Kuribayashi, and Wen-Chun Kao

**Web-Based Interactive 3D Educational Material Development
Framework and Its Authoring Functionalities** 258
Daiki Hirayama, Wei Shi, and Yoshihiro Okada

**Data Relation Analysis Focusing on Plural Data Transition
for Detecting Attacks on Vehicular Network** 270
Jun Yajima, Takayuki Hasebe, and Takao Okubo

**Gait-Based Authentication for Smart Locks Using Accelerometers
in Two Devices** 281
Kazuki Watanabe, Makoto Nagatomo, Kentaro Aburada,
Naonobu Okazaki, and Mirang Park

**Automatic Vulnerability Identification and Security Installation
with Type Checking for Source Code** 292
Shun Hinatsu, Koichi Shimizu, Takeshi Ueda, Benoît Boyer,
and David Mentré

**Blockchain-Based Malware Detection Method Using Shared
Signatures of Suspected Malware Files** 305
Ryusei Fuji, Shotaro Usuzaki, Kentaro Aburada, Hisaaki Yamaba,
Tetsuro Katayama, Mirang Park, Norio Shiratori, and Naonobu Okazaki

**Draft Design of Li-Fi Based Acquisition Layer of DataLake
Framework for IIoT and Smart Factory** 317
ByungRae Cha, Sun Park, Byeong-Chun Shin, and JongWon Kim

**A Novel Hybrid Recommendation System Integrating Content-Based
and Rating Information** 325
Tan Nghia Duong, Viet Duc Than, Tuan Anh Vuong, Trong Hiep Tran,
Quang Hieu Dang, Duc Minh Nguyen, and Hung Manh Pham

**The 14th International Workshop on Network-Based Virtual Reality
and Tele-Existence (INVITE-2019)**

**Proposal of a High-Presence Japanese Traditional Crafts Presentation
System Integrated with Different Cultures** 341
Yangzhicheng Lu, Tomoyuki Ishida, Akihiro Miyakwa,
Yoshitaka Shibata, and Hiromasa Habuchi

**Proposal of Interactive Information Sharing System Using Large
Display for Disaster Management** 350
Ryo Nakai, Tomoyuki Ishida, Noriki Uchida, Yoshitaka Shibata,
and Hiromasa Habuchi

**A TEFL Virtual Reality System for High-Presence
Distance Learning** 359
Steven H. Urueta and Tetsuro Ogi

A Trial Development of 3D Statues Map with 3D Point Server 369
Hideo Miyachi

**Driving Simulator System for Disaster Evacuation Guide Based
on Road State Information Platform** 377
Yoshitaka Shibata and Akira Sakuraba

**The 13th International Workshop on Advanced Distributed
and Parallel Network Applications (ADPNA-2019)**

Estimation Method of Traffic Volume Using Big-Data 387
Kazuki Someya, Ryozo Kiyohara, and Masashi Saito

**Experimental Evaluation of Publish/Subscribe-Based
Spatio-Temporal Contents Management on Geo-Centric
Information Platform** 396
Kaoru Nagashima, Yuzo Taenaka, Akira Nagata, Katsuichi Nakamura,
Hitomi Tamura, and Kazuya Tsukamoto

**DTN Sub-ferry Nodes Placement with Consideration
for Battery Consumption** 406
Kazunori Ueda

**Concept Proposal of Multi-layer Defense Security Countermeasures
Based on Dynamic Reconfiguration Multi-perimeter Lines** 413
Shigeaki Tanimoto, Yuuki Takahashi, Ayaka Takeishi, Sonam Wangyal,
Tenzin Dechen, Hiroyuki Sato, and Atsushi Kanai

Fault Detection of Process Replicas on Reliable Servers 423
Hazuki Ishii, Ryuji Oma, Shigenari Nakamura, Tomoya Enokido,
and Makoto Takizawa

**Deep Recurrent Neural Networks for Wi-Fi Based Indoor
Trajectory Sensing** 434
Hao Li, Joseph K. Ng, and Junxing Ke

**The 10th International Workshop on Heterogeneous Networking
Environments and Technologies (HETNET-2019)**

A Model for Mobile Fog Computing in the IoT 447
Kosuke Gima, Ryuji Oma, Shigenari Nakamura, Tomoya Enokido,
and Makoto Takizawa

**Converting Big Video Data into Short Video: Using 360-Degree
Cameras for Searching Students Location and Judging Students
Learning Style** 459
Noriyasu Yamamoto

**Crowdsourcing Platform for Healthcare: Cleft Lip and Cleft Palate
Case Studies** 465
Krit Khwanngern, Juggapong Natwichai, Suriya Sitthikham,
Watcharaporn Sitthikamtiub, Vivatchai Kaveeta, Arakin Rakchittapoke,
and Somboon Martkamjan

Jaw Surgery Simulation in Virtual Reality for Medical Training 475
Krit Khwanngern, Narathip Tiangtae, Juggapong Natwichai,
Aunnop Kattiyanet, Vivatchai Kaveeta, Suriya Sitthikham,
and Kamolchanok Kammabut

**The 10th International Workshop on Intelligent Sensors
and Smart Environments (ISSE-2019)**

**Fusion Method of Depth Images and Visual Images
for Tire Inspection** 487
Chien-Chou Lin, Chun-Cheng Chang, Ching-Lung Chang,
and Chuan-Yu Chang

**A Research on Constructing the Recognition System for the Dynamic
Pedestrian Traffic Signals Through Machine Vision** 493
Chien-Chung Wu and Yi-Chieh Hsug

**Learning Depth from Monocular Sequence with Convolutional
LSTM Network** 502
Chia-Hung Yeh, Yao-Pao Huang, Chih-Yang Lin, and Min-Hui Lin

Two-Dimensional Inductance Plane Sensor for Smart Home Door Lock 508
Wen-Shan Lin, Chao-Ting Chu, and Chian C. Ho

Wearable EMG Gesture Signal Acquisition Device Based on Single-Chip Microcontroller 517
Wen-Shan Lin, Chao-Ting Chu, and Chian C. Ho

Ring-Based Routing for Industrial Wireless Sensor Networks 526
Ching-Lung Chang, Hao-Ting Lee, and Chuan-Yu Chang

Proposal of Research Information Collection System 537
Takahiro Uchiya, Ryoa Sugisaki, and Ichi Takumi

Agricultural Pests Damage Detection Using Deep Learning 545
Ching-Ju Chen, Jian-Shiun Wu, Chuan-Yu Chang, and Yueh-Min Huang

The 10th International Workshop on Trustworthy Computing and Security (TwCSec-2019)

Analysis of Actual Propagation Behavior of WannaCry Within an Intranet (Extended Abstract) 557
Takanori Oikawa, Masahiko Takenaka, and Yuki Unno

Proposal and Evaluation of Authentication Method Having Shoulder-Surfing Resistance for Smartwatches Using Shift Rule 560
Makoto Nagatomo, Kazuki Watanabe, Kentaro Aburada,
Naonobu Okazaki, and Mirang Park

Evaluation of Manual Alphabets Based Gestures for a User Authentication Method Using s-EMG 570
Hisaaki Yamaba, Shotaro Usuzaki, Kayoko Takatsuka, Kentaro Aburada,
Tetsuro Katayama, Mirang Park, and Naonobu Okazaki

Analysis of the Reasons Affecting the Simulation Results of SAR Imaging 581
Xing-Xiu Song

Finite Element Simulation of Blasting Robote with ANSYS 586
Xing-Xiu Song

The 9th International Workshop on Information Networking and Wireless Communications (INWC-2019)

Performance Evaluation of VegeCare Tool for Tomato Disease Classification 595
Natwadee Ruedeeniraman, Makoto Ikeda, and Leonard Barolli

Performance Analysis of WMNs by WMN-PSODGA Simulation System Considering Load Balancing: A Comparison Study for Exponential and Weibull Distribution of Mesh Clients 604
Seiji Ohara, Admir Barolli, Shinji Sakamoto, and Leonard Barolli

Evaluation of 13.56 MHz RFID System Considering Tag Magnetic Field Intensity 620
Yuki Yoshigai and Kiyotaka Fujisaki

Numerical Analysis of Fano Resonator in 2D Periodic Structure for Integrated Microwave Circuit 630
Hiroshi Maeda, Naoki Higashinaka, and Akihito Ochi

The 8th International Workshop on Advances in Data Engineering and Mobile Computing (DEMoC-2019)

Evaluation of I/O Performance Regulating Function with a Virtual Machine 641
Takashi Nagao, Nasanori Tanabe, Kazutoshi Yokoyama, and Hideo Taniguchi

A Continuous Media Data Broadcasting Model for Base Stations Moving Straight 650
Tomoki Yoshihisa, Yusuke Gotoh, and Akimitsu Kanzaki

A Support System for Second Tourism 658
Yusuke Gotoh

Omotenashi Robots: Generating Funny Dialog Using Visitor Geographical Information 669
Kazuki Haraguchi, Satoshi Aoki, Tomohiro Umetani, Tatsuya Kitamura, and Akiyo Nadamoto

The 8th International Workshop on Web Services and Social Media (WSSM-2019)

Web Service for Searching Halal Compliant Restaurants 683
Imama, Masaki Kohana, and Masaru Kamada

Implementation of Interactive Tutorial for IslayPub by Hooking User Events 692
Daisuke Tanaka, Masaki Kohana, Michitoshi Niibori, Yasuhiro Ohtaki, Shusuke Okamoto, and Masaru Kamada

A Smart Lock Control System with Home Situation 700
Katsumi Ohkura and Masaki Kohana

Card Price Prediction of Trading Cards Using Machine Learning Methods 705
Hiroki Sakaji, Akio Kobayashi, Masaki Kohana, Yasunao Takano, and Kiyoshi Izumi

Kawaii in Tweets: What Emotions Does the Word Describe in Social Media? 715
Jun Iio

Development of Goals Achievement Sharing System Based on Approach Goals in Positive Psychology 722
Yoshihiro Kawano

A Real-Time Programming Battle Web Application by Using WebRTC 731
Ryoya Fukutani, Shusuke Okamoto, Shinji Sakamoto, and Masaki Kohana

Author Index 739

The 22nd International Conference on Network-Based Information Systems (NBiS-2019)

Evaluation of an OI (Operation Interruption) Protocol to Prevent Illegal Information Flow in the IoT

Shigenari Nakamura[1]([envelope]), Tomoya Enokido[2], and Makoto Takizawa[1]

[1] Hosei University, Tokyo, Japan
nakamura.shigenari@gmail.com, makoto.takizawa@computer.org
[2] Rissho University, Tokyo, Japan
eno@ris.ac.jp

Abstract. Various types and millions of nodes including not only computers like servers but also devices like sensors and actuators are interconnected in the IoT (Internet of Things). Here, devices have to be prevented from maliciously accessed. The CapBAC (Capability-Based Access Control) model is proposed to make IoT devices secure. In the CapBAC model, an owner of a device issues a capability token, i.e. a set of access rights to a subject. Here, the subject is allowed to manipulate the device according to the access rights authorized in the capability token. Suppose a subject sb_i is allowed to get data from a device d_2 but not allowed to get data from a device d_1. If another subject can get data from the device d_1 and sends the data to the device d_2, the subject sb_i can get the data of the device d_1 from the device d_2. Here, the data in the device d_1 illegally flows to the subject sb_i. In order to prevent illegal information flow, an OI (Operation Interruption) protocol is proposed in our previous studies. Here, illegal get operations are interrupted. In this paper, we evaluate the OI protocol in terms of the number of illegal get operations. In the evaluation, we show the ratio of the number of illegal get operations to the total number of get operations is kept constant even if the number of subjects increases in the OI protocol.

Keywords: IoT (Internet of Things) · Device security ·
CapBAC (Capability-Based Access Control) model ·
Illegal information flow · Information flow control ·
OI (Operation Interruption) protocol

1 Introduction

Information systems are required to be secure in presence of malicious accesses. For this aim, various types of models and methods are proposed, such as cryptography [12,13] and access control models [1]. Cryptography is used to prevent every information from being forged, stolen, or disclosed by a subject which is granted no permission for the information. In the access control models, only

© Springer Nature Switzerland AG 2020
L. Barolli et al. (Eds.): NBiS-2019, AISC 1036, pp. 3–14, 2020.
https://doi.org/10.1007/978-3-030-29029-0_1

an authorized subject is allowed to manipulate an object in an authorized operation. However, even if a subject is not allowed to get data in an object o_i, the subject can get the data by accessing another object o_j [1]. Here, illegal information flow from the object o_i via the object o_j to the subject occurs. We have to prevent illegal information flow among subjects and objects in the access control models. The LBAC (Lattice-Based Access Control) model [14] is proposed to prevent illegal information flow among subjects and objects. Here, each entity is assigned a security class. Illegal information flow is defined based on the relations among classes and every operation implying the illegal information flow is prohibited. In our previous studies, various types of protocols to prevent illegal information flow are proposed. In the papers [5–7], types of protocols to prevent illegal information flow occurring in distributed database systems are proposed based on the RBAC (Role-Based Access Control) model [15]. In the papers [9,10], protocols to prevent illegal information flow occurring in P2PPSO (Peer-to-Peer Publish/Subscribe with Object concept) systems is proposed based on the TBAC (Topic-Based Access Control) model [11].

The IoT (Internet of Things) is composed of various types and millions of nodes including not only computers but also devices like sensors and actuators. Here, the traditional access control models such as the RBAC [15] and ABAC (Attribute-Based Access Control) [16] models are not adopted for the IoT due to the scalability of the IoT. Hence, the CapBAC (Capability-Based Access Control) model is proposed [3]. Here, an owner of a device issues a capability token to a subject like users and applications. The capability token is defined to be a set of access rights, $\langle d, op \rangle$ for a device d and an operation op. The subject is allowed to manipulate the device d in an operation op only if the capability token including an access right $\langle d, op \rangle$ is issued to the subject.

Suppose a subject sb_i is issued a capability token including a pair of access rights $\langle d_1, get \rangle$ and $\langle d_2, put \rangle$ of a pair of devices d_1 and d_2 by owners of the devices. Suppose the device d_1 is a sensor and the device d_2 is equipped with a pair of sensor and actuator. A sensor just gives sensor data to a subject. On the other hand, an actuator supports an action to store data to the device. A subject sb_j is issued a capability token including an access right $\langle d_2, get \rangle$ by an owner of the device d_2. First, the subject sb_i gets the sensor data dt obtained by the sensor d_1 and then gives the data dt to the device d_2 by using the actuator of the device d_2. Next, the subject sb_j gets the data dt from the sensor of the device d_2. Here, the subject sb_j can obtain the data dt of the sensor d_1 via the device d_2 although the subject sb_j is not issued a capability token including the access right $\langle d_1, get \rangle$. Here, the device d_1 is a source one of information flow. This is illegal information flow from the sensor d_1 to the subject sb_j. In our previous studies [8], an OI (Operation Interruption) protocol is proposed to prevent illegal information flow in the IoT based on the CapBAC model. Here, the legal and illegal information flow relations among devices are defined based on the CapBAC model. In the OI protocol, it is checked whether or not the illegal information flow occurs by performing an operation on a device based on the information flow relations among devices. If the illegal information flow occurs, the operation is interrupted at the device. Hence, every illegal information flow in the IoT is prevented from occurring.

In this paper, we evaluate the OI protocol in terms of the number of illegal get operations interrupted. In the evaluation, we show the ratio of the number of illegal get operations to the total number of get operations is kept constant even if the number of subjects increases in the OI protocol.

In Sect. 2, we discuss the CapBAC model. In Sect. 3, we define types of information flows based on the CapBAC model. In Sect. 4, we discuss the OI protocol to prevent illegal information flow in the IoT based on the definitions of information flows. In Sect. 5, we evaluate the OI protocol in terms of the number of illegal operations interrupted.

2   CapBAC (Capability-Based Access Control) Model

In order to make information systems secure, types of access control models [2, 15, 16] are widely used. Here, a system is composed of two types of entities, subjects and objects. A subject s issues an operation op to an object o. Then, the operation op is performed on the object o. Here, only an authorized subject s is allowed to manipulate an object o in an authorized operation op. Most of the access control models are based on ACLs (access control lists) such as RBAC (Role-Based Access Control) [15] and ABAC (Attribute-Based Access Control) [16] models. An ACL is a list of access rules specified by an authorizer. Each access rule $\langle s, o, op \rangle$ is composed of a subject s, an object o, and an operation op. This means, a subject s is granted an access right of an object o and an operation op. In the ACL system, if a subject s tries to access the data of an object o in an operation op, a service provider has to check whether or not the subject is authorized to manipulate the object o in the operation op by using the ACL, i.e. $\langle s, o, op \rangle$ in the ACL. In scalable systems like the IoT, the ACL gets also scalable and it is difficult to maintain and check the ACL. Hence, the RBAC and ABAC models are not suitable for scalable distributed systems where there is no centralized coordinator and each node is an autonomous process which makes a decision by itself.

The CapBAC (Capability-Based Access Control) model is proposed as an access control model for the IoT [3]. There is an owner of each device. In the CapBAC model, an owner of a device first issues a capability token CAP_i to a subject sb_i. The capability token CAP_i issued to the subject sb_i is defined to be a set of access rights. An access right is a pair $\langle d, op \rangle$ of a device d and an operation op. A subject sb_i is allowed to manipulate a device d in an operation op only if a capability token CAP_i including an access right $\langle d, op \rangle$ is issued to the subject sb_i, i.e. $\langle d, op \rangle \in CAP_i$. Otherwise, the access request is rejected.

In the paper [4], the distributed CapBAC model is proposed. Here, there is no intermediate entity between each pair of a subject and a device to implement the access control. Each subject sb_i issues an access request with a capability token CAP_i to each device d. Figure 1 shows an example of the distributed CapBAC model. First, an owner of a device d issues a capability token CAP_i to a subject

sb_i and the subject sb_i obtains the capability token CAP_i. Next, the subject sb_i sends an access request $\langle d, op \rangle$ with the capability token CAP_i to the device d. The device d checks the validity of the capability token CAP_i and informs the subject sb_i of the authorization decision. If the capability token is valid, the subject sb_i receives the answer of the request. Otherwise, the subject sb_i receives the message which indicates that the request of the subject sb_i is denied.

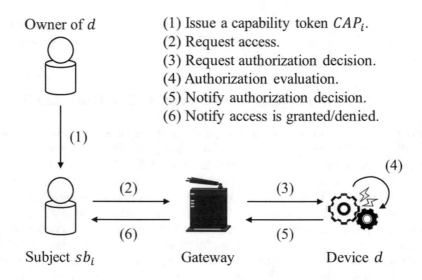

Owner of d

(1) Issue a capability token CAP_i.
(2) Request access.
(3) Request authorization decision.
(4) Authorization evaluation.
(5) Notify authorization decision.
(6) Notify access is granted/denied.

Fig. 1. Distributed CapBAC model.

3   Information Flow

In the IoT, a set D of devices $d_1, \ldots, d_{dn}$ ($dn \geq 1$) are interconnected in networks. In this paper, we consider three types of devices, sensor, actuator, and hybrid device. A sensor device just obtains data collected by sensing events which occur in physical environment. An actuator device acts according to the action request from a subject. Let SB be a set of subjects $sb_1, \ldots, sb_{sbn}$ ($sbn \geq 1$). A subject sb_i gets and puts data from a sensor d_h and to an actuator d_k, respectively. In addition, a subject sb_i issues both get and put operations to a hybrid device d_l like robots and cars.

A subject sb_i can get data from a device d ($\in D$) only if an access right $\langle d, get \rangle$ is granted to the subject sb_i. Let $IN(sb_i)$ be a set of devices whose data a subject sb_i is allowed to get. That is, a subject sb_i is issued a capability token CAP_i including an access right $\langle d, get \rangle$, i.e. $IN(sb_i) = \{d \mid \langle d, get \rangle \in CAP_i\}$ ($\subseteq D$).

Example 1. Let us consider a sensor device d_1 and a hybrid device d_2 as shown in Fig. 2. Suppose the subject sb_i obtains a capability token including a pair of access rights $\langle d_1, get \rangle$ and $\langle d_2, put \rangle$. On the other hand, the subject sb_j obtains a capability token including an access right $\langle d_2, get \rangle$. Here, $IN(sb_i) = \{d_1\}$ for the subject sb_i. $IN(sb_j) = \{d_2\}$ for the subject sb_j. First, the sensor d_1 gets data by sensing events occurring around the sensor d_1 and stores the data in its storage. Next, the subject sb_i gets data from the sensor d_1. Here, the subject sb_i can get the data of the sensor d_1 because the subject sb_i is issued the capability token CAP_i including the access right $\langle d_1, get \rangle$. Then, the subject sb_i puts the data to the hybrid device d_2. Finally, the subject sb_j gets the data from the hybrid device d_2. Here, the subject sb_j can get the data brought to the hybrid device d_2 from the sensor d_1 although the subject sb_j is not allowed to get data from the sensor d_1. Here, the information of the sensor d_1 illegally flows into the subject sb_j.

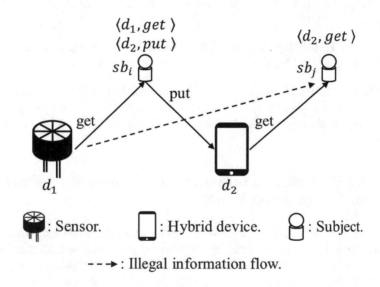

Fig. 2. Illegal information flow in the IoT system.

In this section, we define legal and illegal types of information flow relations on devices and subjects, based on the CapBAC model. Through issuing capability tokens to subjects, data in a device d_1 can be brought to another device d_2 as presented in Example 1. Here, the device d_1 is a *source* device of the device d_2.

Let $sb_i.D$ be a set of devices whose data flows into the subject sb_i. Let $d.D$ be a set of source devices of a device d, whose data flows into the device d by subjects manipulating the device d. For each device d, the set $d.D$ of source devices is manipulated each time a subject runs the device d, which is initially ϕ. If the device d gets data by sensing events, the device d is added to the set $d.D$ of the device d.

Definition 1. Data *flows* from a device d into a subject sb_i ($d \rightarrow sb_i$) iff (if and only if) $d.D \neq \phi$ and $d \in IN(sb_i)$.

If $d \rightarrow sb_i$ holds, data brought to the device d may be brought to the subject sb_i. In Example 1, data in the device d_1 flows into the subject sb_i because $d_1.D$ ($=\{d_1\}$) $\neq \phi$ and $d_1 \in IN(sb_i)$ ($=\{d_1\}$).

Definition 2. Data *legally flows* from a device d into a subject sb_i ($d \Rightarrow sb_i$) iff $d \rightarrow sb_i$ and $d.D \subseteq IN(sb_i)$.

If $d \rightarrow sb_i$, data of the devices in the set $d.D$ may be brought into the subject sb_i. Otherwise, no information flows from the device d into the subject sb_i. The condition "$d.D \subseteq IN(sb_i)$" means that the subject sb_i is allowed to get the data of every device in the set $d.D$. In Example 1, data in the device d_1 legally flows into the subject sb_i ($d_1 \Rightarrow sb_i$) because $d_1 \rightarrow sb_i$ and $d_1.D$ ($=\{d_1\}$) $\subseteq IN(sb_i)$ ($=\{d_1\}$).

Definition 3. Data *illegally flows* from a device d into a subject sb_i ($d \mapsto sb_i$) iff $d \rightarrow sb_i$ and $d.D \nsubseteq IN(sb_i)$.

The condition "$d.D \nsubseteq IN(sb_i)$" means that the subject sb_i is not allowed to get the data in some device of $d.D$. In Example 1, data in the device d_2 illegally flows into the subject sb_j ($d_2 \mapsto sb_j$) because $d_2 \rightarrow sb_j$ and $d_2.D$ ($=\{d_1\}$) $\nsubseteq IN(sb_j)$ ($=\{d_2\}$).

4   An OI (Operation Interruption) Protocol to Prevent Illegal Information Flow

In Sect. 3, the illegal information flow is defined based on the CapBAC model. If the data in a device d_1 brought into a device d_2 is brought into a subject sb_i which is not allowed to get the data in the device d_1, the data in the device d_1 illegally flows from the device d_1 to the subject sb_i. In this section, we discuss the OI (Operation Interruption) protocol [8] to prevent illegal information flow in the IoT based on the definitions of information flows.

In the IoT with the CapBAC model, a subject sb_i is issued a capability token CAP_i which is a set of access rights. If the capability token CAP_i obtained by the subject sb_i includes an access right $\langle d, op \rangle$ of a device d and an operation op, the subject sb_i is allowed to manipulate the device d in an operation op.

$IN(sb_i)$ shows a set of devices whose data is allowed to be got by a subject sb_i, i.e. $IN(sb_i) = \{d \mid \langle d, \ get \rangle \in CAP_i\}$. A set $sb_i.D$ of devices for a subject sb_i shows that data of the devices in $sb_i.D$ is brought to the subject sb_i. A set $d.D$ of source devices of a device d indicates that the data of the devices in $d.D$ is brought to the device d. Data flows into subjects and devices by subjects manipulating devices. For each subject sb_i and device d, the set $sb_i.D$ and $d.D$ are manipulated, which are initially ϕ.

The OI protocol is shown as follows:

[OI (Operation Interruption) protocol]

1. A device d gets data by sensing events occurring around the device d.
   a. $d.D = d.D \cup \{d\}$;
2. A subject sb_i gets data from a device d, i.e. $d \rightarrow sb_i$ holds.
   a. If $d \Rightarrow sb_i$, the subject sb_i gets the data from the device d and $sb_i.D = sb_i.D \cup d.D$;
   b. Otherwise, the get operation is interrupted at the device d;
3. A subject sb_i puts data to a device d.
   a. Data obtained by the subject sb_i is brought to the device d and $d.D = d.D \cup sb_i.D$;

In this paper, we consider three types of devices, sensor, actuator, and hybrid devices. Sensor and hybrid device collect data by sensing events occurring around them. Once a device d gets data by sensing events, the device d is added to the set $d.D$ of the device d. Actuators and hybrid devices get data which is collected by other devices like sensors and hybrid devices. Subjects issue *get* operations to sensors and hybrid devices to get data collected by the sensors and hybrid devices. On the other hand, subjects issue *put* operations to actuators and hybrid devices to store data which the subjects get from sensors and hybrid devices.

When a subject sb_i tries to get data from a device d, the subject sb_i sends an access request to the device d with the capability token CAP_i. On receipt of the access request from the subject sb_i, the device d checks whether or not the legal information flow condition is satisfied according to the capability token CAP_i attached in the access request from the subject sb_i. If the legal condition is satisfied, the device d sends a reply of the access request to the subject sb_i. Otherwise, the access request of the subject sb_i is interrupted at the device d. Here, the device d informs the subject sb_i that the access request is denied.

In Example 1, first, the sensor d_1 obtains data by sensing events occurring around the sensor d_1 and stored the data in its storage, i.e. $d_1.D = \{d_1\}$. Next, the subject sb_i tries to get the data from the sensor d_1. Here, data legally flows from the sensor d_1 into the subject sb_i $(d_1 \Rightarrow sb_i)$ because $d_1 \rightarrow sb_i$ and $d_1.D$ $(=\{d_1\}) \subseteq IN(sb_i)$ $(=\{d_1\})$. Hence, the subject sb_i can get the data from the sensor d_1 and $sb_i.D = sb_i.D$ $(=\phi) \cup d_1.D$ $(=d_1) = \{d_1\}$. Then, the subject sb_i puts the data to the hybrid device d_2. Here, $d_2.D = d_2.D$ $(=\phi) \cup sb_i.D$ $(=\{d_1\}) = \{d_1\}$. Finally, the subject tries to get the data from the hybrid device d_2. Here, data illegally flows from the hybrid device d_2 into the subject sb_j $(d_2 \mapsto sb_j)$ because $d_2 \rightarrow sb_j$ and $d_2.D$ $(=\{d_1\}) \nsubseteq IN(sb_j)$ $(=\{d_2\})$. Hence, the get operation of the subject sb_j is interrupted at the device d_2 as shown in Fig. 3.

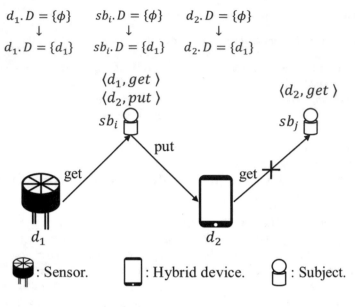

$$d_1.D = \{\phi\} \qquad sb_i.D = \{\phi\} \qquad d_2.D = \{\phi\}$$
$$\downarrow \qquad\qquad\quad \downarrow \qquad\qquad\quad \downarrow$$
$$d_1.D = \{d_1\} \qquad sb_i.D = \{d_1\} \qquad d_2.D = \{d_1\}$$

Fig. 3. OI protocol.

5 Evaluation

We evaluate the OI protocol on a subject set $SB = \{sb_1, \ldots, sb_{sbn}\}$ and a device set $D = \{d_1, \ldots, d_{dn}\}$ in terms of the number of operations interrupted.

In the evaluation, each subject supports *get* and *put* operations. We assume that every subject issues only *get* and *put* operations to sensors and actuators, respectively. On the other hand, every subject issues *get* and *put* operations to hybrid devices. This means, for each d of sensors and actuators, there is one access right $\langle d, get \rangle$ or $\langle d, put \rangle$ and for each hybrid device d, there are a pair of access rights $\langle d, get \rangle$ and $\langle d, put \rangle$. Let sn, an, and hn be the numbers of sensors, actuators, and hybrid devices, respectively. Hence, a set AC of access rights includes $sn + an + 2hn$ access rights. Every capability token CAP_i issued to a subject sb_i is composed of access rights for *get* and *put* operations. Let $mxan$ show the maximum number of access rights to be granted to every subject. acn_i is the number of access rights granted to a subject sb_i. Here, the number acn_i is randomly selected out of numbers $1, \ldots, mxan$. The number acn_i of access rights are randomly granted to each subject sb_i.

Let $ac_{i,j}$ be an access right in a capability token CAP_i issued to a subject sb_i. Every access right $ac_{i,j}$ has validity period $vp_{i,j}$ [tu] which shows how long the access right $ac_{i,j}$ is valid. $mxvp$ show maximum validity period [tu] of every access right. If an access right $ac_{i,j}$ is granted to a subject sb_i, the validity period $vp_{i,j}$ of the access right $ac_{i,j}$ is randomly selected out of numbers $1, \ldots,$ $mxvp$ [tu].

In the evaluation, we consider sbn subjects and one hundred devices ($dn = 100$). First, sbn subjects and dn ($= 100$) devices are randomly generated, i.e. $SB = \{sb_1, \ldots, sb_{sbn}\}$ and $D = \{d_1, \ldots, d_{100}\}$. Every subject is issued capability tokens which are composed of access rights randomly selected from a set AC. The number of access rights granted to each peer in capability tokens is randomly selected out of numbers $1, \ldots, 10$, i.e $mxan = 10$. The validity period $vp_{i,j}$ of each access right $ac_{i,j}$ is randomly selected out of numbers $1, \ldots, 20$, i.e. $mxvp = 20$. After the generation of subjects and devices, the following procedures are performed in the OI protocol:

1. Every subject sb_i selects to either issue an operation or not. If a subject sb_i issues a *get* operation to a device d, data in the device d flows into the subject sb_i. On the other hand, if a subject sb_i issues a *put* operation to a device d, data in the subject sb_i flows into the data d.
2. For every subject sb_i, the validity period $vp_{i,j}$ of every access right $ac_{i,j}$ is decremented by one. If the validity period $vp_{i,j}$ gets 0, the access right $ac_{i,j}$ is revoked from the subject sb_i. If every access right $ac_{i,j}$ included in the capability token CAP_i is revoked from the subject sb_i., the capability token CAP_i is also revoked from the subject sb_i.
3. Every subject sb_i which has no capability token is issued capability tokens which are composed of access rights randomly selected from a set AC. acn_i is randomly selected out of numbers $1, \ldots, mxan$.

The above procedures are assumed to be performed in one time unit [tu]. Let st be a simulation time. This means, the above procedures are iterated st times. Here, $st = 0, 100, 200, 300, 400,$ or 500. We randomly create a subject set SB and a device set D one hundred times. For a pair of given sets SB and D, the above procedures for st [tu] are iterated one hundred times. Finally, we calculate the average number of illegal *get* operations interrupted.

Figure 4 shows the numbers of illegal *get* operations interrupted for the simulation time st [tu] in the OI protocol. Here, every label shows the number of subjects sbn. The number of illegal *get* operations interrupted increases as the simulation time increases.

Figure 5 shows the ratios of the numbers of illegal *get* operations interrupted to the total numbers of *get* operations issued for the simulation time st [tu] in the OI protocol. Here, every label shows the number of subjects sbn. For $st \geq 400$, the ratio of the illegal operations interrupted does not change where $sbn = 30$, 50, 70, or 100. This means, even if the number of subjects sbn increases, the ratio of the illegal *get* operations interrupted is kept at most around 0.32 in the OI protocol.

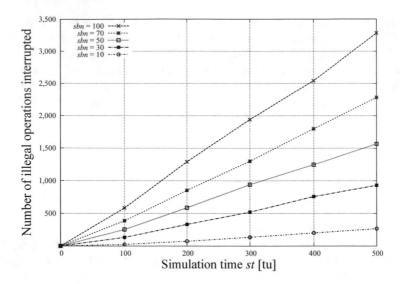

Fig. 4. Number of illegal operations interrupted.

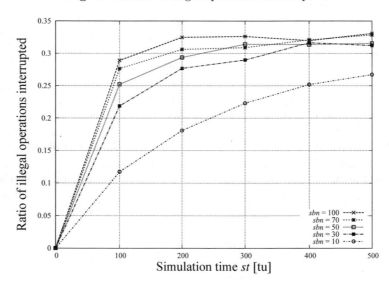

Fig. 5. Ratio of illegal operations interrupted.

6   Concluding Remarks

The IoT is composed of various types and millions of nodes including not only computers but also devices like sensors and actuators. Here, the traditional access control models such as RBAC and ABAC models are not suitable for the IoT due to the scalability. Hence, we take the CapBAC (Capability-Based Access Control) model to make the IoT secure. In this paper, we consider three types of devices, sensor, actuator, and hybrid devices. In the IoT, a subject sb_i can get the data of the device d_1 from another device d_2 to which the data of the device d_1 is brought although the subject sb_i is not allowed to get the data from the device d_1. Here, illegal information flow occurs. In our previous studies, the OI (Operation Interruption) protocol is proposed to prevent illegal information flow in the IoT based on the CapBAC model. In the OI protocol, operations implying illegal information flow are interrupted. In this paper, we evaluated the OI protocol in terms of the number of illegal operations interrupted. In the evaluation, we show the ratio of the number of illegal *get* operations to the total number of *get* operations is kept constant even if the number of subjects increases in the OI protocol.

Acknowledgements. This work was supported by Japan Society for the Promotion of Science (JSPS) KAKENHI Grant Number JP17J00106.

References

1. Denning, D.E.R.: Cryptography and Data Security. Addison Wesley, Boston (1982)
2. Fernandez, E.B., Summers, R.C., Wood, C.: Database Security and Integrity. Adison Wesley, Boston (1980)
3. Gusmeroli, S., Piccione, S., Rotondi, D.: A capability-based security approach to manage access control in the internet of things. Math. Comput. Model. **58**(5–6), 1189–1205 (2013)
4. Hernández-Ramos, J.L., Jara, A.J., Marín, L., Skarmeta, A.F.: Distributed capability-based access control for the internet of things. J. Internet Serv. Inf. Secur. **3**(3/4), 1–16 (2013)
5. Nakamura, S., Duolikun, D., Enokido, T., Takizawa, M.: A flexible read-write abortion protocol to prevent illegal information flow among objects. J. Mob. Multimed. **11**(3&4), 263–280 (2015)
6. Nakamura, S., Duolikun, D., Enokido, T., Takizawa, M.: A write abortion-based protocol in role-based access control systems. Int. J. Adapt. Innov. Syst. **2**(2), 142–160 (2015)
7. Nakamura, S., Duolikun, D., Enokido, T., Takizawa, M.: A read-write abortion protocol to prevent illegal information flow in role-based access control systems. Int. J. Space Based Situated Comput. **6**(1), 43–53 (2016)
8. Nakamura, S., Enokido, T., Barolli, L., Takizawa, M.: Capability-based information flow control model in the IoT. In: Proceedings of the 13th International Conference on Innovative Mobile and Internet Services in Ubiquitous Computing, pp. 63–71 (2019)
9. Nakamura, S., Enokido, T., Takizawa, M.: Information flow control in object-based peer-to-peer publish/subscribe systems. Concurrency and Computation: Practice and Experience (accepted)
10. Nakamura, S., Enokido, T., Takizawa, M.: Causally ordering delivery of event messages in P2 PPSO systems. Cogn. Syst. Res. **56**, 167–178 (2019)
11. Nakamura, S., Ogiela, L., Enokido, T., Takizawa, M.: An information flow control model in a topic-based publish/subscribe system. J. High Speed Netw. **24**(3), 243–257 (2018)

12. Ogiela, L.: Intelligent techniques for secure financial management in cloud computing. Electron. Commer. Res. Appl. **14**(6), 456–464 (2015)
13. Ogiela, M.R., Ogiela, L.: On using cognitive models in cryptography. In: Proceedings of IEEE the 30th International Conference on Advanced Information Networking and Applications (AINA 2016), pp. 1055–1058 (2016)
14. Sandhu, R.S.: Lattice-based access control models. IEEE Comput. **26**(11), 9–19 (1993)
15. Sandhu, R.S., Coyne, E.J., Feinstein, H.L., Youman, C.E.: Role-based access control models. IEEE Comput. **29**(2), 38–47 (1996)
16. Yuan, E., Tong, J.: Attributed based access control (ABAC) for web services. In: Proceedings of the IEEE International Conference on Web Services (ICWS 2005) (2005)

Evaluation of Data and Subprocess Transmission Strategies in the Tree-Based Fog Computing Model

Ryuji Oma[1]([✉]), Shigenari Nakamura[1], Dilawaer Duolikun[1], Tomoya Enokido[2], and Makoto Takizawa[1]

[1] Hosei University, Tokyo, Japan
ryuji.oma.6r@stu.hosei.ac.jp, nakamura.shigenari@gmail.com,
dilewerdolkun@gmail.com, makoto.takizawa@computer.org
[2] Rissho University, Tokyo, Japan
eno@ris.ac.jp

Abstract. In order to increase the performance of the IoT (Internet of Things), the fog computing model is proposed. Here, subprocesses of an application process to handle sensor data are performed on fog nodes in addition to servers. In the TBFC (Tree-Based Fog Computing) model proposed in our previous studies, an application process to handle sensor data is assumed to be a sequence of subprocesses, i.e. linear model. At each level of a TBFC tree, a same subprocess is performed on every node. In this paper, we consider a more general model, GTBFC (General TBFC) model of the IoT where subprocesses of an application process are structured in a tree. Each subprocess in the process tree is performed on fog nodes which are at a same level in the GTBFC tree. Each leaf subprocess is performed on edge nodes which communicate with sensor and actuator devices. We also proposed MEG (Minimum Energy in the GTBFC tree) and SMPRG (Selecting Multiple Parents for Recovery in the GTBFC tree) algorithms to select a new parent node for a child node of a faulty node in the GTBFC tree. In the evaluation, we show the energy consumption of nodes in the SMPRG algorithm as 21% and 31% smaller than the MEG and RD (Random) algorithms.

Keywords: General TBFC (GTBFC) model · Process tree ·
Equivalent nodes · Faults of fog nodes · MEG algorithm ·
SMPRG algorithm

1 Introduction

In the Internet of Things (IoT) [9], not only computers like servers and clients but also millions of sensor and actuator devices are interconnected in networks. In the cloud computing model [4], data collected by sensors is processed by application processes on servers. Networks are congested to transmit the huge volume of sensor data and servers are also overloaded to process the sensor

© Springer Nature Switzerland AG 2020
L. Barolli et al. (Eds.): NBiS-2019, AISC 1036, pp. 15–26, 2020.
https://doi.org/10.1007/978-3-030-29029-0_2

data in realtime manner. The FC (Fog Computing) model [18] is proposed to reduce the communication and processing traffic to handle sensor data in the IoT. Sensor data is processed and the output data is sent to another fog node. On receipt of output data from fog nodes, a fog node further processes the data and sends the processed data to other fog nodes. Thus, servers in clouds finally receive data processed by fog nodes.

It is critical to reduce the electric energy consumed by fog nodes and servers. The power consumption models of a computer are proposed to show the electric power to perform application process [5–8,10]. In order to reduce the energy consumption and execution time of fog nodes and servers, the TBFC (Tree-Based Fog Computing) model [11,12,17] is proposed. Here, fog nodes are hierarchically structured in a height-balanced tree. Nodes at a root level and a bottom level show root node, i.e. a cluster of servers and edge nodes which communicate with sensors and actuators, respectively. Sensors first send data to edge nodes. Each edge node generates output data by processing the input data and sends the output data to a parent node [11]. Thus, each node receives data from child nodes and sends processed data to a parent node. Here, the linear model of an application process is considered, i.e. a sequence of subprocesses. Every node at each level of the TBFC tree is equipped with a same subprocess. In order to be tolerant of node failure, the FTBFC (Fault-tolerant TBFC) [13,15] and MFTBFC (Modified FTBFC) [14,16] models and the subprocess transmission strategy [16] are proposed.

In this paper, we consider a GTBFC (general TBFC) model where subprocesses of an application process are structured in a tree. A subprocess receives input data from child subprocesses and sends output data to a parent subprocess. Each subprocess is supported by fog nodes. A fog node of a subprocess has child nodes where a child subprocess of the subprocess is performed and which sends output data to a parent node which supports a parent subprocess. If a node gets faulty, the child nodes are disconnected in the GTBFC tree. A node which supports the subprocess of the faulty node is an *equivalent* node of the faulty node, which can be a new parent node of the faulty node. In the GTBFC tree, only equivalent nodes of the faulty node can be a new parent node of the disconnected nodes. Output data of each disconnected node has to be sent to an equivalent node. We newly propose an MEG (Minimum Energy in the GTBFC tree) and SMPRG (Selecting Multiple Parents for Recovery in the GTBFC tree) algorithms to select new parent nodes in the equivalent nodes for disconnected nodes so that the energy consumption of new parent nodes can be reduced. In the evaluation, we showed the average energy consumption of new parent nodes selected by the SMPRG algorithm is 21% and 31% smaller than the MEG and random (RD) algorithms, respectively.

In Sect. 2, we propose the GTBFC model for a tree-structured application process. In Sect. 3, we discuss how to select an equivalent node of a faulty node. In Sect. 4, we evaluate the SMPRG algorithm in the GTBFC model.

2   GTBFC Model

2.1   Tree Structure of an Application Process

In this paper, we consider an application process to handle sensor data, which is hierarchically composed of subprocesses. We consider an example [2] where the GTBFC model is composed of eight nodes as shown in Fig. 1. A root node f is a server. A pair of sensors s_1 and s_2 send a pairs of temperature and time data to edge nodes f_{111} and f_{112} and another pair of sensors s_3 and s_4 send a pairs of humidity and time data to edge nodes f_{121} and f_{122} every one second. A subprocess *t-aggregate* of the edge nodes f_{111} and f_{112} calculates an average value of temperature data collected for one minute. Another subprocess *h-aggregate* of the edge nodes f_{121} and f_{122} calculates an average value of humidity data collected for one minute. A parent fog node f_{1i} receives input data from a pair of child fog nodes f_{1i1} and f_{1i2} ($i = 1, 2$). A pair of subprocesses *t-merge* and *h-merge* of parent fog nodes f_{11} and f_{12}, respectively, of an edge fog node f_{1ij} ($i, j = 1, 2$) sort and merge multiple temperature and humidity data in time and sends the merged data to its parent fog node f_1. A subprocess *join* of the fog node f_1 joins data of temperature and humidity from the child fog nodes f_{11} and f_{12}. A subprocess *store* of the root node f receives joined data from the child fog node f_1 and stores the data as a record to the table in the database DB. Thus, the application process p is a hierarchically composed of subprocesses *t-aggregate*, *h-aggregate*, *t-merge*, *h-merge*, *join*, and *store* as shown in Fig. 1(1). Figure 1(2) shows a GTBFC tree of seven fog nodes for the process tree of Fig. 1(1).

An application process P is hierarchically composed of subprocesses. A subprocess p is a root node of a process tree. The root subprocess p has child subprocesses $p_1, ..., p_c$ ($c \geq 1$). Then, a process p_i has child subprocesses $p_{i1}, ..., p_{ic_i}$ ($c_i \geq 1$). Thus, each subprocess p_{Ri} receives data from child subprocesses $p_{Ri1}, ..., p_{Ric_{Ri}}$ ($c_{Ri} \geq 1$) and sends processed data to a parent subprocess p_R as shown in Fig. 2. Here, the label R of a node f_R shows a path from a root subprocess p to the subprocess p_R. Each non-root subprocess in process tree P is supported by one or more than one fog node. A leaf subprocess p_R is supported by an edge node which communicates with receives sensor data from child sensors $s_{R1}, ..., s_{Rsl_R}$ ($sl_R \geq 0$) and sends actions to child actuators $a_{R1}, ..., a_{Ral_R}$ ($al_R \geq 0$). A root subprocess p is supported by a root node, i.e. server.

A node f_R takes a collection D_R of input data $d_{R1}, ..., d_{Rl_R}$ which child nodes $f_{R1}, ..., f_{Rl_R}$ send, respectively. Let $p(f_R)$ be a subprocess supported by a node f_R. A subprocess $p(f_R)$ of the node f_R generates output data d_R by processing input data D_R. Then, the node f_R sends the output data d_R to a parent node $pt(f_R)$. Finally, output data processed by nodes is delivered to a root node, i.e. cloud of servers. A notation $|d|$ shows the size [Byte] of data d. The ratio $|d_R|/|D_R|$ is the *output ratio* ρ_R of a node f_R.

Let $st(f_R)$ show a subtree whose root node is a node f_R in a GTBFC tree. A pair of nodes p_R and p_U are *equivalent* iff the nodes f_R and f_U support a same subprocess and every pair of ancestor nodes of a same level in $as(f_R)$ and $as(f_U)$ support a same subprocess. In Fig. 1(2), a pair of the nodes f_{111} and f_{112}

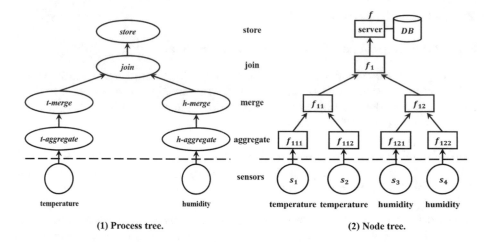

Fig. 1. GTBFC model.

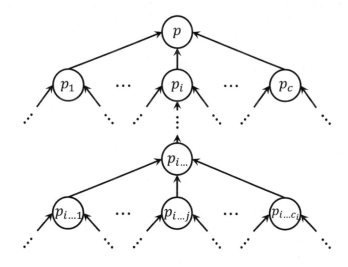

Fig. 2. Process tree.

are equivalent and another pair of the nodes f_{121} and f_{122} are also equivalent. However, the nodes f_{11} and f_{12} are not equivalent since the nodes f_{11} and f_{12} support different subprocesses *t-merge* and *h-merge*, respectively.

2.2   Execution Time of a Fog Node

A node f_R takes input data D_R of size i_R $(=|D_R|)$ from child nodes and sends output data d_R of size o_R $(=|d_R|)$ [11,12,17] to a parent node $pt(f_R)$, where $o_R = \rho_R \cdot i_R$ for the output ratio ρ_R. A node f_R is realized as a sequence of input (I_R), computation (C_R), and output (O_R) modules. The input module I_R

receives input data D_R from the child nodes $f_{R1}, ..., f_{Rc_R}$ and the output module O_R sends output data d_R to the parent node $pt(f_R)$. The computation module C_R is a subprocess $p(f_R)$ which generates the output data d_R by processing the input data D_R. In this paper, we assume the modules are sequentially performed on receipt of input data.

$TI_R(x)$, $TC_R(x)$, and $TO_R(x)$ show the execution time [sec] of the input I_R, computation C_R, and output O_R modules of a node f_R for data of size x, respectively. $TC_R(x)$ depends on the computation complexity of a subprocess $p(f_R)$ of the node f_R. In this paper, $TC_R(x)$ is $ct_R \cdot C_R(x)$ where $C_R(x) = x$ or $C_R(x) = x^2$. Here, ct_R is a constant. A pair of execution time $TI_R(x)$ and $TO_R(x)$ to receive and send data of size x are proportional to x, i.e. $TI_R(x) = rt_R \cdot x$ and $TO_R(x) = st_R \cdot x$. Here, st_R and rt_R are constants.

$$TC_R(x) = ct_R \cdot C_R(x). \tag{1}$$

$$TI_R(x) = rt_R \cdot x. \tag{2}$$

$$TO_R(x) = st_R \cdot x. \tag{3}$$

It takes $TF_R(x)$ [sec] to process input data D_R of size x in each node f_R:

$$TF_R(x) = TI_R(x) + TC_R(x) + \delta_R \cdot TO_R(\rho_R \cdot x). \tag{4}$$

Here, if f_R is a root, $\delta_R = 0$, else $\delta_R = 1$.

The execution time $TI_R(x)$, $TC_R(x)$, and $TO_R(x)$ of a fog node f_R are measured for data size x as shown in Fig. 3. Here, the node f_R is implemented in a Raspberry Pi 3 Model B [1] which is connected with a 10 Gbps network. A pair of fog nodes communicate with each other in UDP [3]. A subprocess $p(f_R)$, i.e. C_R which just selects one value in the input data is realized in the node f_R. A fog node f_R sends and receives data of size x by *send* and *receive* system calls of UDP, respectively. As shown in Fig. 3, $TI_R(x)$ is five times longer than $TO_R(x)$, i.e. $rt_R = 5 \cdot st_R$.

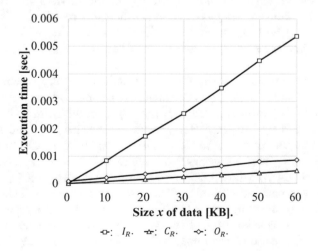

Fig. 3. Execution time of each module.

2.3   Energy Consumption of a Fog Node

$EI_R(x)$, $EC_R(x)$, and $EO_R(x)$ show the electric energy [J] consumed by the input I_R, computation C_R, and output O_R modules [11] of a node f_R for input data of size x, respectively. In this paper, we assume each node f_R follows the SPC (Simple Power Consumption) model [6–8]. The power consumption of a node f_R to perform the computation module C_R is $maxE_R$ [W]. The energy consumption $EC_R(x)$ [J] of the computation module C_R of a node f_R to process input data of size x (>0) is $EC_R(x) = maxE_R$ [W] $\cdot TC_R(x)$ [sec].

A pair of the electric power PI_R and PO_R [W] are consumed by the input I_R and output O_R modules, respectively [6–8]. PI_R and PO_R are $re_R \cdot maxE_R$ and $se_R \cdot maxE_R$, respectively, where $0 < se_R \le re_R \le 1$ in the Raspberry Pi 3 Model B node. The energy consumption $EI_R(x)$ and $EO_R(x)$ [J] to receive and send data of size x (>0) are $EI_R(x) = PI_R \cdot TI_R(x)$ and $EO_R(x) = PO_R \cdot TO_R(x)$, respectively. Each node f_R consumes the energy $EF_R(x)$ to process input data D_R of size x:

$$
\begin{aligned}
EF_R(x) &= EI_R(x) + EC_R(x) + \delta_R \cdot EO_R(\rho_R \cdot x) \\
&= (re_R \cdot TI_R(x) + TC_R(x) + \delta_R \cdot \rho_R \cdot se_R \cdot TO_R(x)) \cdot maxE_R \\
&= (re_R \cdot rt_R \cdot x + ct_R \cdot C_R(x) + \delta_R \cdot \rho_R \cdot se_R \cdot st_R \cdot x) \cdot maxE_R. \quad (5)
\end{aligned}
$$

3   Recovery from Node Faults

A fog node might be faulty in the GTBFC model. If a fog node f_R stops by fault, every child node f_{Ri} of the node f_R is *disconnected*. No disconnected node can deliver the output data to any ancestor node of the node f_R. In the TBFC model [11,12,17], every fog node at the same level supports a same subprocess. On the other hand, in the GTBFC model, every pair of nodes at the same level may not support the same subprocess depending on the process tree. Only a node equivalent to the faulty node f_R can be a candidate parent node of disconnected nodes. Let $en(f_R)$ be a set of equivalent nodes of a node f_R. We assume each child node f_{Ri} knows every equivalent node of the parent node f_R, i.e. f_{Ri} knows $en(f_R)$.

Let f_U be an equivalent node of a faulty node f_R. Every disconnected child node f_{Ri} of the faulty node f_R is reconnected to the node f_U. Here, the node f_U receives data D_R from the nodes $f_{R1}, ..., f_{Rc_R}$ in addition to data D_U from its own child nodes $f_{U1}, ..., f_{Uc_U}$. Here, i_U shows the size $|D_U|$ and i_R shows the size $|D_R|$. Hence, the node f_U has to process both the data D_U and D_R whose total size is $i_U + i_R$ and consumes more energy $EF_U(i_U + i_R)$ than $EF_U(i_U)$. An equivalent node f_U whose energy consumption $EF_U(i_U + i_R)$ is minimum is selected to be a new parent node of the disconnected nodes. This is the MEG (Minimum Energy node in the GTBFC tree) algorithm. However, the energy consumption and execution time of the selected a new parent node f_U increases since the node f_U has to process not only its own input data D_U but also input data D_R.

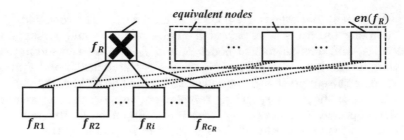

Fig. 4. SMPRG algorithm.

In order to reduce the energy consumption and execution time of a new parent node, we propose the SMPRG (Selecting Multiple Parents for Recovery in the GTBFC tree) algorithm in this paper. Here, an equivalent node f_U is selected for each disconnected node f_{Ri} [Algorithm 1]. Since the output data $d_{R1}, ..., f_{R,c_R}$ of disconnected nodes are distributed to multiple new parent nodes as shown in Fig. 4, the energy consumption and execution time of each new parent node can be smaller than the MEG algorithm.

4   Evaluation

We evaluate the MEG and SMPRG algorithms to select an equivalent node for each disconnected node in the GTBFC model. We consider a height-balanced process tree P of height h as shown in Fig. 5. Here, a root subprocess p has a pair of child subprocesses p_1 and p_2. Each subprocess p_i has one subprocess p_{i1} ($i = 1, 2$). Thus, each subprocess $p_{i1...1}$ has only one child subprocess $p_{i1...11}$ ($i = 1, 2$). A GTBFC tree for the process tree P is a height-balanced four-ary tree with height h (≥ 1), where each non-edge node f_R has four child nodes

---

Algorithm 1. SMPRG algorithm
<hr>

Input: f_R = faulty node;

1    $F = ch(f_R)$; /* set of disconnected nodes. */

2   **while** $F \neq \phi$ **do**

3   **select** a disconnected node f_D in F whose size o_D of output data d_D is maximum;

4        $C = en(f_R)$; /* set of equivalent nodes of f_R. */

5   **while** $C \neq \phi$ **do**

6   **select** an equivalent node f_U where $EF_U(i_U + o_D)$ is minimum in C;

7   **connect** f_D to f_U;

8            $ch(f_U) = ch(f_U) \cup \{f_D\}$; $pt(f_D) = f_U$; /* connect f_U to f_D */

9            $i_U = i_U + o_D$;

10           $C = C - \{f_U\}$;

11       $F = F - \{f_D\}$;

$f_{R1}, ..., f_{Rc}$ $(c = 4)$ and every edge node is at level $h - 1$. The root node f supports the root subprocess p. A pair of the child nodes f_1 and f_2 support the subprocess p_1 and another pair of nodes f_3 and f_4 support the subprocess p_2. Then, the child nodes $f_{11}, ..., f_{14}$ and $f_{21}, ..., f_{24}$ of the nodes f_1 and f_2, respectively, support the subprocess f_{11} as shown in Fig. 6. The child nodes $f_{31}, ..., f_{34}$ and $f_{41}, ..., f_{44}$ of the nodes f_3 and f_4, respectively, support the subprocess p_{21}. Thus, the subprocesses in the process tree P are supported by the nodes in the GTBFC tree. There are totally 4^{h-1} edge nodes in the GTBFC tree.

In the evaluation, we consider two types of data. There are two types of edge nodes, each of which handles one of the two types of data. In the root node f, the two types of data are joined. The nodes $f_{1...}$ and $f_{2...}$ handle one type of the data and the nodes $f_{3...}$ and $f_{4...}$ handle the other type of the data. Let SD be the total size [bit] of sensor data sent by all the sensors, i.e. the sensors totally send sensor data SD [bit] to 4^{h-1} edge nodes. In this paper, we assume SD is 1 [MB]. Each type of the sensor data is $SD/2 = 0.5$ [MB] in size. The size of sensor data which each edge node receives is randomly decided.

In the evaluation, we assume one node f_R is randomly selected to be faulty for each level l $(1 \leq l < h - 1)$. We assume neither a root node f nor every edge node is faulty for simplicity. That is, nodes at levels 0 and $h - 1$ are not faulty. If a node f_R is faulty, c child nodes $f_{R1}, ..., f_{Rc}$ of the faulty node f_R are disconnected.

We consider the RD (Random), MEG, and SMPRG algorithms. In the RD algorithm, an equivalent node is randomly selected to be a new parent node of all the disconnected nodes. In the MEG algorithm, an equivalent node is selected as a new parent node, whose energy consumption is minimum in the equivalent nodes. In the SMPRG algorithm, an equivalent node whose energy consumption is minimum is selected as a new parent node for each disconnected node.

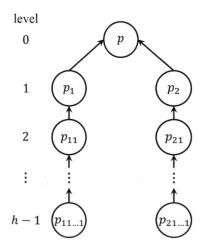

Fig. 5. Process tree in the evaluation.

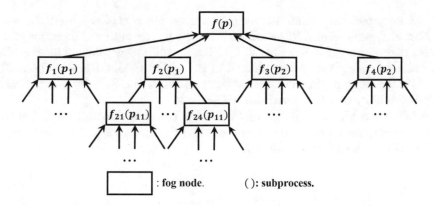

Fig. 6. GTBFC tree.

Each fog node f_R is assumed to be realized by a Raspberry Pi 3 Model B [1]. Here, the maximum electric power consumption $maxE_R$ is 3.7 [W]. In this paper, we assume a pair of the electric power ratios re_R and se_R of a node f_R are 0.729 and 0.676, respectively. Hence, a pair of the power PI_R and PO_R of a node f_R are $0.729 * 3.7 = 2.7$ [W] and $0.676 * 3.7 = 2.5$ [W], respectively. Figures 7 and 8 show the average energy consumed by new parent nodes in the GTBFC tree for the computation complexity $O(x)$ and $O(x^2)$ of subprocesses.

Figures 7 and 8 show the ratios of the energy consumption of the new parent node selected by the SMPRG, MEG, and RD algorithms. The energy consumption of a new parent node selected by the SMPRG algorithm decreases by about 21% and 31% for computation complexity $O(x)$, respectively, and 23% and 37% for $O(x^2)$ in the MEG and RD algorithms, respectively.

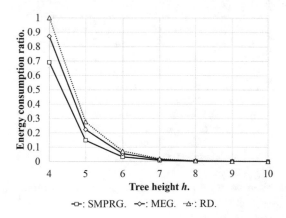

Fig. 7. Energy consumption of new parent nodes of computation complexity $O(x)$ for height h.

In this paper, we assume the execution time ratios rt_R, st_R, and ct_R are 1, 0.222, and 1, respectively. Figures 9 and 10 show the execution time ratios of new parent nodes selected in the SMPRG, MEG, and RD algorithms. The execution time of new parent fog nodes selected in the SMPRG algorithm is about 21% and 31% for computation complexity $O(x)$ and 23% and 37% for $O(x^2)$ shorter than the MEG and RD algorithms, respectively.

As shown in Figs. 7, 8, 9 and 10, the energy consumption and execution time ratios of each new parent node to recover from the faults for disconnected nodes in the GTBFC tree can be reduced.

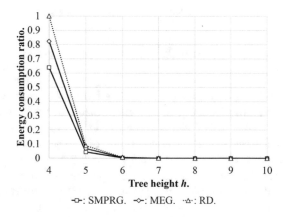

Fig. 8. Energy consumption of new parent nodes of computation complexity $O(x^2)$ for height h.

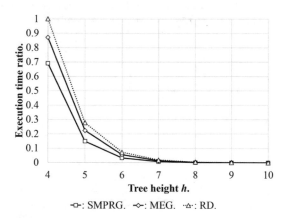

Fig. 9. Execution time of new parent nodes of computation complexity $O(x)$ for height h.

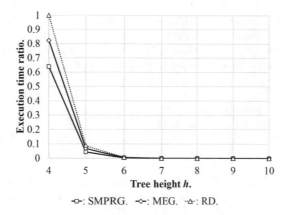

Fig. 10. Execution time of new parent nodes of computation complexity $O(x^2)$ for height h.

5   Concluding Remarks

In this paper, we proposed the GTBFC model where an application process is hierarchically structured. Each subprocess in the process tree is supported by fog nodes. Thus, fog nodes are also structured in a tree according to the process tree. In the GTBFC tree, every pair of nodes at the same level may not support a same subprocess. If a node is faulty, disconnected nodes of the faulty node have to be connected to a node equivalent to the faulty node, i.e. a node which supports a same subprocess as the faulty node. In this paper, we proposed the MEG (Minimum Energy in the GTBFC tree) and SMPRG (Selecting Multiple Parents for Recovery in the GTBFC tree) algorithms to select equivalent nodes for a disconnected node. In the evaluation, the energy to be consumed by equivalent nodes can be reduced in the SMPRG algorithm compared with the RD (Random) and MEG algorithms.

References

1. Raspberry Pi 3 Model B. https://www.raspberrypi.org/products/raspberry-pi-3-model-b/
2. Chida, R., Guo, Y., Oma, R., Nakamura, S., Enokido, T., Takizawa, M.: Implementation of fog nodes in the tree-based fog computing (TBFC) model of the IoT. In: Proceedings of the 7th International Conference on Emerging Internet, Data and Web Technologies (EIDWT 2019), pp. 92–102 (2019)
3. Comer, D.E.: Internetworking with TCP/IP, vol. 1. Prentice Hall, Upper Saddle River (1991)
4. Creeger, M.: Cloud computing: an overview. Queue **7**(5), 3–4 (2009)
5. Duolikun, D., Enokido, T., Takizawa, M.: Dynamic migration of virtual machines to reduce energy consumption in a cluster. Int. J. Grid Util. Comput. (IJGUC) **9**(4), 357–366 (2018)

6. Enokido, T., Ailixier, A., Takizawa, M.: A model for reducing power consumption in peer-to-peer systems. IEEE Syst. J. **4**(2), 221–229 (2010)
7. Enokido, T., Ailixier, A., Takizawa, M.: Process allocation algorithms for saving power consumption in peer-to-peer systems. IEEE Trans. Ind. Electron. **58**(6), 2097–2105 (2011)
8. Enokido, T., Ailixier, A., Takizawa, M.: An extended simple power consumption model for selecting a server to perform computation type processes in digital ecosystems. IEEE Trans. Ind. Inform. **10**(2), 1627–1636 (2014)
9. Hanes, D., Salgueiro, G., Grossetete, P., Barton, R., Henry, J.: IoT Fundamentals: Networking Technologies, Protocols, and Use Cases for the Internet of Things. Cisco Press, Indianapolis (2018)
10. Kataoka, H., Nakamura, S., Duolikun, D., Enokido, T., Takizawa, M.: Multi-level power consumption model and energy-aware server selection algorithm. Int. J. Grid Util. Comput. (IJGUC) **8**(3), 201–210 (2017)
11. Oma, R., Nakamura, S., Duolikun, D., Enokido, T., Takizawa, M.: An energy-efficient model for fog computing in the internet of things (IoT). Internet Things **1–2**, 14–26 (2018)
12. Oma, R., Nakamura, S., Duolikun, D., Enokido, T., Takizawa, M.: Evaluation of an energy-efficient tree-based model of fog computing. In: Proceedings of the 21st International Conference on Network-Based Information Systems (NBiS 2018), pp. 99–109 (2018)
13. Oma, R., Nakamura, S., Duolikun, D., Enokido, T., Takizawa, M.: Fault-tolerant fog computing models in the IoT. In: Proceedings of the 13th International Conference on P2P, Parallel, Grid, Cloud and Internet Computing (3PGCIC 2018), pp. 14–25 (2018)
14. Oma, R., Nakamura, S., Duolikun, D., Enokido, T., Takizawa, M.: Energy-efficient recovery algorithm in the fault-tolerant tree-based fog computing (FTBFC) model. In: Proceedings of the 33rd International Conference on Advanced Information Networking and Applications (AINA 2019), pp. 132–143 (2019)
15. Oma, R., Nakamura, S., Duolikun, D., Enokido, T., Takizawa, M.: A fault-tolerant tree-based fog computing model. Int. J Web Grid Serv. (IJWGS) (2019, accepted)
16. Oma, R., Nakamura, S., Duolikun, D., Enokido, T., Takizawa, M.: Subprocess transmission strategies for recovering from faults in the tree-based fog computing (TBFC) model. In: Proceedings of the 13th International Conference on Complex, Intelligent, and Software Intensive Systems (CISIS-2019) (2019, accepted)
17. Oma, R., Nakamura, S., Enokido, T., Takizawa, M.: A tree-based model of energy-efficient fog computing systems in IoT. In: Proceedings of the 12th International Conference on Complex, Intelligent, and Software Intensive Systems (CISIS 2018), pp. 991–1001 (2018)
18. Rahmani, A.M., Liljeberg, P., Preden, J.S., Jantsch, A.: Fog Computing in the Internet of Things. Springer, Cham (2018)

Performance Evaluation of WMNs by WMN-PSOHC System Considering Random Inertia Weight and Linearly Decreasing Vmax Replacement Methods

Shinji Sakamoto[1](✉), Seiji Ohara[2], Leonard Barolli[3], and Shusuke Okamoto[1]

[1] Department of Computer and Information Science, Seikei University,
3-3-1 Kichijoji-Kitamachi, Musashino-shi, Tokyo 180-8633, Japan
shinji.sakamoto@ieee.org, okam@st.seikei.ac.jp
[2] Graduate School of Engineering, Fukuoka Institute of Technology,
3-30-1 Wajiro-Higashi, Higashi-Ku, Fukuoka 811-0295, Japan
seiji.ohara.19@gmail.com
[3] Department of Information and Communication Engineering, Fukuoka Institute
of Technology, 3-30-1 Wajiro-Higashi, Higashi-Ku, Fukuoka 811-0295, Japan
barolli@fit.ac.jp

Abstract. Wireless Mesh Networks (WMNs) have many advantages such as low cost and increased high-speed wireless Internet connectivity, therefore WMNs are becoming an important networking infrastructure. In our previous work, we implemented a Particle Swarm Optimization (PSO) based simulation system for node placement in WMNs, called WMN-PSO. Also, we implemented a simulation system based on Hill Climbing (HC) for solving node placement problem in WMNs, called WMN-HC. Then, we implemented a hybrid simulation system based on PSO and HC, called WMN-PSOHC. In this paper, we analyse the performance of WMNs by using WMN-PSOHC considering Random Inertia Weight Method (RIWM) and Linearly Decreasing Vmax Method (LDVM). Simulation results show that a good performance is achieved for RIWM compared with LDVM.

1 Introduction

The wireless networks and devices are becoming increasingly popular and they provide users access to information and communication anytime and anywhere [2, 6–9, 12, 17, 23, 28–30]. Wireless Mesh Networks (WMNs) are gaining a lot of attention because of their low cost nature that makes them attractive for providing wireless Internet connectivity. A WMN is dynamically self-organized and self-configured, with the nodes in the network automatically establishing and maintaining mesh connectivity among them-selves (creating, in effect, an ad hoc network). This feature brings many advantages to WMNs such as low up-front cost, easy network maintenance, robustness and reliable service coverage [1]. Moreover, such infrastructure can be used to deploy community networks,

© Springer Nature Switzerland AG 2020
L. Barolli et al. (Eds.): NBiS-2019, AISC 1036, pp. 27–36, 2020.
https://doi.org/10.1007/978-3-030-29029-0_3

metropolitan area networks, municipal and cooperative networks, and to support applications for urban areas, medical, transport and surveillance systems.

We consider the version of the mesh router nodes placement problem in which we are given a grid area where to deploy a number of mesh router nodes and a number of mesh client nodes of fixed positions (of an arbitrary distribution) in the grid area. The objective is to find a location assignment for the mesh routers to the cells of the grid area that maximizes the network connectivity and client coverage. Node placement problems are known to be computationally hard to solve [10,11,34]. In some previous works, intelligent algorithms have been recently investigated [3,5,13,15,18,19,21,26,27].

In [22], we implemented a Particle Swarm Optimization (PSO) based simulation system, called WMN-PSO. Also, we implemented a simulation system based on Hill Climbing (HC) for solving node placement problem in WMNs, called WMN-HC [16,20].

In our previous work, we implemented a hybrid simulation system based on PSO and HC. We called this system WMN-PSOHC. In this paper, we analyse the performance of hybrid WMN-PSOHC system considering Random Inertia Weight Method (RIWM) and Linearly Decreasing Vmax Method (LDVM).

The rest of the paper is organized as follows. The mesh router nodes placement problem is defined in Sect. 2. We present our designed and implemented hybrid simulation system in Sect. 3. The simulation results are given in Sect. 4. Finally, we give conclusions and future work in Sect. 5.

2 Node Placement Problem in WMNs

For this problem, we have a grid area arranged in cells we want to find where to distribute a number of mesh router nodes and a number of mesh client nodes of fixed positions (of an arbitrary distribution) in the considered area. The objective is to find a location assignment for the mesh routers to the area that maximizes the network connectivity and client coverage. Network connectivity is measured by Size of Giant Component (SGC) of the resulting WMN graph, while the user coverage is simply the number of mesh client nodes that fall within the radio coverage of at least one mesh router node and is measured by Number of Covered Mesh Clients (NCMC).

An instance of the problem consists as follows.

- N mesh router nodes, each having its own radio coverage, defining thus a vector of routers.
- An area $W \times H$ where to distribute N mesh routers. Positions of mesh routers are not pre-determined and are to be computed.
- M client mesh nodes located in arbitrary points of the considered area, defining a matrix of clients.

It should be noted that network connectivity and user coverage are among most important metrics in WMNs and directly affect the network performance.

In this work, we have considered a bi-objective optimization in which we first maximize the network connectivity of the WMN (through the maximization of the SGC) and then, the maximization of the NCMC.

In fact, we can formalize an instance of the problem by constructing an adjacency matrix of the WMN graph, whose nodes are router nodes and client nodes and whose edges are links between nodes in the mesh network. Each mesh node in the graph is a triple $v = <x, y, r>$ representing the 2D location point and r is the radius of the transmission range. There is an arc between two nodes u and v, if v is within the transmission circular area of u.

3   Proposed and Implemented Simulation System

3.1   WMN-PSOHC Hybrid Simulation System

3.1.1   Particle Swarm Optimization

In Particle Swarm Optimization (PSO) algorithm, a number of simple entities (the particles) are placed in the search space of some problem or function and each evaluates the objective function at its current location. The objective function is often minimized and the exploration of the search space is not through evolution [14]. However, following a widespread practice of borrowing from the evolutionary computation field, in this work, we consider the bi-objective function and fitness function interchangeably. Each particle then determines its movement through the search space by combining some aspect of the history of its own current and best (best-fitness) locations with those of one or more members of the swarm, with some random perturbations. The next iteration takes place after all particles have been moved. Eventually the swarm as a whole, like a flock of birds collectively foraging for food, is likely to move close to an optimum of the fitness function.

Each individual in the particle swarm is composed of three $\mathcal{D}$-dimensional vectors, where $\mathcal{D}$ is the dimensionality of the search space. These are the current position $\vec{x}_i$, the previous best position $\vec{p}_i$ and the velocity $\vec{v}_i$.

The particle swarm is more than just a collection of particles. A particle by itself has almost no power to solve any problem; progress occurs only when the particles interact. Problem solving is a population-wide phenomenon, emerging from the individual behaviors of the particles through their interactions. In any case, populations are organized according to some sort of communication structure or topology, often thought of as a social network. The topology typically consists of bidirectional edges connecting pairs of particles, so that if j is in i's neighborhood, i is also in j's. Each particle communicates with some other particles and is affected by the best point found by any member of its topological neighborhood. This is just the vector $\vec{p}_i$ for that best neighbor, which we will denote with $\vec{p}_g$. The potential kinds of population "social networks" are hugely varied, but in practice certain types have been used more frequently.

In the PSO process, the velocity of each particle is iteratively adjusted so that the particle stochastically oscillates around $\vec{p}_i$ and $\vec{p}_g$ locations.

3.1.2   Hill Climbing

Hill Climbing (HC) algorithm is a heuristic algorithm. The idea of HC is simple. In HC, the solution s' is accepted as the new current solution if $\delta \leq 0$ holds, where $\delta = f(s') - f(s)$. Here, the function f is called the fitness function. The fitness function gives points to a solution so that the system can evaluate the next solution s' and the current solution s.

The most important factor in HC is to define the neighbor solution, effectively. The definition of the neighbor solution affects HC performance directly. In our WMN-PSOHC system, we use the next step of particle-pattern positions as the neighbor solutions for the HC part.

3.1.3   WMN-PSOHC System Description

In following, we present the initialization, particle-pattern, fitness function and router replacement methods.

Initialization

Our proposed system starts by generating an initial solution randomly, by *ad hoc* methods [35]. We decide the velocity of particles by a random process considering the area size. For instance, when the area size is $W \times H$, the velocity is decided randomly from $-\sqrt{W^2 + H^2}$ to $\sqrt{W^2 + H^2}$.

Particle-Pattern

A particle is a mesh router. A fitness value of a particle-pattern is computed by combination of mesh routers and mesh clients positions. In other words, each particle-pattern is a solution as shown is Fig. 1. Therefore, the number of particle-patterns is a number of solutions.

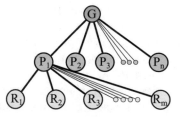

G: Global Solution
P: Particle-pattern
R: Mesh Router
n: Number of Particle-patterns
m: Number of Mesh Routers

Fig. 1. Relationship among global solution, particle-patterns and mesh routers.

Fitness Function

One of most important thing is to decide the determination of an appropriate objective function and its encoding. In our case, each particle-pattern has an own fitness value and compares other particle-pattern's fitness value in order to share information of global solution. The fitness function follows a hierarchical approach in which the main objective is to maximize the SGC in WMN. Thus, we use α and β weight-coefficients for the fitness function and the fitness function of this scenario is defined as:

$$\text{Fitness} = \alpha \times \text{SGC}(\boldsymbol{x}_{ij}, \boldsymbol{y}_{ij}) + \beta \times \text{NCMC}(\boldsymbol{x}_{ij}, \boldsymbol{y}_{ij}).$$

Router Replacement Methods

A mesh router has x, y positions and velocity. Mesh routers are moved based on velocities. There are many router replacement methods in PSO field [4,31–33]. In this paper, we consider RIWM and LDVM.

Random Inertia Weight Method (RIWM)
In RIWM, the ω parameter is changing randomly from 0.5 to 1.0. The C_1 and C_2 are kept 2.0. The ω can be estimated by the week stable region. The average of ω is 0.75 [24,32].
Linearly Decreasing Vmax Method (LDVM)
In LDVM, PSO parameters are set to unstable region ($\omega = 0.9$, $C_1 = C_2 = 2.0$). A value of V_{max} which is maximum velocity of particles is considered. With increasing of iteration of computations, the V_{max} is kept decreasing linearly [25,31].

3.2 WMN-PSOHC Web GUI Tool

The Web application follows a standard Client-Server architecture and is implemented using LAMP (Linux + Apache + MySQL + PHP) technology (see Fig. 2). We show the WMN-PSOHC Web GUI tool in Fig. 3. Remote users (clients) submit their requests by completing first the parameter setting. The parameter values to be provided by the user are classified into three groups, as follows.

- Parameters related to the problem instance: These include parameter values that determine a problem instance to be solved and consist of number of router nodes, number of mesh client nodes, client mesh distribution, radio coverage interval and size of the deployment area.

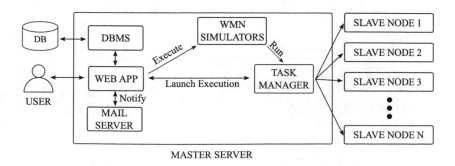

Fig. 2. System structure for web interface.

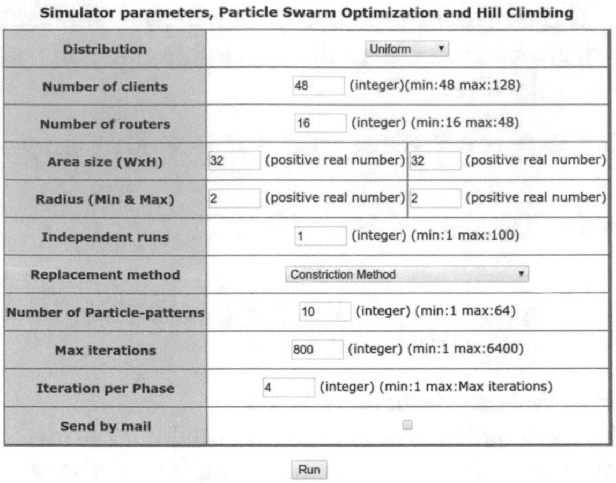

Fig. 3. WMN-PSOHC Web GUI Tool.

- Parameters of the resolution method: Each method has its own parameters.
- Execution parameters: These parameters are used for stopping condition of the resolution methods and include number of iterations and number of independent runs. The former is provided as a total number of iterations and depending on the method is also divided per phase (e.g., number of iterations in a exploration). The later is used to run the same configuration for the same problem instance and parameter configuration a certain number of times.

Table 1. Parameter settings.

Parameters	Values
Clients distribution	Normal distribution
Area size	32.0×32.0
Number of mesh routers	16
Number of mesh clients	48
Total iterations	800
Iteration per phase	4
Number of particle-patterns	9
Radius of a mesh router	2.0
Fitness function weight-coefficients (α, β)	0.7, 0.3
Replacement method	RIWM, LDVM

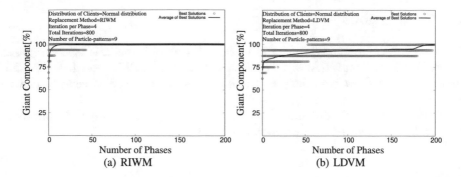

Fig. 4. Simulation results of WMN-PSOHC for SGC.

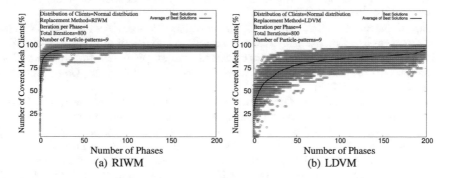

Fig. 5. Simulation results of WMN-PSOHC for NCMC.

4 Simulation Results

In this section, we show simulation results using WMN-PSOHC system. In this work, we consider Normal distributions of mesh clients. The number of mesh routers is considered 16 and the number of mesh clients 48. The total number of iterations is considered 800 and the iterations per phase is considered 4. We consider the number of particle-patterns 9. We conducted simulations 100 times, in order to avoid the effect of randomness and create a general view of results. We show the parameter setting for WMN-PSOHC in Table 1.

We show the simulation results in Figs. 4 and 5. For SGC, both replacement methods reach the maximum (100%). However, RIWM converges faster than LDVM. Also, for the NCMC, RIWM performs better than LDVM. Therefore, we conclude that the performance for RIWM is better compared with LDVM.

5   Conclusions

In this work, we evaluated the performance of a hybrid simulation system based on PSO and HC (called WMN-PSOHC) considering RIWM and LDVM. Simulation results show that the performance is better for RIWM compared with LDVM.

In our future work, we would like to evaluate the performance of the proposed system for different parameters and scenarios.

References

1. Akyildiz, I.F., Wang, X., Wang, W.: Wireless mesh networks: a survey. Comput. Netw. **47**(4), 445–487 (2005)
2. Barolli, A., Sakamoto, S., Barolli, L., Takizawa, M.: Performance analysis of simulation system based on particle swarm optimization and distributed genetic algorithm for WMNs considering different distributions of mesh clients. In: International Conference on Innovative Mobile and Internet Services in Ubiquitous Computing, pp. 32–45. Springer (2018)
3. Barolli, A., Sakamoto, S., Ozera, K., Barolli, L., Kulla, E., Takizawa, M.: Design and implementation of a hybrid intelligent system based on particle swarm optimization and distributed genetic algorithm. In: International Conference on Emerging Internetworking, Data and Web Technologies, pp. 79–93. Springer (2018)
4. Clerc, M., Kennedy, J.: The particle swarm-explosion, stability, and convergence in a multidimensional complex space. IEEE Trans. Evol. Comput. **6**(1), 58–73 (2002)
5. Girgis, M.R., Mahmoud, T.M., Abdullatif, B.A., Rabie, A.M.: Solving the wireless mesh network design problem using genetic algorithm and simulated annealing optimization methods. Int. J. Comput. Appl. **96**(11), 1–10 (2014)
6. Goto, K., Sasaki, Y., Hara, T., Nishio, S.: Data gathering using mobile agents for reducing traffic in dense mobile wireless sensor networks. Mob. Inf. Syst. **9**(4), 295–314 (2013)
7. Inaba, T., Elmazi, D., Sakamoto, S., Oda, T., Ikeda, M., Barolli, L.: A secure-aware call admission control scheme for wireless cellular networks using fuzzy logic and its performance evaluation. J. Mob. Multimed. **11**(3&4), 213–222 (2015)
8. Inaba, T., Obukata, R., Sakamoto, S., Oda, T., Ikeda, M., Barolli, L.: Performance evaluation of a QoS-aware fuzzy-based CAC for LAN access. Int. J. Space Based Situated Comput. **6**(4), 228–238 (2016)
9. Inaba, T., Sakamoto, S., Oda, T., Ikeda, M., Barolli, L.: A testbed for admission control in WLAN: a fuzzy approach and its performance evaluation. In: International Conference on Broadband and Wireless Computing, Communication and Applications, pp. 559–571. Springer (2016)
10. Lim, A., Rodrigues, B., Wang, F., Xu, Z.: k-center problems with minimum coverage. In: Computing and Combinatorics, pp. 349–359 (2004)
11. Maolin, T., et al.: Gateways placement in backbone wireless mesh networks. Int. J. Commun. Netw. Syst. Sci. **2**(1), 44 (2009)
12. Matsuo, K., Sakamoto, S., Oda, T., Barolli, A., Ikeda, M., Barolli, L.: Performance analysis of WMNs by WMN-GA simulation system for two WMN architectures and different TCP congestion-avoidance algorithms and client distributions. Int. J. Commun. Netw. Distrib. Syst. **20**(3), 335–351 (2018)

13. Naka, S., Genji, T., Yura, T., Fukuyama, Y.: A hybrid particle swarm optimization for distribution state estimation. IEEE Trans. Power Syst. **18**(1), 60–68 (2003)
14. Poli, R., Kennedy, J., Blackwell, T.: Particle swarm optimization. Swarm Intell. **1**(1), 33–57 (2007)
15. Sakamoto, S., Kulla, E., Oda, T., Ikeda, M., Barolli, L., Xhafa, F.: A comparison study of simulated annealing and genetic algorithm for node placement problem in wireless mesh networks. J. Mob. Multimedia **9**(1–2), 101–110 (2013)
16. Sakamoto, S., Kulla, E., Oda, T., Ikeda, M., Barolli, L., Xhafa, F.: A comparison study of hill climbing, simulated annealing and genetic algorithm for node placement problem in WMNs. J. High Speed Netw. **20**(1), 55–66 (2014)
17. Sakamoto, S., Kulla, E., Oda, T., Ikeda, M., Barolli, L., Xhafa, F.: A simulation system for WMN based on SA: performance evaluation for different instances and starting temperature values. Int. J. Space Based Situated Comput. **4**(3–4), 209–216 (2014)
18. Sakamoto, S., Kulla, E., Oda, T., Ikeda, M., Barolli, L., Xhafa, F.: Performance evaluation considering iterations per phase and SA temperature in WMN-SA system. Mob. Inf. Syst. **10**(3), 321–330 (2014)
19. Sakamoto, S., Lala, A., Oda, T., Kolici, V., Barolli, L., Xhafa, F.: Application of WMN-SA simulation system for node placement in wireless mesh networks: a case study for a realistic scenario. Int. J. Mob. Comput. Multimedia Commun. (IJMCMC) **6**(2), 13–21 (2014)
20. Sakamoto, S., Lala, A., Oda, T., Kolici, V., Barolli, L., Xhafa, F.: Analysis of WMN-HC simulation system data using Friedman test. In: The Ninth International Conference on Complex, Intelligent, and Software Intensive Systems (CISIS 2015), pp. 254–259. IEEE (2015)
21. Sakamoto, S., Oda, T., Ikeda, M., Barolli, L., Xhafa, F.: An integrated simulation system considering WMN-PSO simulation system and network simulator 3. In: International Conference on Broadband and Wireless Computing, Communication and Applications, pp. 187–198. Springer (2016)
22. Sakamoto, S., Oda, T., Ikeda, M., Barolli, L., Xhafa, F.: Implementation and evaluation of a simulation system based on particle swarm optimisation for node placement problem in wireless mesh networks. Int. J. Commun. Netw. Distrib. Syst. **17**(1), 1–13 (2016)
23. Sakamoto, S., Obukata, R., Oda, T., Barolli, L., Ikeda, M., Barolli, A.: Performance analysis of two wireless mesh network architectures by WMN-SA and WMN-TS simulation systems. J. High Speed Netw. **23**(4), 311–322 (2017)
24. Sakamoto, S., Ozera, K., Barolli, A., Ikeda, M., Barolli, L., Takizawa, M.: Performance evaluation of WMNs by WMN-PSOSA simulation system considering random inertia weight method and linearly decreasing Vmax method. International Conference on Broadband and Wireless Computing, Communication and Applications, pp. 114–124. Springer (2017)
25. Sakamoto, S., Ozera, K., Ikeda, M., Barolli, L.: Performance evaluation of WMNs by WMN-PSOSA simulation system considering constriction and linearly decreasing inertia weight methods. In: International Conference on Network-Based Information Systems, pp. 3–13. Springer (2017)
26. Sakamoto, S., Ozera, K., Oda, T., Ikeda, M., Barolli, L.: Performance evaluation of intelligent hybrid systems for node placement in wireless mesh networks: a comparison study of WMN-PSOHC and WMN-PSOSA. In: International Conference on Innovative Mobile and Internet Services in Ubiquitous Computing, pp. 16–26. Springer (2017)

27. Sakamoto, S., Ozera, K., Oda, T., Ikeda, M., Barolli, L.: Performance evaluation of WMN-PSOHC and WMN-PSO simulation systems for node placement in wireless mesh networks: a comparison study. In: International Conference on Emerging Internetworking, Data and Web Technologies, pp. 64–74. Springer (2017)
28. Sakamoto, S., Ozera, K., Barolli, A., Barolli, L., Kolici, V., Takizawa, M.: Performance evaluation of WMN-PSOSA considering four different replacement methods. In: International Conference on Emerging Internetworking, Data and Web Technologies, pp. 51–64. Springer (2018)
29. Sakamoto, S., Ozera, K., Ikeda, M., Barolli, L.: Implementation of intelligent hybrid systems for node placement problem in WMNs considering particle swarm optimization, hill climbing and simulated annealing. Mob. Netw. Appl. **23**(1), 27–33 (2018)
30. Sakamoto, S., Ozera, K., Barolli, A., Ikeda, M., Barolli, L., Takizawa, M.: Implementation of an intelligent hybrid simulation systems for WMNs based on particle swarm optimization and simulated annealing: performance evaluation for different replacement methods. Soft Comput. **23**(9), 3029–3035 (2019)
31. Schutte, J.F., Groenwold, A.A.: A study of global optimization using particle swarms. J. Glob. Optim. **31**(1), 93–108 (2005)
32. Shi, Y.: Particle swarm optimization. IEEE Connect. **2**(1), 8–13 (2004)
33. Shi, Y., Eberhart, R.C.: Parameter selection in particle swarm optimization. In: Porto, V.W., Saravanan, N., Waagen, D., Eiben, A.E. (eds.) Evolutionary Programming VII, pp. 591–600. Springer, Heidelberg (1998)
34. Wang, J., Xie, B., Cai, K., Agrawal, D.P.: Efficient mesh router placement in wireless mesh networks. In: Proceedings of IEEE International Conference on Mobile Adhoc and Sensor Systems (MASS 2007), pp. 1–9 (2007)
35. Xhafa, F., Sanchez, C., Barolli, L.: Ad hoc and neighborhood search methods for placement of mesh routers in wireless mesh networks. In: Proceedings of 29th IEEE International Conference on Distributed Computing Systems Workshops (ICDCS 2009), pp. 400–405 (2009)

Implementation of a Fuzzy-Based Simulation System and a Testbed for Improving Driving Conditions in VANETs Considering Drivers's Vital Signs

Kevin Bylykbashi[1(✉)], Ermioni Qafzezi[2], Makoto Ikeda[3], Keita Matsuo[3], and Leonard Barolli[3]

[1] Graduate School of Engineering, Fukuoka Institute of Technology (FIT), 3-30-1 Wajiro-Higashi, Higashi-Ku, Fukuoka 811–0295, Japan
bylykbashi.kevin@gmail.com
[2] Sorbonne Université, Université de Technologie de Compiègne, Compiègne, France
ermioni.qafzezi@etu.utc.fr
[3] Department of Information and Communication Engineering, Fukuoka Institute of Technology (FIT), 3-30-1 Wajiro-Higashi, Higashi-Ku, Fukuoka 811-0295, Japan
makoto.ikd@acm.org, {kt-matsuo,barolli}@fit.ac.jp

Abstract. Vehicular Ad Hoc Networks (VANETs) have gained a great attention due to the rapid development of mobile internet and Internet of Things (IoT) applications. With the evolution of technology, it is expected that VANETs will be massively deployed in upcoming vehicles. In addition, ambitious efforts are being done to incorporate Ambient Intelligence (AmI) technology in the vehicles, as it will be an important factor for VANET to accomplish one of its main goals, the road safety. In this paper, we propose an intelligent system for improving driving condition using fuzzy logic. The proposed system considers in-car environment data such as the ambient temperature and noise, and driver's vital signs data, i.e. heart and respiratory rate, to make the decision. Then, it uses the smart box to inform the driver and to provide a better assistance. We aim to realize a new system to support the driver for safe driving. We evaluated the performance of proposed system by computer simulations and experiments. From the evaluation results, we conclude that the driver's heart rate and respiratory rate, noise level and vehicle's inside temperature have different effects to the driver's condition.

1 Introduction

Traffic accidents, road congestion and environmental pollution are persistent problems faced by both developed and developing countries, which have made people live in difficult situations. Among these, the traffic incidents are the most serious ones because they result in huge loss of life and property. For decades, we have seen governments and car manufacturers struggle for safer roads and car

© Springer Nature Switzerland AG 2020
L. Barolli et al. (Eds.): NBiS-2019, AISC 1036, pp. 37–48, 2020.
https://doi.org/10.1007/978-3-030-29029-0_4

accident prevention. The development in wireless communications has allowed companies, researchers and institutions to design communication systems that provide new solutions for these issues. Therefore, new types of networks, such as Vehicular Ad Hoc Networks (VANETs) have been created. VANET consists of a network of vehicles in which vehicles are capable of communicating among themselves in order to deliver valuable information such as safety warnings and traffic information.

Nowadays, every car is likely to be equipped with various forms of smart sensors, wireless communication modules, storage and computational resources. The sensors will gather information about the road and environment conditions and share it with neighboring vehicles and adjacent roadside units (RSU) via vehicle-to-vehicle (V2V) or vehicle-to-infrastructure (V2I) communication. However, the difficulty lies on how to understand the sensed data and how to make intelligent decisions based on the provided information.

As a result, Ambient Intelligence (AmI) becomes a significant factor for VANETs. Various intelligent systems and applications are now being deployed and they are going to change the way manufacturers design vehicles. These systems include many intelligence computational technologies such as fuzzy logic, neural networks, machine learning, adaptive computing, voice recognition, and so on, and they are already announced or deployed [1]. The goal is to improve both vehicle safety and performance by realizing a series of automatic driving technologies based on the situation recognition. The car control relies on the measurement and recognition of the outside environment and their reflection on driving operation.

On the other hand, we are focused on the in-car information and driver's vital information to detect the danger or risk situation and inform the driver about the risk or change his mood. Thus, our goal is to prevent the accidents by supporting the drivers. In order to realize the proposed system, we use some Internet of Things (IoT) devices equipped with various sensors for in-car monitoring.

In this paper, we propose a fuzzy-based system for improving driving conditions considering four parameters: Heart Rate (HR), Respiratory Rate (RR), Noise Level (NL) and Vehicle's Inside Temperature (VIT) to decide the Driver's Situation Condition (DSC).

The structure of the paper is as follows. In Sect. 2, we present an overview of VANETs. In Sect. 3, we present a short description of AmI. In Sect. 4, we describe the proposed fuzzy-based system and its implementation. In Sect. 5, we discuss the simulation and experimental results. Finally, conclusions and future work are given in Sect. 6.

2   Vehicular Ad Hoc Networks (VANETs)

VANETs are a type of wireless networks that have emerged thanks to advances in wireless communication technologies and the automotive industry. VANETs are considered to have an enormous potential in enhancing road traffic safety and traffic efficiency. Therefore, various governments have launched programs dedicated to the development and consolidation of vehicular communications and networking and both industrial and academic researchers are addressing many related challenges, including socio-economic ones, which are among the most important [2].

The VANET technology uses moving vehicle as nodes to form a wireless mobile network. It aims to provide fast and cost-efficient data transfer for the advantage of passenger safety and comfort. To improve road safety and travel comfort of voyagers and drivers, Intelligent Transport Systems (ITS) are developed. The ITS manages the vehicle traffic, support drivers with safety and other information, and provide some services such as automated toll collection and driver assist systems [3].

The VANETs provide new prospects to improve advanced solutions for making reliable communication between vehicles. VANETs can be defined as a part of ITS which aims to make transportation systems faster and smarter, in which vehicles are equipped with some short-range and medium-range wireless communication [4]. In a VANET, wireless vehicles are able to communicate directly with each other (i.e., emergency vehicle warning, stationary vehicle warning) and also served various services (i.e., video streaming, internet) from access points (i.e., 3G or 4G) through roadside units.

3   Ambient Intelligence (AmI)

The AmI is the vision that technology will become invisible, embedded in our natural surroundings, present whenever we need it, enabled by simple and effortless interactions, attuned to all our senses, adaptive to users and context and autonomously acting [5]. High quality information and content must be available to any user, anywhere, at any time, and on any device.

In order that AmI becomes a reality, it should completely envelope humans, without constraining them. Distributed embedded systems for AmI are going to change the way we design embedded systems, as well as the way we think about such systems. But, more importantly, they will have a great impact on the way we live. Applications ranging from safe driving systems, smart buildings and home security, smart fabrics or e-textiles, to manufacturing systems and rescue and recovery operations in hostile environments, are poised to become part of society and human lives.

The AmI deals with a new world of ubiquitous computing devices, where physical environments interact intelligently and unobtrusively with people. AmI environments can be diverse, such as homes, offices, meeting rooms, hospitals, control centers, vehicles, tourist attractions, stores, sports facilities, and music devices.

In the future, small devices will monitor the health status in a continuous manner, diagnose any possible health conditions, have conversation with people to persuade them to change the lifestyle for maintaining better health, and communicates with the doctor, if needed [6]. The device might even be embedded into the regular clothing fibers in the form of very tiny sensors and it might

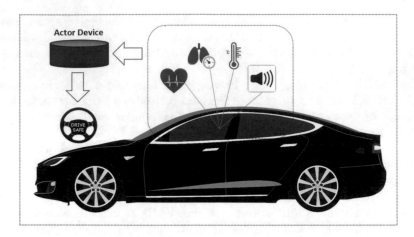

Fig. 1. Proposed system architecture.

communicate with other devices including the variety of sensors embedded into the home to monitor the lifestyle. For example, people might be alarmed about the lack of a healthy diet based on the items present in the fridge and based on what they are eating outside regularly.

The AmI paradigm represents the future vision of intelligent computing where environments support the people inhabiting them [7–9]. In this new computing paradigm, the conventional input and output media no longer exist, rather the sensors and processors will be integrated into everyday objects, working together in harmony in order to support the inhabitants [10]. By relying on various artificial intelligence techniques, AmI promises the successful interpretation of the wealth of contextual information obtained from such embedded sensors and will adapt the environment to the user needs in a transparent and anticipatory manner.

4   Proposed System

In this work, we use fuzzy logic to implement the proposed system. Fuzzy sets and fuzzy logic have been developed to manage vagueness and uncertainty in a reasoning process of an intelligent system such as a knowledge based system, an expert system or a logic control system [11–16]. In Fig. 1, we show the architecture of our proposed system.

4.1   Proposed Fuzzy-Based Simulation System

The proposed system called Fuzzy-based System for improving Driver's Situation Condition (FSDSC) is shown in Fig. 2. For the implementation of our system, we consider four input parameters: Heart Rate (HR), Respiratory Rate (RR), Noise Level (NL) and Vehicle's Inside Temperature (VIT) to determine the Driver's

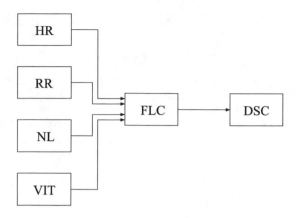

Fig. 2. Proposed system structure.

Table 1. Parameters and their term sets for FSDSC.

Parameters	Term sets
Heart Rate (HR)	Slow (S), Normal (No), Fast (F)
Respiratory Rate (RR)	Slow (Sl), Normal (Nm), Fast (Fa)
Noise Level (NL)	Quiet (Q), Noisy (N), Very Noisy (VN)
Vehicle's Inside Temperature (VIT)	Low (L), Medium (M), High (H)
Driver's Situation Condition (DSC)	Extremely Bad (EB), Very Bad (VB), Bad (B), Normal (Nor), Good (G), Very Good (VG), Extremely Good (EG)

Situation Condition (DSC). These four input parameters are not correlated with each other, for this reason we use fuzzy system. The input parameters are fuzzified using the membership functions showed in Fig. 3(a), (b), (c) and (d). In Fig. 3(e) are shown the membership functions used for the output parameter. We use triangular and trapezoidal membership functions because they are suitable for real-time operation. The term sets for each linguistic parameter are shown in Table 1. We decided the number of term sets by carrying out many simulations. In Table 2, we show the Fuzzy Rule Base (FRB) of FSDSC, which consists of 81 rules. The control rules have the form: IF "conditions" THEN "control action". For instance, for Rule 1: "IF HR is S, RR is Sl, NL is Q and VIT is L, THEN DSC is VB" or for Rule 38: "IF HR is No, RR is Nm, NL is Q and VIT is M, THEN DSC is EG".

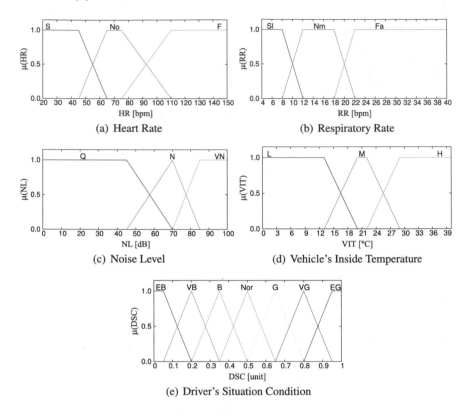

Fig. 3. Membership functions.

4.2   Testbed Description

In order to evaluate the proposed system, we implemented a testbed and carried out experiments in a real scenario. A snapshot of testbed is shown in Fig. 4. The testbed is composed of sensing and processing components. The sensing system consists of two parts. The first part is implemented in the Arduino Platform while the second one consists of a Microwave Sensor Module (MSM) called DC6M4JN3000. We set-up sensors on Arduino Uno to measure the environment temperature and noise and used the MSM to measure the driver's respiratory and heart rate. Then, we implemented a processing device to get the sensed data and to run our fuzzy system. The sensing components are connected to the processing device via USB cable. We used Arduino IDE and Processing language to get the sensed data from the first module, whereas the MSM generates the sensed data in the appropriate format itself. Then, we use FuzzyC [17] to fuzzify these data and to determine the Driver's Situation Condition which is the output of our proposed system. Based on the DSC an appropriate task can be performed.

Table 2. FRB of FSDSC.

No	HR	RR	NL	VIT	DSC	No	HR	RR	NL	VIT	DSC	No	HR	RR	NL	VIT	DSC
1	S	Sl	Q	L	VB	28	No	Sl	Q	L	Nor	55	F	Sl	Q	L	EB
2	S	Sl	Q	M	B	29	No	Sl	Q	M	VG	56	F	Sl	Q	M	B
3	S	Sl	Q	H	VB	30	No	Sl	Q	H	Nor	57	F	Sl	Q	H	EB
4	S	Sl	N	L	EB	31	No	Sl	N	L	B	58	F	Sl	N	L	EB
5	S	Sl	N	M	VB	32	No	Sl	N	M	G	59	F	Sl	N	M	VB
6	S	Sl	N	H	EB	33	No	Sl	N	H	B	60	F	Sl	N	H	EB
7	S	Sl	VN	L	EB	34	No	Sl	VN	L	VB	61	F	Sl	VN	L	EB
8	S	Sl	VN	M	EB	35	No	Sl	VN	M	Nor	62	F	Sl	VN	M	EB
9	S	Sl	VN	H	EB	36	No	Sl	VN	H	VB	63	F	Sl	VN	H	EB
10	S	Nm	Q	L	Nor	37	No	Nm	Q	L	VG	64	F	Nm	Q	L	B
11	S	Nm	Q	M	G	38	No	Nm	Q	M	EG	65	F	Nm	Q	M	G
12	S	Nm	Q	H	Nor	39	No	Nm	Q	H	VG	66	F	Nm	Q	H	B
13	S	Nm	N	L	B	40	No	Nm	N	L	G	67	F	Nm	N	L	VB
14	S	Nm	N	M	G	41	No	Nm	N	M	EG	68	F	Nm	N	M	Nor
15	S	Nm	N	H	B	42	No	Nm	N	H	G	69	F	Nm	N	H	VB
16	S	Nm	VN	L	VB	43	No	Nm	VN	L	Nor	70	F	Nm	VN	L	VB
17	S	Nm	VN	M	B	44	No	Nm	VN	M	VG	71	F	Nm	VN	M	B
18	S	Nm	VN	H	VB	45	No	Nm	VN	H	Nor	72	F	Nm	VN	H	VB
19	S	Fa	Q	L	VB	46	No	Fa	Q	L	Nor	73	F	Fa	Q	L	EB
20	S	Fa	Q	M	B	47	No	Fa	Q	M	VG	74	F	Fa	Q	M	B
21	S	Fa	Q	H	VB	48	No	Fa	Q	H	Nor	75	F	Fa	Q	H	EB
22	S	Fa	N	L	EB	49	No	Fa	N	L	B	76	F	Fa	N	L	EB
23	S	Fa	N	M	VB	50	No	Fa	N	M	G	77	F	Fa	N	M	VB
24	S	Fa	N	H	EB	51	No	Fa	N	H	B	78	F	Fa	N	H	EB
25	S	Fa	VN	L	EB	52	No	Fa	VN	L	VB	79	F	Fa	VN	L	EB
26	S	Fa	VN	M	EB	53	No	Fa	VN	M	Nor	80	F	Fa	VN	M	EB
27	S	Fa	VN	H	EB	54	No	Fa	VN	H	VB	81	F	Fa	VN	H	EB

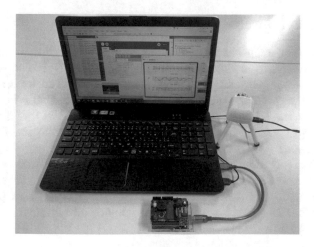

Fig. 4. Snapshot of testbed.

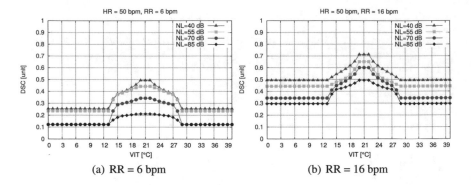

(a) RR = 6 bpm   (b) RR = 16 bpm

Fig. 5. Simulation results for HR = 50 bpm.

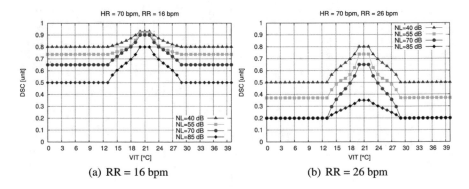

(a) RR = 16 bpm   (b) RR = 26 bpm

Fig. 6. Simulation results for HR = 70 bpm.

5 Proposed System Evaluation

5.1 Simulation Results

In this subsection, we present the simulation results for our proposed system. The simulation results are presented in Figs. 5, 6 and 7. We consider the HR and RR as constant parameters. The NL values considered for simulations are from 40 to 85 dB. We show the relation between DSC and VIT for different NL values. We vary the VIT parameter from 0 to 40 °C.

In Fig. 5, we consider the HR value 50 bpm and change the RR from 6 to 16 bpm. This heart rate is not a big concern as it is not a very low rate, but it can be seen that any abnormal respiratory rate, noise over 70 dB or temperature other than 17 and 25 °C range could affect very much the driver. Therefore, a good situation for the driver with his heart beating 50 times per minute is when he breathes 16 times per minute, there is not any annoying noise and the ambient temperature is between 17 and 25 °C.

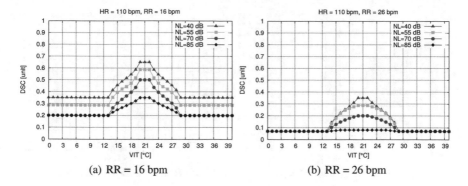

(a) RR = 16 bpm   (b) RR = 26 bpm

Fig. 7. Simulation results for HR = 110 bpm.

In Fig. 6, we present the simulation results for HR 70 bpm. In Fig. 6(a) is considered the scenario with a normal respiratory rate. We can see that the DSC values are much better than all the other considered scenarios. This is due to the driver's vital signs, which indicate a very good status of the driver's body, so the driver's situation condition is much better, and we can see that he could endure situations when a noise might be present or the temperatures are not the best ones. As the breath rate increases (see Fig. 6(b)), it can be seen that the DSC values are decreased and the effect of noise and temperature is more intense.

In Fig. 7, we increase the value of HR to 110 bpm. If the driver breathes normally we can see that there are some cases when the DSC value is normal such as when the ambient is quiet or when the ambient temperature is between 17 and 25 °C. On the other hand, when the driver breathes rapidly, there is not any situation that can be considered as a normal situation.

In the cases where the driver's situation is decided as bad or very bad continuously for relatively long time, the system can perform a certain action. For example, the system may limit the vehicle's maximal speed, suggest him to have a rest, or to call the doctor if he breathes abnormally and/or his heart beats at very low/high rates.

5.2   Experimental Results

The experimental results are presented in Figs. 8, 9 and 10. In Fig. 8(a) are shown the results of DSC when HR and RR is "Slow". As we can see, there is not any DSC value that is decided as a normal situation by the system. From Fig. 8(b), we can see a number of good and normal DSC values. These values are achieved when the ambient is quiet and the temperature is within 18–24 °C range. All other values are bad and very bad.

The results of DSC for normal heart rate are presented in Fig. 9, with Fig. 9(a) and (b) presenting the experimental results for normal and fast respiratory rate, respectively. Here the driver is in better conditions and when he breathes normally, many values are decided as good or very good. Several values are decided

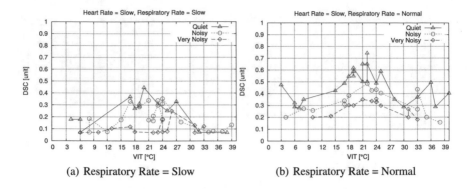

(a) Respiratory Rate = Slow   (b) Respiratory Rate = Normal

Fig. 8. Experimental results for slow heart rate.

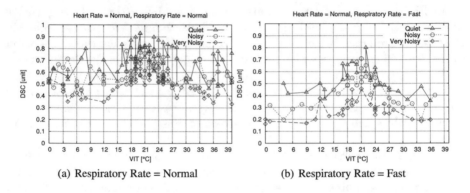

(a) Respiratory Rate = Normal   (b) Respiratory Rate = Fast

Fig. 9. Experimental results for normal heart rate.

as extremely good as well. As we explained in the simulation results (Fig. 6), we get better DSC values due to the driver's vital signs which indicate a very good status of the driver's body.

In Fig. 10 are shown the results of DSC for fast heart rate. The results are almost the same with that of Fig. 7 where the good and normal values happen to be only when the driver breathes normally and the ambient is quiet or the temperature is between 17 and 25 °C. When he breathes rapidly any situation is decided as bad, very bad or extremely bad. In these situations, the driver might have been experiencing forms of anxiety which increases the risk of a potential accident. Therefore, the system decides to perform the appropriate action in order to provide the driving safety.

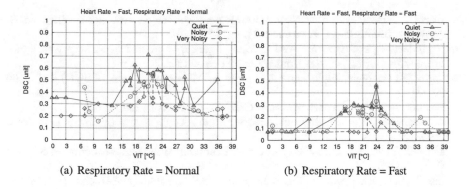

(a) Respiratory Rate = Normal   (b) Respiratory Rate = Fast

Fig. 10. Experimental results for fast heart rate.

6 Conclusions

In this paper, we proposed a fuzzy-based system to decide the driver's situation condition. We took into consideration four parameters: heart rate, respiratory rate, noise level and vehicle's inside temperature. We evaluated the performance of proposed system by simulations and experiments. From the evaluation results, we conclude that the driver's heart rate and respiratory rate, noise level and vehicle's inside temperature have different effects on the decision of the driver's condition.

In the future, we would like to make extensive simulations and experiments to evaluate the proposed system and compare the performance with other systems.

References

1. Gusikhin, O., Filev, D., Rychtyckyj, N.: Intelligent vehicle systems: applications and new trends. In: Cetto, J.A., Ferrier, J.L., Costa dias Pereira, J., Filipe, J. (eds.) Informatics in Control Automation and Robotics, pp. 3–14. Springer, Heidelberg (2008)
2. Santi, P.: Mobility Models for Next Generation Wireless Networks: Ad Hoc, Vehicular and Mesh Networks. Wiley, Chichester (2012)
3. Hartenstein, H., Laberteaux, L.: A tutorial survey on vehicular ad hoc networks. IEEE Commun. Mag. **46**(6), 164–171 (2008)
4. Karagiannis, G., Altintas, O., Ekici, E., Heijenk, G., Jarupan, B., Lin, K., Weil, T.: Vehicular networking: a survey and tutorial on requirements, architectures, challenges, standards and solutions. IEEE Commun. Surv. Tutor. **13**(4), 584–616 (2011)
5. Lindwer, M., Marculescu, D., Basten, T., Zimmennann, R., Marculescu, R., Jung, S., Cantatore, E.: Ambient intelligence visions and achievements: linking abstract ideas to real-world concepts. In: 2003 Design, Automation and Test in Europe Conference and Exhibition, pp. 10–15, March 2003
6. Acampora, G., Cook, D.J., Rashidi, P., Vasilakos, A.V.: A survey on ambient intelligence in healthcare. Proc. IEEE **101**(12), 2470–2494 (2013)
7. Aarts, E., Wichert, R.: Ambient intelligence. In: Bullinger, H.J. (ed.) Technology Guide, pp. 244–249. Springer, Heidelberg (2009)

8. Aarts, E., De Ruyter, B.: New research perspectives on ambient intelligence. J. Ambient. Intell. Smart Environ. **1**(1), 5–14 (2009)
9. Vasilakos, A., Pedrycz, W.: Ambient Intelligence, Wireless Networking, and Ubiquitous Computing. Artech House, Inc., Norwood (2006)
10. Sadri, F.: Ambient intelligence: a survey. ACM Comput. Surv. (CSUR) **43**(4), 36 (2011)
11. Kandel, A.: Fuzzy Expert Systems. CRC Press, Boca Raton (1991)
12. Zimmermann, H.-J.: Fuzzy Set Theory and Its Applications. Springer, Dordrecht (1991)
13. McNeill, F.M., Thro, E.: Fuzzy Logic: A Practical Approach. Academic Press, San Diego (1994)
14. Zadeh, L.A., Kacprzyk, J.: Fuzzy Logic for the Management of Uncertainty. Wiley, New York (1992)
15. Klir, G.J., Folger, T.A.: Fuzzy sets, uncertainty, and information (1988)
16. Munakata, T., Jani, Y.: Fuzzy systems: an overview. Commun. ACM **37**(3), 69–77 (1994)
17. Inaba, T., Sakamoto, S., Oda, T., Barolli, L., Takizawa, M.: A new FACS for cellular wireless networks considering QoS: a comparison study of FuzzyC with MATLAB. In: Proccedings of the 18th International Conference on Network-Based Information Systems (NBiS 2015), pp. 338–344 (2015)

Multi-cloud System for Content Sharing Using RAID-Like Fragmentation

Takumi Murakami and Shinji Sugawara[✉]

Chiba Institute of Technology, Narashino 275-0016, Japan
murakami@sugawara-lab.org, shinji.sugawara@it-chiba.ac.jp

Abstract. This paper proposes a multi-cloud content sharing system using a RAID-like storage and a redundancy avoidance technique aiming at high speed data transmission and capacity efficiency. The proposed method reduces files' upload and download times by not only using replications of fragmented files but also using RAID-like file fragmentation and parity files. The effectiveness of the proposal is evaluated by actual experiments over the Internet.

1 Introduction

Along with the great development of computer devices, communication terminals and networks, wide variety of digital contents with big capacity have been exchanged over the Internet [1], and cloud systems with diverse characteristics have been applied for content sharing services [2]. In addition, multi-cloud systems constructed by a combination of multiple clouds have become common, mainly in order to avoid "vendor lock-in," [3] or to keep privacy.

However in many cases, it is difficult for the conventional multi-cloud systems to achieve both high speed data transmission and efficient use of storage capacity at the same time. In the related research, at the time of uploading, contents are divided into small fragments, and each fragment of a file is equally redundantly scattered to some selected clouds, which causes that the downloading speed is greatly improved, as well as keeping privacy successfully. However, this method does not take into account the cost of keeping the large capacity of the content replicas. Because fragments of files are allocated redundantly, consuming a large amount of capacity of the cloud storage is a serious problem.

Therefore in the previous research of us, we proposed a multi-cloud content sharing system using RAID technology and redundancy avoidance techniques aiming at high speed data transmission and efficient use of storage. But still a room to be improved remains especially in the lengths of upload and download times.

In this paper, we propose an improved method to reduce the times better compared with the previous research, and newly show more effective way of content sharing with a multi-cloud system.

The rest of this paper consists of four more sections shown as follows. The previous studies is explained in Sect. 2 and improved method is proposed in

© Springer Nature Switzerland AG 2020
L. Barolli et al. (Eds.): NBiS-2019, AISC 1036, pp. 49–61, 2020.
https://doi.org/10.1007/978-3-030-29029-0_5

Sect. 3. Then, the proposed method is evaluated by actual experiments over the Internet, and discussed on the method's effectiveness to be expected in Sect. 4. Finally, this paper is concluded in Sect. 5.

2   Related Works

2.1   MyCloud

MyCloud [4] was proposed as a high-performance and secure distributed storage system by using a plural number of clouds in parallel. However from the viewpoint of this research, the most interesting function of MyCloud is its redundant chunk allocation.

In this method, like Shamir's method [5], a content to be shared is divided into some small chunks and they are redundantly scattered to the clouds. Firstly, a non-negative integer number, called "redundancy level" from zero to the number, i.e., the total number of clouds minus 1, should be set. When the number is zero, any content cannot be recreated from the chunks even if only one cloud is failed. When the redundancy level is set to 2, any content can be recreated even if one or two clouds are failed.

Figure 1 shows the case of using 5 clouds and 5 chunks with redundancy level 2. In this case, even if any two clouds are down, any chunk can be retrieved. For example, when Cloud A and B are down simultaneously, every chunk can be retrieved only from Clouds C, D, and E, and the original content file will be recreated.

As mentioned above, although each content is divided into some chunks, and each of them is redundantly allocated to the clouds, this redundant allocation enlarges the range of selecting clouds for downloading, and chunks are retrieved at high-speed by choosing the normally working clouds with the fastest communication links.

The behaviors of content allocation and retrieval are as follows.

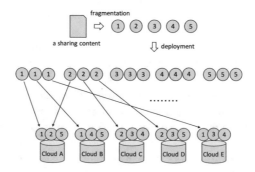

Fig. 1. Concept of chunk allocation in MyCloud.

2.1.1   Content Allocation

1. A client sends a content to the user interface.
2. The interface divides and encoded the content into fragments according to the redundancy level. (For example, if the number of clouds is five and the redundancy level is 2, each content should be divided into 5 fragments)
3. The interface allocates the fragments redundantly to the clouds with avoiding the overlapping of the same fragments in the same cloud storage.

2.1.2   Content Retrieval

1. A client sends a content request to the user interface.
2. The interface requests corresponding fragments to all of the clouds.
3. When all the fragments are retrieved by the interface, the interface recreates the original content from the fragments.

2.2   Method Proposed in Previous Research

The previous research of us [6] proposed a method to share contents using multiple cloud storages which are controlled by an administration server. In this method, contents are allocated to plural cloud storages in the way that is just like a RAID [7] system. Furthermore, when a content is uploaded in a cloud storage, the content is judged if there are any replicas of it in the network, and if there exist, they are eliminated in order to save the storage capacity. Hash function is used for the replication judgment and hash values of the contents stored in the clouds are maintained in the administration server and compared with the value of the content which is going to be stored.

2.2.1   Outline of the Method

This is a method to share contents using multiple cloud storages which are controlled by an administration server. The basic concept of this method is shown in Fig. 2. In this method, contents are allocated to plural cloud storages in the way that is just like a RAID system. Furthermore, when a content is uploaded in a cloud storage, the content is judged if it is duplicated with the existing content and duplicated ones are eliminated in order to save the storage capacity. Hash function is used for the duplication judgment and hash values of the contents stored in the clouds are maintained in the administration server and compared with the one of the content which is going to be stored.

The algorithm of the proposed method consists of following four parts.

- Behavior of Contents Deployment
- Behavior of Contents Retrieval
- Redundant Contents Deployment
- Contents' Redundancy Avoidance

For the specific explanations of these parts, refer to Ref. [6].

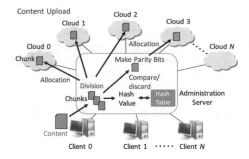

Fig. 2. Concept of previous research method.

3   Proposed Method

3.1   Basic Concept

The method improved and proposed in this paper is a combination of ideas of MyCloud, the method proposed in the previous research of us, and RAID6. A file of a content to be shared in the system is divided into small sized fragments and they are deployed with their parity codes to a plural number of clouds just like MyCloud and the method of our previous research, however, the fragments and parity codes are not redundantly replicated, and two kinds of parity codes are needed as is the case with RAID6.

Generally, when the cloud storage is working in a normal way, parity files are not used. If there are cloud storages which store only parity files, they are usually not used and their efficiency of use is low. In the proposed method, all of the cloud storages are used for downloading chunks in parallel for short download time. The total amount of data capacity stored in the cloud storages are suppressed by using RAID6 redundancy control. This causes upload time to be shortened and efficiency of using cloud storages to be increased.

3.2   Redundant Fragments Allocation Procedure

The redundancy of the contents in the proposed method is achieved not by replication of the fragments but making parity files. Two parity files are made for every three fragments. As shown in Fig. 3, in the case of using five clouds and five fragments (i.e., from fragment 1 to 5) of an original content, a parity code is made for every three fragments, thus two parity codes such as the parity code for fragment 1, 2, 3, and that for fragment 4, 5 will be made. By this way, just like RAID6, the redundancy level of this system becomes 2 and it is possible to cope with the situation where two cloud storages go down simultaneously.

Figure 3 illustrates a working example of the improved method. In the case that a sharing content is divided into 5 fragments from #1 to #5, two kinds of parity codes for a group of files from #1 to #3 and another group of files of #4 and #5 are made. Five fragments and four parity codes are deployed on five clouds as shown in the figure.

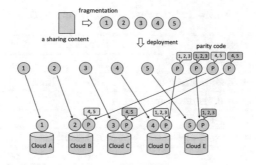

Fig. 3. A working example of proposed method.

In this case, if two clouds such as Cloud A and B are failed for example, fragments #1 and #2 can be recovered with fragment #3 and two parity codes made by the fragment group of #1, #2 and #3. Because this method does not replicate any of the fragments and parity codes, which causes each cloud keeps only one fragment, upload and download time can be more shortened than the conventional methods.

As shown in the Fig. 3, there are two kinds of parity codes used in this system. Like RAID6, there are about two ways of making the two kinds of parity codes, i.e., P+Q and 2D-XOR. Both ways can be used in the proposed method, however the specific explanations of how to make the parity codes and how to rebuild the lost fragments with those two kinds of parity codes is omitted in this paper, because especially the way of rebuilding fragments is complicated and this topic is not essential for the proposal in this paper.

4   Evaluation

4.1   Evaluation Environment

In order to evaluate the effectiveness of the proposed method, we actually construct a simple prototype of the contents sharing system which works according to the proposed procedure over the Internet, execute the test content's uploading and downloading as experiments, and compared its performance with that of the other prototypes such as MyCloud, the method proposed in the previous research, and RAID6-like method. The cloud storages used for the construction of multi-cloud environment are cloudme, box, opendrive, teracloud, and hidrive. The specific environment of the experiment is shown in Table 1 shown below.

Uploading and downloading times of the test content over the Internet are actually measured, and compared among the methods. Especially, downloading times are measured both in the cases of normal and the situation where two of five clouds are down simultaneously.

Table 1. Environment of experiment.

Parameters	Values
Number of clouds	5
Number of simulation runs	50–100 times
Number of division to fragments	50
Redundancy level	2
Capacity of test content	4.4 [MB]
Processor specification	Intel Core i5 2.60 [GHz]
OS	Windows Pro 10

Transmission of the test content fragments are executed over HTTP. The procedure of the experiment is shown below. The perspective of the system for the evaluation is illustrated in Fig. 4.

1. Division of the test content.
2. Making two kinds of parity codes. (Measuring the elapsed uploading time begins)
3. Uploading the fragments to the clouds in parallel. (Measuring the elapsed uploading time ends)
4. Downloading the fragments to the client in parallel. (Measuring the elapsed downloading time begins)
5. Rebuilding the test content from the retrieved fragments.

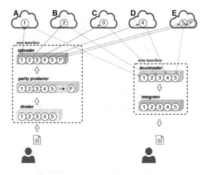

Fig. 4. Perspective of system for evaluation.

4.2 Evaluation Procedures

As mentioned above, the methods executed in the experiment are MyCloud, the method proposed in the previous research, and RAID6-like method, as well as proposed method. Their procedures of downloading and uploading test content of are shown below.

4.2.1 MyCloud

- *Normal Situation without Clouds' Failures*

  Because redundancy is ensured by replication of fragments, making parity codes is omitted. According to the experimental condition shown in Table 1, redundancy level is set to 2, then the number of replicas of each fragment is set to three, and the replicated fragments are uploaded in the five clouds. The situation is illustrated in Fig. 5.

- *Situation Where Two Clouds Fail*

  Because retrievals of fragments from two clouds are impossible in this situation, all fragments from the resting three clouds are downloaded. For example, in Fig. 6, when cloud A and B are failed, fragments 2, 4, and 5 are downloaded in parallel at first, and then, fragments 1 and 3 are done next from clouds C, D, and E.

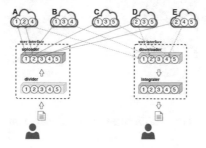

Fig. 5. Evaluation of MyCloud.

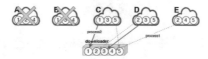

Fig. 6. Behavior of MyCloud during clouds are down.

4.2.2 Previous Research Method

- *Normal Situation without Clouds' Failures*

  In this method, parity codes are created from every four fragments. Because the number of division of the test content is five, two parity codes for fragments 1 to 4 and for fragment 5 are created like the illustration in Fig. 7. The fragments are replicated in a same way of MyCloud, however, the number of replicated fragments is two for redundancy level 2 to cope with two clouds' down at the same time because of using the parity codes. One of the five clouds is exclusively used only for parity codes, and the downloading is executed from the four other clouds in parallel.

- *Situation Where Two Clouds Fail*
  For example in Fig. 8, When clouds A and B are failed, all the fragments can be downloaded from clouds C and D, and the original test content can be rebuild without using parity nodes. However, only two clouds are used, firstly two fragments, secondary two, and lastly a single are downloaded in three steps and it takes long time. When the two clouds including cloud E which holds parity codes, all the fragments can be downloaded from the remaining three clouds and it does not take very long time for fragments retrieval.

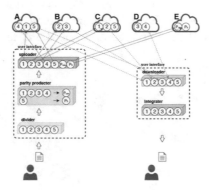

Fig. 7. Evaluation of previous research method.

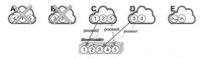

Fig. 8. Behavior of previous research method during clouds are down.

4.2.3   RAID6-like Method

- *Normal Situation without Clouds' Failures*
  In the RAID6-like system, when the number of clouds is five, the number of fragments should be three because two kinds of parity codes are needed. Downloading is executed from three clouds in parallel as shown in Fig. 9.
- *Situation Where Two Clouds Fail*
  The most hard situation is where two of three clouds holding fragments are failed. Even in this case, from a single fragment and two kinds of parity codes, the original test content can be rebuilt. This situation is illustrated in Fig. 10.

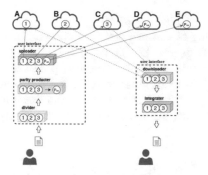

Fig. 9. Evaluation of RAID6-like method.

Fig. 10. Behavior of RAID6-like method during clouds are down.

4.2.4   Proposed Method

- *Normal Situation without Clouds' Failures*

  According to the environment shown in Table 1, test content is divided into five fragments and parity codes are created with every three fragments in two ways. With considering an efficient upload and download, the fragments and two parity codes (i.e., created from fragments 1, 2, and 3, and from fragments 4 and 5) each of which exists two kinds are allocated in the way shown in Fig. 11.

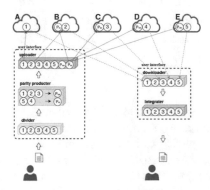

Fig. 11. Evaluation of proposed method.

- *Situation Where Two Clouds Fail*
  The way of rebuilding of the original test content is just the same with that of RAID6, as shown in Fig. 12. That is, fragments 3, 4, and 5 held in the normally working clouds are downloaded in parallel, and then, two kinds of parity codes created from fragments 1, 2, and 3 are downloaded in parallel. Finally, fragments 1 and 2 are rebuilt from the downloaded parity codes and fragment 3. At this time, all the fragments are collected as a whole original content.

Fig. 12. Behavior of proposed method during two clouds are down.

4.3 Results of the Evaluations and Discussions

The average values and standard deviations of the times of uploading and downloading test content data are shown in Table 2, Figs. 13 and 14. In these figures, the upload and download times are shown as normal distributions of the probability.

Table 2. Result of experiment.

	MyCloud	Previous	RAID6	Proposed
Average of upload time [s]	14.7	14.1	7.8	8.9
Standard deviation of upload time [s]	5.8	6.2	4.4	4.3
Average of download time [s]	35.9	39.8	56.6	37.0
Standard deviation of download time [s]	16.7	22.2	16.5	14.7
Total storage capacity [MB]	13.1	10.2	7.3	7.3

As shown in Fig. 13, proposed method as well as RAID6 achieves the shortest upload time, because those two do not make any replicas of fragments for keeping redundancy and total capacity to be uploaded is the smallest.

On the other hand, download time of the proposed method has almost the same distribution characteristic with MyCloud, and achieves the short average download time with the smallest standard deviation, which is illustrated in Fig. 14. Then, in both upload and download situations, we can see that the proposed method keeps the best performance.

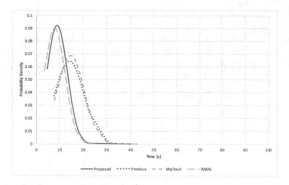

Fig. 13. Distribution of upload time.

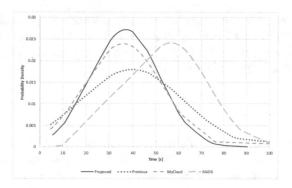

Fig. 14. Distribution of download time.

Finally, distributions of download times are shown in Fig. 15. In this experiment, cloud A and B fail. Proposed method seems to achieve much better performance than any other methods. However, in the experiment, cloud A and B are down and cloud C has the slowest link among the normally working clouds. Under this condition, the proposed method is the most advantageous because it only requires downloading one fragment from each cloud (one fragment from cloud C, and one parity code plus one fragment from each of cloud D and E). On the contrary, previous method must download the fragments from only two clouds of C and D in two times (There is no need to use parity codes held in cloud E), and MyCloud needs to download fragments from all of three clouds, but two fragments are downloaded from the slowest cloud C. RAID6-like method is just the same way of downloading fragment and parity codes and rebuilding original content with the proposed, but because the number of division of the content is small, the capacity of each fragment is large and it takes longer time for downloading than the proposed. Even if the other combination of two clouds is failed at the same time, the proposed method is expected to achieve almost the same result.

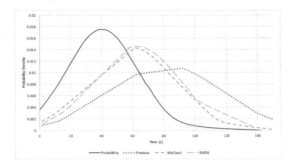

Fig. 15. Distribution of download time during two clouds are down.

From this discussion, because downloading time will be changed according the combinations of failed clouds, allocation of the fragments, and the order of the downloading fragments, the proposed method cannot always achieve the best performance when the clouds are failed.

5   Conclusion

We proposed a novel content sharing method using multi-cloud system with RAID6-like redundancy control. And it was found that the proposed method with improved utilization of cloud storages exhibits high performance and its effectiveness from the experiment by the experiments over the Internet. Especially in the average upload and download time, the proposed method demonstrated the better results than the others. However, when some clouds are failed, we found that downloading time will be changed according the combinations of failed clouds, allocation of the fragments, and the order of the downloading fragments. Then, the proposed method cannot always achieve the best performance in this situation.

As a future work, when some clouds are failed, the way to keep the shortest downloading time should be explored.

Acknowledgement. This work was partially supported by JSPS KAKENHI Grant Number JP17K00134.

References

1. Cisco Visual Networking Index: Forecast and Trends, 2017–2022. White paper, Cisco public. https://www.cisco.com/c/en/us/solutions/collateral/service-provider/visual-networking-index-vni/white-paper-c11-741490.pdf
2. Sugawara, S.: Survey of cloud-based content sharing research: taxonomy of system models and case examples. IEICE Trans. Commun. **E100-B**(04), 484–499 (2017)
3. Abu-Libdeh, H., Princehouse, L., Weatherspoon, H.: RACS: a case of cloud storage diversity. In: Proceedings SoCC 2010. ACM, June 2010

4. Horiuchi, K.: MyCloud: a high-performance, secure, distributed storage system combining several vendors' cloud storage services. Comput. Softw. JSSST **29**(2), 158–167 (2010)
5. Shamir, A.: How to share a secret. Commun. ACM **22**(11), 612–613 (1979)
6. Sugino, H., Nakano, T., Sugawara, S.: Efficient content sharing by multiple users with RAID based multi-cloud storages. In: Proceedings of IEEE International Conference on Cloud Networking (CloudNet), October 2018
7. Patterson, D.A., Gibson, G., Katz, R.H.: A case for redundant arrays of inexpensive disks (RAID). ACM SIGMOD Rec. **17**, 109–116 (1988)

A VR System for Alleviating a Fear of Heights Based on Vital Sensing and Placebo Effect

Iku Kitanosono[1(✉)], Toshiyuki Haramaki[2], Tsuneo Kagawa[2], and Hiroaki Nishino[2]

[1] Graduate School of Engineering, Oita University, Oita, Japan
v18e3009@oita-u.ac.jp
[2] Faculty of Science and Technology, Oita University, Oita, Japan
{haramaki,t-kagawa,hn}@oita-u.ac.jp

Abstract. Acrophobia is a phobia that persons fear for heights and tend to have negative images like falling down from their standing positions. Persons in acrophobia avoid going high places, so they have little experience in such places. Making them experience high places as much as possible is an effective treatment to improve the symptoms. Placebo effect is a well-known phenomenon that person's psychological belief influences his/her physical and mental status. In this paper, we propose a VR simulation system to overcome acrophobia based on a placebo effect by making persons virtually experience high places. We show a method for effectively presenting the heights to viewers according to their fear strength and elaborate our implementation approach of the proposed system.

1 Introduction

Acrophobia is one of phobia that persons feel a fear of heights. Such persons often have negative images like falling down from the high places despite almost no such danger. Persons who have acrophobia can never go to high places by themselves. Therefore, they have poor experience for standing at high places, walking around the area, and looking down at the window. Making them experience high places is an effective treatment because it will convince them that high places are not so dangerous.

A placebo is a psychological effect that persons' strong belief influences their physical and mental conditions. For instance, some patients get real benefits and effects of therapy even if they take fake drugs because they strongly believe the drugs are real and effective ones. Various medical treatments actually use the placebo effect. It may be more effective than medicine especially in the case of mental illness treatments.

Current virtual reality (VR) technology provides many powerful functions to make users feel immersion. The users can experience various virtual contents through VR technology. If VR technology reproduces a realistic view in high places, persons who have acrophobia can safely experience the high places. The experiences with the VR simulation make them believe that they are really stand at the high places and such places are not so dangerous. Placebo effects generated by VR technology have potential power to solve the problem and provide a good therapeutic effects for acrophobia persons as shown in Fig. 1.

© Springer Nature Switzerland AG 2020
L. Barolli et al. (Eds.): NBiS-2019, AISC 1036, pp. 62–72, 2020.
https://doi.org/10.1007/978-3-030-29029-0_6

We propose a VR simulation system immersing users in high places to remedy acrophobia based on placebo effects. The proposed system simulates the users' experiences of high places and makes them believe that high places are safe even in some unusual and dangerous situations. The system uses various sensors for obtaining their physical and vital data to estimate their emotional states, especially for their fears. We propose a method to effectively present the view of high places based on the estimated strength of users' "fear of heights" feeling. It achieves VRET (Virtual Reality Exposure Therapy) meaning that it stimulates the patients by directly presenting the principal subject of fear in a VR space.

Fig. 1. Proposed VR system concept to make a user believe that high places are not dangerous.

2 Related Work

There are some precedent studies for adopting the VR technology for treatments of acrophobia [1, 2]. These works make users to continuously immerse in high places in a virtual environment as a countermeasure against phobia. The users, however, tend to take long time to overcome acrophobia and have stressful feelings during treatments.

There are some precedent studies for implementing VR systems designed for overcoming phobia. Donker et al. developed a smartphone application because virtual reality exposure therapy is an effective manner in treating certain phobias [3]. The smartphone application is compatible with VR using Google's card board. They developed an inexpensive smartphone application based on the card board function because acrophobia treatments with VR exposure therapy still cost as high as regular exposure therapy. Additionally, they conducted experiments to verify the usefulness of the proposed application. Despite the use of less realistic red and blue anaglyph images, they proved their approach is useful for treating phobia as it significantly reduces users' fear of spiders. Their study shows that VR is a cost-effective way for treating phobias. Chardonnet et al. developed a simulator for train workers who are doing their work at heights using VR [4]. They tried to improve the sense of immersion for making the workers get used to heights. An effort to improve the immersion is to reproduce gust while the workers are performing in the VR simulator with ladders and lifelines. They also added a tactile feedback function when the workers touch a virtual power line.

Compared with the above-mentioned trials, our proposed system utilizes the placebo effect to further enhance VR exposure that has proven to be useful in some treatments. The proposed system also judges and predicts the strength of user's fear for smoothly presenting the virtual contents in the acrophobia treatment.

3   Proposed System

3.1   System Configuration

Figure 2 shows the proposed system overview. The system consists of three parts. The vital data acquisition part measures the users' physical and mental states using multiple sensors. The fear state estimation part judges and predicts the users' emotion especially the strength of fear. The VR simulation part generates and displays computer graphics scenes of the heights according to the user's fear state.

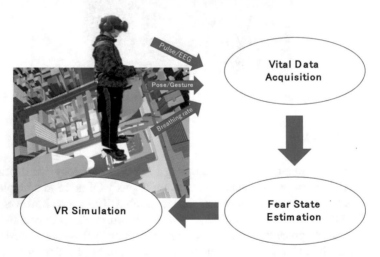

Fig. 2. Overview of proposed system consisting of three parts, Vital Data Acquisition, Fear State Estimation and VR Simulation.

3.2   Vital Data Acquisition

The system uses the following three vital sensors for obtaining the user's physical and vital conditions. Then, the system generates appropriate VR contents according to the user's state estimated from the acquired sensing data.

1. **Position detection sensors:** A user attaches six wireless position detection sensors to recognize his/her poses and gestures such as "looking down from high places". Figure 3 shows the body parts to attach the six sensors that are head on an HMD display, both hands with motion controllers, waist, and both feet.

2. **A wristband-type activity sensor:** A user wears like a watch for measuring his/her heart rate and respiratory volume.
3. **An electroencephalography (EEG) sensor:** A user wears as a headset for continuously obtaining his/her brain wave.

The system tracks the user's motions with a set of position detection sensors and monitors the user's vital activities by using worn vital sensors as shown in Fig. 4. However, currently, breathing sensor have not been construct automatic sensing, we must have log data after VR simulation.

Fig. 3. Full body motion tracking by using six position detection sensors attached on head, both hands, waist, and both feet.

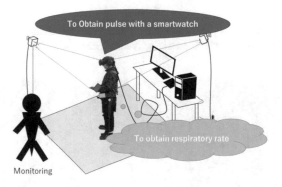

Fig. 4. Vital sensors to estimate the user's emotional state.

3.3   Fear State Estimation

The system estimates the user's fear state after data acquisition. Firstly, the system acquire a set of data used for estimating the user's fear state from the activity and EGG sensors. Next, the system maps the data set onto the Russell's emotional ring model as shown in Fig. 5. The Russell's emotional model can be used as a classification framework to map the sensing data onto a specific human emotion arranged in the two dimensional coordinate consisting of two orthogonal axes such as "Activation-Deactivation" and "Pleasant-Unpleasant" axes [5, 6]. Then, the system predicts the user's next state from the mapped time series data. It effectively handles the user's vital data to judge the user's current status and predict the next status.

A dedicated operator is necessary to estimate the users' fear state and provide the user appropriate instructions during the simulation. If the operator judges that the user become familiar with the current floor by observing the sensor values, then ask him/her to get on the elevator and go up to higher floors.

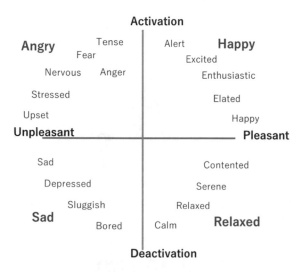

Fig. 5. Russell's emotional model for mapping various human emotions on a two dimensional space constructed by two orthogonal axes: "activation-deactivation" and "pleasant-unpleasant".

Fig. 6. An example image of high places presented in VR simulation.

3.4   VR Simulation

Figure 6 shows an example image of high places presented in the VR simulation. In a virtual view, there are some tall buildings in a city, and the user moves up and down the floors in a tall building. The system takes different approaches to make the user experience higher floors according to his/her fear state. A typical approach is to guide the user for gradually going up from the lower floor to the upper in the building. It enables the user to gradually adjust to the height.

After the system started, the user stands at the scene in front of the elevator as shown in Fig. 7. He/she recognizes to go up to higher floors using the elevator in the building. After going up lower floors, he/she looks some scenes from windows. On each floor, he/she experiences the three different types of rendered image in a specific scene from phase 1 to 3 as described in Sect. 4. The user is expected to gradually overcome acrophobia through experiencing these scenes.

We adopted a VR device called HTC Vive which includes an HMD, a pair of motion sensors for tracking the HMD position and direction, and a pair of motion controllers. We developed computer graphics rendering and control software modules using Unity and C#.

Fig. 7. Start image of VR simulation where a user stands in front of an elevator and feel reality to go up to higher floors with the elevator.

4 Placebo Effect

Acrophobia can be divided into severe and mild phobia. The severe patient needs to take counseling by a doctor for treating his/her trauma. Our study focuses on the mild case. Unconsciously making negative images for high places, such as "The floor may be broken", "The window may be broken", and "The fence may be broken and it will fall", will cause mild acrophobia. Conversely, persons who are not afraid of the height tend to have positive feelings such as "The view from window looks very nice" and "landscape in higher places are wide and magnificent". If they can change their negative emotions to positive ones, they can overcome acrophobia. Therefore, the system should make the users with mild acrophobia strongly belief that the high places are enjoyable and fearless, helping them to have good images about high places.

We implemented a visualization method in the VR simulation for helping users to have good impression about the high places. After a user arrives at a floor, the system gradually makes some objects to be transparent in the floor such as roof, windows, and walls. The VR simulation visualizes unusual and dangerous altitude scenes in a virtual environment. The scenes are designed for making them strongly feel that they are secure although they go to high altitude. Firstly, the system takes the user to a normal room in a tall building. After that, the system gradually changes the same room scene as follows (shown in Fig. 8):

Phase 1:   After the user arrive at a room in a floor, the system makes the ceiling and window grasses of the room transparent. The user can see a slightly more open landscape than usual view.

Phase 2:   The system removes the walls from phase 1 image. The user can see the surrounding landscape further open.

Phase 3:   The system finally removes all floorings from phase 2 image. The user may feel he/she is floating in the air. If the user stays in the higher floor and looks down in this visualization phase, he/she should feel a fear of heights. The right image in Fig. 6 is an example of 20th floor rendered in this mode.

(a) Phase 1: Roof and windows are removed from the normal room.

(b) Phase 2: Walls are removed from the phase 1 room scene.

(c) Phase 3: Floors are removed from the phase 2 room scene.

Fig. 8. Visualization method of virtual room in a building.

5   Experiments

5.1   Experimental Task

We prototyped the system and verified the effectiveness of the proposed treatment method with placebo effect to improve acrophobia. We employed five subjects with mild acrophobia who are basically scared for high places in the experiment. They experienced the virtual high altitude with the following three methods in the VR simulation:

Method 1:   We apply placebo effects in each floor for making the subjects experience transparent phase 1 through phase 3. In this method, the subjects gradually go up the floors with an elevator starting from the first floor, next the second floor, and then the third, the fifth, and finally the 20th floor.

Method 2:   This is basically the same with the method 1 except not applying placebo effects. The subjects always stay a normal room (no transparent phase applied) on each floor.

Method 3:   This method takes the subjects directly to the highest 20th floor without applying placebo effects (no transparent visualization). This case is a similar method normally used in VRET treatments.

In this experiment, visualization reality and sense of immersion are very important to make the subjects feel in high places. Before starting the simulation process, we explain about the simulation procedure and acrophobia. After that, we show a consent form to the subjects and asked them to sign it for enhancing the reality as a treatment. After finishing the simulation, we asked the following questions to evaluate the system:

Q1. How much are you comfortable at high places?
We asked the subjects to answer in five ranks: five means the subjects are comfortable at high places and one means they aren't comfortable.

Q2. How much do you feel a fear of heights in this system?
We asked the subjects to answer in three ranks: three means the subjects are scary and one means they aren't scary.

Q3. Which is the most effective one among three visualization method?
Beside the questionnaire, we further interviewed the subjects about the reasons to select their answers.

5.2   Experimental Results

Table 1 shows the answers for the above-mentioned questions from each subject. According to the results in Q2 in Table 1, the subjects seemed to feel reality well in the VR simulation because most of them felt strong fear of heights. The subjects' opinions about the reasons of their evaluations include "I suddenly felt like I was falling from a high place when the wall suddenly disappears at a higher floor", "I conceived strong negative images when looking down from a higher floor". There were some contrary opinions not to feel like staying at high places in the simulation. The results of Q3 in Table 1 shows the placebo effects designed and implemented in the system is useful. Some subjects pointed the usefulness of the gradual floor rising procedure such as "It was less scary to move up the floor from the lower floor" and "It was hard to feel a fear of heights because I got used to the heights by starting from a lower floor". However, not so strong supporters on the transparency visualization function.

Table 2 shows the two subjects' heart rate measured in the experiment. Their heart rate increase is minimum in the case of method 1, showing that the gradual floor going up with placebo effects is effective for suppressing their fear of heights.

Table 1. Results of experiment.

	User A	User B	User C	User D	User E	Average
Q1.	1	1	1	2	1	1.2
Q2.	3	3	3	2	3	2.8
Q3.	1	1	1	1	1	1.0

Table 2. Subjects' heart rate measured in the experiment (bpm).

Subject		User A	User E
The normal heart rate		77.33	74.23
Method 1	Heart rate in high place	79.00	90.50
	Difference	1.67	16.27
Method 2	Heart rate in high place	92.00	97.00
	Difference	14.67	22.77
Method 3	Heart rate in high place	91.50	99.00
	Difference	14.17	24.77

Table 3. Subjects' breathing rate measured in the experiment.

	Floor	User A	User B	User C	User D	User E	Average
Method 1	1F	15.60	17.30	16.40	16.35	16.20	16.37
	2F	15.20	16.80	15.90	16.00	15.32	15.84
	3F	16.30	17.20	14.95	13.30	17.30	15.81
	5F	16.20	17.30	14.80	14.87	17.50	16.13
	20F	17.50	17.90	18.90	14.32	18.40	17.40
Method 2	1F	16.70	17.20	18.30	18.05	13.20	16.69
	2F	16.50	17.60	18.70	17.60	14.50	16.98
	3F	17.00	17.90	18.90	15.40	16.70	17.18
	5F	17.20	17.20	19.00	14.50	17.60	17.10
	20F	17.90	17.30	19.10	13.05	19.80	17.43
Method 3	1F	15.90	17.20	17.50	17.20	15.00	16.56
	20F	13.25	18.00	14.00	20.30	22.00	17.51

Table 3 shows the subjects' breathing rate measured in the experiment. We could not find any significant relationship between subjects' breathing rate and floor heights. In the experiment, we manually obtained the vital data and could not determine their fear status.

5.3 Discussion

Based on the experimental results, the VR simulation indicated a good sign for the treatment of acrophobia. According to the results of Q3, all subjects supported the method 1 was the most effective in terms of qualitative assessment. From the results shown in Table 2, the method 1 achieved the smallest difference in heart rate between lower and higher floors than other methods. However, the effectiveness of the placebo effects designed and implemented in the system still is unclear because only one subject mentioned the advantage of the transparent visualization. Therefore, the influence of the transparent visualization may not so strong. We must consider more attractive simulation contents for giving users stronger impressions that high places are not dangerous.

Individual difference in vital sensing is a problem to deal with. We should, therefore, need to conduct other evaluations for reducing the individual difference by improving the sense of immersion in the VR simulation. Fine and user-oriented sensing data measurement is required.

The psychological burden of patient is the most important issue in the treatment by VRET with placebo effects. We would like to define and implement the fear state estimation function based on the proposed method in the near future. If the user's fear state is jumping, the system should be able to restart the procedure from milder contents with transparency phase 1 in the building hierarchy.

6 Conclusions

In this paper, we proposed a VR system for improving and dealing with acrophobia by using placebo effects. The proposed system collects multiple sensing data and quantitatively analyzes the user's fear state about the heights. We also designed the VR contents rendering method by giving the user a scenario for taking an elevator to go up to higher floors in a tall building and gradually making the user to accustom with altitude. We conducted an experiment to verify the effectiveness of the proposed system by employing subject with mild acrophobia. Although the experiment still was a preliminary one, the results were promising.

As future work, we would like to implement the function for predicting the user's fear state based on the vital data. We need to define a mapping method between a set of EEG and heart rate data onto a specific portion in the Russell's emotional model. Designing a prediction module based on deep learning techniques may be a potential approach. Reconsidering to present placebo effects in the VR simulation is another possible future improvement like incorporating Miller hypnosis.

References

1. Krijn, M., Emmelkamp, P.M., Biemond, R., de Wilde de Ligny, C., Schuemie, M.J., van der Mast, C.A.: Treatment of acrophobia in virtual reality: the role of immersion and presence. Behav. Res. Ther. **42**(2), 229–239 (2004)
2. Rothbaum, B.O., Hodges, L.F., Kooper, R., Opdyke, D., Williford, J.S., North, M.: Virtual reality graded exposure in the treatment of acrophobia: a case report. Behav. Ther. **26**(3), 547–554 (1995)
3. Donker, T., Van Esveld, S., Ficher, N., Van Straten, A.: 0Phobia – towards a virtual cure for acrophobia: study protocol for a randomized controlled trial. Trials **19**(433), 1–11 (2018)
4. Chardonnet, J.-R., Di Loreto, C., Ryard, J., Rousseau, A.: A virtual reality simulator to detect acrophobia in work-at-height situations. In: Proceedings 2018 IEEE International Conference on Virtual Reality and 3D User Interfaces, pp. 747–748 (2018)
5. Yamamoto, J., Kawazoe, M., Nakazawa, J., Takashio, K., Tokuda, H.: MOLMOD: analysis of feelings based on vital information for mood acquisition. In: MobileHCI 2009 Measuring Mobile Emotions, Bonn, Germany, 15 September 2009. ACM (2009). ISBN 978-1-60558-281-8
6. Ikeda, Y., Okada, Y., Horie, R., Sugaya, M.: Estimate emotion method to use facial expressions and biological information. In: 10th International Conference on Foundations of Augmented Cognition: Neuroergonomics and Operational Neuroscience, AC 2016 and Held as Part of 18th International Conference on Human-Computer Interaction, HCI International 2016, Toronto, Canada (2016)

Exploring Bit Arrays for Join Processing in Spatial Data Streams

Wendy Osborn$^{(\boxtimes)}$

University of Lethbridge,
4401 University Drive West, Lethbridge, AB T1K 3M4, Canada
wendy.osborn@uleth.ca

Abstract. In this paper, the use of bit arrays for processing spatial joins in spatial data streams is explored. Although spatial joins between objects have been explored in other contexts, such as centralized and distributed systems, they have not been explored in great detail in spatial data streams. This work explores the use of a Bloom-filter (i.e., bit array) inspired representation of a spatial object. Strategies for both mapping objects to bit arrays, and processing spatial joins using the bit arrays in a data stream environment are presented. The strategies are evaluated and compared with spatial (non-bit) join approaches. Performance improvements are identified, and areas of improvement are also identified.

1 Introduction

Nowadays, applications exist that generate data in a continuous stream, where the amount of data cannot be stored in its entirety, and also may lose its validity after a certain amount of time [4,7]. Data streams require different strategies for handling them, as conventional strategies were designed for stored data [4]. This also extends to spatial data streams, where the streaming data consists of points or objects of non-zero area [15]. One type of strategy that needs to be re-visited is the spatial join [18] on two or more spatial data sets. Conventional spatial join processing requires that all objects are present for the join. Now, spatial objects are arriving as spatial data streams and therefore must be handled as they arrive. In addition, decisions must be made as to which objects to keep [7].

The spatial join has been applied previously in the research literature. Existing strategies can be classified into the following categories: spatial joins in a centralized system [2,3,9,10,17,20,21], a distributed system [1,6,8,11–14,16,19] and a streaming data system [15]. Although a significant amount of study has gone into applying spatial joins to queries centralized and distributed systems the main limitation in most scenarios is that both sets of spatial objects must available in their entirety when applying a spatial join between them. The only existing work that studied spatial joins in a spatial data stream environment is that of Kwon and Li [15]. They proposed a progressive join strategy where objects are joined multiple times using various levels of granularity in order to

© Springer Nature Switzerland AG 2020
L. Barolli et al. (Eds.): NBiS-2019, AISC 1036, pp. 73–85, 2020.
https://doi.org/10.1007/978-3-030-29029-0_7

Fig. 1. Sensor network

reduce the cost of the spatial join. A limitation of this work is the multiple times the same two objects are processed.

Therefore, this paper proposes several strategies for processing a spatial join in a spatial data stream environment. In addition to proposing a strategy that uses a conventional spatial join, two additional strategies are proposed that utilize a bit array representation of a spatial object, in order to improve the time it takes to perform a spatial join, at the expense of some spurious results being generated. Two of the three strategies also apply a spatial join on a restricted portion of the space containing objects. An empirical evaluation and comparison versus simplified versions of the strategy in [15] shows that improvements in running time and accuracy can be achieved with the new strategies. In addition, some shortcomings of the new strategies lead to future research directions.

Section 2 provides some background that is required for the work to be proposed. In Sect. 3, the three strategies are proposed, and two simplified versions of the work of [15] are presented. Section 4 presents the methodology and results of the performance evaluation and comparison of the proposed strategies. Finally, Sect. 5 concludes the paper and provides some future research directions.

2   Background

In this section, the required background notably, spatial data stream system, spatial join and Bloom filter are summarized. Figure 1 depicts an example data stream. Here, multiple sensors (in the example, two) are generating data continuously, which is sent to a stream data processor. From here, any results are sent on a result stream to the required destination [7]. In a spatial data stream, the data being streamed is spatial data, which can be either point data, object data, or a combination.

A data stream processor can only store a certain amount of information from the streams. Several strategies have been proposed for choosing which data to keep [7]. One strategy is a sliding window, which is a memory of limited size that takes streaming data as it arrives and stores it for a limited time. When the sliding window is full, decisions are made as to which objects to remove. For the strategies in this paper, a first-in-first-out (FIFO) strategy is assumed.

A spatial join [18] takes two sets of spatial objects S_1 and S_2 and relates an object from each of the two sets using a spatial predicate, such as overlap, containment, and adjacency. The strategies proposed below assume the use of the overlap predicate, but with slight modification any other predicate can be utilized. Also assumed is the use of a nested-loop strategy for processing a join of a portion of each set spatial objects. Two different representations of a set of spatial objects are used in the proposed strategies below: a geometric (i.e., vector) representation, where each polygon is represented using points connected with lines; and a bit array representation, where a sequence of **1** and **0** represents the entire set of spatial objects.

The bit array utilized in the proposed strategies below is inspired by the Bloom filter [5]. A Bloom filter is a bit array that is generated by hashing some of the bit locations to **1**, while the other bit locations are left as **0**. They can be used to provide a compact representation of a set of values, such as an attribute in a database relation. If the Bloom filters of two or more attributes are created using the same hashing functions, then the filters can be bit-wise-anded to find potential common attribute values. Because hashing may produce the same bit addresses for two or more attribute values, some false positives may occur in other words, the resulting Bloom filter may indicate that a value exists in both attributes, when in fact this is not the case. Therefore, additional testing is necessary to eliminate false positives. The bit array approach that is utilized by the proposed strategies is presented in Sect. 3.

3   Stream Spatial Join Strategies

In this section, several strategies are proposed for join processing in spatial data streams two which utilize a bit array representation, and one that utilizes the original polygon representation. In addition, two polygon-based strategies will be summarized that will be used for comparison in Sect. 4. First, the approach for creating the bit array will be proposed. Then, the strategies Bit1, Bit2, and Spatial2 will be proposed, followed by a summarization of Spatial1 and Join.

For all strategies, where appropriate, the following notation will be utilized: S_1 and S_2 are the spatial data streams that provide objects or bit arrays to the data stream processor; E_1 and E_2 are the regions that contain objects from S_1 and S_2; OR is denoted as an "overall region", which contains some combination of E_1 and E_2; m and n are the grid dimensions used to partition OR; RS is the result stream; SW is the sliding window; $numobj(SW)$ is the current number of objects or bit arrays maintained in SW; b_1 and b_2 are bit arrays representing objects o_1 and o_2 respectively; and o_{swx}, $x = 1$ to $numobj(SW) - 1$ are the objects or bit arrays in SW.

3.1   Bit Array Mapping

As mentioned above, the bit array is inspired by the Bloom filter [5]. However, the following differences are applied to their creation here:

– A separate bit array is created for each spatial object, instead of one bit array for the entire object set. Because each spatial object is being transmitted from a sensor one at a time, and may not remain in the sliding window for the duration of the join operation, it makes sense to create individual bit arrays over one bit array for the entire object set.
– The location for the **1** bits are chosen based the location of the object on a grid, instead of using hashing.

Therefore, the following is the proposed strategy for taking an object in polygon form and mapping it to a bit array representation. First, the region OR is formed by combining E_1 and E_2 and determining the minimum extent that encompasses both. OR is then partitioned into a grid of $n \times m$ cells. Following this, a bit array is created and initialized to all zeros. The size of the bit array is $m * n$, which means that each bit in the array corresponds to one location in the grid. For an object o_i, if any portion overlaps a cell, a 1 bit will be assigned to the corresponding bit in the bit array; otherwise the bit is left as 0.

To determine if potential overlap exists between two objects using their bit arrays, this can be done by performing a simple bit-wise-and of the bit arrays. If the result of the bit-wise-and is "positive" (i.e., b_1 & $b_2 != 0$), then potential overlap exists between the objects.

Figure 2 depicts the mapping of the two rectangles (pictured as red and blue) into their respective bit arrays, and the result of a bit-wise-and operation. It should be noted that the mapping takes place in this example in column-major format however, this is not strictly required and using row-major format will not affect the result. It can be seen that overlap exists between the red and blue rectangles since the bit arrays have 1 bits in the same locations, which would produce a non-zero bit-wise-and result.

However, it is possible that the result of a mapping and bit-wise-and computation will produce a "positive" result that is a false positive (i.e., a true outcome that is actually false). Figure 3 depicts such a result. Both the red and blue rectangles map to bit arrays that produce a "positive" outcome when a bit-wise-and operation is applied to them. But we see that the rectangles do not overlap. This is referred to in this paper as a *spurious tuple* one that is generated even though, officially, it is not part of the result. Therefore, a trade-off exists, between (hopefully) achieving a faster result versus having extra tuples formed.

Now, the spatial data stream join processing strategies will be presented, beginning with the bit array strategies, then followed by the polygon strategies.

3.2 Bit1 Strategy

The **Bit1** strategy utilizes a bit array representation of every object that arrives from the spatial data streams. The sensors that generate objects must coordinate in order to determine the overall region OR so that all objects are properly mapped to bit arrays and can be compared by the spatial data stream processor. Then, the spatial stream module processes the spatial join in the following manner:

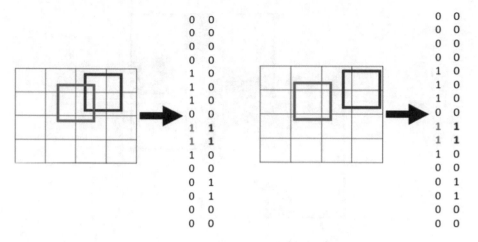

Fig. 2. Bit mapping - accurate result   **Fig. 3.** Bit mapping - false positive

1. Bit arrays b_1 and b_2, representing objects o_1 and o_2 are received one from stream S_1 and the other from stream S_2.
2. First, if required, room must be made for the new bit arrays if the sliding window is full. Although any strategy can be used for selecting which bit arrays to remove, this work assumed a First-in-First-out strategy, which reflects that "older" bit arrays should be removed from the window.
3. The bit array b_1 is added to the sliding window. One at a time, b_1 is bit-wise-anded with all other bit arrays o_{swx} that are from S_2 and currently in SW. Any bit-wise-and operations that produce a "positive" result (i.e., b_1 & $o_{swx}! = 0$) are placed onto RS.
4. Next, the bit array b_2 is added to the sliding window. One at a time, b_2 is bit-wise-anded with all other bit arrays o_{swx} that are from S_1 and currently in SW. Any bit-wise-and operations that produce a "positive" result (i.e. b_2 & $o_{swx}! = 0$) are placed onto RS.
5. This process repeats from Step 1 until no more bit arrays are sent.

3.3   Bit2 Strategy

The *Bit2* Strategy also utilizes a bit array representation of objects. However, it improves upon the *Bit1* strategies in the following way. Figure 4 contains two overall regions E_1 and E_2, both with potential stream objects. Note the overlapped area (denoted from here as OR), which is highlighted in gray. One major improvement can be made to the *Bit1* strategy if we observe the following: Only objects that overlap with OR can potentially be part of the final join result. Therefore, only those objects that overlap OR need to be mapped into bit arrays! One other difference that must be noted is that the stream data processor performs the mapping of objects into bit arrays however, the mapping will only be with respect to OR, and it is expected that fewer objects needs to be mapped.

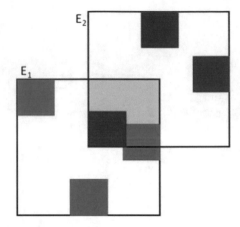

Fig. 4. Overlap of two overall regions

Given these modification, the spatial stream module of the **Bit2** Strategy processes the spatial join in the following manner:

1. First, the overall regions E_1 and E_2 are received from S_1 and S_2 by the stream query processor in order to determine OR.
2. Then, two objects o_1 and o_2 are received from S_1 and S_2.
3. If required, room must be made for the objects if the sliding window is full, using the First-in-First-out strategy.
4. If the object o_1 overlaps OR, it is mapped into b_1 and added to the sliding window. One at a time, b_1 is bit-wise-anded with all other bit arrays o_{swx} that are from S_2 and currently in SW. Any bit-wise-and operations that produce a "positive" result (i.e. b_1 & $o_{swx}! = 0$) are placed onto RS.
5. If the object o_2 overlaps OR, it is mapped into b_2 and added to the sliding window. One at a time, b_2 is bit-wise-anded with all other bit arrays o_{swx} that are from S_1 and currently in SW. Any bit-wise-and operations that produce a "positive" result (i.e. b_2 & $o_{swx}! = 0$) are placed onto RS.
6. This process repeats from Step 2 until no more bit arrays are sent.

3.4   Spatial2 Strategy

The *Spatial2* strategy is also proposed in this paper. Its main purpose here is to provide a comparison between utilizing the actual objects from a spatial data stream versus utilizing bit array representations. However, to the best of the author's knowledge, this approach has not be proposed before.

Spatial2 is an adaptation of *Bit2* that tests arriving objects o_1 and o_2 for overlap with OR, and instead of mapping o_1 and o_2 to bit arrays, the objects themselves are stored in SW. The strategy proceeds as follows:

1. First, the overall regions E_1 and E_2 are received from S_1 and S_2 by the stream query processor in order to determine OR.

2. Then, two objects o_1 and o_2 are received from S_1 and S_2.
3. If required, room must be made for the objects if SW is full, using the First-in-First-out strategy.
4. If the object o_1 overlaps OR, it is added to SW. One at a time, o_1 tested for overlap with all other objects o_{swx} that are from S_2 and currently in SW. Any overlap operations that produce a "positive" result (i.e., $o_1 \cap o_{swx}! = 0$) are placed onto RS.
5. If the object o_2 overlaps OR, it is added to SW. One at a time, o_2 tested for overlap with all other objects o_{swx} that are from S_1 and currently in SW. Any overlap operations that produce a "positive" result (i.e., $o_2 \cap o_{swx}! = 0$) are placed onto RS.
6. This process repeats from Step 2 until no more bit arrays are sent.

3.5   Spatial1 and Join Strategies

Finally, two strategies *Spatial1* and *Join* are presented. Their main purpose is for comparison. *Spatial1* is an adaptation of *Bit1* where, instead of receiving bit arrays from the data streams, the actual objects are received, and are managed and processed similarly to the *Bit1* strategy. *Join* is an adaptation of *Spatial1* where a sliding window of unlimited size is utilized for processing the spatial join. *Spatial1* and *Join* can be viewed as an adaptation of the progressive join approach proposed in [15], where only one level of object resolution is utilized.

4   Evaluation

In this section, the empirical evaluation of the Bit1, Bit2, and Spatial2 strategies is presented, along with a comparison against the Join and Spatial1 strategies. The framework and evaluation methodology is presented first. Then, the results of the evaluation and resulting discussion is presented.

4.1   Framework and Methodology

For all experiments, an environment was utilized that contained two simulated spatial data streams, with each containing a bit mapper. The central stream query processor utilizes a sliding window for maintaining a subset of objects or bit arrays that have arrived from the data streams. It also contains a bit mapper.

All of the strategies presented in Sect. 3 are implemented in C++ on a PC running Linux Centos 7. They were evaluated using several simulated spatial data streams that utilized synthetic sets of 10×10 rectangles. This approach was chosen so that certain characteristics such as the overall coverage area of a stream, as well as the overlap between the coverage areas of two streams, could be controlled. Altogether, eight sets of rectangles were utilized in pairs for spatial joins each pair contained 500, 1000, 1500 and 2000 rectangles, respectively. Each set of n rectangles is drawn from a region of space of dimension $(\sqrt{n} * 10) \times (\sqrt{n} * 10)$. For example, the rectangles in each of the 1000-rectangle set were drawn from a 310×310 region of space. In addition, each pair of rectangles were created so that 25% of the overall region of space between them had overlap.

For all strategies covered in Sect. 3, three sets of tests were carried out that varied: (1) the number of objects sent through each data stream, (2) the size of the sliding window, and (3) the size of the grid used for mapping the bit arrays. For each set of tests:

- *Varying the number of objects.* Four tests were carried out, for the 500×500, 1000×1000, 1500×1500 and 2000×2000 spatial join pairs respectively. The grid size was set to 11×11 (i.e., the closest to 128 bits, without going past that value), and the window size was set at 16000 bytes (i.e., 1000 rectangles or 1000 128-bit arrays).
- *Varying the window size.* Four tests were carried out using 8000, 16000, 24000 and 32000 bytes (i.e., 500, 1000, 1500 or 2000 rectangles or 128-bit arrays) respectively. The 1000×1000 spatial join query was utilized, along with an 11×11 grid size.
- *Varying the grid size.* Finally, four tests were carried out, for the 8×8, 11×11, 13×13 and 16×16 grid sizes (i.e., equivalent to 64, 128, 192, and 256 bits) respectively. The 1000×1000 spatial join query was utilized, along with a 16000 byte window size.

For all tests, in addition to the final spatial join stream, two performance factors were recorded:

- The CPU time over the entire join at the stream query processor.
- The number of joined tuples in the final result. Given this value for each strategy and the final spatial joins streams, accuracy was determined by calculating the following:
  - The number of tuples in the Join result that also existed in the Spatial1, Spatial2, Bit1 and Bit2 results, respectively (i.e., true positives).
  - The number of tuples in the Bit1 and Bit2 results that did not exist in the Join result (i.e., spurious tuples, or false positives).

4.2   Results and Discussion

The running time results will be presented first, followed by the accuracy results.

Figures 5, 6 and 7 present the running time results from varying the data size, sliding window size, and grid size respectively. With respect to the size of the spatial data streams on the running time (Fig. 5), it is observed that although the running time of all proposed strategies increase with the amount of data, the increase is not as significant when compared to the Join result. The best performing strategies are *Spatial1* and *Bit2*, which both have the lowest increases. *Bit2* has a slightly higher running time, which is due to the extra time required for mapping polygons to bit arrays. However, this difference over *Spatial2* is not significant. Both *Spatial1* and *Bit1* have the least desirable performance out of the four strategies. However, *Bit1* has a lower running time due to the use of the bit-wise-and operation, which does run faster than an overlap comparison of two polygons (even as simple rectangles).

With respect to varying streaming window size (Fig. 6), it is observed that once again *Spatial2* and *Bit2* have the best and fairly consistent performance, independent of the sliding window size. This is likely due to the restriction to the shared overlap area for identifying candidate objects and bit array for spatial joins not as many objects are being processed, and also added to and removed from the sliding window. Again, *Bit2* had a higher running due to having to perform bit array mapping. Surprisingly, *Spatial1* and *Bit1* have poor performance as the sliding window size increases. An increase in the size of the sliding window also leads to an increase in the time to search it whenever new objects or bit arrays arrive at the spatial stream server. It was expected that a larger sliding window would result in decreases in time, but was not the case.

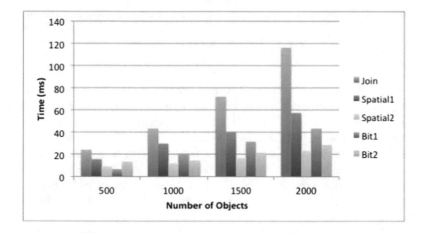

Fig. 5. Time for varying number of objects

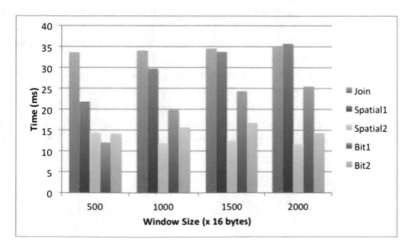

Fig. 6. Time for varying window sizes

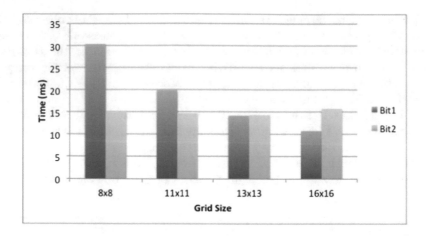

Fig. 7. Time for varying grid sizes

Finally, with respect to varying the grid size on the running time of *Bit1* and *Bit2* (Fig. 7), it is observed that for *Bit1* than an increase in the grid granularity actually leads to a decrease in running time! Again, the opposite was expected, but there is a reasonable explanation for this. As the grid size increases, the size of the bit array increases, which means that fewer bit arrays fit in the sliding window which leads to a lower running time for repeatedly searching the window. This trend is not observed for *Bit2*. However, the reason for this could be that the focus area is much smaller and therefore fewer bit arrays are being dealt with in general.

Tables 1, 2, and 3 present the accuracy results from varying the data size, sliding window size and grid size respectively. In all tables, the strategy names indicate the number of tuples in the final result, while %join indicates the accuracy of the strategy, and #spur indicates the number of spurious tuples that were generated by *Bit1* and *Bit2*. The accuracy will be discussed first, followed by the issue of spurious tuples.

With respect to the size of the spatial data streams on the accuracy of the results (Table 1) it is observed that, the best results were *Spatial2* and *Bit2*, both with 100% accuracy, with a slight decrease beginning at 2000 objects. This is a result of the combination of the window size (1000 objects or bit arrays, or 16000 bytes) and only consulting the restricted overlapped space. Very few if any

Table 1. Accuracy for varying data sizes

Data size	Join	Sp1	%join	Sp2	%join	Bit1	%join	#spur	Bit2	%join	#spur
500	421	421	100	421	100	2841	100	2420	1054	100	633
1000	952	712	74.79	952	100	8406	74.79	7694	2918	100	1966
1500	1455	845	58.08	1455	100	11268	58.08	10423	5613	100	4158
2000	1930	841	43.58	1922	99.59	14969	43.58	14128	7923	99.59	6002

Table 2. Accuracy for varying window sizes

Window size	Join	Sp1	%join	Sp2	%join	Bit1	%join	#spur	Bit2	%join	#spur
500	952	403	42.33	934	98.10	4832	42.33	4429	2862	98.10	1928
1000	952	712	74.79	952	100	8406	74.79	7694	2918	100	1966
1500	952	894	93.91	952	100	10428	93.91	9534	2918	100	1966
2000	952	952	100	952	100	11083	100	10131	2918	100	1966

Table 3. Accuracy for varying grid sizes

Grid size	Join	Bit1	%join	#spur	Bit2	%join	#spur
8×8	952	18581	100	17629	4201	100	3249
11×11	952	8406	74.79	7694	2918	100	1966
13×13	952	4790	54.94	4267	2630	100	1678
16×16	952	2838	42.33	2435	2250	98.21	1321

objects or bit arrays need to be removed in order to make room for new ones. Unfortunately, the same cannot be said for *Spatial1* and *Bit2*. As the number of objects in the spatial data stream increases, the accuracy decreases significantly down to just over 40%. This is a result of considering all objects in both streams participating in the spatial join.

With respect to varying the sliding window size on the accuracy of the results (Table 2), a similar outcome is seen again for *Spatial2* and *Bit2*, which is almost 100% accuracy, regardless of the size of the sliding window. For *Spatial1* and *Bit1*, it is observed that a larger sliding window significantly helps to improve accuracy, with 100% being achiever for the 2000×2000 case. With more objects being considered here, many are being moved out of the sliding window before future objects have arrived that require them for a join.

Finally, with respect to the grid size on the accuracy of the results (Table 3), we see a similar result again for *Bit2* as before. For *Bit1*, it is observed that the accuracy significantly decreases as the grid size increases. This is a result of the larger grid producing a larger bit array, which means fewer of them can be stored in the same sized sliding window. Therefore, more bit arrays would need to be removed from the sliding window in order to make room for more, and having some that were required later on not longer being available.

With respect to *Bit1* and *Bit2*, both are generating an unacceptable number of spurious tuples. Given that *Bit2* operates on a reduced overlapped space, the number being generated by it is significantly lower than the number being generated by *Bit1*. The issue is with the grid size. A very fine grained (i.e., large) grid will result in a lower number of spurious tuples, since the cells will be smaller. However, the drawback is that the bit array is larger and fewer of them will fit into the same sized sliding window as the actual objects themselves.

Overall, it is found that *Spatial2* and *Bit2* has the best performance, and the most consistent performance for both running time and accuracy, regardless of the number of objects, sliding window size, and grid size. However, for *Spatial1* and *Bit1*, it is observed that for an increase in data size, there is an increase in running time but a significantly decrease in accuracy. Given an increase in the sliding window size, there is an increase in accuracy and running times, while an increase in grid size results in a decrease in both accuracy and running time. Therefore, reducing the space that is looked at between two spatial data streams results in significant performance improvements.

5   Conclusion

In this paper, three spatial join strategies for spatial data streams are proposed. Two utilize a bit array mapping strategy to speed up comparisons between spatial objects, while one utilizes a conventional spatial join between objects. Two of the three strategies operate on a reduced space where both spatial data streams overlap. An experimental evaluation and comparison versus modified versions of an existing progressive spatial join strategy shows that all three algorithms outperform the existing strategies with respect to running time, with the two that operate on reduced spatial outperforming all others with almost 100% accuracy regardless of the size of the data stream, sliding window, or grid.

This work has resulted in many future research directions. One very important one is determining how to relate the bit array back to the actual geometric object it is representing in a data stream environment, which is not being done at the moment. Other directions include: (1) reducing the number of spurious tuples being generated by all strategies, (2) With respect to the *Bit2* strategy, one improvement that could help further improve its running time is to have the bit arrays mapped on the sensor generating the spatial data, (3) considering spatial data streams where the coverage area is not known in advance, (4) considering spatial data streams where the arrival times of objects differ, (5) comparing the strategies with larger spatial data streams and against the original progressive join algorithm proposed by [15].

References

1. Abel, D., Ooi, B., Tan, K.L., Power, R., Yu, J.: Spatial join strategies in distributed spatial DBMS. In: Proceedings of the 4th International Symposium on Advances in Spatial Databases (1995)
2. Aji, A., Wang, F., Vo, H., Lee, R., Liu, Q., Zhang, X., Saltz, J.: Hadoop-GIS: a high-performance spatial data warehousing system over MapReduce. Proc. VLDB **6**, 1009–1020 (2013)
3. Arge, L., Procopiuc, O., Ramaswamy, S., Suel, T., Vitter, J.: Scalable sweeping-based spatial join. In: Proceedings of the 24th International Conference on Very Large Databases, pp. 570–581 (1998)
4. Babu, S., Widom, J.: Continuous queries over data streams. SIGMOD Rec. **30**(3), 109–120 (2011)
5. Bloom, B.H.: Space/time trade-offs in hash coding with allowable errors. Commun ACM **13**(7), 422–426 (1970)

6. Farruque, N., Osborn, W.: Efficient distributed spatial semijoins and their applica-
    tion in multiple-site queries. In: Proceedings of the 28th IEEE International Con-
    ference on Advanced Information Networking and Applications. IEEE Computer
    Society (IEEE) (2014)
7. Han, J., Kamber, M., Pei, J.: Data Mining: Concepts and Techniques. Morgan
    Kaufmann, Boston (2011)
8. Hua, Y., Xiao, B., Wang, J.: BR-tree: a scalable prototype for supporting multiple
    queries of multidimensional data. IEEE Trans. Comput. **58**(12), 1585–1598 (2009)
9. Huang, Y.W., Jing, N., Rundensteiner, E.: Integrated query processing strate-
    gies for spatial path queries. In: Proceedings of the 13th International Conference
    on Data Engineering, pp. 477 –486 (1997). https://doi.org/10.1109/ICDE.1997.
    582010
10. Jacox, E., Samet, H.: Spatial join techniques. ACM Trans. Database Syst. **32**(1),
    1–44 (2007). Article No. 7
11. Kalnis, P., Mamoulis, N., Bakiras, S., Li, X.: Ad-hoc distributed spatial joins on
    mobile devices. In: Proceedings of the 20th IEEE International Parallel and Dis-
    tributed Processing Symposium (2006)
12. Kang, M.S., Ko, S.K., Koh, K., Choy, Y.C.: A parallel spatial join processing for
    distributed spatial databases. In: Proceedings of the 5th International Conference
    on Flexible Query Answering Systems, FQAS 2002, London, UK, pp. 212–225.
    Springer (2002). http://portal.acm.org/citation.cfm?id=645424.652610
13. Karam, O.: Optimizing distributed spatial joins using R-trees. Ph.D. thesis, Tulane
    University (2001)
14. Karam, O., Petry, F.: Optimizing distributed spatial joins using R-trees. In: Pro-
    ceedings of the 43rd ACM Southeast Conference (2006)
15. Kwon, O., Li, K.J.: Progressive spatial join for polygon data stream. In: Pro-
    ceedings of the 19th ACM SIGSPATIAL International Conference on Advances in
    Geographic Information Systems. ACM (2011)
16. Osborn, W., Zaamout, S.: Using spatial semijoins over multiple sites in distributed
    spatial query processing. Can. J. Electr. Comput. Eng. **39**(2), 71–81 (2016)
17. Patel, J., DeWitt, D.: Partition based spatial-merge join. In: Proceedings of the
    1996 ACM SIGMOD International Conference on Management of Data, vol. 25,
    pp. 259–270 (1996)
18. Shekhar, S., Chawla, S.: Spatial Databases: A Tour. Prentice Hall, Upper Saddle
    River (2003)
19. Tan, K.L., Ooi, B., Abel, D.: Exploiting spatial indexes for semijoin-based join
    processing in distributed spatial databases. IEEE Trans. Knowl. Data Eng. **12**(6),
    920–937 (2000)
20. Zhong, Y., Han, J., Zhang, T., Li, Z., Fang, J., Chen, G.: Towards parallel spatial
    query processing for big spatial data. In: Proceedings of the 26th IEEE Interna-
    tional Parallel and Distributed Processing Symposium Workshops, pp. 2085–2094
    (2012)
21. Zhou, X., Abel, D., Truffet, D.: Data partitioning for parallel spatial join process-
    ing. Geoinformatica **2**, 175–204 (1998)

2D Color Image Enhancement Based on Conditional Generative Adversarial Network and Interpolation

Yen-Ju Li, Chun-Hsiang Chang, Chitra Meghala Yelamandala, and Yu-Cheng Fan[✉]

Department of Electronic Engineering,
National Taipei University of Technology, Taipei, Taiwan
skystar@ntut.edu.tw

Abstract. In the rapid development of autonomous driving technology. Precise detection of objects might assist self-driving cars to drive as safely as human. The object detection is frequently uses point clouds and produces high quality environment color images to match. However, at night or when the light is dim, it affects the quality of color images. In order to overcome this, the existing image enhancement is focused on the histogram equalization method [1] and Retinex algorithm [2]. This paper proposes to use the Conditional Generative Adversarial Network (cGAN) [3] to train the intrinsic images for quickly decomposed shadow layer, and then use the interpolation method to achieve the image contrast enhancement.

1 Introduction

In recent years, the multimedia industry has proposed various applications such as color cameras, depth sensors, and LIDAR in which LIDAR is evolving rapidly. The environmental sensing of the self-driving vehicles is equipped with high quality sensor LIDAR combined with a color camera to detect obstacles or pedestrians to avoid collision.

In addition to environmental detection, self-driving vehicles have begun to focus on using the Internet to connect with other vehicle. LIDAR is installed on each vehicle and LIDAR data is shared with other vehicles via the network. The established model will expand the environment detection area and accuracy to achieve the perfect object detection.

There will be color and brightness distortion problems. To overcome this, advanced image decomposition method is used for image contrast enhancement and it requires long-term convergence time due to complex algorithms, but selects a simple linear image contrast algorithm for brightness enhancement.

The cGAN is used to achieve fast and accurate intrinsic image decomposition, and the image shadow layer is enhanced by interpolation. Finally, the image enhancement is optimized by combining the shadow layer and the reflection layer.

L. Barolli et al. (Eds.): NBiS-2019, AISC 1036, pp. 86–95, 2020.
https://doi.org/10.1007/978-3-030-29029-0_8

2   Method

2.1   Architecture

The architecture proposed in this paper is shown in the Fig. 1. It is divided into three parts. The first part is training cGAN. It is necessary to collect the original image and the Ground Truth reflection layer image for training, and pre-process the reflection layer picture to avoid unnecessary feature interference training.

The second part uses the cGAN for image decomposition. In which the original image is the input, and the corresponding reflection layer image is the output. Then the intrinsic image decomposition formula can be used to obtain the shadow layer.

In the third part, the shadow images are interpolated and compensated for image enhancement, and then multiplied and combined with the reflection layer generated by the cGAN to obtain the ideal two-dimensional color picture output.

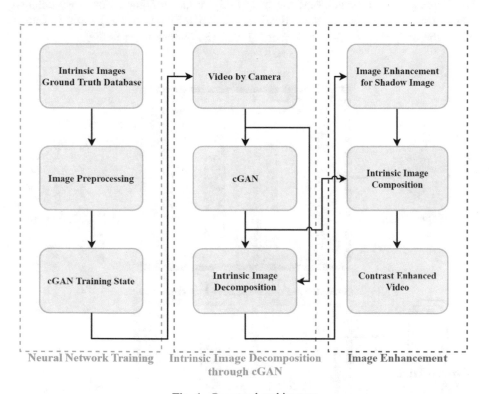

Fig. 1. Proposed architecture

2.2   cGAN

The cGAN architecture proposed in this paper, as shown in Fig. 2, improves the architecture of the literature [3], and divides it into a generation network and a discriminating network. The input of the generated network is the original picture, and the output is the generated reflection layer picture. The detailed structure can be divided into two parts, first collecting features, then magnifying the features of the small pictures and arranging them into new pictures.

The input of the discriminating network architecture is divided into two pictures, which are the image generated by generating network and the ground truth ideal picture, and the output is the similar probability of the two pictures. Initially, the two input images are concatenated, and then the features of each region of the image are compressed. Through the loss function of cGAN, the similarity correction can be made for each area of each picture. The similar probability of two input images is longer as the training time increases, and the probability value is higher. For other detailed training parameters, we use the following, the Batch Size is 5, the Learning Rate is 10^{-4}, the optimizer is AdamOptimizer [4], and the Epochs is set to 200.

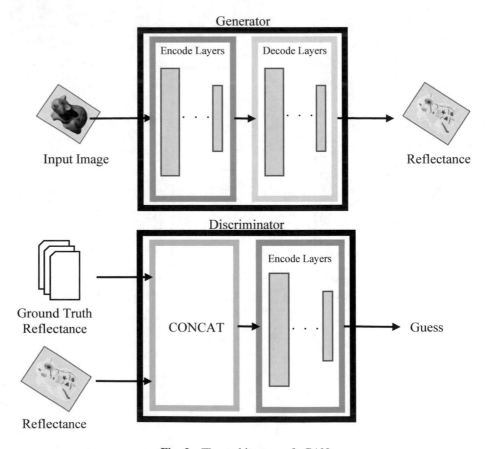

Fig. 2. The architecture of cGAN

2.3 Interpolation

After obtaining the reflection layer image from the previous step, we use the internal image decomposition formula [5] to decompose the shadow layer.

$$O = R \times S \tag{1}$$

Where O is the original picture, R is the reflection layer image, and S is the shadow layer image. First we will use simple difference compensation, because the reflection layer has the characteristics of the most realistic picture of the image. So when we subtract the original image pixel value O(y, x) and the reflection layer image pixel value R(y, x), the larger the error value Err(y, x) is, the larger the exposure difference in the picture. The place may be too bright or too dark. The formula is as follows.

$$Err(y, x) = O(y, x) - R(y, x) \tag{2}$$

When we get the difference, we interpolate the shadow layer to compensate the image. Here we use Contrast Limited Adaptive Histogram Equalization (CLAHE) [6] to obtain a continuous and detailed shadow image of gray-scale values by interpolating between large blocks. First, the selected picture is divided into several blocks, and the average error value of each block is calculated as the interpolation value of the subsequent interpolation compensation algorithm.

$$Err(k) = \frac{1}{row \times col} \sum (O(i, j) - R(i, j)) \tag{3}$$

Where k denotes k blocks, row denotes the number of columns in the region, and columns denotes the number of rows in the region. Secondly, the interpolation method is shown in Fig. 3. Where the red region denotes four corners using nearest neighbor interpolation, the yellow region compensates for edges using linear interpolation, and the middle blue region is compensated using bilinear interpolation.

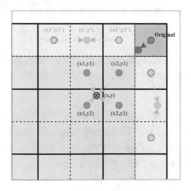

Fig. 3. Interpolation rule

After the above-mentioned interpolation method performs error compensation on each area, the strong edge is likely to cause brightness imbalance after compensation. Figure 4(a) is the original picture, Fig. 4(b) is the hypothetical corresponding reflection layer picture, and Fig. 4(c) is the result obtained by the above interpolation compensation method. It can be seen that there is a phenomenon at the edge of Fig. 4(c) that is opposite to the enhancement result.

When the edge position is interpolated, the average error value of the adjacent areas is very different. The formula is calculated to cause interference and offset, resulting in a lower compensation value than expected, and there is no brightness imbalance at the weak edge, as shown in Fig. 5(a), (b) and (c).

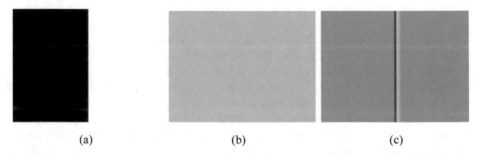

(a)   (b)   (c)

Fig. 4. Vertical strong edge (a) Original image; (b) The hypothetical corresponding reflection layer picture; (c) The result obtained by the above interpolation compensation method.

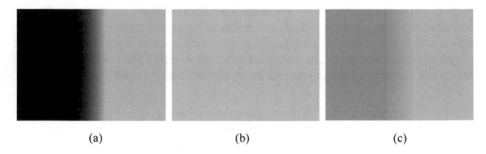

(a)   (b)   (c)

Fig. 5. Vertical weak edge (a) Original image; (b) The hypothetical corresponding reflection layer picture; (c) The result obtained by the above interpolation compensation method.

In order to correct the problem of the edge of the brightness, we adjust the weight of interpolation for the area around the strong edge. If there are no blocks in the block that are marked as having strong edges, then the weight values remain linearly interpolated values. If there are strong edge blocks around, we need to adjust the weight value to 0 or 1, and then return the expression to the original interpolation method to get better interpolation compensation results, as shown in Fig. 6(a), (b) and (c).

When we get the final shadow layer enhancement image $Shadow_{EnhancedEdge}$, we can use the following internal image decomposition formula to merge with the reflection layer R obtained by the previous cGAN to get the final image enhancement color image $I_{Enhanced}$.

$$I_{Enhanced} = Shadow_{EnhancedEdge} \times R \tag{4}$$

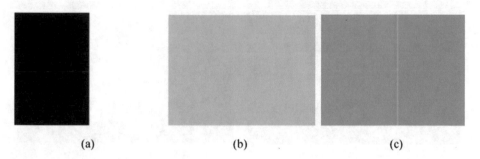

(a)   (b)   (c)

Fig. 6. Vertical strong edge (a) Original image; (b) The hypothetical corresponding reflection layer picture; (c) The result obtained by the modified interpolation compensation method.

3 Experimental Results

The MIT dataset [7] and the Intrinsic Image in the Wild (IIW) dataset [8] are used as a training data. There are 19 pictures in the MIT dataset, 5230 pictures in the IIW dataset, and then the datasets are divided into training materials and test data in a ratio of 9:1. A large amount of training data is used for image decomposition training through the cGAN architecture, and the remaining test data is used to test the correctness of the training results.

The enhanced results of the interpolation compensation algorithm are shown in Fig. 7(a)–(f). This paper makes the following comparisons. First, we compare the correctness of the data of the reflective layer. We compare it with two papers, the first paper adopts Clustering method [9], and the second paper presents L1 Flattening scheme [10]. From the experimental data, we can see that the WHDR (Weighted Human Disagreement Rate) value of the proposed method is 0.2650, which means that the error rate is about 26.5%. Compared with other literatures, the reflection layer produced by cGAN is not much different from other algorithms (Table 1).

Table 1. WHDR value

Clustering	L1	Proposed
0.2658	0.3038	0.2650

The internal image decomposition processing time of cGAN is compared as shown in Table 2. We propose the internal image decomposition of cGAN, as long as the neural network architecture is set up with parameters training, the image reflection layer can be generated immediately. It can be seen from Table 2. The algorithm proposed in

[9–11] requires a long operation time. It can be seen that the internal image decomposition proposed by us is feasible, and not only the decomposition effect is excellent, but also the speed is greatly improved.

(a)

(b)

(c)

(d)

(e)

(f)

Fig. 7. NPEA Dataset-Rail (a) Original image; (b) Reflection layer image generated by cGAN; (c) Proposed method; (d) NPEA; (e) CRM; (f) LIME.

Table 2. Internal image decomposition processing time

Reference	Processing Time	Resolution	Environment
[9]	5–15 s	Based on clusters	Intel Core i5-2500 CPU at 3.30 GHz
[10]	25.600 s	320 × 240	Intel Core i7-4790 CPU at 3.60 GHz
	182.612 s	640 × 480	
[11]	12 s	398 × 600	CPU at 3.40 GHz
Proposed	0.051 s	320 × 240	Intel Core i7-7700K NVIDIA GeForce GTX 1080 Ti
	0.110 s	640 × 480	

Next, we compare the image enhancement data, we use the NPEA [12] dataset. We have two data comparisons, the first one is Information Entropy [13], the formula is as follows

$$H(x) = \sum_{i=1}^{k} p(i) \log_2 \frac{1}{p(i)} \tag{5}$$

i represents the grayscale value of 0–255, $p(i)$ is the probability of grayscale value. It can be known from the formula that $H(x)$ is calculating the concentration of each grayscale value. When the probability of partial grayscale value is too high, the value of $H(x)$ will be too low, which means that the more concentrated the grayscale value is, the worse the contrast is. The second comparison data is Lightness Order Error (LOE) [12]. The larger the LOE is, the more the dissimilarity is. The brightness consistency before and after image enhancement will make the LOE value smaller. So the smaller the LOE value, the better.

We have three papers in our comparison. The first paper proposes the Naturalness Preserved Enhancement Algorithm (NPEA), which decomposes the shadow layer and the reflection layer through the Bright-pass Filter, and then uses the bi-log algorithm to enhance the details of the shadow layer of the image. The second paper proposes Low-light Image Map Estimation (LIME) [14], using the algorithm to solve the illumination mapping, and then doing gamma correction on the shadow layer [15]. The third paper proposes Camera Response Model (CRM) [16], which uses the camera to take different exposure photos to obtain the camera response module, and then uses the brightness estimation algorithm to obtain the exposure rate map. Finally, the two photos are combined to analyze the photos and adjusted the normal exposure to achieve image enhancement.

The experimental results are shown in Figs. 8 and 9. It is found from the experimental results that the performance of our algorithm on LOE is not optimal, because the brightness of the reflection layer is darker than that of the original image, resulting in the interpolation compensation algorithm reducing the shadow layer. The grayscale value, which is easy to cause the grayscale value hierarchy to change, the original grayscale is higher and higher, and the LOE result is larger, especially when the light and dark contrast is enhanced. However, our algorithm produces another good result, that is, the contrast is greatly improved.

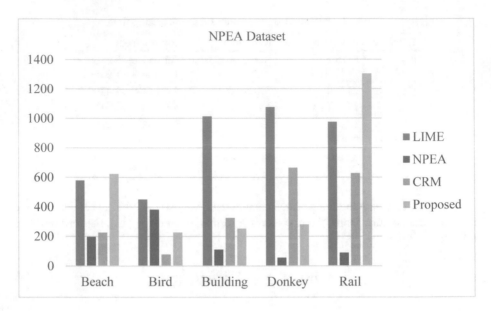

Fig. 8. LOE comparison

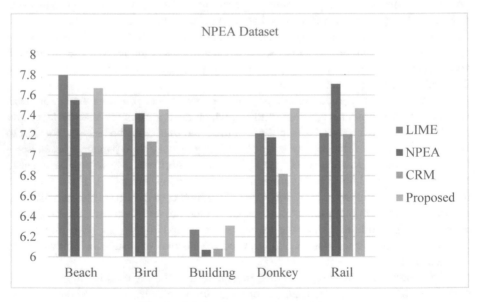

Fig. 9. Information entropy comparison

From the Information Entropy data, we can see that the contrast of our enhanced image is better than other algorithms, because the algorithm does not blindly increase the brightness, but reflects the bright and dark real correction according to the reflection layer image, so the contrast get a better upgrade.

4 Conclusion

cGAN is proposed for intrinsic image decomposition in which the processing time is greatly reduced. Then the original image and the reflection layer image are substituted into the interpolation compensation algorithm for reinforcement. As a result, the image quality has a good contrast and naturalness. In the future, we believe that our proposed method will have good results in brightness and contrast enhancement on multimedia devices.

References

1. Gonzalez, R.C., Woods, R.E.: Digital Image Processing, 2nd edn. Prentice-Hall, Upper Saddle River (2002)
2. Land, E.H., McCann, J.J.: Lightness and Retinex theory. J. Opt. Soc. Amer. **61**(1), 1–11 (1971)
3. Isola, P., Zhu, J.-Y., Zhou, T., Efros, A.A.: Image-to-image translation with conditional adversarial networks. In: CVPR, pp. 2–5 (2017)
4. Kingma, D., Ba, J.: Adam: a method for stochastic optimization. In: International Conference on Learning Representations (2015)
5. Barrow, H.G., Tenenbaum, J.M.: Recovering intrinsic scene characteristics from images. In: Computer Vision Systems. Academic, New York (1978)
6. Zuiderveld, K.: Contrast limited adaptive histogram equalization. In: Graphics Gems IV, Academic Press (1994)
7. Grosse, R., Johnson, M.K., Adelson, E.H., Freeman, W.T.: Ground truth dataset and baseline evaluations or intrinsic image algorithms. In: International Conference on Computer Vision (2009)
8. Bell, S., Bala, K., Snavely, N.: Intrinsic images in the wild. ACM Trans. Graph. **33**(4) (2014)
9. Garces, E., Munoz, A., Lopez-Moreno, J., Gutierrez, D.: Intrinsic images by clustering. In: Computer Graphics Forum (Eurographics Symposium on Rendering), vol. 31, no. 4 (2012)
10. Bi, S., Han, X., Yu, Y.: An L1 image transform for edgepreserving smoothing and scene-level intrinsic decomposition. ACM Trans. Graph. (TOG) **34**(4), 78 (2015)
11. Mittal, A., Soundararajan, R., Bovik, A.C.: Making a 'Completely Blind' image quality analyzer. IEEE Signal Process. Lett. **20**(3), 209–212 (2013)
12. Wang, S., Zheng, J., Hu, H.M., Li, B.: Naturalness preserved emhancement algorithm for non-uniform illumination images. IEEE Trans. Image Process. **22**(9), 3538–3578 (2013)
13. Ye, Z., Mohamadian, H., Ye, Y.: Discrete entropy and relative entropy study on nonlinear clustering of underwater and arial image. In: Proceedings of IEEE International Conference on Control Applications, pp. 318–323, October 2007
14. Guo, X., Li, Y., Ling, H.: Lime: low-light image enhancement via illumination map estimation. IEEE Trans. Image Process. **26**(2), 982–993 (2017)
15. Poynton, C.A., Kaufmann, M.: Digital Video and HDTV: Algorithm and Interfaces, pp. 260, 630 (2003)
16. Ying, Z., Li, G., Ren, Y., Wang, R., Wang, W.: A new lowlight image enhancement algorithm using camera response model. In: 2017 IEEE International Conference on Computer Vision Workshops (ICCVW), pp. 3015–3022, October 2017

Position Follow-up Control for Hand Delivery of Object Between Moving Robot Arms of Remote Robot Systems with Force Feedback

Qin Qian[1], Yutaka Ishibashi[1(⊠)], Pingguo Huang[2], and Yuichiro Tateiwa[1]

[1] Nagoya Institute of Technology, Nagoya, Japan
q.qian.924@stn.nitech.ac.jp, {ishibasi,tateiwa}@nitech.ac.jp
[2] Seijoh University, Tokai, Japan
huangpg@seijoh-u.ac.jp

Abstract. In this paper, we propose position follow-up control for deal-ing with hand delivery of an object between moving robot arms of remote robot systems with force feedback, and we investigate the effect of the control by experiment. We also examine influences of the network delay on the hand delivery of the object under the control. In each system, a user can operate the moving robot arm having a force sensor by using a haptic interface device while watching video. An electric hand which can hold the object is attached to the tip of the robot arm. In the position follow-up control, the position of one robot arm is automatically moved close to the other moving robot arm, and then smooth hand delivery is realized in combination with manual operation. In the experiment, we make a comparison between the case in which the position follow-up con-trol is carried out and the case in which the control is not performed. Experimental results show that the average operation time is greatly decreased under the control, and the average operation time increases as the network delay becomes larger.

1 Introduction

In recent years, remote robot systems with force feedback have actively been researched [1–9]. Especially, a number of researchers focus on cooperative work among multiple remote robot systems with force feedback in which multiple robot arms move an object by holding the object together or one robot arm moves and hand-delivers the object to another robot arm [10,11]. It is possible for users to perceive the shape, weight, and softness of a remote object through haptic interface devices by using force feedback. Therefore, the efficiency and accuracy of the cooperative work are expected to be improved largely. However, when force and/or position information is transmitted over a network like the Internet, which does not guarantee the quality of service (QoS) [14], the quality of experience (QoE) [15] such as the operability of the haptic interface device may seriously be degraded due to the network delay, delay jitter, and packet

© Springer Nature Switzerland AG 2020
L. Barolli et al. (Eds.): NBiS-2019, AISC 1036, pp. 96–107, 2020.
https://doi.org/10.1007/978-3-030-29029-0_9

loss [3,4]. Also, instability phenomena in the remote robot systems with force feedback such as vibrations of the robot arm and haptic interface device may largely affect the remote operation. To solve the problems, it is necessary to carry out stabilization control and QoS control [5,7]. In particular, when performing cooperative work among multiple robot arms, it becomes more difficult to solve the problems [12].

In [10], two remote robot systems with force feedback are adopted. A user at the master terminals remotely controls a robot arm at the slave terminal of each system to move an object together by operating two haptic interface devices with both hands while watching video. The authors investigate influences of the network delay on the cooperative work by experiment. In [11], the authors clarify the influences of the network delay on the cooperative work in which a user hands over (or receives) an object with an electric hand of a robot arm to (or from) that of the other robot arm. As a result, it is demonstrated that the average operation time increases as the network delay becomes larger.

To solve the problem, in this paper, we propose position follow-up control for hand delivery of an object between robot arms while moving the robot arms. We investigate the effect of the control, and examine influences of the network delay on the hand delivery of the object under the control.

The organization of this paper is as follows. Section 2 describes the remote robot systems with force feedback. In Sect. 3, the position follow-up control is proposed. Then, the experiment method is explained in Sect. 4, and the experimental results are presented in Sect. 5. We conclude the paper in Sect. 6.

2   Remote Robot Systems with Force Feedback

2.1   System Configuration

Figure 1 shows the configuration of the remote robot systems (called *systems 1 and 2* here) with force feedback used in this paper. Each system consists of the master terminal and the slave terminal. In system 1, the master terminal consists of PC for haptic interface device and PC for video. The haptic interface device called 3D Systems Touch [16] is connected to PC for haptic interface device. The slave terminal consists of PC for industrial robot and PC for video. PC for industrial robot is directly connected to the industrial robot via an Ethernet cable (100 BASE-TX). A web camera (5WH-00003 by Microsoft Corp.) is connected to PC for video, and the camera is set in front of the industrial robot. The video resolution is 1920×1080 pixels. PC for haptic interface device and PC for industrial robot are linked to each other by switching hubs over a network. The industrial robot consists of a robot arm (RV-2F-D [17] by Mitsubishi Electric Corp.), a robot controller (CR750-Q [17]), a force interface unit (2F-TZ561 [18]), and a force sensor (1F-FS001-W200 [18]) which is attached to the surface of the flange of the robot arm. In our experiment, an electric hand [19] is attached to the tip of the force sensor. As shown in Fig. 2, the robot arm is installed on a metal platform.

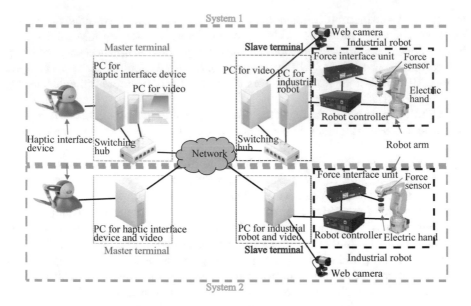

Fig. 1. Configuration of remote robot systems with force feedback.

System 2 is almost the same as system 1. In system 2, there is only one PC in each of the master and the slave terminals, and it has roles of PC not only for haptic interface device or industrial robot but also for video.

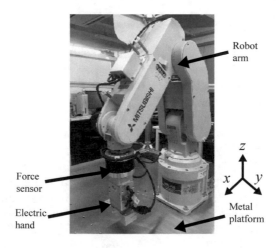

Fig. 2. Configuration of robot arm.

2.2   Remote Operation

In each system, a user at the master terminal can remotely operate the robot arm at the slave terminal by using the haptic interface device. The initial position of the haptic interface device is the original position which corresponds to the initial position of the industrial robot (i.e., the electric hand attached to the tip of the robot arm).

The master terminal acquires the position information from the haptic interface device every millisecond, calculates the reaction force, and outputs the force via the device. The position information is transmitted to the slave terminal by UDP. At the slave terminal, PC for industrial robot employs the real-time control function [20] and real-time monitor function [20] to get the position information and the information about the force sensor from the robot controller every 3.5 ms (the value is equal to the control period of the industrial robot). The two types of information are transmitted as different UDP packets between the robot controller and PC for industrial robot. Then, PC for industrial robot forwards the position information of the robot arm and force information to the master terminal. Also, it sends the information of instruction based on the position information of the haptic interface device to the industrial robot every 3.5 ms.

The reaction force $\boldsymbol{F}_t^{(\mathrm{m})}$ outputted at time t (ms) ($t \geq 1$) against the haptic interface device of the master terminal is calculated as follows:

$$\boldsymbol{F}_t^{(\mathrm{m})} = K_{\mathrm{scale}} \boldsymbol{F}_{t-1}^{(\mathrm{s})} \tag{1}$$

where $\boldsymbol{F}_t^{(\mathrm{s})}$ is the force received from the slave terminal at time t. K_{scale} is the mapping ratio about the force between the slave terminal and the master terminal ($K_{\mathrm{scale}} = 0.1$ [13] in this paper). Moreover, if the absolute value of reaction force exceeds the maximum allowable reaction force of 3.3 N, 3.3 N is outputted.

At the slave terminal, the robot arm is operated on the basis of the position information transmitted from the master terminal. The position vector about the tip of the robot arm $\boldsymbol{S}_t$ ($t \geq 1$) is calculated as follows:

$$\boldsymbol{S}_t = \boldsymbol{M}_{t-1} \tag{2}$$

where $\boldsymbol{M}_t$ is the position vector of the haptic interface device that is received from the master terminal at time t. It should be noted that the mapping ratio from the work space of the robot arm at the slave terminal to that of the haptic interface device at the master terminal is set to 1:1 [13].

3   Position Follow-up Control

In the position follow-up control, a position near the location of the hand delivery (called the *target position* here) is determined from the current position information of the robot arm (called *robot arm 1*) of system 1, and the robot arm (*robot arm 2*) of system 2 is automatically moved to the target position (the determination method of the target position will be described in Sect. 4). Therefore, the position information of robot arm 1 is transmitted from PC for industrial robot of system 1 to PC for industrial robot of system 2. The movement in the x-axis direction by the position follow-up control is calculated by the following equation instead of Eq. (2) (the equations in the y-axis and z-axis are the same as that in the x-axis).

$$S_t^{(x)} = \begin{cases} S_{t-1}^{(x)} + V_{\text{follow}-\text{up}}^{(x)} & (S_{t-1}^{(x)} < S_{\text{target}}^{(x)}) \\ S_{t-1}^{(x)} - V_{\text{follow}-\text{up}}^{(x)} & (S_{t-1}^{(x)} > S_{\text{target}}^{(x)}) \\ S_{\text{target}}^{(x)} & (S_{t-1}^{(x)} = S_{\text{target}}^{(x)}) \end{cases} \tag{3}$$

where $S_t^{(x)}$ is the position of robot arm 2 at time t, and $S_{\text{target}}^{(x)}$ is the target position based on the position information of robot arm 1. Also, $V_{\text{follow}-\text{up}}^{(x)}$ is the follow-up speed of robot arm 2, which is set to 0.1 mm/ms in this paper. When robot arm 2 moves, the stylus of the haptic interface device of system 2 also moves in the same way as robot arm 2. The hand delivery of an object is performed in three axes in combination with automatic and manual operations after robot arm 2 has reached the target position. Specifically, robot arm 2 automatically follows up robot arm 1 in the x-axis direction, and is manually moved in the other two axes.

4   Experiment Method

To investigate the effect of the position follow-up control, we dealt with hand delivery of an object between moving robot arms (the effect of that between fixed robot arms are referred to [13]) with force feedback under the control and no control (the control is not performed) by measuring the average operation time of the cooperative work under the different moving speed of the robot arms. We also examined influences of the network delay on the hand delivery of the object with the work efficiency by using a network emulator (NIST Net [21]). The same constant delay in both directions was added to each packet transferred between the master and slave terminals of each system. The network delay was not add to each packet transferred between systems 1 and 2. We need to examine the influence of the delay on the hand delivery[1]; this is for further study.

In our experiment, we performed two types of cooperative work (called *work A* and *work B* here) in which a wooden stick of 30 cm was hand-delivered between the two moving robot arms while watching video. In work A, a user (called *user 2* here) operated robot arm 2 to move the wooden stick which was held by the electric hand toward robot arm 1. The other user (*user 1*) operated robot arm 1

---

[1] PC for industrial robot of system 2 can also know the position of robot arm by using the video camera without sending the position information. However, this is outside the scope of this paper.

to grasp the stick with closing the electric hand, and then pulled it. Then, user 2 opened the electric hand to release the stick. In work B, the wooden stick was held by robot arm 1 at the beginning of the work. User 2 moved the electric hand of robot arm 2 toward the wooden stick and closed the electric hand to grasp the stick. Then, user 1 opened the electric hand and hand-delivered the stick to user 2.

In the experiment, to make the two robot arms move in the same way as each other, we moved robot arm 1 forward and back from -6 cm to 6 cm in the x-axis direction horizontally, and the position follow-up control was applied for robot arm 2 to follow up and approach robot arm 1. The movement in the x-axis direction of robot arm 1 is calculated by the following equation instead of Eq. (2).

$$
S_t^{(x)} = \begin{cases} S_{t-1}^{(x)} + V_{\text{moving}}^{(x)} T & \text{(move forward)} \\ S_{t-1}^{(x)} - V_{\text{moving}}^{(x)} T & \text{(move backward)} \end{cases} \tag{4}
$$

where -6 cm $\leq S_t^{(x)} \leq 6$ cm, and $V_{\text{moving}}^{(x)}$ is the moving speed of robot arm 1, and $T = \lceil 3.5 \rceil$ ($=4$) ms (i.e., the control period of each robot arm). For simplicity, the directions of the two electric hand were set to the same as each other. Also, in this paper, the target position is the position at which the electric hand and the wooden stick are at the same height, at a distance of 0 cm in the x-axis direction and 1 cm in the y-axis direction. When robot arm 2 reaches the target position and moves together with robot arm 1, the hand delivery is started. Furthermore, we set the moving speed of robot arm 1 to 0.005 mm/ms, 0.015 mm/ms, and 0.025 mm/ms. We investigate the effect of the control on the work efficiency by measuring the average operation times of work A and work B at different moving speed. The average operation time is defined as the average time from the moment the work is started until the instant the stick is hand-delivered. One of the authors operates robot arm 2, and another person does robot arm 1. We carried out the experiment 10 times for each combination of the type of work and moving speed.

We also examined influences of the network delay on the two types of work when the moving speed of robot arm 1 is 0.025 mm/ms. In the experiment, the additional delay was selected from among 0 ms, 100 ms, and 200 ms. The additional delay was selected for systems 1 and 2 in random order, and we also carried out the experiment 10 times for each combination of the type of work and additional delay.

5   Experimental Results

We show the average operation times of the two types of work as a function of the moving speed of robot arm 1 under the position follow-up control and no control in Fig. 3. We also plot the average operation times of work A and work B versus the addition delay under the control when the additional delay in both systems is 0 ms, 100 ms, and 200 ms in Fig. 4. The 95% confidence intervals were further shown in Figs. 3 and 4. In addition, we only show the position and force

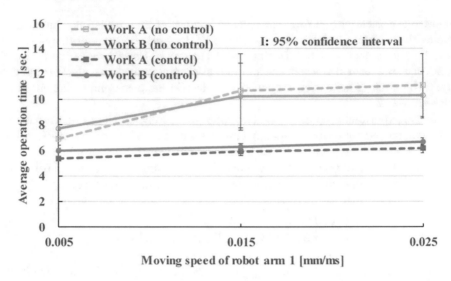

Fig. 3. Average operation time versus moving speed of robot arm 1.

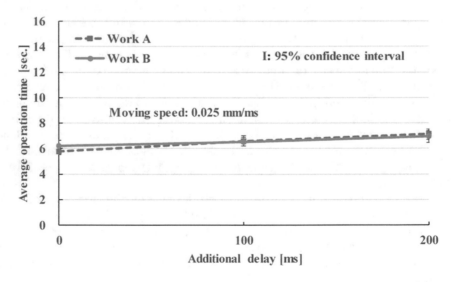

Fig. 4. Average operation time versus additional delay.

of robot arm 2 in the y-axis (the left and right direction) versus the elapsed time from the beginning of work B to the end of the work under the control and no control (additional delay is 0 ms, and the moving speed of robot arm 1 is 0.025 mm/ms) in Figs. 5 and 6 because of space limitations.

In Fig. 3, we see that the average operation times under the control are shorter than those under no control for all the moving speeds considered here. In addition, the average operation time increases as the speed becomes faster when the position follow-up control is not used, and the increase of the average operation time can be suppressed under the control. Therefore, the two types of work is difficult to be carried out under no control, and especially when the moving speed of robot arm 1 is fast.

From Fig. 4, we observe that the average work times increase as the additional delays become larger in both work A and work B under the position follow-up control. Also, the average operation time in work A is almost the same as that in work B.

From Fig. 5(a), we see that the position of robot arm 2 increases linearly from 0 to about 0.5 s; this indicates that the robot arm automatically moves to the target position under the control. Then, the position keeps constant until about 2.7 s. Then, it starts to increase gradually until about 4.7 s. Finally, it keeps constant; this means that the start of the hand delivery of the object after robot arm 2 has reached the target position. It is should be noted that from around 5.5 s to 6.2 s, the electric hand of robot arm 2 is closed and that of robot arm 1 is opened. From Fig. 5(b), we can see that at 0 s, relatively large force is generated due to the inertia of the start of the position follow-up control, and then the force is almost zero because the user does not operate robot arm 2. After the start of the hand delivery at around 2.7 s, relatively small force is generated since robot arm 2 moves together with robot arm 1. At around 5.5 s, the force starts to changes; this means that the object is hand-delivered from robot arm 1 to robot arm 2.

In Fig. 6(b), we notice that large force is generated at about 6.0 s and 9.0 s. This is because the electric hand of robot arm 2 hit the wooden stick held by robot arm 2 twice. At about 6.0 s, the electric hand could not reach a suitable location to grasp the stick, but hit the stick. Also, at about 9.0 s, although the electric hand of robot arm 2 reached a suitable location, but hit the stick while grasping the stick; this means that the stick was not located at the center of the electric hand. We can confirm in Fig. 6(a) that the position largely changes at those times. Therefore, the work becomes difficult to be carried out under no control, and it cannot be performed smoothly under no control.

From the above discussions, we can say that the position follow-up control is effective.

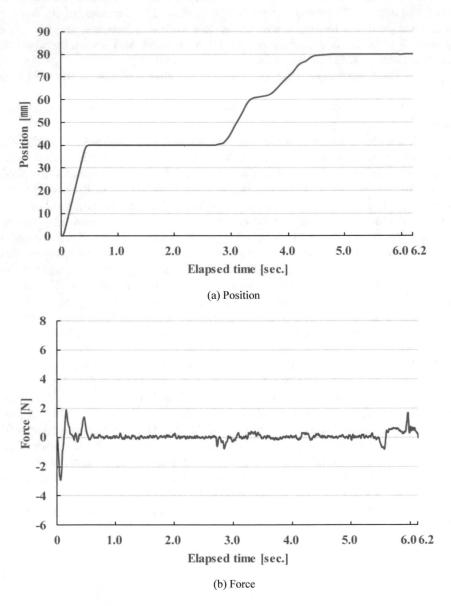

(a) Position

(b) Force

Fig. 5. Position and force of robot arm 2 versus elapsed time under control in work B.

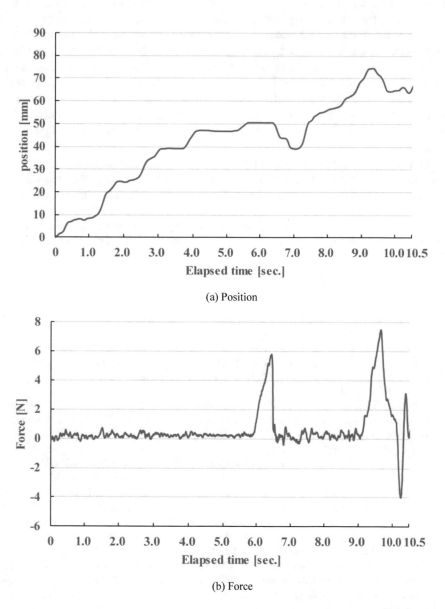

(a) Position

(b) Force

Fig. 6. Position and force of robot arm 2 versus elapsed time under no control in work B.

6   Conclusion

In this paper, we proposed the position follow-up control for hand delivery of an object between the two moving robot arms of the remote robot systems with force feedback, and investigated the effect of the control by comparing the average operation time of the two types of cooperative work under the control and no control. We also examined the influences of the network delay on the hand delivery under the control. As a result, we found that the average operation time under the control is smaller than that under no control, and smooth hand delivery can be realized between the moving robot arms under the control. Also, we saw that the average operation time increases as the network delay becomes larger.

As the next step of our research, we need to examine influences of the network delay between systems 1 and 2 on hand delivery of an object. We will also deal with the cooperative work when the robot arms move in the three axes.

Acknowledgment. The authors thanks Yuichi Toyoda for his support to construct the experiment system. This work was supported by JSPS KAKENHI Grant Number 18K11261.

References

1. Ohnishi, K.: Real world haptics: its principle and future prospects. IEEJ **133**(5), 268–269 (2013). (in Japanese)
2. Kawai, T.: Haptics for surgery. IEEJ **133**(5), 282–285 (2013). (in Japanese)
3. Miyoshi, T., Terashima, K.: A stabilizing method for non-passive force-position teleoperating system. In: SICE Symposium on Control Theory, vol. 35, pp. 127–130 (2006). (in Japanese)
4. Miyoshi, T., Maeda, Y., Morita, Y., Ishibashi, Y., Terashima, K.: Development of haptic network game based on multi-lateral tele-control theory and influence of network delay on QoE. Trans. Virtual Real. Soc. Jpn. Spec. Issues Haptic Contents **19**(4), 559–569 (2014). (in Japanese)
5. Ishibashi, Y., Huang, P.: Improvement of QoS in haptic communication and its future. IEICE Trans. Commun. (Jpn. Ed.) **J99-B**(10), 911–925 (2016)
6. Huang, P., Miyoshi, T., Ishibashi, Y.: Stabilization of bilateral control in remote robot system. IEICE Technical report, CQ2016-125, March 2017. (in Japanese)
7. Huang, P., Toyoda, Y., Taguchi, E., Miyoshi, T., Ishibashi, Y.: Improvement of haptic quality in stabilization control of remote robot system. IEICE Technical report CQ2017-79, November 2017. (in Japanese)
8. Arima, R., Huang, P., Ishibashi, Y., Tateiwa, Y.: Softness assessment of objects in remote robot system with haptics: comparison between reaction force control upon hitting and stabilization control. IEICE Technical report, CQ2017-98, January 2018. (in Japanese)
9. Suzuki, K., Maeda, Y. Ishibashi, Y., Fukushima, N.: Improvement of operability in remote robot control with force feedback. In: Proceedings of IEEE Global Conference on Consumer Electronics (GCCE), pp. 16-20, October 2015
10. Taguchi, E. Ishibashi, Y., Huang, P., Tateiwa, Y.: Experiment on collaborative work between remote robot systems with haptics. In: IEICE Global Conference, B-11-17, March 2018. (in Japanese)

11. Toyoda, Y., Ishibashi, Y., Huang, P., Tateiwa, Y., Watanabe, H.: Influence of network delay on efficiency of cooperative work with human in remote robot control with haptic sense. IEICE Technical report, CQ2018-9, April 2018. (in Japanese)
12. Taguchi, E., Ishibashi, Y., Huang, P., Tateiwa, Y., Miyoshi, T.: Comparison of stabilization control in cooperation between remote robot systems with force feedback. In: Proceedings of International Conference on Future Computer and Communication (ICFCC), February 2019
13. Toyoda, Y., Ishibashi, Y., Huang, P., Tateiwa, Y., Watanabe, H.: Follow-up control of robot position for hand delivery of object between remote robot systems with force feedback. IEICE Technical report, CQ2018-90, January 2019. (in Japanese)
14. ITU-T Rec. I. 350: General aspects of quality of service and network performance in digital networks, March 1993
15. ITU-T Rec. G. 100/P. 10 Amendment 1: New appendix I – Definition of quality of experience (QoE) (2007)
16. 3D Systems Touch. https://www.3dsystems.com/haptics-devices/touch
17. RV-2F-D Series Standard Specifications Manual. http://dl.mitsubishielectric. co.jp/dl/fa/members/document/manual/robot/bfp-a8900/bfp-a8900x.pdf.   (in Japanese)
18. CR750/CR751 Controller Force Sense Function Instruction Manual. http://dl. mitsubishielectric.co.jp/dl/fa/members/document/manual/robot/bfp-a8947/bfp-a8947b.pdf. (in Japanese)
19. Electric gripper body (2 claws) Function Instruction Manual. http://www.taiyo-ltd.co.jp/products/electrically-powered/docs/Manual_esg1-2f_201508.pdf.   (in Japanese)
20. CR750/CR751 series controller, CR800 series controller Ethernet Function Instruction Manual. http://dl.mitsubishielectric.co.jp/dl/fa/members/document/ manual/robot/bfp-a3379/bfp-a3379b.pdf. (in Japanese)
21. Carson, M., Santay, D.: NIST Net - a Linux-based network emulation tool. ACM SIGCOMM **33**(3), 111–126 (2003)

A SOA Based SLA Negotiation and Formulation Architecture for Personalized Service Delivery in SDN

Shuraia Khan[✉] and Farookh Khadeer Hussain

School of Computer Science,
University of Technology Sydney, Ultimo, Australia
Shuraia.khan@student.uts.edu.au,
Farookh.hussain@uts.edu.au

Abstract. Supporting end-to-end personalized Quality of Services (QoS) delivery in existing network architecture is an ongoing issue. Software Defined Networking (SDN) model has emerged in response to the limitations of traditional network. Integrating Software Defined Network (SDN) architecture with Service Oriented Architecture (SOA) brings new concept for future service oriented delivery in SDN services. Researchers from both academic and industry are working to resolve the QoS limitations of service delivery, however; most of the proposed solutions are application oriented and unable to provide a reliable personalized QoS delivery in future service oriented SDN. This research propose a reliable Service Level Agreement (SLA) oriented Service Negotiation framework that would be able to provide reputation based personalized service delivery and assist in QoS management in SDN for informed decision making. Moreover, potential benefits of the proposed framework are also discussed in this paper in social, scientific and business aspects.

Keywords: Service negotiation · Personalized service delivery · SLA

1 Introduction

Highly diverse and dynamic network demanded by current and emerging applications bring new challenges of service provisioning in future networks. There was an argument about future internetworking architecture that able to function with SOA principles, such as; the network will be service contact oriented, loose coupling, abstraction, reusability, independence, statelessness, discoverability and compos ability [1]. SDN introduce itself as an effective solution for current and future network that capable to support intelligent applications [2]. Thus future networking architecture demand of allowing service provision according to current business needs with full potentiality in SDN is essential [1]. Applications of SOA principle in SDN addressing the challenging issues of end to end QoS provisioning; such as driving the current networking technology into a Network-as-a-service (NaaS) paradigm, which makes the future carrier networks look more like Clouds [1].

In service oriented computing, personalization is described as an advance feature of service selection which means meeting the consumer's needs more effectively and

© Springer Nature Switzerland AG 2020
L. Barolli et al. (Eds.): NBiS-2019, AISC 1036, pp. 108–119, 2020.
https://doi.org/10.1007/978-3-030-29029-0_10

efficiently where the "one size fits all" idea is not applicable. Therefore, rather than receiving a whole service pack; the consumer can customize the service request based on their needs and receiving the services as a pay as you go basis. By offering personalization, companies empowered consumers to make their services feel like their own. The service requirements are demarcated in service contract stage which formally made in early stage of Service Level Agreement (SLA).

In this research, we propose a SLA Negotiation Framework with the impression of SOA principals that enable to provide personalized service delivery in SDN. SLA negotiation framework is a unique framework that combines the advantages of consumer selection as well as provider selection while selecting the required services. This framework offer a reputation based selection process that facilitate in intelligent decision making. In addition, the framework is grounded with SOA concept that able to fulfill various service requirements of the respected consumer's and create a scope of delivering services in personalized manner. Moreover, it uses logic driven selection and decision making analogy to determine a successful service negotiation and develop SLA. The implementation details however are not in the scope of this paper.

This paper is organized as follows. In Sect. 2, the related work is summarized. Section 3 explains the proposed SLA negotiation framework for SDN with SOA principles followed by the potential benefits of the proposed framework in Sect. 4. The paper is concluded with future recommendation in Sect. 5.

2 Related Work

Service Level Agreement (SLA) driven Service negotiation frameworks have been proposed from various service oriented domains previously, however in this paper, SLA driven Service negotiation framework in SDN are only discussed. A SLA driven service provisioning architecture is proposed [3] that enables flexible and quantitative SLA negotiations for network services. The architecture is an agent-mediated network service prototype where dynamic SLA negotiations scheme is proposed between a user agent and a network provider agent. There are four SLA negotiation scheme is proposed such as Bandwidth Negotiation at Resource Limit (BNRL), Guaranteed Session Start Time Negotiation with Delay (STN-D), Guaranteed Session Duration Negotiation for both Session Cut Short (SDN-CS) and Temporary Session Bandwidth Drop off (SDNTBD). In this approach, the negotiation take place when the network performances of the above negotiation schemes are degrade. This research analyzes the impact of negotiation schemes on three dimensions, such as; service availability, network utilization or revenues and means user satisfaction to reduce the rejection probability.

To identify the best suited protocol target to fulfill client's requirements among the offered protocols by provider, a protocol similarity model and a similarity of virtual networks negotiation model is proposed [4]. The work focus on two major aspects; VN negotiation: where the model is capable to categorize the protocols that customize to VN/SDN. In second aspect, a metric is proposed that can evaluate the similarity between same kinds of protocols.

A SLA re-negotiation based approach is proposed [5] to adjust the allocation of network resources as close as possible according to Virtual Network's (VN) consumer's demand. Therefore, this approach prevents resource idleness, as well as the loss of QoS experience. However, the design architecture and implementation technique of this proposed approach remains unstated.

A dynamic Flow negotiation based approach in SDN has been proposed [6] where the flows are able to pass through to meet the QoS requirements. In this approach, the flows connect devices passing through a wired or wireless domain in-contrast with separate SDN controllers. The approach also propose a QoS negotiation mechanism among the domain and SDN controllers.

A real-life implementation based technique in SDN is proposed by Korner et al. [7] that enable to apply QoS requirements using a conjoin approach (WS-Agreement standard and Open flow standard). This approach allows to implement the following capabilities such as: defining QoS requirements of the network, negotiate service level objectives based on current network utilization, create a SLA for cloud-based network and establish a QoS overlay in Open-flow network based on SLA. In addition, developing an SLA framework named WSAG4J (Part of this work) enhance the capabilities of the approach and allow to implement 'WS-agreement' and 'WS-negotiation' as protocol. Consequently, the experimental result depicts that, this framework able to assist through simpler network management and network utilization calculation and predict the available capacity in relation to all end-to-end overlays.

A SDN based self-customizable architecture is proposed [8] where the architecture is comprises with a cloud based front-end portal and SDN based back end APIs. The architecture is beneficial for subscriber to improve streaming video (YouTube) quality and video conferencing such as skype. This is the only paper that offer personalizing network experience in cloud based SDN. However, this framework is not able to provide personalizing service delivery according to the term "Personalization". The concept is quite different what they have offered. No SOA principles have been considered in this framework.

As evident from the above literature survey, there is very little research on Service Level Agreement (SLA) driven service provisioning in SDN. As a result, there is no research that able to define how the Service negotiation strategies can take place amongst service provider and service consumer with the presence of third party (Service agent) that leads to Service Oriented Architecture (SOA) Principles. Therefore, personalize service delivery in SDN remain anonymous features in existing researches.

3   Proposed Framework - Service Negotiations and SLA Formulation

"Service negotiation and SLA formulation" is an intelligent framework where the criteria of the required services will be specified before the SLA formulated. Therefore, the framework enables to provide the opportunity of service consumer to personalize their services. Moreover, the consumer will have opportunity to select service provider based on their reputation. In this framework, the consumer would be able to request and receive services based on their current needs or current business demand. The framework is framed with four key tasks such as, Service Specification, Service Provider Selection, Suitable Consumer Selection and Service Level Agreement Formulation. There are five steps in the framework, as shown in Fig. 1.

Step 1: Consumer Send a request for a service (To the registry)

Step 2: Service Provider ((Intermediate) registry/broker) receives service request.

Step 3: Service providers reply with their suitability decision making.

Step 4: Service Provider Selection (Accept or Reject).

Step 5: Service Negotiation based on Service Providers offer or Service Level Agreement Formulation.

Step 1: Consumer sends a request for a service (to the registry)

The consumer build an initial draft of service request with their requirements in this step. The service request draft may include specific service criteria such as: service name or modified service items, service availability, service duration, service specification etc. The consumer sends the service request draft to the registry or broker or service agent. As an example; send an open tender for a service.

Step 2: Service Provider (Intermediate) registry/broker)) receives service request

Service agent/Registry/Broker (Intermediate agent) receives consumer's service request that is specified in service request draft. As soon as the intermediate agent receives the service request draft, they send that request to the interested tender recipients (service providers). In this stage, the generic scenario can be, intermediate agent/broker send the service request to their registered service provider company with specification and waiting for their response. In this stage, the intermediate agent also sends an acknowledgement to the service requester (consumer) by mentioning that their request has been received with service requirements clarification.

Step 3: Service providers reply with their suitability decision making

Interested service providers receive the service request with service specification and then they take some intelligent decision based on their suitability. The interested service providers measure their suitability by considering some important factors and then send their interest of accepting the service request to the requester (consumer). To accept the service request, the factors that service provider will consider while calculating their suitability are;

Factor 1 or F1: Reliability of the Consumer
Factor 2 or F2: Duration of the Services they require
Factor 3 or F3: Risk Propensity

F1: Reliability of the Consumer

Reliability of the consumer is the first factor that provider will consider prior to agreeing of providing services to the consumer. Reliability of the consumer can be measured using requester's (consumer's) historical data. The historical data includes the company reputation data from last few years. The reputation can be consider with two variations such as; company's overall reputation rating and other provider's satisfaction rating about this company, as we assume that, the service requester company is not a new company and the company has worked with other service providers before.

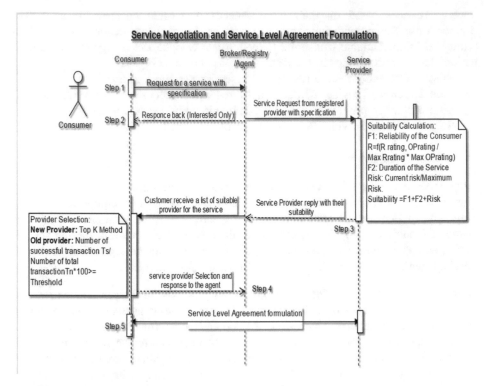

Fig. 1. Negotiation and service level agreement formulation process (sequence diagram)

Reliability R can be calculated by taking the two parameters; Reputation Rating (R rating) and other providers reputation rating (OPR rating) and dividing them by Maximum reputation rating (Max R rating) and Maximum other reputation rating (Max OPR rating), This comparison calculation perform using some calculation function ($\int$) where the $\int$ = (+) or (−) or (*) or (/).

$$R = \int (Reputation\ rating,\ Other\ Providers\ Reputation\ rating) \qquad (1)$$

$$R = \int \left(\frac{Reputation\ rating,\ Other\ Providers\ Reputation\ rating}{Max\ Reputation\ rating,\ Max\ Other\ Reputation\ rating} \right) \qquad (2)$$

$$R = \int \left(\frac{R\ rating,\ OPR\ rating}{Max\ R\ rating,\ Max\ OPR\ rating} \right) \qquad (3)$$

Reputation rating is considered from 1 to 5 where 1 represents lowest rating and 5 represents highest rating. Max R rating is the maximum rating score within the range that the consumer company can achieve from their Consumer, which are 5 in this case. Max OPR rating may consider with similar proportion.

Factor 2 or F2: Duration of the Services Consumer Require

The suitability calculation is also considered the duration of the services that the requester (consumer) needs. A rating formula has been created that can represent the duration of the services. The service provider able to change the rating based on their best fit for their company. The Duration of the requested service can be represent as D and duration rating can be defined in the following ways.

0 to 3 months = 6
More than 3 months to less than 1 year = 5
More than 1 year to less than 3 years = 4
More than 3 years to less than 5 years = 3
More than 5 years to less than 10 years = 2
More than 10 years = 1

0 to 3 months duration request can be consider as less suitable because the contract can be establish for short-term period. On the other hand, more than 10 years' service request looks more reliable and suitable. As a result, longer duration service request can be considered as the most preferable request and define as high priority.

Factor 3 or F3: Risk Propensity of Accepting a Contract

Risk Propensity is an important factor that needs to take into account while measuring the service provider's suitability. There are several circumstances take into account while measuring Risk propensity of service provider. As an example, a service provider company has resources to provide services to the consumer. A Consumer requires x amount of services for next one year. The service provider company doesn't have that much hardware and software resources to provide x amount of services. Or they have enough hardware and software resources to provide x amount of services however they have lack of technical skill staff to provide x amount of services in required level. In this case they need to hire expertise. In those cases, the service provider company needs to calculate their financial solidity to purchase additional hardware or Software or expertise for next one year. Based on this circumstance, the service provider will calculate their financial investment prior to take the contract.

Here we also create a rating formula that can represent additional investment rate and this risk propensity rating value is used to measure the suitability of the provider. The rating value is also changeable based on organization's 'demand. The risk rating can be defined in following:

Table 1. Possible risk rating

Risk rating	Resource used	Expense range
0	No additional resource require	$0.00–$n1
1	20% of their resource require	$n1–$n2
2	50% of their resource require	$n2–$n3
3	70% of their resource require	$n3–$n4
4	90% of their resource require	$n4–$n5
5	100% of their resource require & may need extra	$n5–$n6 or more

The maximum risk that a provider company may take is rated as 5, Risk rating and the maximum risk rate is changeable.

To accept the service request, the provider company need to calculate their resource availability or necessity of additional expenses and those calculate can be define as the probable Risk Propensity for the provider company. To find the probable Risk propensity, we evaluate the current risk level for the provider (Table 1) of accepting the service request and then the current risk level divided by Maximum Risk (Table 1). Then we perform some calculation function ($\int$) where the $\int = (+)$ or $(-)$ or $(*)$ or $(/)$.

$$\text{Risk R} = \int \left(\frac{Current\ Risk\ Level}{Maximum\ Risk} \right) \tag{4}$$

Suitability Calculation
As the provider company will take decision of whether or not accepting the consumer service request after considering the above three factors. In this stage, we build an intelligent system that able to determine the suitability for the provider. In this intelligent system, fuzzy interface method used to map input variables to the output value. Takagi-Sugeno method (T-S method) is a very popular Fuzzy method that is used to calculate the suitability of the Consumer where the values of the previous three factors are considering as inputs. The basic equation can be as following:

$$Suitability = Reliability\ \&|\ Duration\ \&|\ Risk\ Propensity \tag{5}$$

Step 4: Service Provider Selection (Accept or Reject)
The possible outcomes of previous step (Step 3) are a list of interested provider who would like to provide the services to the requester consumer. In this step, the consumer also has opportunity to select their suitable service provider. In other words, the provider who can be the best fit for their requested services and in one word is "provider selection". We categorize Service providers in two groups based on their durations of continuing providing services.

(A) Old providers - who are already in the market and running their business successfully from few years.
(B) New providers - who are new and just have started their business.

There are two distinct approaches have been proposed for old providers and new providers to find the suitable providers.

For Old Providers
Information about all existing SDN providers is stored in a provider profile repository. This provider repository is accessible to the consumer and able to see the status of all interested service provider. All of the old providers who have previous transaction records are listed in the repository. To select the suitable provider, the consumer considers the number of successful transaction. If a provider A, has performed Ts number of successful transaction among Tn number of total transaction, then the suitability of the provider can be calculated using following equation [9].

$$T_s/Tn \times 100 \geq Threshold \tag{6}$$

The Threshold is the acceptable percentage of success ratio that will be determined by the consumer. This acceptable percentage of success ratio can be varied upon to different consumers.

For New Providers

The new provider does not have any previous transaction history that may assist to analyze the status about provider. The consumer will use top-K nearest neighbors profile patterns and will take a predictive decision. This method is also used to take a predictive decision to select a provider who are relatively new and not have any previous transaction history [9]. We select the top-K nearest neighbors from a set of providers with the maximum number of similarities to a new provider. The top-K nearest neighbors comprises with three sub-steps:

(a) Selection of Similar Neighbors
(b) Transaction Trend of new provider
(c) Decision to approve or reject new provider

Selection of Similar Neighbors

To find nearest similar neighbors, we use UCF method which is widely used in user-based recommender systems [9]. UCF is capable to determine the likely transaction trend of a new consumer based on its nearest top-K resource usage profile. Selection of nearest neighbors depends on some criteria and the equation are providing below.

$$NN = \frac{\sum_i^n [sim(r, n_i) \times rs_i \{n_i \in N | rs \in R\}]}{n} \tag{7}$$

The nearest neighbors NN is determined by considering the requesting provider r, where ni is the ith nearest neighbor from a set of all neighbors N and rsi is the ith resources used by the neighbors and R is a set of all resources.

We also need to determine the strength of the similarity between a new provider and existing providers. To calculate the strength of similarity, we will use Pearson Correlation Coefficient (PCC) equation [9] and the equation is presenting below.

$$sim(r, r_a) = \frac{n(\sum rr_a) - (\sum r)(\sum r_a)}{\sqrt{[n(\sum r^2 - (\sum r)^2)][n \sum r_a^2 - (\sum r_a)^2}} \tag{8}$$

In the above Equation, sim(r, ra) represents the strength of similarity between requesting provider r and all of it's neighbors ra. Here r represents the requesting provider and ra is a set of nearest neighbors. To improve the prediction accuracy, we will choose only those neighbors whose PCC value is positive.

To calculate the transaction trend, we also need to determine the enhanced top-K nearest neighbors. The equation [9] is following;

$$TKN_{enh}(r) = \{r_a | r_a \in T_k(r), sim1(r_a, r) > 0, r_a \neq r\} \qquad (9)$$

TKNenh are the enhanced top-K nearest neighbors where r is requesting new provider. Tk(r), sim1(ra, r) represents the similarity between a provider r and a set of traditional top-K nearest neighbors.

Transaction Trend of New Providers
To calculate the transaction trends of new provider, we consider all top-K nearest neighbors, their transaction trends and the degree of similarity they have to a new consumer. The transaction trend calculation [9] can be performed using following equation:

$$T_{trend}(rn) = \frac{1}{n}[\sum\nolimits_{i=1}^{n} TKN_{enh}(i)\{PCC(i) \times T_{trend}(i) \qquad (10)$$

Here Ttrend(rn) represents a transaction trend of the requesting new provider where TKNenh(i) is the ith enhanced top-K nearest neighbors and PCC(i) is the Pearson Correlation Coefficient value for ith nearest neighbors. Ttrend(i) is a transaction trend for ith nearest neighbors that starts from 1 and moves to n where n is the total number of top-K nearest neighbors.

Decision to Approve or Reject New Provider
The consumer take decision to approve or reject new provider based on the transaction trend of the new provider that has been calculated from previous equation (Eq. 10). The consumer will select the new provider if the provider's transaction trend is greater than or equal to the threshold value [9]. This threshold value can be signified by the consumer on their choice.

$$\text{Accept request} = T_{trend}(rn) \geq threshold \qquad (11)$$

Step 5: Service Level Agreement Formulation
In this step, the consumer and the provider company make a formal commitment. The service provider who were agreed to provide the requested services have a clear understanding about the consumer company as well as their previous history of required services and transaction records with service requirements. In the same way, the consumer already has detail knowledge about the service provider company and their previous transaction history. In this stage, the consumer has the opportunity to signify their service requirements or Quality of Service requirements in a formal document. In this document, the consumer able to specify several service requirements based on their choice or company demand with very fine granular specifications. Moreover, consumer can indicate differentiated service request (services for different applications with various requirements) and specify their expectation level of each services. The provider company will review the service requirements document where they can have very comprehensive understanding about the requested services.

Since, the provider already has Consumer Company's background, their service requirements and expectations. Therefore, the provider company enables to provide personalized service delivery as the personalization features are demarcated in the document. In addition, the service provider company and the consumer company able to negotiate several other things such as: service cost, SLA violation penalties etc. After agreeing from both parties, a SLA document can be formulated and finalized that is signed by both parties.

4 Potential Benefits of Proposed Framework

The benefit of following Service Level Agreement (SLAs) to the service-oriented Business Model is crucial as this is becoming an increasingly popular way of managing and maintaining the service quality. Service Oriented SDN architecture enhance of delivering the network application functionality as a service. The SLA contains not only the terms of the services, but also includes the non-functional requirements of the services that also specified as Quality of service (QoS) terms such as obligations, service pricing, penalties in case of agreement violations [10]. In order to avoid costly SLA violations and timely react to failures and environmental changes, advanced SLA enactment strategies are necessary. SLA enactment strategies include suitable service negotiation and SLA formulation that are proposed in this paper for early intervention before SLA violation.

There is very little research that offer SLA negotiation oriented service delivery in SDN. However, there is no research that able to offer personalize service delivery in SDN network. Moreover, there is no research that offers the SLA negotiation in SOA based SDN architecture. This proposed framework brings several potential benefits for SDN researchers, SDN service providers and consumers. The benefits are as follows.

1. Able to provide a structured and intelligent service negotiation process for entire networks instead of application based solution.
2. Assists of providing differentiated services in the network which is the most advanced method for managing traffic in terms of Class of Services (CoS).
3. Enable to provide personalized service delivery that help to builds loyalty between service provider and consumer.
4. Leverage the service oriented architecture in SDN to bring service oriented future SDN.
5. Assisting SDN service providers in decision making based on service requirements and requester's reputation.
6. Assisting SDN service consumers in intelligent decision making based on providers previous reputation.
7. Highly improve the trust and satisfaction among providers and consumers, as both parties obtain very comprehensive knowledge about the other party before articulate the SLA.
8. Service Level Agreement negotiation assists to save the business from facing a big financial loss for both service provider and consumer side.

5 Conclusion and Future Works

In this paper, we introduced a SLA based service negotiation framework that enables to provide personalized service delivery in SOA oriented SDN service delivery. After reviewing related literature find that there is very little or almost no research on SLA negotiation in SDN however a SLA negotiation framework in SDN is proposed[3] where the negotiation take place after the performance of network parameters are contaminated. Moreover, most of the framework not developed for service delivery purpose. Therefore, personalize service delivery in SDN which is an advance feature of SOA, cannot be achieved.

This SLA negotiation framework in SDN service delivery has open pathways for service negotiation between service consumer and provider. This research federated with three important tasks such as; consumer selection, provider selection and SLA formulation in the presence of third party agent and perform those tasks in five distinguish steps. This research offers great flexibility in reputation based service provider selection as selecting a trustworthy service provider is an ongoing challenge for service consumer. In addition the proposed framework has the capability to provide personalize service delivery in SOA based SDN that offers a list of potential benefits for SDN researchers, consumers and providers are discussed in this paper. In future, this proposed framework can be combined with some Artificial Intelligence (AI) systems that may be able to find QoS parameters for a service in a SDN network. This combination would bring opportunity in predictive decision making as service oriented QoS parameters identification can add a new advancement in SLA formulation that offers QoS guarantee in SDN service delivery.

References

1. Duan, Q., Ansari, N., Toy, M.: Software-defined network virtualization: an architectural framework for integrating SDN and NFV for service provisioning in future networks. IEEE Netw. 30(5), 10–16 (2016)
2. Martini, B., Paganelli, F.: A service-oriented approach for dynamic chaining of virtual network functions over multi-provider software-defined networks. Future Internet 8(2), 24 (2016)
3. Chieng, D., Marshall, A., Parr, G.: SLA brokering and bandwidth reservation negotiation schemes for QoS-aware internet. IEEE Trans. Netw. Serv. Manag. 2(1), 39–49 (2005)
4. Gomes, R.L., Bittencourt, L.F., Madeira, E.R.: A similarity model for virtual networks negotiation. In: Proceedings of the 29th Annual ACM Symposium on Applied Computing. ACM (2014)
5. Gomes, R.L., Bittencourt, L.F., Madeira, E.R.: SLA renegotiation according to traffic demand. In: 2nd Workshop on Network Virtualization and Intelligence for the Future Internet (WNetVirt) (2013)
6. Ghalwash, H., Huang, C.: A QoS framework for SDN-based networks. In: 2018 IEEE 4th International Conference on Collaboration and Internet Computing (CIC) (2018)
7. Körner, M., Stanik, A., Kao, O.: Applying QoS in software defined networks by using WS-agreement. In: 2014 IEEE 6th International Conference on Cloud Computing Technology and Science (CloudCom). IEEE (2014)

8. Gharakheili, H.H., et al.: Personalizing the home network experience using cloud-based SDN. In: Proceeding of IEEE International Symposium on a World of Wireless, Mobile and Multimedia Networks 2014. IEEE (2014)
9. Walayet Hussain, F.K.H., Hussain, O.K., Chang, E.: Provider-based optimized personalized viable SLA (OPV-SLA) framework to prevent SLA violation. Comput. J. Adv. Access **59**, 1760–1783 (2016). Section A
10. Emeakaroha, V.C., et al.: Low level metrics to high level SLAs-LoM2HiS framework: bridging the gap between monitored metrics and SLA parameters in cloud environments. In: 2010 International Conference on High Performance Computing and Simulation (HPCS). IEEE (2010)

An Energy-Efficient Process Replication Algorithm with Multi-threads Allocation

Tomoya Enokido[1($\boxtimes$)], Dilawaer Duolikun[2], and Makoto Takizawa[2]

[1] Faculty of Business Administration, Rissho University, Tokyo, Japan
eno@ris.ac.jp
[2] Department of Advanced Sciences, Faculty of Science and Engineering,
Hosei University, Tokyo, Japan
dilewerdolkun@gmail.com, makoto.takizawa@computer.org

Abstract. In order to realize energy-efficient information systems, it is necessary to not only achieve performance objectives but also reduce the total electric energy consumption of a system. In our previous studies, the RECLB (redundant energy consumption laxity-based) algorithm is proposed to select multiple virtual machines in a server cluster for redundantly performing each application processes in presence of server faults so that the total electric energy consumption of a server cluster can be reduced. Here, one thread on a CPU is bounded to a virtual machine in a server and replicas of each application process are performed on the virtual machine by using only one thread even if some threads are not used in the server. In this paper, the RECLB-MT (RECLB with multi-threads allocation) algorithm is proposed to furthermore reduce the total electric energy consumption of a server cluster by allocating more number of threads to each virtual machine. We evaluate the RECLB-MT algorithm in terms of the total electric energy consumption of a server cluster compared with the RECLB algorithm.

Keywords: Energy-efficient information systems ·
Process replication · Green computing systems · Virtual machines ·
Load balance

1 Introduction

Distributed applications have to be implemented on scalable, high performance, and fault-tolerant computing systems like cloud computing systems [1] to provide available and reliable application services. Server cluster systems [2–7] equipped with virtual machine technologies [8] are widely used to implement these distributed applications. In order to provide available and reliable application services in presence of server faults [9], multiple replicas [10] of each application process can be redundantly performed on multiple virtual machines in a server cluster. Here, since replicas of each application process are performed on multiple virtual machines which are performed on multiple servers, a larger amount

© Springer Nature Switzerland AG 2020
L. Barolli et al. (Eds.): NBiS-2019, AISC 1036, pp. 120–131, 2020.
https://doi.org/10.1007/978-3-030-29029-0_11

of electric energy is consumed in a server cluster than non-redundant models. It is necessary to design and implement not only fault-tolerant but also energy-efficient server cluster systems equipped with virtual machines as discussed in Green computing [1].

In our previous studies, the *redundant energy consumption laxity based* (*RECLB*) algorithm [7] is proposed to select multiple virtual machines in a server cluster for redundantly performing replicas of each *computation type application process* (*computation process*) [6] which mainly consumes CPU resources of a virtual machine. We assume every process is deterministic [11]. In the RECLB algorithm, each time a load balancer receives a request process, the load balancer selects a subset of virtual machines in a server cluster, where the total amount of electric energy laxity of a server cluster is the minimum. Then, the load balancer forwards the request process to every selected virtual machines and replicas of the request process are performed on selected virtual machines. As the results, the total electric energy consumption [J] of a server cluster to redundantly perform the computation processes can be reduced in the RECLB algorithm. On the other hand, only one thread on a CPU is bounded a virtual machine in a server even if some other threads are not used by the other virtual machines. This means, computation resource of a server is not efficiently utilized and the server consumes more electric energy since it takes longer time to perform processes.

In this paper, the *RECLB with multi-threads allocation* (*RECLB-MT*) algorithm is newly proposed to furthermore reduce the total electric energy consumption [J] of a server cluster and the average computation time [sec] of each computation process by allocating idle threads to each virtual machine which computation processes are performed. In the simulation, we evaluate the RECLB-MT algorithm. The evaluation results show the total electric energy consumption of a server cluster and the average computation time of each process in the RECLB-MT algorithm can be more reduced than the RECLB algorithm.

In Sect. 2, we discuss the computation model of a virtual machine and power consumption model of a server. In Sect. 3, we discuss the RECLB-MT algorithm. In Sect. 4, we evaluate the RECLB-MT algorithm compared with the RECLB algorithm.

2 Computation and Power Consumption Models

2.1 Computation Model of a Virtual Machine

A system is composed of multiple servers s_1, ..., s_n ($n \geq 1$). In this paper, we assume a server s_t is equipped with one CPU composed of multiple homogeneous cores. Let nc_t be the total number of cores in a server s_t ($nc_t \geq 1$). Let C_t be a set of cores c_{1t}, ..., $c_{nc_t t}$ in a server s_t. We assume the Hyper-Threading Technology [12] is enabled on a CPU. Let ct_t be the number of threads on each core c_{ht} in a server s_t. Let nt_t be the total number of threads in a server s_t, i.e. $nt_t = nc_t \cdot ct_t$. Let TH_t be a set of threads th_{1t}, ..., $th_{nt_t t}$ ($nt_t \geq 1$) in a server s_t. Threads $th_{(h-1) \cdot ct_t + 1}$, ..., $th_{h \cdot ct_t}$ ($1 \leq h \leq nc_t$) are bounded to a core c_{ht}. Let V_t be a set of virtual machines VM_{1t}, ..., $VM_{nt_t t}$ in a server s_t. A virtual machine VM_{vt} is performed on threads in a server s_t. Let $VT_{vt}(\tau)$ be a set of threads bounded to a virtual machine VM_{vt} at time τ. $nVT_{vt}(\tau)$ is $|VT_{vt}(\tau)|$ and $1 \leq nVT_{vt}(\tau) \leq nt_t$. A virtual machine VM_{vt} is referred to as *active* iff (if and only if) at least one process is performed on the virtual machine VM_{vt}. Otherwise, the virtual machine VM_{vt} is *idle*. A thread th_{kt} is referred to as *active* iff at least one virtual machine VM_{vt} is active on the thread th_{kt}. Otherwise, the thread th_{kt} is *idle*. A core c_{ht} is referred to as *active* iff at least one thread th_{kt} is active on the core c_{ht}. Otherwise, the core c_{ht} is *idle*. In this paper, we assume each thread th_{kt} in a server s_t is not bounded to multiple virtual machines at the time τ. Each active virtual machine VM_{vt} can be exclusively performed on at least one thread th_{kt} in a server s_t at time τ.

In this paper, we consider *computation processes*, where CPU resources are mainly consumed and deterministic. A term *process* stands for a computation process in this paper. A notation pt_{kt}^i stands for a process p^i performed on a thread th_{kt} in a server s_t. Let Th_{kt}^i be the total computation time [msec] of a process pt_{kt}^i which the process p^i is performed on a thread th_{kt}. Let $minTh_{kt}^i$ be the minimum computation time of a process pt_{kt}^i where the process pt_{kt}^i is exclusively performed on one thread th_{kt} on a core c_{ht} and the other threads on the core c_{ht} are *idle* in a server s_t. We assume $minTh_{1t}^i = minTh_{2t}^i = \cdots = minTh_{nt_t t}^i$ in a server s_t. $minTh^i = minTh_{kt}^i$ on the fastest server s_t. We assume one virtual computation step [vs] is performed for one time unit [msec] on one thread th_{kt} in the fastest server s_t. That is, the maximum computation rate $Maxf_{kt}$ of a thread th_{kt} on a core c_{ht} in the fastest server s_t where only the thread th_{kt} is active on the core c_{ht} is 1 [vs/ms]. $Maxf_{ku} \leq Maxf_{kt}$ on the slower server s_u. We assume $Maxf_{1t} = Maxf_{2t} = \cdots = Maxf_{nt_t t}$ in a server s_t. $Maxf = max(Maxf_{k1}, ..., Maxf_{kn})$. A process p^i is considered to be composed of VS^i virtual computation steps. $VS^i = minTh^i \cdot Maxf = minTh^i$ [vs]. The maximum computation rate $maxf_{kt}^i$ of a process pt_{kt}^i is $VS^i / minTh_{kt}^i$ $(0 \leq maxf_{kt}^i \leq 1)$ where the process p_{kt}^i is exclusively performed on a thread th_{kt} on a core c_{ht} and only the thread th_{kt} is active on the core c_{ht} in a server s_t.

The computation rate $FT_{kt}(\tau)$ of a thread th_{kt} on a core c_{ht} in a server s_t at time τ is given as follows:

$$FT_{kt}(\tau) = Maxf_{kt} \cdot \beta_{kt}(at_{kt}(\tau)). \tag{1}$$

Here, $at_{kt}(\tau)$ is the number of active threads on the core c_{ht} at time τ where the thread th_{kt} is bounded to the core c_{ht}. Let $\beta_{kt}(at_{kt}(\tau))$ be the *performance degradation ratio* of a thread th_{kt} on a core c_{ht} at time τ $(0 \leq \beta_{kt}(at_{kt}(\tau)) \leq 1)$ where multiple threads are active on the same core c_{ht}. $\beta_{kt}(at_{kt}(\tau)) = 1$ if $at_{kt}(\tau) = 1$. $\beta_{kt}(at_{kt}(\tau_1)) \leq \beta_{kt}(at_{kt}(\tau_2)) \leq 1$ if $at_{kt}(\tau_1) \geq at_{kt}(\tau_2)$.

Suppose a virtual machine VM_{vt} is performed on a set $VT_{vt}(\tau)$ of threads in a server s_t at time τ. The computation rate $FV_{vt}(\tau)$ of the virtual machine VM_{vt} at time τ is given as follows:

$$FV_{vt}(\tau) = \sum_{th_{kt} \in VT_{vt}(\tau)} FT_{kt}(\tau). \tag{2}$$

In this paper, replicas of each process are performed on virtual machines installed in each server s_t. A notation p_{vt}^i stands for a replica of a process p^i performed on a virtual machine VM_{vt} in a server s_t. Replicas which are being performed and already terminate on a virtual machine VM_{vt} at time τ are *current* and *previous*, respectively. Let $CP_{vt}(\tau)$ be a set of current replicas on a virtual machine VM_{vt} at time τ and $NC_{vt}(\tau)$ be $|CP_{vt}(\tau)|$. In this paper, we assume the computation rate $FV_{vt}(\tau)$ of a virtual machine VM_{vt} at time τ is uniformly allocated to every current replica on the virtual machine VM_{vt}.

The computation rate $f_{vt}^i(\tau)$ of a replica p_{vt}^i performed on a virtual machine VM_{vt} at time τ is given as follows:

$$f_{vt}^i(\tau) = \begin{cases} \alpha_{vt}(\tau) \cdot FV_{vt}(\tau) \ / \ NC_{vt}(\tau), & \text{if } NC_{vt}(\tau) > nVT_{vt}(\tau), \\ FT_{kt}(\tau), & \text{otherwise.} \end{cases} \quad (3)$$

Here, $\alpha_{vt}(\tau)$ is the *computation degradation ratio* of a virtual machine VM_{vt} at time τ ($0 \le \alpha_{vt}(\tau) \le 1$). $\alpha_{vt}(\tau_1) \le \alpha_{vt}(\tau_2) \le 1$ if $NC_{vt}(\tau_1) \ge NC_{vt}(\tau_2)$. $\alpha_{vt}(\tau) = 1$ if $NC_{vt}(\tau) \le 1$. $\alpha_{vt}(\tau)$ is assumed to be $\varepsilon_{vt}^{NC_{vt}(\tau)-1}$ where $0 \le \varepsilon_{vt} \le 1$.

Suppose that a replica p_{vt}^i starts and terminates on a virtual machine VM_{vt} at time st_{vt}^i and et_{vt}^i, respectively. Let T_{vt}^i be the total computation time of a replica p_{vt}^i performed on a virtual machine VM_{vt}. Here, $T_{vt}^i = et_{vt}^i - st_{kt}^i$ and $\sum_{\tau=st_{vt}^i}^{et_{vt}^i} f_{vt}^i(\tau) = VS^i$. At time st_{vt}^i a replica p_{vt}^i starts, the computation laxity $lc_{vt}^i(\tau)$ is VS^i [vs]. The computation laxity $lc_{vt}^i(\tau)$ [vs] of a replica p_{vt}^i at time τ is given as follows:

$$lc_{vt}^i(\tau) = VS^i - \sum_{x=st_{vt}^i}^{\tau} f_{vt}^i(x). \quad (4)$$

2.2 Power Consumption Model of a Server

Let $E_t(\tau)$ be the electric power [W] consumed by a server s_t at time τ. Let $maxE_t$ and $minE_t$ be the maximum and minimum electric power [W] of a server s_l, respectively. Let $ac_t(\tau)$ be the number of active cores in a server s_t at time τ. $minC_t$ shows the electric power [W] where at least one core c_{ht} is active on a server s_t. Let cE_t be the electric power [W] consumed by a server s_t to make one core active.

The *PCSV* (*power consumption of a server with virtual machines*) model [6] is proposed where computation processes are performed on virtual machines in a server s_t. According to the PCSV model, the electric power consumption $E_t(\tau)$ [W] of a server s_t to perform current replicas on virtual machines at time τ is given as follows [6]:

$$E_t(\tau) \ = \ minE_t \ + \ \sigma_t(\tau) \ \cdot \ (minC_t \ + \ ac_t(\tau) \ \cdot \ cE_t) \quad (5)$$

Here, $\sigma_t(\tau) = 1$ if at least one core c_{ht} is active on a server s_t at time τ. Otherwise, $\sigma_t(\tau) = 0$. The electric power consumption $E_t(\tau)$ of a server s_t depends on the number $ac_t(\tau)$ of active cores at time τ.

The total electric energy $TE_t(\tau_1, \tau_2)$ [J] consumed by a server s_t from time τ_1 to τ_2 is given as follows:

$$TE_t(\tau_1, \tau_2) = \sum_{\tau=\tau_1}^{\tau_2} E_t(\tau). \tag{6}$$

The processing power $PE_t(\tau)$ [W] of a server s_t at time τ is $E_t(\tau)$ - $minE_t$. The total processing electric energy $TPE_t(\tau_1, \tau_2)$ [J] consumed by a server s_t from time τ_1 to τ_2 is given as follows:

$$TPE_t(\tau_1, \tau_2) = \sum_{\tau=\tau_1}^{\tau_2} PE_t(\tau). \tag{7}$$

3   Process Replication Algorithm

3.1   System Model

A cluster S is composed of multiple servers $s_1, ..., s_n$ ($n \geq 1$). One thread th_{kt} is bounded to a virtual machine VM_{vt} in a server s_t as shown in Fig. 1. Servers in a server cluster S might stop by fault. Let nF be the maximum number of servers which concurrently stop by fault in the cluster S. Let rd^i be the *redundancy* of a process p^i ($nF + 1 \leq rd^i \leq n$). If a server s_t stops by fault, every virtual machine VM_{vt} performed on the server s_t also stops. Hence, each process p^i has to be redundantly performed on rd^i virtual machines performed on different servers. Each time a load balancer K receives a request process p^i from a client cl^i, the load balancer K selects a subset VMS^i ($|VMS^i| = rd^i$) of virtual machines in the server cluster S and forwards the process p^i to every virtual machines VM_{vt} in the set VMS^i. On receipt of a process p^i, a virtual machine VM_{vt} creates and performs an instance p_{vt}^i of the process p_{vt}^i. An instance p_{vt}^i of the process p^i performed on a virtual machine VM_{vt} is a *replica* of the process p^i. On termination of a replica p_{vt}^i, the virtual machine VM_{vt} sends a reply r_{vt}^i to the load balancer K. The load balancer K takes only the first reply r_{vt}^i and ignores every other reply. This means, a load balancer K can receive at least one reply r_{vt}^i from a virtual machine VM_{vt} even if nF servers stop by fault in a server cluster S.

3.2   Total Processing Electric Energy Laxity

The total processing electric energy laxity $tpel_t(\tau)$ [J] shows how much electric energy a server s_t has to consume to still perform every current replica on every active virtual machine in the server s_t at time τ. Let $\mathbf{CP}_t(\tau)$ be a family $CP_{1t}(\tau), ..., CP_{nt_t t}(\tau)$ of current replica sets of all the virtual machines $VM_{1t}, ..., VM_{nt_t t}$ in a server s_t at time τ. The total processing electric energy

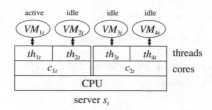

Fig. 1. Threads and virtual machines in a server s_t.

laxity $tpel_t(\tau)$ of a server s_t at time τ is obtained by the following **Elaxity**(s_t, τ) procedure [7]:

Elaxity(s_t, τ) {
    if $CP_{vt}(\tau) = \phi$ for every replica p_{vt}^i in $\mathbf{CP}_t(\tau)$ **return**(0);
    for each core c_{ht} in a server s_t, {
        nv = number of active virtual machines on c_{ht} at time τ;
        if $nv \geq 1$, $ac_t(\tau) = ac_t(\tau) + 1$;
        for each VM_{vt} on a core c_{ht}, {
            for each replica $p_{vt}^i \in CP_{vt}(\tau)$, {
                $lc_{vt}^i(\tau + 1) = lc_{vt}^i - f_{vt}^i(\tau)$;
                if $lc_{vt}^i(\tau + 1) \leq 0$, $CP_{vt}(\tau) = CP_{vt}(\tau) - \{p_{vt}^i\}$;
            } /* for end. */
        } /* for end. */
    } /* for end. */
    $tpel_t(\tau) = E_t(\tau) - minE_t$; /* processing power */
    return ($tpel_t(\tau) + $ **Elaxity**(s_t, $\tau + 1$));
}

3.3 The RECLB Algorithm

The RECLB (*redundant energy consumption laxity-based*) algorithm [7] is proposed to select multiple virtual machines for redundantly performing each process p^i so that the total processing electric energy consumption of a server cluster S can be reduced. Let $TPEL_{vt}(\tau)$ be the total amount of processing electric energy laxity (TPEL) [J] of a server cluster S where a replica p_{vt}^i of a new request process p^i is allocated to a virtual machine VM_{vt} at time τ. In the RECLB algorithm, a subset VMS^i of virtual machines in a server cluster S, where the total amount of processing electric energy laxity $TPEL_{vt}(\tau)$ of a server cluster S is the minimum, is selected for each new request process p^{new} at time τ by the following **RECLB**(p^{new}, τ) procedure:

RECLB(p^{new}, τ) { /* a term VM stands for a virtual machine. */
    $VMS^i = \phi$;
    while ($rd^i > 0$) {
        for each VM_{vt} in a server cluster S, {

$$CP_{vt}(\tau) = CP_{vt}(\tau) \cup \{p^{new}\};$$
$$TPEL_{vt}(\tau) = \sum_{t=1}^{n} \textbf{ELaxity}(s_t, \tau);$$
$\}$ /* for end. */
$vm = $ a virtual machine VM_{vt} where $TPEL_{vt}(\tau)$ is the minimum;
$VMS^i = VMS^i \cup \{vm\};$
$S = S - \{s_t\};$
/* the server s_t which performs the virtual machine vm is removed. */
$rd^i = rd^i - 1;$
$\}$ /* while end. */
$\textbf{return}(VMS^i);$
$\}$

3.4   The RECLB-MT Algorithm

In the RECLB algorithm, only one thread is bounded to a virtual machine in a server. Hence, some threads may be idle while other threads are active to perform many replicas. This means, a server consumes more electric energy since it takes longer time to perform replicas if more number of replicas are performed. We newly propose the *RECLB-MT* (*RECLB with multi-threads allocation*) algorithm to furthermore reduce the total processing electric energy consumption (TPE) of a server cluster S in this paper. In the RECLB-MT algorithm, the total processing electric energy consumption of a server cluster S and the average computation time of each process can be reduced by allocating idle threads to active virtual machines in each server s_t. In the RECLB-MT algorithm, each time a load balancer K receives a new request p^{new}, a virtual machine VM_{vt} where the total processing energy laxity $TPEL_{vt}(\tau)$ of a server cluster S is the minimum is selected for the request p^{new} at time τ by the $\textbf{RECLB}(p^{new}, \tau)$ procedure.

Suppose a server s_t is equipped with a dual-core CPU as shown in Fig. 1. Four threads th_{1t}, th_{2t}, th_{3t}, and th_{4t} are bounded to virtual machines VM_{1t}, VM_{2t}, VM_{3t}, and VM_{4t}, respectively. Suppose only the virtual machine VM_{1t} is active in the server s_t at time τ. Hence, only the thread th_{1t} is active in the server s_t at time τ. In the RECLB algorithm [7], every current replica is performed on the virtual machine VM_{1t} by using only one thread th_{1t} even if the other threads are idle in the server s_t at time τ. Here, only the core c_{1t} is active in the server s_t at time τ, i.e. $ac_t(\tau) = 1$. Here, even if current replicas are performed on the virtual machine VM_{1t} by using a pair of threads th_{1t} and th_{2t} on the same core c_{1t} at time τ, the number $ac_t(\tau)$ of active cores in the server s_t at time τ is one, i.e. $ac_t(\tau) = 1$. In the PCSV model, the electric power consumption $E_t(\tau)$ of a server s_t depends on the number $ac_t(\tau)$ of active cores in the server s_t at time τ. At time τ, the electric power consumption $E_t(\tau)$ of the server s_t where current replicas are performed on the virtual machine VM_{1t} with one thread th_{1t} is the same as the electric power where current replicas are performed on the virtual machine VM_{1t} with a pair of threads th_{1t} and th_{2t} on the same core c_{1t} since $ac_t(\tau) = 1$.

In the RECLB-MT algorithm, if a thread $th_{k't}$ on a core c_{ht} bounded to virtual machine $VM_{v't}$ is idle, i.e. the virtual machine $VM_{v't}$ is idle, the thread $th_{k't}$ is used for another active virtual machine VM_{vt} performed on the same core c_{ht} in each server s_t at time τ. Then, the computation time of each replica performed on the virtual machine VM_{vt} can be reduced in the RECLB-MT algorithm than the RECLB algorithm since the computation rate $FV_{vt}(\tau)$ of the virtual machine VM_{vt} increases. As a result, the total processing electric energy consumption of each server s_t can be reduced in the RECLB-MT algorithm than the RECLB algorithm.

Let $idle_{ht}(\tau)$ be a set of idle threads on a core c_{ht} in a server s_t at time τ. Let $avm_{ht}(\tau)$ be a set of active virtual machines on a core c_{ht} at time τ. At time τ, idle threads on a core c_{ht} are allocated to each active virtual machine VM_{vt} performed on the core c_{ht} in a server s_t by the following **Thread_Alloc()** procedure:

Thread_Alloc(τ) {
    for each core c_{ht} in a server s_t, {
        $avm_{ht}(\tau) = $ a set of active VMs on a core c_{ht} at time τ;
        if $idle_{ht}(\tau) \geq 1$ and $avm_{ht}(\tau) \geq 1$, {
            while $|idle_{ht}(\tau)| > 0$, {
                $th = th_{kt} \in idle_{ht}(\tau)$; /* a thread th_{kt} is randomly selected in $idle_{ht}(\tau)$. */
                $VM_{vt} = $ a VM where $CP_{vt}(\tau)$ is the maximum in $avm_{ht}(\tau)$;
                $nVT_{vt}(\tau) = nVT_{vt}(\tau) \cup \{th\}$; /* a thread th is bounded to VM_{vt}. */
                $idle_{ht}(\tau) = idle_{ht}(\tau) - \{th\}$;
                $avm_{ht}(\tau) = avm_{ht}(\tau) - \{VM_{vt}\}$;
                if $avm_{ht}(\tau) = \phi$, $avm_{ht}(\tau) = $ a set of active VMs on c_{ht} at time τ;
            } /* while end. */
        } /* if end. */
    } /* for end. */
}

4   Evaluation

4.1   Environment

We evaluate the RECLB-MT algorithm in terms of the total processing electric energy consumption [J] of a homogeneous server cluster S and the average computation time of each process p^i compared with the RECLB [7] algorithm.

We consider a homogeneous cluster S composed of five servers s_1, ..., s_5 ($t = 5$) in this evaluation. We assume the fault probability fp_t for every server s_t is the same $fp = 0.1$. Every server s_t ($1 \leq t \leq 5$) follows the same power consumption model as shown in Table 1. The parameters of each server s_t are obtained from the experiment [6]. Every server s_t is equipped with a dual-core CPU ($nc_t = 2$). A pair of threads are bounded for each core in each server s_t, i.e. $ct_t = 2$. Hence, the number nt_t of threads in each server s_t is four. $minE_t = 14.8$ [W], $minC_t = 6.3$ [W], $cE_t = 3.9$ [W], and $maxE_t = 33.8$ [W]. The total number of threads in the server cluster S is twenty. Each virtual machine VM_{vt} holds at least one virtual CPU and at least one thread th_{kt} is bounded to each virtual machine VM_{vt} in each server s_t ($v = 1, ..., 4$ and $t = 1, ..., 5$). Hence, there are twenty virtual machines in the server cluster S. Every virtual machine VM_{vt} follows the same computation model as shown in Table 2. The parameters of each virtual machine VM_{vt} are obtained from the experiment [6]. The parameter ε_{vt} in the computation degradation ratio $\alpha_{vt}(\tau)$ of a virtual machine VM_{vt} is 1. The maximum computation rate $Maxf_{kt}$ of each thread th_{kt} in every server s_t is 1 [vs/ms]. The performance degradation ratios $\beta_{kt}(1)$ and $\beta_{kt}(2)$ are 1 and 0.6, respectively, for every thread th_{kt} in the server cluster S. We assume no virtual machine migrates to another server in the server cluster S.

Table 1. Homogeneous cluster S.

server	nc_t	ct_t	nt_t	$minE_t$	$minC_t$	cE_t	$maxE_t$
s_t	2	2	4	14.8 [W]	6.3 [W]	3.9 [W]	33.8 [W]

Table 2. Parameters of virtual machine.

virtual machine	ε_{vt}	$Maxf_{kt}$	$\beta_{kt}(1)$	$\beta_{kt}(2)$
VM_{vt}	1	1 [vm/ms]	1	0.6

The number m of processes $p^1, ..., p^m$ ($0 \le m \le 140{,}000$) are issued to the server cluster S. The starting time of each process p^i is randomly selected in a unit of one millisecond where the average inter-request time is 0.5 [ms] in the total simulation time x [msec], i.e. $x/m = 0.5$ [ms]. We assume the redundancy rd^i for each process p^i is the same rd ($= \{1, 2, 3, 4, 5\}$) and $nF = rd - 1$ holds. The minimum computation time $minTh^i$ of every process p^i is assumed to be 1 [ms].

4.2 Average Computation Time of Each Process

The computation time T^i [msec] for each process p^i is the computation time T_{vt}^i of a replica p_{vt}^i which commits earliest in the replicas of the process p^i. The computation time T^i for each process p^i is measured five times for each redundancy rd ($=\{1, 2, 3, 4, 5\}$) and each total number m of processes ($0 \le m \le 140{,}000$). Let $T_{tm}^m(rd)$ be the computation time T^i [msec] obtained in the tm-th simulation to perform the total number m of processes with redundancy rd. The average computation time $AT^m(rd)$ [msec] of each process to perform the total number m of processes with redundancy rd is $\sum_{tm=1}^{5} \sum_{i=1}^{m} T_{tm}^{i,m} / (m \cdot 5)$.

Figure 2 shows the average computation time [msec] of each process to perform the total number m of processes with redundancy rd in the RECLB and RECLB-MT algorithms. In Fig. 2, RECLB(rd) and RECLB-MT(rd) stand for the average computation time [msec] of each process in RECLB and RECLB-MT algorithms, respectively, to perform the total number m of processes with redundancy rd. The average computation time in the RECLB-MT algorithm can be more reduced than the RECLB algorithm for each redundancy rd. In the RECLB-MT algorithm, idle threads in each server s_t are bounded to active virtual machines in the server s_t if the total processing electric energy of the server s_t does not increase. Then, the computation rate of each active virtual machine to perform replicas increases. As a result, the average computation time of each process can be more reduced in the RECLB-MT algorithm than the RECLB algorithm since the computation resources in the server cluster S can be more efficiently utilized in the RECLB-MT algorithm than the RECLB algorithm.

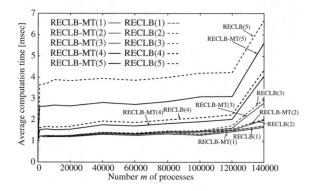

Fig. 2. Average computation time [msec] of each process ($fp = 0.1$).

4.3 Total Processing Electric Energy of a Server Cluster

Let $TE_{tm}^m(rd)$ be the total processing electric energy [J] to perform the number m of processes ($0 \le m \le 140{,}000$) with redundancy rd ($= \{1, 2, 3, 4, 5\}$) in the server cluster S obtained in the tm-th simulation. The total processing electric energy $TE_{tm}^m(rd)$ is measured five times for each number m of processes and each redundancy rd. The average total processing electric energy $ATE^m(rd)$ [J] to perform the total number m of processes with redundancy rd is calculated as $\sum_{tm=1}^{5} TE_{tm}^m(rd)/5$.

Figure 3 shows the average total processing electric energy [J] of the server cluster S to perform the number m of processes with redundancy rd in the RECLB and RECLB-MT algorithms. In Fig. 3, RECLB(rd) and RECLB-MT(rd) stand for the average total processing electric energy of the RECLB and RECLB-MT algorithms, respectively, to perform the total number m of processes with redundancy rd. In the RECLB and RECLB-MT algorithms, the average total processing electric energy of the server cluster S increases as the number m of processes and redundancy rd increases. In the RECLB-MT algorithm, the average computation time of each process can be more reduced than the RECLB algorithm as shown in Fig. 2 since idle threads in each server s_t are bounded to active virtual machines in the server s_t. Then, the active time of each virtual machine can be more reduced in the RECLB-MT algorithm than the RECLB algorithm. As a result, the average total processing electric energy of the server cluster S can be more reduced in the RECLB-MT algorithm than the RECLB algorithm for $rd \geq 1$.

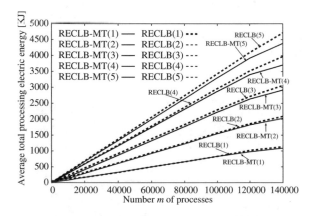

Fig. 3. Average total processing electric energy [KJ] of a server cluster S ($fp = 0.1$).

The evaluation results show the average total processing electric energy consumption of the server cluster S and the average computation time of each process can be more reduced in the RECLB-MT algorithm than the RECLB algorithm since idle threads are more efficiently utilized in the RECLB-MT algorithm than the RECLB algorithm. Following the evaluation, the RECLB-MT algorithm is more useful than the RECLB algorithm.

5   Concluding Remarks

In this paper, we newly proposed the RECLB-MT (RECLB with multi-threads allocation) algorithm to select multiple virtual machines for redundantly performing each computation process so that the total processing electric energy consumption of a server cluster and the average computation time of each process can be reduced by allocating idle threads to active virtual machines in each server. In the RECLB-MT algorithm, threads in each server are more efficiently utilized than the RECLB algorithm. In the evaluation, we showed the total processing electric energy consumption of a server cluster and the average computation time of each process can be more reduced in the RECLB-MT algorithm than the RECLB algorithm. Following the evaluation, the RECLB-MT algorithm is more useful than the RECLB algorithm.

References

1. Natural Resources Defense Council (NRDS): Data center efficiency assessment - scaling up energy efficiency across the data center industry: Evaluating key drivers and barriers (2014). http://www.nrdc.org/energy/files/data-center-efficiency-assessment-IP.pdf
2. Enokido, T., Aikebaier, A., Takizawa, M.: Process allocation algorithms for saving power consumption in peer-to-peer systems. IEEE Trans. Ind. Electron. **58**(6), 2097–2105 (2011)
3. Enokido, T., Aikebaier, A., Takizawa, M.: A model for reducing power consumption in peer-to-peer systems. IEEE Syst. J. **4**(2), 221–229 (2010)
4. Enokido, T., Aikebaier, A., Takizawa, M.: An extended simple power consumption model for selecting a server to perform computation type processes in digital ecosystems. IEEE Trans. Ind. Inf. **10**(2), 1627–1636 (2014)
5. Enokido, T., Takizawa, M.: Integrated power consumption model for distributed systems. IEEE Trans. Ind. Electron. **60**(2), 824–836 (2013)
6. Enokido, T., Takizawa, M.: Power consumption and computation models of virtual machines to perform computation type application processes. In: Proceedings of the 9th International Conference on Complex, Intelligent and Software Intensive Systems (CISIS-2015), pp. 126–133 (2015)
7. Enokido, T., Takizawa, M.: An energy-efficient process replication algorithm in virtual machine environments. In: Proceedings of the 11th International Conference on Broadband and Wireless Computing, Communication and Applications (BWCCA-2016), pp. 105–114 (2016)
8. KVM: Main Page - KVM (Kernel Based Virtual Machine) (2015). http://www.linux-kvm.org/page/Mainx_Page
9. Lamport, R., Shostak, R., Pease, M.: The byzantine generals problems. ACM Trans. Programing Lang. Syst. **4**(3), 382–401 (1982)
10. Schneider, F.B.: Replication Management Using the State-Machine Approach. Distributed Systems, 2nd edn, pp. 169–197. ACM Press, Reading (1993)
11. Kshemkalyani, A.D., Singhal, M.: Distributed Computing - Principles, Algorithms, and Systems. Cambridge University Press, Cambridge (2008)
12. Intel Xeon processor 5600 series: The next generation of intelligent server processors (2010). http://www.intel.com/content/www/us/en/processors/xeon/xeon-5600-brief.html

Application of Perceptual Features for User Authentication in Distributed Systems

Marek R. Ogiela[1(✉)] and Lidia Ogiela[2]

[1] Cryptography and Cognitive Informatics Research Group, AGH University of Science and Technology, 30 Mickiewicza Avenue, 30-059 Kraków, Poland
mogiela@agh.edu.pl
[2] Department of Cryptography and Cognitive Informatics, Pedagogical University of Krakow, Podchorążych 2 Street, 30-084 Kraków, Poland
lidia.ogiela@gmail.com

Abstract. This paper will present selected approaches based on cognitive solutions for creation security procedures and cryptographic protocols. In particular will be described the ways of using perceptual and cognitive features in creation of secure protocols, oriented on user authentication and data management. The new areas of cognitive cryptography will be also described, which offer a new possibilities in security areas oriented on application of perceptual features and personal characteristics.

Keywords: Cognitive authentication · Security protocols · Cryptographic protocols

1 Introduction

Recent solutions in cryptography and security areas are often connected with application of personal features, behavioral patterns or cognitive abilities [1, 2]. One of the most important among them are protocols oriented for user authentication. For this purpose have been already proposed many secure procedures, but it seems that in distributed systems it will be also possible to create a new user oriented procedures, which allow to authenticate only selected group of persons [3, 4]. So the general objective of our research were connected with development of two types of cryptographic procedures.

The first ones have been intended for direct user authentication, using visual codes. Such protocols should require from participants a very specialized knowledge in a specific information area, as well as special perception abilities for proper semantic evaluation of presented patterns [5, 6].

The second class of protocols was connected with data division tasks, using a personal threshold approach, and considering personal characteristics of participants in sharing protocol. Information division procedures can be performed in distributed structures and systems, and enable generation of different secret parts taking into account personal characteristics [7, 8]. Both types of protocols are oriented for application in distributed network architectures and Cloud infrastructures.

L. Barolli et al. (Eds.): NBiS-2019, AISC 1036, pp. 132–136, 2020.
https://doi.org/10.1007/978-3-030-29029-0_12

2 Cognitive-Based Security Protocols

Cognitive-based security protocols can be proposed thanks to the application of cognitive systems and perceptual features. Cognitive vision systems have already been defined by the authors [9, 10], and various classes of such systems have been proposed. Among them, image analysis systems, and systems for identification of personal characteristics were described, but also systems supporting management processes. A common feature of different classes of such systems, is the ability to determine the semantics of the analyzed data. In case of personal data analysis, such systems allow to extract very specific personal characteristics and behavioral patterns. So this type of features we can use in creation of new authentication procedures that take the form of cognitive verification codes. Authentication using such procedures require specific knowledge or perception abilities, necessary for proper evaluation of the meaning of recognized visual patterns. Such codes allow to perform very selective authentication, especially granting users representing specific expertise area, and having particular perception skills.

Cognitive systems allow to develop of secure protocols for sharing and management data in Cloud computing and distributed systems. Sharing procedure consider the number of participants of the division protocol at different levels of the cloud or edge infrastructure, as well as consider personal characteristics in sharing procedure.

3 User Authentication and Data Sharing Protocols

Cognitive user authentication protocols can be developed base on application of expert knowledge, and perceptual features of authenticated users. For this purpose, cognitive information systems may be used, which allow to create collections of visual patterns for authentication procedure. Such patterns should be selected according semantic content, representing the selected area or domain, in which user are qualified. After creation of sets containing visual patterns, it can be used in the authentication procedure, carried out in two independent ways.

The first one is the authentication based on specific expert knowledge, which require from user indication of correct pattern subsets depending on the verification questions.

Second approach is conducted with presentation of a single visual pattern, having many semantic elements, which should be determined during verification, performed in an iterative way (see Fig. 1). On each stage in verification procedure are asked questions with increasing degree of details. The questions should concern elements placed in different areas of the pattern and presented in fuzzy images with various degrees of fuzziness. Such filtered patterns can be recognized only by users who have knowledge related to the subject of patterns, and having special perceptual abilities that allow to see and correctly interpret fuzzy patterns. The main problem in efficient application of such authentication procedures is generation of collections containing visual patterns with specific semantic meaning and representing a selected area of expert knowledge.

Figure 1 presents an example of multi-level expert knowledge authentication procedure, in which answers depend on the displayed structures or marked regions.

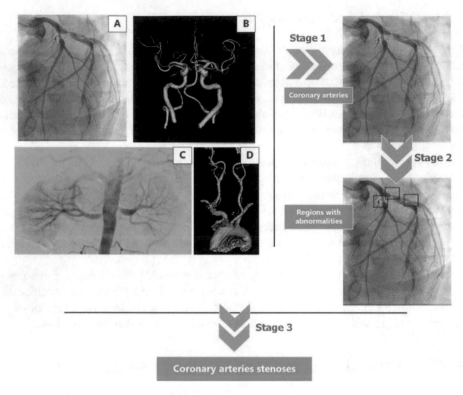

Fig. 1. Authentication protocol based on visual patterns with expert knowledge verification. Image presents a set of selected medical visualization. In particular it presents: (A) coronary arteries, (B) brain vessels, (C) renal arteries, (D) carotid arteries. The first verification question can be connected with finding coronary structures among these visualizations (stage 1). Having detected coronary arteries next questions can be connected with finding places of pathologies (stage 2). Having detected by user visible pathologies, next question can be asked about the name of vessel, type of pathologies etc.

During verification in this procedure it is possible to ask several questions about visible patterns, screening modality, lesion details etc. Proper understanding the content of this patterns allow user to be successfully authenticated. In such protocols is possibilities to quickly verify if authenticated person represents a particular area of expertise and has knowledge connected with presented patterns. If patterns are blurred or softened it is also possible to test users' perception skills, who answer several questions with different detail levels.

Second class of security procedures based on cognitive systems is protocols for data sharing, distribution and reconstruction in Cloud infrastructure. In such procedures, cognitive systems can be applied to extract selected personal or behavioral features, which next can be applied in information sharing processes oriented to individual users. It is worth noting that personal characteristics may have a form of biometrics, but also can have a form of parameters defining non-standard personal features of participants. Such features may have form of behavioral parameters (e.g. gestures or movements), morphometric parameters, or diagnostic values (e.g., cerebral perfusion).

After selection of feature and evaluation of personal characteristics, they can be involved in the process of generation of personalized parts of divided secret.

Generated shares should be distributed among the participants of the sharing protocols in such manner that distribution should consider the hierarchical relationships and the distributed nature of the structures, in which the secret division takes place [11, 12]. This means that data are divided in different ways depending on the number of participants of the protocol at a given level, and also considering different personal characteristics of different users.

4   Conclusions

In this paper has been presented general ideas and approaches for creation of user authentication protocols, and sharing data procedures, with application of cognitive systems and perceptual features. Presented solution allow to perform user authentication tasks with relation to theirs expertise and knowledge, as well as perceptual skills and cognition abilities. Described approaches perform authentication tasks in several iterations, presenting sequentially blurred patterns with different stages of fuzziness. Such patterns can be recognized by users at different perception thresholds. Because it is required advanced semantic content interpretation during verification, such authentication protocols fulfil security features. Presented solutions are oriented for groups of users having particular perception skills, and expertise knowledge.

Also sharing procedure are based on application of cognitive systems. Such protocols can involve in division processes selected personal features, and characteristics, which allow sharing procedure to make more personalized and oriented for particular user.

Presented protocols enrich classical cryptography towards new classes of security procedures connected with cognitive cryptography, which join traditional encryption approaches [13, 14] with cognitive functions and application of personal features.

Acknowledgments. This work has been supported by the National Science Centre, Poland, under project number DEC-2016/23/B/HS4/00616. This work has been supported by the AGH University of Science and Technology research Grant No 16.16.120.773.

References

1. Ogiela, M.R., Ogiela, L.: On using cognitive models in cryptography. In: IEEE AINA 2016 - The IEEE 30th International Conference on Advanced Information Networking and Applications, Crans-Montana, Switzerland, 23–25 March, pp. 1055–1058 (2016)
2. Ogiela, M.R., Ogiela, L.: Cognitive keys in personalized cryptography. In: IEEE AINA 2017 The 31st IEEE International Conference on Advanced Information Networking and Applications, Taipei, Taiwan, 27–29 March, pp. 1050–1054 (2017)

3. Meiappane, A., Premanand, V.: CAPTCHA as Graphical Passwords - A New Security Primitive: Based on Hard AI Problems. Scholars' Press (2015)
4. Osadchy, M., Hernandez-Castro, J., Gibson, S., Dunkelman, O., Perez-Cabo, D.: No bot expects the DeepCAPTCHA! Introducing immutable adversarial examples, with applications to CAPTCHA generation. IEEE Trans. Inf. Forensics Secur. **12**(11), 2640–2653 (2017)
5. Ogiela, M.R., Ogiela, U., Ogiela, L.: Secure information sharing using personal biometric characteristics. In: Kim, T.-H., et al. (eds.) Computer Applications for Bio-technology, Multimedia and Ubiquitous City. CCIS, vol. 353, pp. 369–373. Springer, Heidelberg (2012)
6. Ogiela, M.R., Ogiela, L., Ogiela, U.: Biometric methods for advanced strategic data sharing protocols. In: The Ninth International Conference on Innovative Mobile and Internet Services in Ubiquitous Computing (IMIS-2015), July 8–10, Blumenau, Brazil, pp. 179–183 (2015). https://doi.org/10.1109/imis.2015.29
7. Ogiela, L., Ogiela, M.R.: Bio-inspired cryptographic techniques in information management applications. In: IEEE AINA 2016 - The IEEE 30th International Conference on Advanced Information Networking and Applications, Crans-Montana, Switzerland, 23–25 March, pp. 1059–1063 (2016)
8. Ogiela, U., Ogiela, L.: Linguistic techniques for cryptographic data sharing algorithms. Concurr. Comput. Pract. E. **30**(3), e4275 (2018). https://doi.org/10.1002/cpe.4275
9. Ogiela, L., Ogiela, M.R.: Insider threats and cryptographic techniques in secure information management. IEEE Syst. J. **11**, 405–414 (2017)
10. Ogiela, M.R., Ogiela, U.: Secure information management in hierarchical structures. In: Kim, T.-H., et al. (eds.) AST 2011. CCIS, vol. 195, pp. 31–35. Springer, Heidelberg (2011)
11. Ogiela, L., Ogiela, M.R., Ogiela, U.: Efficiency of strategic data sharing and management protocols. In: The 10th International Conference on Innovative Mobile and Internet Services in Ubiquitous Computing (IMIS-2016), 6–8 July, Fukuoka, Japan, pp. 198–201 (2016). https://doi.org/10.1109/imis.2016.119
12. Ogiela, L.: Advanced techniques for knowledge management and access to strategic information. Int. J. Inf. Manag. **35**(2), 154–159 (2015)
13. Easttom, Ch.: Modern Cryptography: Applied Mathematics for Encryption and Information Security. McGraw-Hill Education, New York (2015)
14. Schneier, B.: Applied Cryptography. Wiley, Indianapolis (2015)

User Oriented Protocols for Data Sharing and Services Management

Lidia Ogiela[1(✉)], Makoto Takizawa[2], and Urszula Ogiela[1]

[1] Pedagogical University of Krakow, Podchorążych 2 Street,
30-084 Kraków, Poland
lidia.ogiela@gmail.com, uogiela@gmail.com
[2] Department of Advanced Sciences, Hosei University,
3-7-2, Kajino-cho, Koganei-shi, Tokyo 184-8584, Japan
makoto.takizawa@computer.org

Abstract. In this paper will be presented a new user-oriented approach for creation of cryptographic protocols dedicated for efficient and secure data sharing and services management. Such new procedures will be proposed with application of mathematical linguistic formalisms and application of personal characteristics, involved in creation of the division and management procedures. Such new protocols can be created with application of different data representation classes. Some possible application will also be described.

Keywords: Data sharing · Service management · User oriented protocols

1 Introduction

Data sharing processes can be considered with different security aspects. Very often data protection protocols are used not only to secure secret information, but also to management such data [1–4, 10].

The most important task carried out by the protocols for securing data, especially these from the group of threshold schemes, are the splitting processes included distribution stage of all secret parts. In process of data securing and management is also possible to secure service data/information by share them. Secret parts are distributed between the selected trusted users or participants. The main idea of these protocols were described in [5, 7, 8].

Classical data sharing protocols perform shadow distribution tasks between protocol participants without determining the manner, in which they belong to the secret holder. New classes in this area are biometric data sharing protocols. They provide the option of distributing shadows using the personal marking of each secret part. Secret marking takes place using a selected (by arbitrator or random) type of biometrics. A special feature of biometric threshold schemes is the unambiguity of assigning the secret (or part of it) to its owner.

The main idea of these algorithms consists the secure processes realization at the one (selected) stages. In this paper we present the new aspects of guaranteeing service security and possibilities of management in multilevel structures. Multilevel secret

© Springer Nature Switzerland AG 2020
L. Barolli et al. (Eds.): NBiS-2019, AISC 1036, pp. 137–141, 2020.
https://doi.org/10.1007/978-3-030-29029-0_13

management will allow to implement data protection procedures at different levels and taking into account independent data protection procedures at different levels.

2 User Oriented Protocols in Threshold Schemes

User oriented protocols denote the threshold schemes, which include personal features for marking service-secret parts. In this class of protocols can be use different types of personal features – standard and non-standard. The standard personal features are for example: DNA code, finger print, eye scan, face features, etc. At the non-standard group is possible select the following personal features: movements, gestures, rhythm and/or repetitive behaviour, etc. In this paper we propose use for service-data security non-standard characteristic.

New idea of user oriented threshold schemes are used them to multilevel secret labelling. Figure 1 presents an idea of user oriented threshold schemes for service-secret parts labelling process at multilevel threshold schemes.

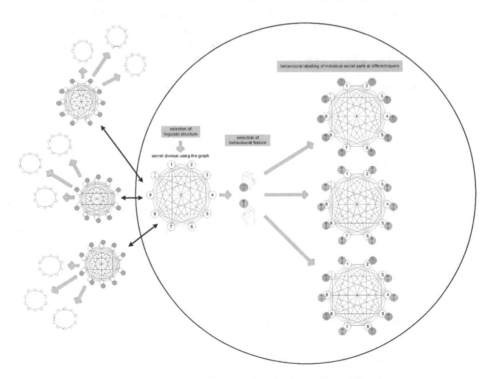

Fig. 1. An idea of user oriented threshold scheme in multilevel structure.

Figure 1 presents user oriented threshold scheme created by application of behavioural features – gesture B1 and B2 (means O.K by left and right hands). The main idea of proposed algorithm is selection proper linguistic structure to describe service-data – in this example is a graph structure with 10 edges. Service-data was share between 10 secret holders. All of them receive one secret part and mark this part by non-standard personal feature. Figure 1 presents three possible options – mark secret parts by the first gesture B1, mark all secret parts by the second non-standard feature B2, and mark secret parts by different features B1 or B2. Selection of the solution is dependent on the use of threshold schemes. This algorithm is a part of multilevel structure, so it is possible to apply this solution on many levels of the network. In each part of the network it is possible to apply an independent solution. In this way a network of independent solutions is created with the possibility of various applications of user oriented threshold schemes.

3 User Oriented Protocols for Service Management

The proposed user oriented protocols for service-data sharing, are also used to support the management processes [6, 9]. Proposed techniques can be used at different aspects or configurations. First of all, they can be used for:

- supporting data management processes,
- secret management,
- management of individual parts of the secret,
- management of different groups of secrets,
- secret management in basic structure at selected level,
- secret management in complex structures at various levels – multilevel management.

This paper focuses on the last example regarding the multilevel management of secret data created in complex structure at various levels. Figure 2 presents an example of proposed solution.

Figure 2 presents the user oriented protocols dedicated to multilevel management processes. These processes are realise at different levels – organization level, Fog level and Cloud level. Also, the applied threshold schemes can be different – Fig. 2 presents example of non-standard patterns (gesture B1 or B2). The network of connections between particular levels of structures indicates the connections between the introduced solutions. Each of the relationships can be implemented independently, but the value of the proposed algorithm is that it is possible to implement all the specific management processes together from different levels. Occurring connections between particular levels indicate the necessity of using different threshold techniques and methods for described service-data.

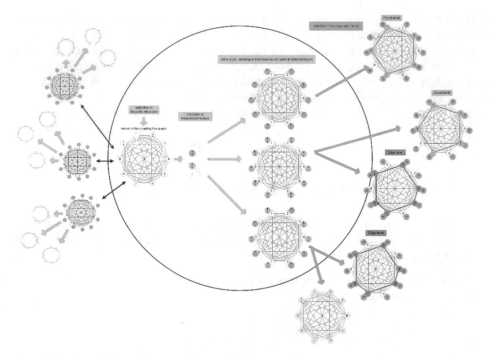

Fig. 2. Scheme of user oriented protocols for multilevel management processes.

4   Conclusions

In this paper secret management methods based on the user oriented protocols were presented. The possibilities of applying these methods enrich the previously known solutions. Secret network management in multi-level data structures is possible by using threshold techniques. Threshold techniques have been extended by the biometric data marking stage implemented by use non-standard biometric patterns. Biometric features as individual data tags uniquely assign data sets (shadows of a shared secret) to their holder.

Proposed solutions due to their universality can be used both in an independent way dedicated to specific data, as well as for the entire data network – which was the main theme of this work. In this paper was presented the possibility of using the user oriented protocols and discussed solution in data management processes.

Acknowledgments. This work has been supported by the National Science Centre, Poland, under project number DEC-2016/23/B/HS4/00616.

This work was supported by JSPS KAKENHI grant number 15H0295.

References

1. Gregg, M., Schneier, B.: Security Practitioner and Cryptography Handbook and Study Guide Set. Wiley, Hoboken (2014)
2. Laudon, K.C., Laudon, J.P.: Management Information Systems – Managing the Digital Firm, 7th edn. Prentice-Hall International Inc., Cliffs (2002)
3. Menezes, A., van Oorschot, P., Vanstone, S.: Handbook of Applied Cryptography. CRC Press, Waterloo (2001)
4. Ogiela, L.: Advanced techniques for knowledge management and access to strategic information. Int. J. Inf. Manage. **35**(2), 154–159 (2015)
5. Ogiela, L.: Cryptographic techniques of strategic data splitting and secure information management. Pervasive Mob. Comput. **29**, 130–141 (2016)
6. Ogiela, L., Ogiela, M.R.: Insider threats and cryptographic techniques in secure information management. IEEE Syst. J. **11**(2), 405–414 (2017)
7. Ogiela, M.R., Ogiela, U.: Secure information splitting using grammar schemes. In: Nguyen, N.T., Katarzyniak, R.P., Janiak, A. (eds.) New Challenges in Computational Collective Intelligence. Studies in Computational Intelligence, vol. 244, pp. 327–336. Springer, Heidelberg (2009)
8. Ogiela, M.R., Ogiela, U.: Secure information management in hierarchical structures. In: Kim, T., Adeli, H., Robles, R.J., Balitanas, M. (eds.) Advanced Computer Science and Information Technology. Communications in Computer and Information Science, vol. 195, pp. 31–35. Springer, Heidelberg (2011)
9. Ogiela, U., Takizawa, M., Ogiela, L.: Classification of cognitive service management systems in cloud computing. In: Barolli, L., Xhafa, F., Conesa, J. (eds.) Advances on Broad-Band Wireless Computing, Communication and Applications. BWCCA 2017. Lecture Notes on Data Engineering and Communications Technologies, vol. 12, pp. 309–313. Springer, Heidelberg (2018). https://doi.org/10.1007/978-3-319-69811-3_28
10. Yan, S.Y.: Computational Number Theory and Modern Cryptography. Wiley, New York (2013)

Development of Borehole Imaging Method with Using Visual-SLAM

Tsuneo Kagawa[✉]

Oita University, 700 Dannoharu, Oita City, Oita, Japan
t-kagawa@oita-u.ac.jp

Abstract. Borehole imaging is a method to investigate inside vertical bored hole with an optical camera. It is very important for construction and civil engineering that can accurately investigate the inner ground and geology at low cost. Recently borehole imaging method has also become important in terms of natural disaster prevention, such as landslides. However, easily surveyed cameras are difficult to control their position or pose in the hole because they are always rolling and rotating in the ground. In this study, our purpose is to realize a high-precision survey using cameras at low cost. We consider the introduction of visual SLAM (Simultaneous Localization and Mapping), which is used for autonomous robot navigation, self-driving car or smart phone augmented reality application. In this paper we discuss about development of borehole imaging method with using visual-SLAM.

1 Introduction

Borehole imaging is one of most effective geological investigation methods. It is logging and data-processing methods that are used to produce centimeter-scale images of the borehole wall and the rocks that make it up. An investigator obtains the core sample by drilling a vertical hole to observe the condition of the inside structural information about stratum by an optical camera or an acoustic sensor. An investigator can detect fractures underground easily. In addition, current image processing technique provide accuracy investigation for the inner ground and geology at low cost.

To analyze the geological structure or condition in detail from the in-hole image, accurate information is required, such as the depth, the azimuthal direction and so on. Ordinally, sophisticated borehole camera system is utilized for investigation because these types of camera are well-controlled even inside of deep hole. Such cameras, however, must be installed on a well-maintained area near the hole, and it is necessary to adjust such as calibration to improve the survey accuracy, which requires a large cost. Furthermore, such cameras are expensive and large, which makes them difficult to handle in the actual field.

Recently, borehole imaging investigations has also become important in terms of natural disaster prevention such as landslides. In the case of natural disaster prediction, many information, such as size, depth and azimuthal direction of fractures in the hole, are required. It is strongly necessary to conduct investigations at difficult points in the installation of cameras such as undeveloped land and slopes. Furthermore, surveys and analysis may be conducted at multiple points, so that there is no cost associated with

© Springer Nature Switzerland AG 2020
L. Barolli et al. (Eds.): NBiS-2019, AISC 1036, pp. 142–153, 2020.
https://doi.org/10.1007/978-3-030-29029-0_14

moving and installing the camera as much as possible, and the actual survey as fast and easy as possible. However, easily surveyed cameras are difficult to control their position or pose because they are always rolling and rotating in the ground. If a big fracture is found in the hole wall, it is completely impossible to know the fracture is in which azimuthal direction of north, south, east, or west when the camera is rotated.

In this study, our purpose is to realize a high-precision survey using cameras at low cost. It is necessary to acquire camera information such as the position and pose of the camera to improve the resolution of video captured borehole wall image. Furthermore, in order to analyze images with high accuracy, it is necessary not only to accumulate and combine a series of images, but also to capture the position and posture of the borehole camera as accurately as possible. We consider the introduction of visual SLAM (Simultaneous Localization and Mapping), which is used for autonomous robot navigation, self-driving car or smart phone augmented reality application. SLAM is a method that simultaneously estimates the position of a sensor and the position of landmarks in the external world using distance data. In addition, visual SLAM uses similar three-dimensional information obtained by processing such as image measurement from surrounding images. In this paper, we discuss about a method to estimate the position and orientation of a borehole camera using the image features of the hole walls image. We introduce development of borehole imaging method with using visual-SLAM.

2 Borehole Camera

2.1 Borehole Camera

Borehole camera is an optical camera that can record the wall inside of a hole excavated by drilling. The borehole camera can capture movies while the camera is hanged down to the bottom of the hole. An investigator uses the movie to investigate geological structure and detect fractures in the stratum. However, this investigation wastes man-power and time-consuming task because the movie is basically very long. Recently, there has been proposed that automatic fracture detection method [6–9]. There are basically two types of borehole camera, such as borehole imaging processing system (BIPS) and forward vision camera system (FVCS).

Borehole imaging processing system (BIPS) is a camera system that can accurately investigate hole wall images. The camera of the system is horizontally fixed and can move vertically without any rotation and rolling. Furthermore, the investigators operate camera automatically, the speed of camera is stable when capturing bore hole wall. Therefore, it can capture high accuracy images and measure various property of the fracture, such as size, depth, position and azimuthal direction. However, generally, this type of camera system cost money, manpower and time for calibration and operation. Furthermore, wide and clean environment is necessary to set the system and operate camera correctly. These types of camera system are not suitable for disaster prevention because many holes and quick operations are required to estimate natural disaster. In addition, the investigator must operate in very wild and inconvenient environments.

In contrast, forward vision camera system (FVCS) in which has a cylindrical body is lowered from the top as shown in Fig. 1. FVCS is less expensive and easier to install than BIPS. It has less space constraints. Quick surveys are possible even in unclean environments such as steep slopes, which is very useful in disaster prevention. However, the camera itself may rotate violently due to the cable, and there are always widely differences between the image center and actual hole center because of camera is rolling in the depth of borehole. Many FVCS manually drop the camera into borehole. The behavior of camera in the deep borehole, such as rotating, rolling and descent speed becomes unstable, which makes it difficult to analyze the borehole wall image with image processing technique. A camera shown in Fig. 1 is utilized for our work, this camera system has a depth measuring device, which can measure length of camera cable as camera depth. Figure 2 is an example of frame image.

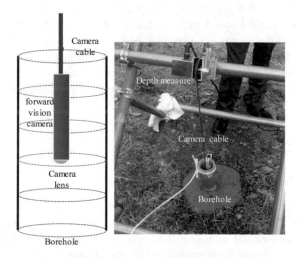

Fig. 1. Borehole camera in a forward vision camera system. It is a simple camera and it is hanged with camera cable in borehole. A depth measure is a device which measures length of delivered cable as camera depth position.

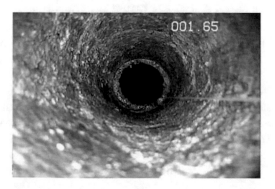

Fig. 2. An example of borehole wall image with using borehole camera.

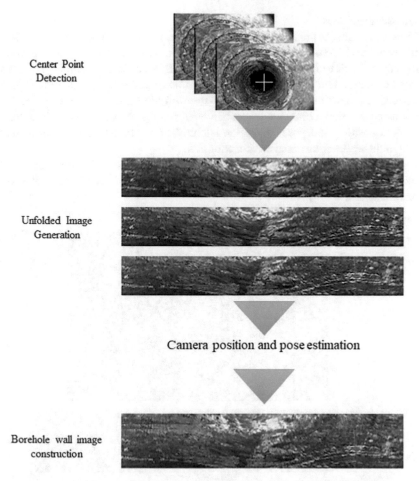

Fig. 3. Borehole imaging method. At first, center point of the hole is detected from each frame image in the movie. After that, these images are expanded for feature extraction and measurement. Based on these parameters, bore hole camera position and camera pose are estimated. Finally, these images are stitched to integrate for single borehole wall image reconstruction with these parameters.

3   Borehole Imaging Method

3.1   Borehole Imaging Method

Figure 3 shows an abstract of borehole imaging investigation. Borehole imaging is basically carried out in the following [1–4]:

1. Center points detection of borehole.
   A center point in each frame is estimated with image processing.

2. Unfolding images.

   In order to analyze the images, such an image is expanded in the azimuthal direction to generate a hole wall image. FVCS captures image with only low resolution, and it is very difficult to investigate fractures in detail, since each frame in the movie has large deviation derived from large camera rotation and rolling, it is difficult to obtain azimuthal alignment between frames. After unfolded image generation, features are extracted to find correspondence between frames in the movie. 3-dimensional camera position and pose estimation is simultaneously estimated from features.

3. Borehole wall texture image reconstruction.

   Based on camera pose and position estimation, images which are extracted from movie data are synthesized with camera parameter and borehole wall texture is reconstructed.

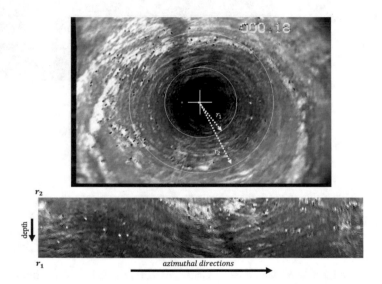

Fig. 4. Unfolding borehole image.

3.2   Center Points Detection

At first, each frame is extracted from borehole movie and is expanded with polar axis to generate an unfolded image as shown in the Fig. 4. In unfolded image, the vertical axis represents depth of hole, and the horizontal axis represents azimuth direction. When combining with the borehole wall image of the front and back frames, the depth difference is shifted in the vertical direction, and the rotation angle difference is shifted in the horizontal direction.

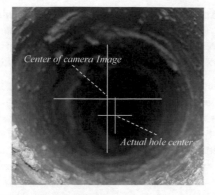

Fig. 5. The gap between an image center in each frame and the actual borehole center.

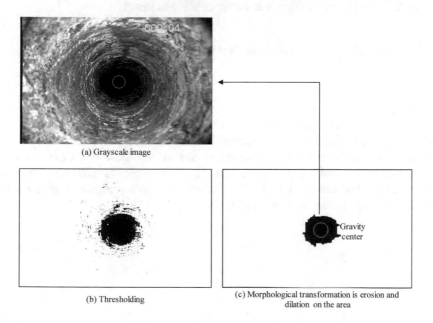

(a) Grayscale image

(b) Thresholding

(c) Morphological transformation is erosion and dilation on the area

Fig. 6. Center detection method.

In FVCS, camera is only hanged from the top of the hole. By camera rolling in the hole, there occurs the gap between an image center and the actual borehole center as shown in Fig. 5. The borehole center should be determined as accurately as possible because it greatly affects the results of the unfolded image generation. Therefore, some estimation method is required. The detection process of the hole center is shown in the Fig. 6. Usually, the center area of the borehole is often darker than the surrounding area because hole center is the deepest part. Our method detects hole center as below:

1. The original frame is converted to grayscale image (a)
2. The grayscale frame image is binarized by simple threshold processing (b).

3. By morphological transformation (c). Morphological transformation is erosion and dilation on the area, it reduces the noisy small area around larger area.
4. The outline of the central area is extracted.
5. The gravity center of the obtained area is applied as the hole center candidate.

However, in this method, when black rock or sediment exists in the hole, the area including the hole center will not be circular and the gravity center will be far apart from the actual hole center. In this way, accurate detection of the central area becomes very difficult because it is susceptible to noise. As will be described later, an appropriate number of candidates are randomly set around the center point as center point candidates. After that, an unfolded image is generated based on each center point candidate.

In practice, in order to estimate the position and orientation of the camera, $r_1 = 100$ pixels, $r_2 = 200$ pixels in Fig. 4, and to match the previous and subsequent frames, to minimize the influence of distortion of the developed image.

4 Camera Position and Pose Estimation

4.1 Visual SLAM

In this method, the camera position and pose estimation is required while acquiring information from borehole imaging to analyze movie data in detail with a simple FVCSs. We apply SLAM (Simultaneous Localization and Mapping) used for robot navigation. Using the video by the borehole camera, we estimate the camera's exact rotation orientation and depth in each frame, and also estimate the three-dimensional position of the borehole. Let C_t be the position and pose parameters of the camera calculated at each frame as shown in Fig. 7.

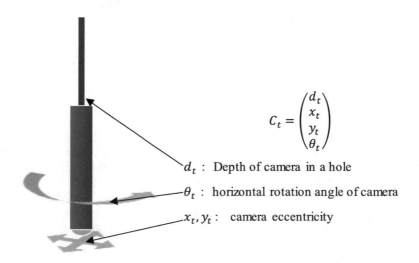

$$C_t = \begin{pmatrix} d_t \\ x_t \\ y_t \\ \theta_t \end{pmatrix}$$

d_t : Depth of camera in a hole

θ_t : horizontal rotation angle of camera

x_t, y_t : camera eccentricity

Fig. 7. Camera position and pose estimation parameters.

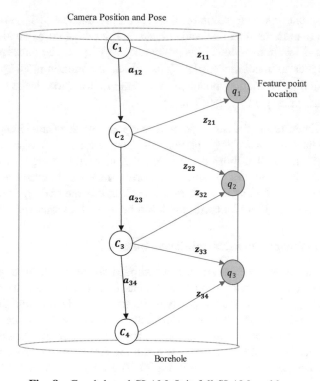

Fig. 8. Graph-based SLAM. It is full SLAM problem.

$$C_t = \begin{pmatrix} d_t \\ x_t \\ y_t \\ \theta_t \end{pmatrix}$$

Where, d_t is the vertical depth of camera in the borehole at frame t in the video, (x_t, y_t) is the eccentric displacement of the camera center from the hole center, θ_t represents the rotation angle of the camera. At this time, it is assumed that the camera does not rotate in the depth direction. The surrounding environment of the target camera is a tubular 3-dimensional shape with a constant radius, and the camera basically travels from top to bottom, that is, it is a general robot because it is limited to movement only in the height direction The computational cost can be reduced compared to the navigation of robotics.

We adopt graph-based SLAM shown in Fig. 8 to realize visual SLAM [5, 6]. Here, C_t represents the position and orientation of the camera. **a** represents the camera motion. $\mathbf{q}_i$ and z_t^i represent estimated landmarks position and their observations respectively. In this case, however, they are a set of image feature points from borehole wall texture. The feature points of the image and their correspondence will be described in detail later. Since the video of the borehole wall is processed as input, it is regarded as a full SLAM problem in which all the data are prepared   in advance. In this method, we estimate C_t and $\boldsymbol{m}$ which maximize the probability density $\mathrm{p}(C_{0:t}, \boldsymbol{m}|z_{1:t}, \boldsymbol{a}_{1:t}, \boldsymbol{c}_{1:t})$. $\boldsymbol{m}$ represents a map in the borehole, and $\boldsymbol{c}_{1:t}$ is a time series of correspondence between landmarks and measurement data. This probability density can be transformed into the product of the measurement model $\mathrm{p}(z_t^i|C_t, \boldsymbol{m}, c_t^i)$ and the motion model $\mathrm{p}(C_t|C_{t-1}, a_t)$, Here, assuming that the motion model has a uniform distribution, make it constant and think about maximizing $\prod_t \prod_i \mathrm{p}(z_t^i|C_t, \boldsymbol{m}, c_t^i)$.

Instead of determining the center point as one point in each frame image, a plurality of point candidates is listed using a particle filter, and expanded images corresponding to each are generated. The position and orientation of the camera is estimated by calculating the likelihood of finding the correspondence between the respective developed images and finding C_t to be maximized. In this method, optical flow is used for feature point extraction and correspondence between previous and next frames.

4.2   Correspondence Between Developed Images

The Shi-Thomas corner detection method applies to detect feature points in the borehole image. Corner detection in Shi-Thomas is stable method, it can detect the same place among frames as feature points while scoring and excluding those with low accuracy. It is effective to take correspondence between each frame in video. In addition, Lucas-Kanade optical flow method is applied to associate each frame images.

Lucas-Thomasi optical flow is used for matching feature points. An example of the finally obtained optical flow is shown in Fig. 9. These optical flows are the important information to connect frame images, both depth and rotation differences are calculated with optical flow information as shown in Fig. 10.

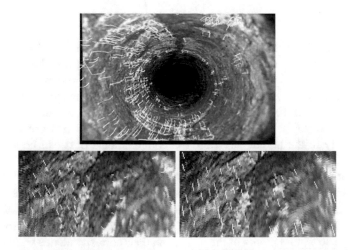

Fig. 9. Optical flow is calculated for find correspondence of two frames. If optical flow line is vertical in the image, camera is estimated to move down the hole. If also optical flow is horizontal, camera maybe rotated.

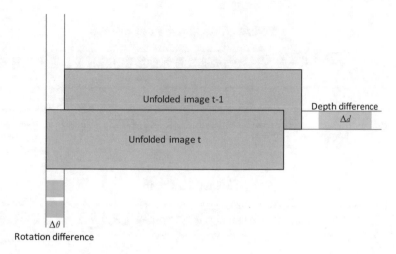

Fig. 10. Image merging. Both depth and rotation differences are calculated with optical flow information

In this method, as described in the previous section, the fitness of the center point candidate in each frame is calculated using the consistency of the optical flow. At present, the variance of the flow direction displayed in the image is simply calculated, and it is considered that the degree of conformity is high if it is estimated that it is small and that many of them are pointing in the same direction.

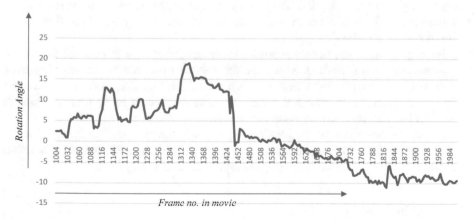

Fig. 11. Rotation estimation.

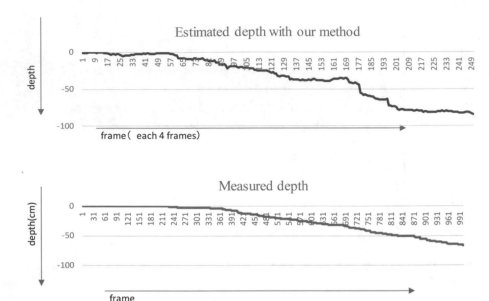

Fig. 12. Depth estimation.

5 Experimental Results

We applied this method to a movie taken on the actual site and estimated the depth and rotation angle. The hole depth is about 100 cm, about 1,000 frames images are utilized for estimation. The results are shown in the Fig. 11 and the Fig. 12. Overall accuracy of both data is not good.

For depth estimation, feature points are extracted for every 4 frames and finds their correspondence in this method. At the site, the depth is measured by the cable length of the downed camera, and the value is shown below the Fig. 12. As for the estimated value, it starts to fall at almost the same time, and it seems to be grasping almost correctly, but also rises halfway and is unstable.

As for the estimation of the rotation angle, the estimation was performed for the frame of the section with large rotation. The fluctuation of the estimated value is stable compared to the depth. For accuracy, a detailed analysis is required.

6 Conclusions

In this paper, we proposed a method to investigate the structure of a borehole well using a simple camera. In order to obtain the movement sequence of the camera, we estimate the position and orientation of the camera using visual SLAM. In the experiment, although the corresponding result is obtained, it has unstable fluctuation and needs detailed analysis.

However, there is much room for re-examination, such as the setting of particle filters and the matching of inter-frame features. In addition, this method is premised on the existence of the texture pattern with the characteristic of the hole wall. For example, it is necessary to consider application to walls without features such as concrete. A lot of research has been done on the method of SfM (Structure from Motion) to obtain 3D structure from monocular images, and we will continue to investigate and study it.

Finally, we assume off-line batch processing using movies, but in the future, we will work on the development of an online method that can investigate pore walls in real time. In addition, analysis method that applies VR (Virtual Reality) technology is also considered by finally reconstructing the hole wall image as a high definition 360° image. Future works include accuracy improvement of camera position estimation, improvement of rotation angle accuracy, application of existing methods such as LSD-SLAM, and examination of generation method of high-definition hole wall image.

References

1. Chite, P.P., Rana, J.G., Bhambare, R.R., More, V.A., Kadu, R.A., Bendre, M.R.: IRIS recognition system using ICA, PCA, Daugman's rubber sheet model together. Int. J. Comput. Technol. Electron. Eng. **2**(1), 16–23 (2012)
2. Wang, S., Luo, S., Huang, Y., Zheng, J.Y., Dai, P., Han, Q.: Railroad online: acquiring and visualizing route panoramas. Vis. Comput. **30**(9), 1045–1057 (2014)
3. Rousso, B., Peleg, S., Finci, I., Rav-Acha, A.: Universal mosaicing using pipe projection. In: Proceedings of Sixth International Conference on Computer Vision, pp. 945–950 (1998)
4. Cao, M., Deng, Z., Rai, L., Teng, S., Zhao, M., Collier, M.: Generating panoramic unfolded image from borehole video acquired through APBT. Multimedia Tools Appl. **77**(19), 25149–25179 (2018)
5. Sebastian, T., Wolfram, B., Dieter, F., Ryuichi, U.: Probablistic Robotics (in Japanese), MyNavi Shuppan, ISBN 978-4839952983, (2016)
6. Chuanying, W., Xianjian, Z., Zengqiang, H., Jinchao, W., Yiteng, W.: The automatic interpretation of structural plane parameters in borehole camera images from drilling engineering. J. Pet. Sci. Eng. **154**, 417–424 (2017)
7. Xi-Ning, L., Jin-Song, S., Wu-Yang, Y., Zhen-Ling, L.: Automatic fracture-vug identification and extraction from electric imaging logging data based on path morphology. Pet. Sci. **16**, 58–76 (2018)
8. Rommed, A.Q.C., Diego, C.C., Renato, M.S., Evandro, J.R.P., Fabiana, R.L., Esteban, W.G. C.: Improving accuracy of automatic fracture detection in borehole images with deep learning and GPUs. In: 30th International Conference on Graphics, Patterns and Images (SIBGRAPI), pp. 345–350 (2017)
9. Mariusz, M., Magdalena, H., Norbert, S.: The application of the automatic search for visually similar geological layers in a borehole in introscopic camera recordings. Measurement **85**, 142–151 (2016)

A Mechanism of Window Switching Prediction Based on User Operation History to Facilitate Multitasking

Keizo Sato[✉], Shotaro Imada, Shinya Mazume,
and Makoto Nakashima

Oita University, 700 Dannoharu, Oita-shi 870-1192, Japan
{k-sato,v18e3002,v1553058,nakasima}@oita-u.ac.jp

Abstract. Multitasking on a PC has long been a common practice for users. However, recent rapid technological advancements often complicate multi-tasking. At this moment, users can utilize many applications on one or more wide displays simultaneously. Thus, the number of tasks and their windows are dramatically increased. Under these circumstances, window switching costs cannot be underestimated. In this paper, we propose a window switching prediction mechanism to facilitate finding the desired window. The mechanism predicts and suggests which window will be needed for the next task based on user operation history. Prediction precision and efficiency of multitasking were investigated by collecting and analysing logs of actual multitasking on subjects' PCs. The experimental results show that the mechanism was able to reduce the time costs of window switching. We also found that the subjects tended to use both traditional and the proposed window switching methods.

1 Introduction

Multitasking on a PC currently is a common practice for many PC users. Although, in the past, the number of tasks and their windows were strictly limited because of the specs of PCs, recent rapid technological advancements have eased the restrictions. Additionally, wide displays are available for office and personal use. Therefore, a user can utilize many applications (i.e., a word processor application, a spread sheet application and/or a graphic tool) at the same time.

Such environmental changes, however, create new issues for efficient multitasking. Task management on a PC has become more complicated since hardware improvements enable the user to perform many tasks in parallel. When the number of tasks increases, the number of total windows naturally increases. Under these circumstances, the user usually uses Windows Flip or the taskbar or visually inspects the displays. However, these methods have low efficiency in finding a specific window when many are opened on a display. The reason is that these methods rely on a user's eyes because the number of the items shown is not reduced. A more efficient method for the user is required.

To assist task management on a PC, the window grouping approach is most frequently used. Virtual Desktop [8] enables a user to expand working space on a PC

L. Barolli et al. (Eds.): NBiS-2019, AISC 1036, pp. 154–166, 2020.
https://doi.org/10.1007/978-3-030-29029-0_15

according to the number of tasks and arrange windows based on the tasks. However, a user will suffer from the strict task management when the number of tasks is increased. It is also difficult to determine the space necessary for some common applications like a mailer or a web browser since they are often used for multiple tasks.

Push-and-pull switching [14] and docking window framework [11] are for grouping windows according to their visibility or the user operation. The former system automatically makes groups so that mutually un-overlapped windows belong to the same group. A user can switch the window groups to perform any task. The latter system is used to combine two or more windows which are arranged close to each other. Once the windows are combined, a group can be moved around as a unit. However, the user has to frequently switch the groups if the number of groups is high and determining which group includes the desired window is difficult.

Baeza-Yates et al. [1] proposed an app prediction method based on the long-term multiple user logs on app usage on smartphones. However, tasks and their windows are frequently changed, therefore the long-term log is not fit to ease daily multitasking since the relation among windows is not suitable.

We here propose a window switching prediction mechanism to automatically predict the window to be used next from many alternative ones. In our mechanism, two values, the *WindowRelation* and the *WindowImportance*, are applied to predict the next window (called the *target window*). *WindowRelation* determines the relationship between windows belonging to the same task. When two windows are used for the same task, they are shown at the same time or are frequently switched with each other. *WindowImportance* determines the most frequently used windows in a task instead of making strict window groups. Even when the relationship between a task and a window group is unclear, the most frequently used window can be easily identified because it should be possible to identify at least one window that is frequently used in a specific task. The prediction is performed by examining the *WindowRelation* and the *WindowImportance* of each window against the active window. We also developed a fan-shaped user interface where the chosen windows based on the WindowRelation or the WindowImportance are shown in the form of banners including the application icons, window texts and the partial images of windows.

The efficiency of the mechanism was investigated using a two stage evaluation: a computational evaluation and a user experience evaluation. In the former evaluation, we collected user operation history data for window use including any mouse/keyboard events on the user's PC (with permission) and simulated the prediction mechanism. In total, 15 subjects' operation histories were collected. The results indicated that the window switching efficiency was improved with a high prediction precision. In the latter evaluation, we prepared the user interface and asked the subjects to switch the windows based on the proposed window prediction mechanism while existing methods were also made available to be used on demand. The subjects used both existing and proposed methods.

The remaining sections are organized as follows: Sect. 2 surveys related work. Section 3 describes the details of our window prediction mechanism. In Sect. 4 we detail the preliminary study that was conducted with the cooperation of 15 university students. Section 5 explains how the user interface for the prediction mechanism is developed. Section 6 shows the resultant efficiency of the window prediction mechanism and its user interface in actual multitasking. Finally, Sect. 7 summarizes the proposed method and discusses a future research direction.

2   Related Work

In a research paper about display space usage [4], it was reported that 78.1% of the subjects used 8 or more windows at the time of the study in 2004. Frequently used traditional window switching methods are the taskbar, shortcut keys (a combination of alt key and tab key) and directly clicking a window. A common problem with these methods is the cost of picking out the target window from many windows, since it is assumed that the target window is invisible or only partially visible when other windows are used. Oliver et al. realized three prototypes displaying thumbnails of available windows [9]. In this approach, a user has to visually identify the thumbnail of the target window just as with the existing shortcut key method when the number of windows is increased.

Another possible solution is to classify windows into several groups which correspond to tasks. Push-and-pull switching [14] and docking window framework [11] are proposed to facilitate managing windows. Although windows must be classified based on the task they belong to under the grouping approach, it is difficult to accurately and automatically make window groups since the relation between windows can be evaluated by various measures. Additionally, when the number of windows is high, it is difficult to find the target window from the pool of in the same task.

Taskposé [2] shows windows in the form of thumbnail images while closely related windows are arranged physically close to each other. Warr et al. [13] performed a study on window switching. Three types of window switching interfaces–Cards, Exposé and Mosaic–were examined from the point of view of selection time and total errors. These interfaces require a user's effort to find a specific window when the number of windows is high.

Lischke et al. proposed the interaction technique for window management on large high-resolution displays [6]. Multiple windows are shown on multiple displays so that a user can easily manage multiple windows with the wide view. The interaction technique is effective when the number of windows is low compared with the size of total visible area. However, there is still the cost of preparing multiple large high-resolution displays for a user and it is difficult to manage many windows because of the limitations of visible ratios of windows.

Fitchett et al. [3] and Baeza-Yates et al. [1] took different approaches to finding the target window. These approaches utilize the user operation history so that the target window is predicted based on the history. The long-term operational history is applied to rank possible actions and predict the next action and/or the next window. However, the long-term history is not suitable for determining the usage of a window which is utilized for multiple purposes.

Ishak et al. [5] and Waldner et al. [12] used a technique of the window transparency to allow visualization of the partially overlapped window. Areas of a window can be divided into two types based on the amount of the information. The part of the underlying window which is visible through the window above is an area with less information in order to be able to maximize the amount of information on a desktop. However, the combined information in the same window causes information overload problems. FoXpace [15, 16] realized an autonomous window arrangement so that the total amount of information shown on a desktop is maximized. However, the calculation for arranging windows is complicated and the usefulness in actual multitasking is unclear.

Our prediction mechanism uses short-term user operation history to choose the candidate windows for the target window. Chosen windows are shown in the form of a banner where a partial image of the window, the window text and icon of the corresponding application are shown.

3   The Window Switching Prediction Mechanism

In this section, the basic idea and the procedure of window switching prediction are introduced.

3.1   The Basic Idea

A user uses many windows to perform tasks while switching windows on demand under a multitasking environment. The main purpose of the window switching prediction mechanism is to reduce the time to find the next active window (called the *target window*) so that a user can easily shift to the next process in multitasking.

Window switching activities can be divided into two types: window switching within a task and window switching among tasks. Window switching within a task is done to switch the active window to another window which is strongly related to the active window and the relationship is determined by examining whether these two windows are used in parallel. Window switching among tasks is done to switch the active window belonging to a task to the window belonging to another task. In this case, window switching is done by searching the window which is the most frequently used in a task.

In the prediction mechanism, first, the relationships between the active window and other windows are examined for finding the windows which have a high possibility of being used in the immediate future. Second, the degree of window use for each window is examined to find the windows which are regarded as the main window of a task. The prediction is done by specifying the windows with a high degree of relationship or a high frequency of use.

3.2   WindowRelation and WindowImportance

We here apply the information (with timestamps) on (i) window switching, (ii) the rectangular area of windows and (iii) the numbers of mouse/keyboard events on each window to determine the relationship among windows (called the *WindowRelation*) and the degree of window use (called the *WindowImportance*).

These three types of information are here collectively described as the user operation history and are used for calculating the *WindowRelation* and *the WindowImportance*.

WindowRelation

Two windows, which are used in the same task, are arranged to be visualized simultaneously or are frequently switched with each other. The former and the latter relations are calculated by formula 1 and 2, respectively. Two windows are here defined as w_i, $w_j \in W$. W indicates the all available windows. The value $s(w_i, w_j)$ is the number of times that a user switches window w_i to w_j. Formula 1 was inspired by Liu et al. [7]. The value $t(w_i, w_j)$ is the time that both w_i and w_j are visible to a user.

$$WindowRelation_1\left(w_i, w_j\right) = \frac{s\left(w_i, w_j\right) + s\left(w_j, w_i\right)}{s(w_i, w) + s(w, w_i) + s\left(w_j, w\right) + s\left(w, w_j\right)} \tag{1}$$

$$WindowRelation_2\left(w_i, w_j\right) = \frac{t\left(w_i, w_j\right)}{total\ working\ time} \tag{2}$$

WindowImportance

The degree of window usage can be calculated by examining how many mouse/keyboard events have happened on a window. This value is calculated based on the information (iii). To normalize this value, we define formula 3 as follows.

$$WindowImportance(w_i) = \frac{number\ of\ mouse/keyboard\ events\ on\ w_i}{number\ of\ mouse/keyboard\ events\ on\ W} \tag{3}$$

3.3   Prediction Mechanism and Procedure

A window switching prediction mechanism consists of three major modules: collecting, predicting and switching modules. Figure 1 shows the detailed structure of a window switching prediction mechanism. The collecting module collects the user operation history to be used to predict the candidate windows for the target window (called the predicted windows). The prediction module has three sub modules: ranking modules 1, 2 and 3. These three sub modules are used for ranking the inactive windows according to *WindowRelation* and *WindowImportance*. Here the predicting module chooses *WindowRelation₁* or *WindowRelation₂* based on the prediction precisions up until the time of the previous window switching. If the prediction precision of *WindowRelation₁* is higher than that of *WindowRelation₂*, *WindowRelation₁* is used to rank the inactive windows – and vice versa. Top-k windows of the chosen *WindowRelation* are dealt as the predicted windows within a task. Top-k windows of *WindowImportance* are also dealt as the predicted windows among tasks. Window switching history is collected and stored as a part of the user operation history to evaluate the ranking precisions of *WindowRelation₁* and *WindowRelation₂*.

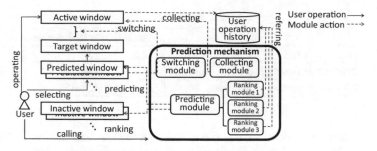

Fig. 1. The detailed structure of the prediction mechanism.

When a user wants to switch the window, the user performs the pre-defined mouse gesture (described in Sect. 5). Then the mechanism calculates the *WindowRelation* between the active window and each of the inactive windows. The windows which indicate a high value of *WindowRelation* are shown to the user as the predicted windows in the form of a kind of banner. In the case of a user changing the task, *WindowImportance* is used to suggest the predicted windows to a user. The user selects one of the predicted windows and then the focus of the active window is switched to the selected window. In order to define the parameter k, we performed preliminary experiments where the relationship between the precision of the prediction mechanism and the value k was examined. After k was defined, we also designed the fan shaped user interface to switch the windows with the prediction mechanism.

4 Preliminary Experiments

The precision of the prediction mechanism and the proper value of the parameter k were examined through the preliminary experiments where the prediction mechanism was simulated based on previously collected user operation history. In order to collect the user operation history for the simulation, we asked university students to permit logging data about their actual multitasking with their PC. The subjects ranged from fourth-grade students to the second year of graduate school.

The detailed procedures of the preliminary experiments are as follows:

i. With the subjects' permission, user operation history was collected including the elapsed time for switching windows on a subjects' PCs.
ii. Prediction simulation was performed based on the collected history.
iii. When the target window was included in top-k windows of *WindowRelation* or *WindowImportance*, the target window was considered to be correctly predicted for a user.

Table 1 shows the statistics for the performed tasks on the subjects' PCs. In this experiment, the numbers of tasks were generally low.

Table 1. The statistics for the tasks.

	Average	Standard deviation
Number of tasks	2.3	1.0
Number of times of window switching	241.4	154.4
Maximum number of windows	13.2	5.3

The resultant precisions of the prediction mechanism are shown in Fig. 2. As shown in the figure, when the parameter k is set to 3 or more, the prediction precision is highly maintained. Totally 91.3% of window switching within a task and totally 80.2% of window switching among tasks were successfully predicted respectively when $k = 3$.

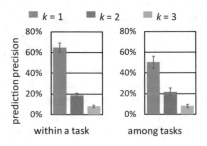

Fig. 2. Prediction precisions.

5   Fan Shaped User Interface Design

An appropriate visualization of the predicted windows is required to facilitate window switching with the prediction mechanism. Note that the windows themselves are not suitable for visualizing the predicted windows when each of the windows occupies a major part of the desktop space and/or the windows overlap with each other.

The predicted windows are here shown as a banner where the application icon, the window text and a partial image of the window is shown in a box. In cases in which no mouse event is detected, the thumbnail images of the predicted windows are shown on banner areas. Figure 3 shows the banner design. Figure 3(a) is the banner if no mouse event has occurred on the window after the prediction mechanism launch. Figure 3(b) is the banner if a mouse event has occurred on the window. Here the partial image of the window is applied instead of the thumbnail of the window since there are often multiple windows open of the same application and it is difficult to identify a particular window of the same application. The partial image is clipped from the original window image with a central focus on the coordinates of the latest clicked point.

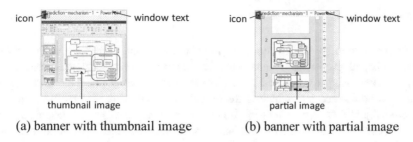

icon   window text   icon   window text

thumbnail image   partial image

(a) banner with thumbnail image   (b) banner with partial image

Fig. 3. Banner design.

Figure 4 shows the arrangement of the banners. The factors for deciding the location of the banners are the size and number of banners in addition to the direction of the mouse gesture. The number of banners is set to 3 so that a user can easily identify the target window from the displayed banners. The banners are arranged as a fan-shape so that a user can choose any banner with the same mouse movement cost. The radius R between the mouse cursor and the banner is set to 200 pixels. Layer I indicates the predicted windows for window switching within a task. Layer II indicates the predicted windows for window switching among tasks. This two-layers arrangement is applied since the frequency of window switching within a task is usually higher than that among tasks. Layer I is arranged preferentially close to the mouse cursor. The width S for banners in layer I is 300 pixels and the width S' for banners in layer II is 336 pixels based on major web banner size [10].

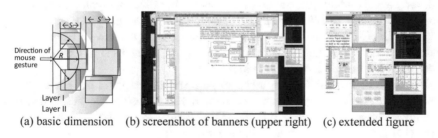

Direction of mouse gesture

Layer I
Layer II

(a) basic dimension   (b) screenshot of banners (upper right)   (c) extended figure

Fig. 4. Banner arrangement.

The fan-shaped banner arrangement is invoked when a user performs the mouse gesture – a right click of the mouse with a movement of the mouse to the right or left. The mouse gesture is applied so that a user can call the fan-shaped arrangement from any location on the desktop while avoiding mistaken application operations. The direction of the mouse gesture is selectable, left to right or right to left, even when the mouse cursor located on the extreme right or left as shown in Fig. 4(b) and (c). The target window selection is performed by moving the mouse cursor to the banner and releasing the right button of the mouse. The target window is automatically brought to the foreground on the desktop. Note that the size and the X-Y coordinates of the target window are not modified to allow a user to easily resume their task.

6 Evaluations

We conducted two types of evaluations for examining the usability and the efficiency of the proposed prediction mechanism. The usability evaluation was performed by interview after using the prediction mechanism on the same PC. The efficiency evaluation was performed by collecting the actual user operation history for each subject from the subject's PC while the subject actually performed tasks. We asked subjects to use the prediction mechanism and any traditional window switching method on demand for more than 7 days. After the experimental period, we investigated the logged data concerning any user operation, especially in view of the time to switch the window and in view of the ratio of the number of times the prediction mechanism was used to total window switching.

6.1 Usability Evaluation

We first performed the usability evaluation to clarify whether the user interface of the prediction mechanism was easily used to switch windows. In the usability evaluation, subjects were asked to perform the three types of tasks. In the task 1, each subject searched historical documents and collected images of the documents found. After collecting the images, the subjects assigned names to the collected images in the spreadsheet application. In the task 2, each subject researched prepared terms and entered the information about the given terms. In the task 3, each subject accessed indicated web sites and clipped an area previously defined with a snipping tool. The clipped images were saved to a folder and were pasted into Microsoft Word files named with the corresponding web site titles. To begin the tasks, two windows of Microsoft Word, one window of Microsoft Excel, one window of Google Chrome, one window of a free snipping tool and two windows of Windows Explorer were launched. In this evaluation, one 30-in. monitor (2,560 × 1,600 pixels) was used to examine usability under the same environmental conditions.

Ten university students were asked to perform the common tasks and then they were asked to answer the questionnaire shown in Table 2 using 7-level Likert scales (1: Strongly disagree, 2: Disagree, 3: Somewhat disagree, 4: Neither agree or disagree, 5: Somewhat agree, 6: Agree, 7: Strongly agree).

Table 2. Questionnaires for the usability evaluation.

	Questionnaire item
Q1	The fan-shaped interface was easy to use
Q2	The usage of the fan-shaped interface was easier than that of traditional methods
Q3	The burden of the fan-shaped interface was smaller than that of traditional methods
Q4	The fan-shaped interface really worked to find the target window
Q5	The locations of displayed banners were appropriate
Q6	I want to continue to utilize the prediction mechanism in future work

Figure 5 shows the number ratios of answers for each questionnaire item. Except for Q2, the number of positive answers (level 5 to 7) were statistically higher than that of negative answers (level 1 to 3) (binominal test, $p < 0.05$). According to the figure, the prediction mechanism gained the acceptance of the subjects as an additional window switching method. On the other hand, many of the subjects tended to keep using the traditional window switching methods.

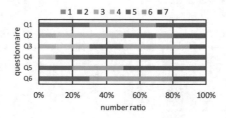

Fig. 5. Number ratios of answers from subjects.

6.2   Efficiency Evaluation

We next examined how the subjects used the prediction mechanism and how the efficiency of window switching was improved in actual multitasking on subjects' PCs. The subjects were asked to use the prediction mechanism with the traditional window switching methods. Table 3 shows the statistics for the subjects and their evaluation environments.

Table 3. Statistics for subjects and evaluation environments.

Subject	Number of days	Grade	Monitor information	
			First monitor	Second monitor
A	12	Fourth-year student	30-in., (2,560 × 1,600 pixels)	—
B	9	Fourth-year student	30-in., (2,560 × 1,600 pixels)	19-in., (1,280 × 1,024 pixels)
C	7	Second-year graduate student	30-in., (2,560 × 1,600 pixels)	24-in., (1,920 × 1,200 pixels)
D	7	Second-year graduate student	24-in., (1,920 × 1,200 pixels)	24-in., (1,920 × 1,200 pixels)

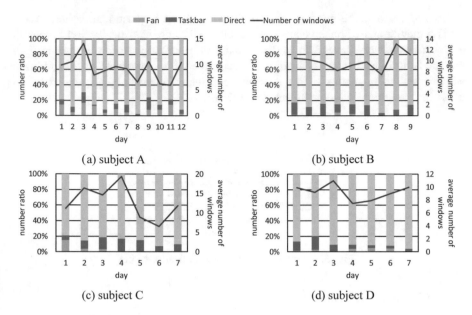

Fig. 6. Number ratios of the used window switching methods.

Figure 6 shows the number ratios of usage of each window switching method. The label "Fan" means the proposed method. The label "Direct" means visual inspection by the subject. Keyboard shortcut for window switching was not used in the evaluations. There was a correlation between the window number and the number ratio for the fan-shaped window switching method for each subject (A 0.58 correlation coefficient was obtained for subject A. For the subject B, C and D, −0.31, 0.05, −0.2 correlation coefficients were obtained, respectively.). The main task of subject A was gathering information about the websites of university libraries in Japan. There were multiple Microsoft Excel windows launched to record the gathered information with a single monitor. In the limited workspace, the proposed method worked especially well.

The elapsed time for performing window switching was also measured to investigate whether the efficiency of window switching was improved according to the prediction mechanism. The results are shown in Fig. 7. In the figure, the X-axis indicates the window switching count from the working start time each day. The Y-axis indicates the elapsed time of window switching. The diagrams indicate that the elapsed time using the proposed method was kept low in many cases though users often spent a great deal of time switching windows by the traditional methods.

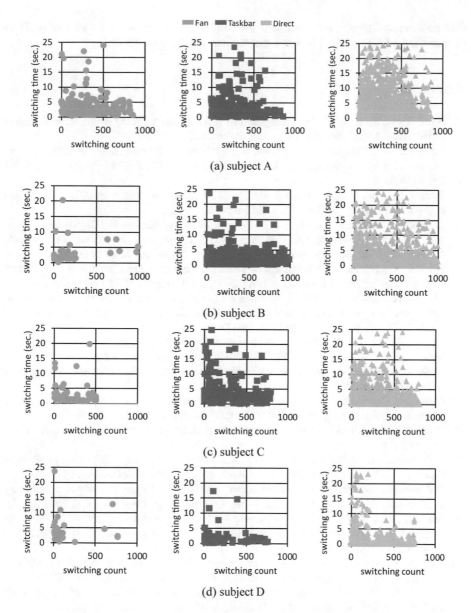

Fig. 7. Scatter diagrams of the elapsed time for window switching.

7 Conclusions

In this paper, we proposed a mechanism of predicting the window to be used based on user operation history. Through the preliminary experiments and the evaluations, it is clear that the proposed mechanism was accepted as an additional window switching method by the subjects and enabled the costs of window switching to be reduced.

Our next research issue is to perform the long-term evaluation for more subjects to examine the effects of the proposed mechanism on daily multitasking.

References

1. Baeza-Yates, R., Jiang, D.: Predicting the next app that you are going to use. In: Proceedings of the WSDM 2015, pp. 285–294. ACM Press (2015)
2. Bernstein, M., Shrager, J., Winograd, T.: Taskposé: exploring fluid boundaries in an associative window visualization. In: Proceedings of the UIST 2008, pp. 231–234. ACM Press (2008)
3. Fitchett, S., Cockburn, A.: AccessRank: predicting what users will do next. In: Proceedings of the CHI 2012, pp. 2239–2242. ACM Press (2012)
4. Hutchings, D.R., Smith, G., Meyers, B., Czerwinski, M., Robertson, G.: Display space usage and window management operation comparisons between single monitor and multiple monitor users. In: Proceedings of the AVI 2004, pp. 32–39. ACM Press (2004)
5. Ishak, E.W., Feiner, S.K.: Interacting with hidden content using content-aware free-space transparency. In: Proceedings of the UIST 2004, pp. 189–192. ACM Press (2004)
6. Lischke, L., Mayer, S., Hoffmann, J., Kratzer, P., Roth, S., Wolf, K., Woniak, P.: Interaction techniques for window management on large high-resolution displays. In: Proceedings of the MUM 2017, pp. 241–247. ACM Press (2017)
7. Liu, S., Tajima, K.: WildThumb: a web browser supporting efficient task management on wide displays. In: Proceedings of the IUI 2010, pp. 159–168. ACM Press (2010)
8. Multiple desktops in Windows 10. https://support.microsoft.com/en-us/help/4028538/. Accessed 23 Apr 2019
9. Oliver, N., Czerwinski, M., Smith, G., Roomp, K.: RelAltTab: assisting users in switching windows. In: Proceedings of the IUI 2008, pp. 385–388. ACM Press (2008)
10. Standard Banner Sizes List. https://blog.bannersnack.com/banner-standard-sizes/. Accessed 17 Apr 2019
11. Shibata, H., Omura, K.: Docking window framework: supporting multitasking by docking windows. In: Proceedings of the APCHI 2012, pp. 227–236. ACM Press (2012)
12. Waldner, M., Steinberger, M., Grasset, R., Schmalstieg, D.: Importance-driven compositing window management. In: Proceedings of the CHI 2011, pp. 959–968. ACM Press (2011)
13. Warr, A., Chi, E.H., Harris, H., Kuscher, A., Chen, J., Flack, R., Jitkoff, N.: Window shopping: a study of desktop window switching. In: Proceedings of the CHI 2016, pp. 3335–3338. ACM Press (2016)
14. Xu, Q., Casiez, G.: Push-and-Pull switching: window switching based on window overlapping. In: Proceedings of the CHI 2010, pp. 1335–1338. ACM Press (2010)
15. Yoshida, K., Ozono, T., Shintani, T.: FoXpace: manipulating windows based on the user's work history. In: Proceedings of the IIAI-AAI 2016, pp. 698–703. IEEE (2016)
16. Yoshida, K., Ozono, T., Shintani, T.: Developing an automatic window manipulation system considering content on application windows and user's behavior. Int. J. Smart Comput. Artif. Intell. 1(2), 59–75 (2017)

Opportunistic Communication
by Pedestrians with Roadside Units
as Message Caches

Tomoyuki Sueda[1] and Naohiro Hayashibara[2(✉)]

[1] Graduate School of Frontier Informatics, Kyoto Sangyo University, Kyoto, Japan
i1888088@cc.kyoto-su.ac.jp
[2] Faculty of Computer Science and Engineering, Kyoto Sangyo University,
Kyoto, Japan
naohaya@cc.kyoto-su.ac.jp

Abstract. Opportunistic communication is one of the critical technologies in the area of advertisement, information sharing, disaster evacuation guidance in delay-tolerant networks (DTNs), vehicular ad hoc networks (VANETs) and so on. The efficiency of opportunistic communication is correlated to the movement pattern. Random walks are often used as the movement patterns of a pedestrian. Even amongst those, Lévy walk that is a family of random walks is attracted attention as a human movement pattern. There are lots of works of Lévy walk in the context of target detection in swarm robotics, analyzing human walk patterns, and modeling the behavior of animal foraging in recent years. According to these results, it is known as an efficient method to search and come across one another in a two-dimensional plane. However, all these works assume a continuous plane and hardly any results on graphs are available. In this paper, we focus on message delivery based on opportunistic communication by pedestrians who move based on Lévy walk and Random walk movement patterns on the road network of a city. Moreover, we introduce roadside units located in the city, which play a role in the distributed message cache. So, we evaluate the impact of roadside units on message delivery in delay-tolerant networks consists of mobile devices of human pedestrians. We assume Random walk and Lévy walk as a pedestrian mobility model. Our simulation results show that the roadside units have a significant impact on message delivery with a small number of pedestrians.

1 Introduction

It is well-known that random walks become an important building block for designing and analyzing protocols in mobile ad hoc networks [9,10,25], searching and collecting information [4,6] in P2P networks, and solving global optimization problems [27,28]. In particular, Lévy walk (also called Lévy flight) has recently attracted attention due to the optimal search in animal foraging [11,26] and the statistical similarity of human mobility [19]. It is a mathematical fractal which

© Springer Nature Switzerland AG 2020
L. Barolli et al. (Eds.): NBiS-2019, AISC 1036, pp. 167–177, 2020.
https://doi.org/10.1007/978-3-030-29029-0_16

is characterized by long segments followed by shorter hops in random directions. The pattern has been proposed by Paul Lévy in 1937 [16], but the similar pattern has also been evolved and sophisticated as a naturally selected strategy that gives animals an edge in the search for sparse targets to survive [11].

Although most of the works on Lévy walk have been done on a continuous plane, we assume unit disk graphs as the underlying environment. Unit Disk Graphs are widely used for routing, topology control, analysis of virus spreading in ad hoc networks (e.g., [2, 15, 25]). Roughly speaking, the difference between a graph and a continuous plane is the freedom of movement. It means that the link structure of the graph restricts the movement. We suppose an agent as a mobile entity. Agents can only move to a neighbor node of the current node on the graph, although agents move anywhere on the continuous plane. It means that agents are restricted their movement by the link structure of a graph. There are several research results on random walks on graphs [14, 18] but there are a few result on Lévy walk on graphs [21].

We assume roadside units (RSUs) as a distributed message cache, which are located in a city. We also assume a delay-tolerant network (DTN) formed by human pedestrians who have mobile devices such as smartphones. Thus, messages are sent, carried, and forwarded by devices of pedestrians who move based on Random walk and Lévy walk. The structure of the road network that is often modeled as a graph restricts the movement of pedestrians in a city.

In this paper, we analyze the impact of RSUs as message caches on the efficiency of message delivery by opportunistic communication in DTNs formed by mobile devices of human pedestrians. More precisely, we measure the average delivery ratio and the latency in our simulations in the presence of RSUs.

2   Related Work

We introduce several research works on Lévy walk and random walks.

Birand et al. proposed the Truncated Levy Walk (TLW) model based on real human traces [5]. The model gives heavy-tailed characteristics of human motion. Authors analyzed the properties of the graph evolution under the TLW mobility model.

Valler et al. analyzed the impact of mobility models including Lévy walk on epidemic spreading in MANET [25]. They adopted the scaling parameter $\lambda = 2.0$ in the Lévy walk mobility model. From the simulation result, they found that the impact of velocity of mobile nodes does not affect the spread of virus infection.

Thejaswini et al. proposed the sampling algorithm for mobile phone sensing based on Lévy walk mobility model [23]. Authors showed that proposed algorithm gives significantly better performance compared to the existing method in terms of energy consumption and spatial coverage.

Fujihara et al. proposed a variant of Lévy walk which is called Homesick Lévy Walk (HLW) [12]. In this mobility model, agents return to the starting point with a homesick probability after arriving at the destination determined by the power-law step length. As their result, the frequency of agent encounter obeys the power-law distribution.

Mizumoto et al. measured the encounter rate of Lévy walk and Brownian walk regarding mating of sexual dimorphism (e.g., male and female) on the one-dimensional and the two-dimensional infinite and borderless space with limited lifespans [17]. They also analyzed the relationship of the scaling parameter of Lévy walk, the encounter rate and the lifespan based on the simulation result. From the result, Lévy walk is efficient regarding the encounter rate in a short lifespan, whereas Brownian walk becomes efficient by the increase of the lifespan.

There are several papers on random walks on finite graphs. Ikeda et al. showed the impact of local degree information of nodes in a graph [14]. They proposed a random walk, called the β random walk, which uses the degree of neighboring nodes and moves to a node with a small degree in high probability compared to the ones with a high degree. The hitting time becomes $O(n^2)$ and the cover time becomes $O(n^2 log\ n)$ in the β random walk despite the hitting and cover times are both $O(n^3)$ in the simple random walk without local degree information.

Nonaka et al. presented the hitting time and the cover time in the Metropolis walk which obeys the transition probability produced by the Metropolis-Hastings algorithm [18]. It is a typical random walk used in Markov chain Monte Carlo methods. They showed that the hitting time is $O(n^2)$ and the cover time $O(n^2 log\ n)$.

Despite of many articles on Lévy walk, they evaluated it on continuous fields, and hardly any results on graphs have been available. Shinki et al. newly defined the algorithm for Lévy walk on a unit disk graph and evaluated the efficiency of message dissemination compared with the random walk, β random walk and the metropolis walk [21]. According to the evaluation, the Lévy walk movement pattern is efficient for message dissemination compared to other random walks.

3   System Model

In this paper, we assume delay-tolerant networks (DTNs) consists of human pedestrians who have mobile devices that have a short-range communication capability such as Bluetooth, ad hoc mode of IEEE 802.11, Near Field Communication (NFC), infrared transmission, and so on. Moreover, we assume a city road network and roadside units (RSUs) located randomly in the road network. They also exchange messages with pedestrians and help to distribute messages by storing some messages temporally.

There exist several messaging protocols in DTN, such as Epidemic [24], Spray and Wait [22] and RAPID [3]. We use the Epidemic protocol in our simulation. It is a flooding-style messaging protocol where every pedestrian sends copies of all messages that are stored in the mobile device of the pedestrian to others in the communication range. To prevent exhausting storage resources in the network, we assume TTL (Time-to-Live) for every message.

4   The Movement Patterns of Human Pedestrians

The efficiency of message broadcasting by opportunistic communication depends on the movement pattern of agents [13]. We focus on Lévy walk and Random walk as the movement pattern of pedestrians. Both are often used as human mobility models. Lévy walk in particular statistically resembles human mobility traces [19]. It has a parameter λ that determines the behavior of agents. In general, the ballistic trajectory frequently appears on agents' movement with a decrease of λ. Each pedestrian is assumed to move at the same velocity.

4.1 Lévy Walk

As we mentioned in Sect. 3, we assume a city road network modeled as a unit disk graph. So, we use the algorithm for Lévy walk on unit disk graphs proposed by Shinki et al. [21].

Lévy walk is a variation of random walks where each node selects a direction uniformly from within $[0, 2\pi)$ and a step length of a walk is determined the Lévy probability distribution described as follows.

$$p(d) \propto d^{-\lambda} \tag{1}$$

d is a step length and λ is the scaling parameter to draw the different shape of the probability distribution. Now we show the behavior of Lévy walk with $\lambda \in \{1, 2, 1.5, 2.0\}$ in a 2D plane (see Figs. 1, 2 and 3).

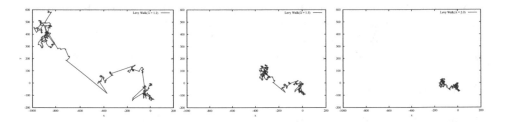

Fig. 1. Lévy walk with $\lambda =$ 1.2 on 2D plane. **Fig. 2.** Lévy walk with $\lambda =$ 1.5 on 2D plane. **Fig. 3.** Lévy walk with $\lambda =$ 2.0 on 2D plane.

According to the figures, the maximum step length is getting longer as λ decreases. In general, Lévy walk becomes similar to the random walk in terms of its behavior when λ is greater than 3.0 [7].

We describe the algorithm for Lévy walk on unit disk graphs.

The main difference between continuous planes and graphs is freedom of mobility. Agents can only move to the neighbor node of the current node in graphs. It means that the movement of agents in a graph is more restricted than the one in a continuous plane.

We explain the algorithm for Lévy walk on unit disk graphs proposed in [21].

In general, an agent selects a destination node v from a set of neighbors $N(u)$ of the current node u randomly and moves to v in random walks on graphs. In contrast, an agent determines the orientation o from $[0, 2\pi)$ at random, and the step length d by the power law distribution (see Eq. 1) at the beginning of Lévy walk. Then, it selects v according to the information and moves to it d times. Note that d is the distance (hops) in a graph.

We now define *a walk* as the sequence of the movement in a direction and *a step* as the movement with the step length $d = 1$. Thus, a walk consists of a sequence of steps in Lévy walk though a walk equals to a step in random walks.

The algorithm described in Algorithm 1 is for a walk of Lévy walk. It means that the algorithm is repeated until the termination condition is satisfied.

Algorithm 1. Lévy walk on unit disk graphs

1: **Initialize:**
$\quad\quad c \leftarrow u$ ▷ the walk starts from u.
$\quad\quad o \leftarrow 0$ ▷ orientation of the walk.
$\quad\quad PN(c) \leftarrow \emptyset$ ▷ possible neighbors to move.
2: d is determined by the power-law distribution
3: o is randomly chosen from $[0, 2\pi)$
4: **while** $d > 0$ **do**
5:   $PN(c) \leftarrow \{x | abs(\theta_{ox}) < \delta, x \in N(c)\}$
6:   **if** $PN(c) \neq \emptyset$ **then**
7:    $d \leftarrow d - 1$
8:    move to $v \in PN(c)$ where v has the minimum abs(θ_{ov})
9:    $c \leftarrow v$
10:   **else**
11:    **break**   ▷ walk failed.
12:   **end if**
13: **end while**

In every walk, each agent determines the step length d by the power-law distribution described in Eq. 1, and selects the orientation o of a walk randomly from $[0, 2\pi)$.

Each agent can obtain a set of neighbors $N(c)$ and a set of possible neighbors $PN(c) \subseteq N(c)$, to which agents can move, from the current node c. In other words, a node $x \in PN(c)$ has the link with c that the angle θ_{ox} between o and the link is smaller than δ which is a given parameter, called a *permissible error*.

Each agent selects the next node $v \in PN(c)$ which has the minimum θ_{ov} and move to v. Thus, each agent moves towards the determined o with δ, d times in a walk.

4.2 Random Walk

We now introduce several random walks which have been used to compare with Lévy flight.

The random walk originally came from the description of the random motion of particles. There are lots of variations; here we describe a simple implementation of random walks. In the simple random walk, agents move to a neighbor of the current node at uniformly random. Thus, the probability $Pr(u, v)_{RW}$ that each agent, which is located at the node u, moves to its neighbor v is defined as follows.

$$Pr_{RW}(u, v) = \begin{cases} \frac{1}{deg(u)} & (v \in N(u)) \\ 0 & (otherwise) \end{cases} \tag{2}$$

The hitting time and the cover time of the simple random walk is $O(n^3)$ [1]. This random walk is also called as Brownian walk.

5   Performance Evaluation

The performance evaluation aims to clarify the impact of RSUs as message caches on the efficiency of message delivery. We have conducted simulations by using The ONE simulator, which is a simulator for opportunistic network environments implemented in Java [8].

We assume a message is broadcasted by someone periodically and evaluate the average delivery ratio.

We assume the Helsinki city scenario (see Fig. 4) as a simulation field included in The ONE simulator.

Fig. 4. Helsinki city scenario in The ONE simulator

5.1   Parameters

We configure the following parameters for our simulations.

Communication range: The communication range of each pedestrian is 10 m. Two pedestrians can communicate each other when the distance between them is less than 10 m.

TTL (Time-to-Live): TTL is introduced to prevent exhausting storage resources in the network because the Epidemic protocol we assume is a flooding-style protocol which sends copies of all stored messages to all neighbors. We set TTL 300 min.

Num. of pedestrians n: It is the number of pedestrians located in a field. Each pedestrian is located at a unique position and the position of it is determined at random. We set $n = [10, 320]$.

Num. of RSUs r: The number of RSUs $r \in \{0, 50, 100, 200\}$. $r = 0$ means that there is no RSUs that temporally hold messages.

Scaling parameter λ: It is a parameter for the Lévy walk to determine the trajectory of agents. The ballistic trajectory is emphasized if λ increases. We set $\lambda = 1.2$ because it is efficient for a search on unit disk graphs [20] and it can cover a wider area on a graph.

5.2 Results

We measured the average delivery ratio and the average latency of message delivery by pedestrians based on Lévy walk and Random walk movement pattern. We analyze the impact of RSUs as message caches on the delivery ratio and the latency. We also compare Lévy walk and Random walk from these points of view. Each result is plotted as an average value of 100 simulation runs with 95% confidence interval.

5.2.1 Average Delivery Ratio

Figures 5 and 6 show the average delivery ratio on messages with the number of pedestrians based on Lévy walk and Random walk, respectively. Each line corresponds to the number of RSUs from 0 to 200.

Figures 7 and 8 show the average delivery ratio with the number of RSUs. Each line corresponds to the number of pedestrians $n = \{10, 20, 40, 80, 160, 320\}$.

According to the results shown in Figs. 5 and 6, they clarify the impact of RSUs as message caches on the average delivery ratio. The average delivery ratio significantly improves with RSUs in the case of $n \leq 40$ in both Lévy walk and Random walk.

Now, we compare the result with and without RSUs. In Fig. 5, it improves by 89.1%, 62.8% and 22.4% with 10, 20 and 40 pedestrians based on Lévy walk, respectively by comparing 50 RSUs and no RSU. On the other hand, Fig. 6 shows that it improves by 55.0%, 52.9% and 20.4% with 10, 20, and 40 pedestrians based on Random walk. According to the result, the impact of RSUs on the average delivery ratio is significant with few pedestrians. On the impact of the number of RSUs, the average delivery ratio improves from 5% to 20% by increasing the number of RSUs on both movement patterns.

Contrary to this, RSUs does not have an impact on the average delivery ratio with $n \geq 80$. It means that there are enough pedestrians to disseminate messages.

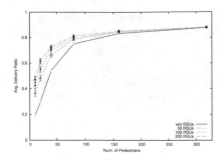
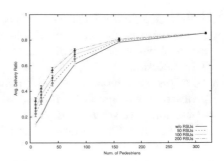

Fig. 5. Average delivery ratio v.s. the number of RSUs with various number of pedestrians based on Lévy walk.

Fig. 6. Average delivery ratio v.s. the number of RSUs with various number of pedestrians based on Random walk.

Figures 7 and 8 also prove that the impact of RSUs depends on the number of pedestrians. In the case of Lévy walk, the average delivery ratio improves by 36.8% on average by comparing (10, 20) pedestrians. It also improves by 41.2%, 20.7% on average by comparing (20, 40) and (40, 80) pedestrians. In the case of Random walk, the difference of improvement on the delivery ratio by the number of pedestrians is more significant than Lévy walk.

Moreover, these figures clarify the impact of the movement pattern on the average delivery ratio. Lévy walk outperforms by 49.2%, 46.0%, 37.5%, 17.7%, 5.8% and 2.9% in average regarding the average delivery ratio in comparison with Random walk with 10, 20, 40, 80, 160 and 320 pedestrians, respectively. The efficiency of message delivery by Lévy walk is emphasized with few pedestrians compared with Random walk. This result corresponds to the result by Shinki et al. [21].

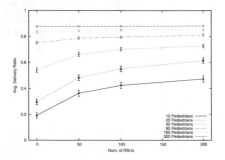
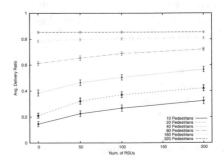

Fig. 7. Average delivery ratio v.s. the number of pedestrians with various number of RSUs

Fig. 8. Average delivery ratio v.s. the number of pedestrians with various number of RSUs

5.2.2 Latency of Message Delivery

We have measured the latency of message delivery. Figures 9 and 10 show the average latency on messages by pedestrians based on Lévy walk and Random walk, respectively. Number of pedestrians n is set $n = \{10, 20, 40, 80, 160, 320\}$. Each result is plotted as an average latency with 95% confidence interval.

Both figures have the same tendency to increase the latency and then reduce it according to the increment of n. However, there are overlaps among configurations of the number of RSUs. It means that it is difficult to say which one is better regarding the latency because of the few pedestrians (i.e., $n \leq 20$).

On the impact of RSUs on the latency of message delivery, Lévy walk movement pattern is significant compared to Random walk, especially with $n = \{40, 80\}$. The average latency of $n = 40$ and 80 with 50 RSUs improve 7.4% and 10.5% compared with the one with no RSU in the case of Lévy walk. On the other hand, they improve 4.8% and 4.2% in the case of Random walk.

On a smaller number of pedestrians (i.e., $n \leq 20$), it is difficult to identify one with another on the impact of RSUs on the latency.

The difference between configurations of the number of RSUs gets smaller according to the increment of the number of pedestrians. In particular, the impact of RSUs becomes negligible with $n \geq 160$ because a sufficient number of pedestrians for message delivery exist in the field.

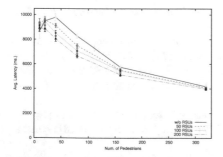

Fig. 9. Latency v.s. the number of pedestrians based on Lévy walk with various number of RSUs

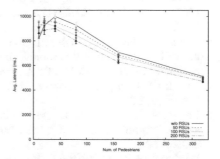

Fig. 10. Latency v.s. the number of pedestrians based on Random walk with various number of RSUs

6 Conclusion

In this paper, we assumed RSUs as message caches located in a city and evaluated the impact of RSUs on the message delivery ratio and the latency of messages delivery by opportunistic communication in DTNs that consist of human pedestrians. The impact of RSUs is significant on the average delivery ratio on both Lévy walk and Random walk, especially with a small number of pedestrians (e.g., $n \leq 40$). It means that RSUs help to distribute messages efficiently.

On the other hand, RSUs show the efficiency on the latency of message delivery with $40 \leq n < 160$. In particular, the difference of the latency with and without RSUs in Lévy walk is more significant than the one in Random walk.

We believe that this result would be useful for the advertisement for shops, events and public services in a city with mobile devices. We plan to develop an application specific protocol for such an advertisement service with RSUs in the future.

References

1. Aleliunas, R., Karp, R.M., Lipton, R.J., Lovasz, L., Rackoff, C.: Random walks, universal traversal sequences, and the complexity of maze problems. In: Proceedings of the 20th Annual Symposium on Foundations of Computer Science (SFCS 1979), pp. 218–223 (1976)
2. Alzoubi, K.M., Wan, P.J., Frieder, O.: Message-optimal connected dominating sets in mobile ad hoc networks. In: Proceedings of the 3rd ACM International Symposium on Mobile Ad Hoc Networking & Computing, MobiHoc 2002, pp. 157–164. ACM, New York (2002). https://doi.org/10.1145/513800.513820. http://doi.acm.org/10.1145/513800.513820
3. Balasubramanian, A., Levine, B., Venkataramani, A.: Dtn routing as a resource allocation problem. SIGCOMM Comput. Commun. Rev. **37**(4), 373–384 (2007). https://doi.org/10.1145/1282427.1282422. http://doi.acm.org/10.1145/1282427.1282422
4. Baldoni, R., Beraldi, R., Quema, V., Querzoni, L., Tucci-Piergiovanni, S.: Tera: Topic-based event routing for peer-to-peer architectures. In: Proceedings of the 2007 International Conference on Distributed Event-based Systems, pp. 2–13 (2007)
5. Birand, B., Zafer, M., Zussman, G., Lee, K.W.: Dynamic graph properties of mobile networks under levy walk mobility. In: Proceedings of the 2011 IEEE Eighth International Conference on Mobile Ad-Hoc and Sensor Systems, MASS 2011, pp. 292–301. IEEE Computer Society, Washington, DC (2011). https://doi.org/10.1109/MASS.2011.36
6. Bisnik, N., Abouzeid, A.A.: Optimizing random walk search algorithms in p2p networks. Comput. Netw. **51**(6), 1499–1514 (2007). https://doi.org/10.1016/j.comnet.2006.08.004
7. Buldyrev, S.V., Goldberger, A.L., Havlin, S., Peng, C.K., Simons, M., Stanley, H.E.: Generalized lévy-walk model for dna nucleotide sequences. Phys. Rev. E **47**(6), 4514–4523 (1993)
8. Desta, M.S., Hyytiä, E., Keränen, A., Kärkkäinen, T., Ott, J.: Evaluating (geo) content sharing with the one simulator. In: Proceedings of the 14th ACM Symposium Modeling, Analysis and Simulation of Wireless and Mobile Systems (MSWiM) (2013)
9. Dolev, S., Schiller, E., Welch, J.L.: Random walk for self-stabilizing group communication in ad hoc networks. IEEE Trans. Mob. Comput. **5**(7), 893–905 (2006). https://doi.org/10.1109/TMC.2006.104
10. Draief, M., Ganesh, A.: A random walk model for infection on graphs: Spread of epidemics & rumours with mobile agents. Discrete Event Dyn. Syst. **21**(1), 41–61 (2011). https://doi.org/10.1007/s10626-010-0092-5
11. Edwards, A.M., Phillips, R.A., Watkins, N.W., Freeman, M.P., Murphy, E.J., Afanasyev, V., Buldyrev, S.V., da Luz, M.G.E., Raposo, E.P., Stanley, H.E., Viswanathan, G.M.: Revisiting lévy flight search patterns of wandering albatrosses, bumblebees and deer. Nature **449**, 1044–1048 (2007)

12. Fujihara, A., Miwa, H.: Homesick lévy walk and optimal forwarding criterion of utility-based routing under sequential encounters. In: Proceedings of the Internet of Things and Inter-cooperative Computational Technologies for Collective Intelligence 2013, pp. 207–231 (2013)
13. Helgason, Ó., Kouyoumdjieva, S.T., Karlsson, G.: Opportunistic communication and human mobility. IEEE Trans. Mob. Comput. **13**(7), 1597–1610 (2014)
14. Ikeda, S., Kubo, I., Yamashita, M.: The hitting and cover times of random walks on finite graphs using local degree information. Theor. Comput. Sci. **410**(1), 94–100 (2009)
15. Kuhn, F., Wattenhofer, R.: Constant-time distributed dominating set approximation. In: Proceedings of the Twenty-second Annual Symposium on Principles of Distributed Computing, PODC 2003, pp. 25–32. ACM, New York (2003). https://doi.org/10.1145/872035.872040. http://doi.acm.org/10.1145/872035.872040
16. Lévy, P.: Théorie de L'addition des Variables Aléatoires. Gauthier-Villars, Paris (1937)
17. Mizumoto, N., Abe, M.S., Dobata, S.: Optimizing mating encounters by sexually dimorphic movements. J. Roy. Soc. Interface **14**(130), 20170086 (2017)
18. Nonaka, Y., Ono, H., Sadakane, K., Yamashita, M.: The hitting and cover times of metropolis walks. Theor. Comput. Sci. **411**(16–18), 1889–1894 (2010)
19. Rhee, I., Shin, M., Hong, S., Lee, K., Kim, S.J., Chong, S.: On the levy-walk nature of human mobility. IEEE/ACM Trans. Netw. **19**(3), 630–643 (2011). https://doi.org/10.1109/TNET.2011.2120618
20. Shinki, K., Hayashibara, N.: Resource exploration using lévy walk on unit disk graphs. In: The 32nd IEEE International Conference on Advanced Information Networking and Applications (AINA-2018). Krakow, Poland (2018)
21. Shinki, K., Nishida, M., Hayashibara, N.: Message dissemination using lévy flight on unit disk graphs. In: The 31st IEEE International Conference on Advanced Information Networking and Applications (AINA 2017). Taipei, Taiwan ROC (2017)
22. Spyropoulos, T., Psounis, K., Raghavendra, C.S.: Efficient routing in intermittently connected mobile networks: The multiple-copy case. IEEE/ACM Trans. Netw. **14**, 77–90 (2008)
23. Thejaswini, M., Rajalakshmi, P., Desai, U.B.: Novel sampling algorithm for human mobility-based mobile phone sensing. IEEE Internet Things J. **2**(3), 210–220 (2015)
24. Vahdat, A., Becker, D.: Epidemic routing for partially-connected ad hoc networks. Technical Report CS-2000-06, Duke University (2000)
25. Valler, N.C., Prakash, B.A., Tong, H., Faloutsos, M., Faloutsos, C.: Epidemic spread in mobile ad hoc networks: Determining the tipping point. In: Proceedings of the 10th International IFIP TC 6 Conference on Networking - Volume Part I, NETWORKING 2011, pp. 266–280. Springer-Verlag, Heidelberg (2011). http://dl.acm.org/citation.cfm?id=2008780.2008807
26. Viswanathan, G.M., Afanasyev, V., Buldyrev, S.V., Murphy, E.J., Prince, P.A., Stanley, H.E.: Lévy flight search patterns of wandering albatrosses. Nature **381**, 413–415 (1996)
27. Yang, X.S.: Cuckoo search via lévy flights. In: Proceedings of World Congress on Nature & Biologically Inspired Computing (NaBIC 2009), pp. 210–214 (2009)
28. Yang, X.S.: Firefly algorithm, lévy flights and global optimization. Res. Dev. Intell. Syst. **XXVI**, 209–218 (2010)

VLSI Implementation of K-Best MIMO Detector with Cost-Effective Pre-screening and Fast Sorting Design

Jheng-Jhan He and Chih-Peng Fan$^{(\boxtimes)}$

Department of Electrical Engineering, National Chung Hsing University,
Taichung 402, Taiwan, R.O.C.
cpfan@dragon.nchu.edu.tw

Abstract. For MIMO detections, the K-Best algorithm has been widely applied for multiple-antenna wireless communications. In this paper, to raise the throughput by the cost-effective architecture, the efficient pre-screening and fast sorting schemes are used for the proposed K-Best detector. At first, the pre-screening based enumeration decreases almost half number of leaf nodes for searching, and the searching number of leaf nodes is reduced by the pre-screening based scheme. Next, the applied fast sorting method reduces the hardware complexity in the sorting process. For VLSI realization, the developed MIMO detector is implemented by TSMC 90 nm CMOS technology. The throughput of proposed 4 × 4 K-Best detector achieves up to 4.4 Gbps at the 64QAM mode. Compared with previous K-Best hardware designs, the proposed design provides larger throughputs and performs higher hardware efficiency.

1 Introduction

Lately, the multiple-input multiple-output (MIMO) technology has been widely used for high-throughput and high-performance wireless communications. To boost the data throughput rate, many researchers applied multiple antennas at the transmitter and receiver sides, and the spectral efficiency was raised by spatial signal processing. Figure 1 illustrates the system model of MIMO communication systems. In the receiver, the MIMO detection technologies are applied to reconstruct and predict the received signals from the transmitter. Due to the limitation of hardware complexity and bit error rate (BER), the MIMO detector required to achieve a high throughput performance. In [1], the zero forcing (ZF) method only requires low computational load for signal detections. However, the ZF method will cause the noise effect in computational process, and the performance will be decreased. By the concept of the ZF method, the minimum mean square error (MMSE) method [1] is also developed to detect received signals by signal-to-noise ratios.

For MIMO signal detections, the maximum likelihood (ML) algorithm provides the optimal BER performance. However, to obtain the optimal maximum likelihood in signal detections, the computational complexity is very large when the high constellation modes are used for modulations. Besides, the computing load increases exponentially with the number of antennas. If the detector uses the full searching scheme for

© Springer Nature Switzerland AG 2020
L. Barolli et al. (Eds.): NBiS-2019, AISC 1036, pp. 178–188, 2020.
https://doi.org/10.1007/978-3-030-29029-0_17

signal detections, the hardware cost will be high. Therefore, the sphere decoding algorithm in [2, 3] are developed to reduce the highly computational complexity of the ML algorithm. By the searching boundary with the spherical radius R, the sphere decoding scheme decreases the searching nodes effectively. But, owing to flexible throughputs and long searching paths, the sphere decoding scheme is not suitable to be used for hardware design. To conquer the shortcoming of sphere decoding algorithm, the K-Best algorithm [4] is developed to detect the MIMO signals, and it uses the tree-based breadth-first searching procedure. The K-Best algorithm can be realized with the pipelined architecture since the algorithm utilizes a one-way searching flow and does not require any reverse searching path.

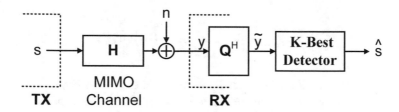

Fig. 1. The system model of MIMO communications

For K-Best algorithm, the applied K values mainly affects the complexity of the algorithm to compute the Euclidean distance (PED) and sorting process. When the K value becomes large, the number of leaf nodes, which are required for searching, also be increased largely, and the hardware complex is raised. In this paper, to develop a cost-effective and high-throughput K-Best detector by applying the pre-screening enumeration skill and fast sorting scheme, the hardware cost is decreased to achieve a high throughput. The rest of the paper is organized as follows. In Sect. 2, the well-known K-Best algorithm is briefly reviewed, and the applied pre-screening based enumeration and fast sorting scheme are described. In Sect. 3, the system simulation and architecture design of the proposed K-Best detector are revealed. The VLSI implementation results and comparison are shown in Sect. 4. Finally, a conclusion is given.

2   Proposed Architecture Design for K-Best Detection

2.1   Brief Review of K-Best Detection

The MIMO detector [4] at the receiver side is presented by

$$\|y - Hs\|^2 = \|\tilde{y} - Rs\|^2 = \sum_{i=M}^{1} \left\| \tilde{y} - \sum_{j=i}^{M} r_{i,j}s_j \right\|^2, \tag{1}$$

where y is the received vector signal, H is the MIMO matrix from the channel estimation in Fig. 1, R is the up triangular matrix by the QR decomposition, s is the transmitted vector signal, $\hat{s}$ is the estimation vector signal after the MIMO detection, and M is the total number of detection layers. In each layer, the partial Euclidean distance (PED) of the candidate nodes will be calculated. Define that $PED_L(s^i)$ is the PED in the layer L, $|e_i(s^i)|^2$ means the distance increment, and $PED_{M+1}(s^{M+1}) = 0$, where

$$e_i(s) = \left\| \tilde{y}_i \quad \sum_{j=i}^{M} r_{i,j} s_j \right\|^2 , \tag{2}$$

and Eq. (2) can be written as

$$e_i(s) = \left\| \tilde{y}_i - \sum_{j=i+1}^{M} r_{i,j} s_j - r_{i,i} s_i \right\|^2 \tag{3}$$

In the K-Best algorithm, the K candidate nodes in each layer can be sorted out by the PED values. When the K value becomes large, the BER performance of the K-Best detection approaches that of the ML algorithm. But, the K-Best detector with large K values needs to keep a large number of candidate child nodes and the corresponding PED values. Thus, how to reduce the computational complexity and hardware cost is an important issue when the large K is used in the K-Best algorithm design.

2.2 Proposed Cost-Effective Pre-screening and Sorting Based K-Best Detector

In the K-Best algorithm, when the number of candidate leaf nodes is increased, the sorting process will require a large number of computations, and the hardware cost is also raised. Therefore, the sorting process needs more PED computations. To decrease the number of candidate nodes and keep the suitable BER performance, the pre-screening based enumeration method [10] is applied for the design of proposed K-Best detector. In the direct K-Best processing, the PEDs of all child nodes in each layer need to be calculated. To search candidate child nodes, the PED values of these selected child nodes must be less than those of the other nodes. Then the selected candidate nodes are extended to be as the parent nodes of the next detection layer. The computation in (3) can be described as

$$L_i = \tilde{y}_i - \sum_{j=i+1}^{M} r_{i,j} s_j, \; F_i = r_{i,i} s_i \tag{4}$$

By (4), the center point C_i for detections is located at

$$C_i = \frac{L_i}{r_{i,i}}, \tag{5}$$

where $r_{i,i}$ is the diagonal element of the R matrix, which is generated by QR decomposition. By (4) and (5), the system pre-screens several child nodes to constrain the searching extension to the next layer.

In previous designs, the computational order of PEDs is decided by the direct enumeration. By the direct 1-D enumeration, all child nodes are required to be selected for PED computations. Table 1 lists the conditions for the pre-screening based enumeration at 64QAM. In Table 1, it means that the far child nodes from the center point do not need to be considered as candidates. By using the pre-screening and the 1-D enumeration technologies, more impossible child nodes can be omitted by the enumeration order. By the effective pre-screening based enumeration scheme, the system efficiently predicts the number of the child nodes, which are extended from the parent node, and the number of the candidate child nodes is decreased effectively. Figure 2 reveals the simulation results of the BER performances in 4 × 4 MIMO systems at 16QAM with and without the pre-screening based enumeration. In Fig. 2, "PD-2Q" means that the number of the pre-screening nodes is 2, and "PD-4Q" means that the number of the pre-screening nodes is 4. By simulations, the BER performance of the proposed PD-2Q design at the 16QAM mode is close to that of the direct K-Best design without the pre-screening scheme. Generally, when the complexity of the modulation mode becomes large, the number of pre-screening nodes can also be raised. By the applied pre-screening scheme, the detector only enumerates a few child nodes, which own the high possibility to be extended. Therefore, almost half of required visiting child nodes are reduced for detections, and the following sorting process reduces the corresponding operations to cut down the hardware cost and enhance the hardware efficiency.

Table 1. The conditions for the pre-screening based enumeration at 64QAM

64QAM	
Conditions	Order of enumerations
$C_i \leq -6$	$\{-7,-5,-3,-1,1,3,5,7\}$
$-6 < C_i \leq -5$	$\{-5,-7,-3,-1,1,3,5,7\}$
$-5 < C_i \leq -4$	$\{-5,-3,-7,-1,1,3,5,7\}$
$\vdots$	
$-2 < C_i \leq -1$	$\{-1,-3,1,-5,3,-7,5,7\}$
$-1 < C_i \leq 0$	$\{-1,1,-3,3,-5,5,-7,7\}$
$C_i > 6$	$\{7,5,3,1,-1,-3,-5,-7\}$
$6 \geq C_i > 5$	$\{5,7,3,1,-1,-3,-5,-7\}$
$5 \geq C_i > 4$	$\{5,3,7,1,-1,-3,-5,-7\}$
$\vdots$	
$2 \geq C_i > 1$	$\{1,3,-1,5,-3,7,-5,-7\}$
$1 \geq C_i > 0$	$\{1,-1,3,-3,5,-5,7,-7\}$

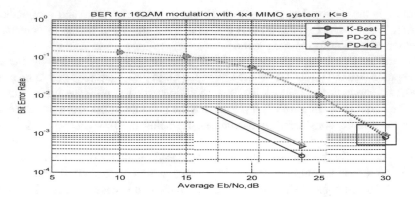

Fig. 2. Simulations of BER performances at 16QAM in 4 × 4 MIMO

In [6], Batcher's Odd-Even Sorting Algorithm (BOESA) is utilized for fast sorting process. In each layer of K-Best detection, the K sub-nodes are needed to be sorted out as the parent nodes of the next layer for successive searching extensions. To sort out the K minimum outputs, the detector may do the whole sorting operations to generate the correct sorted outputs by the BOESA scheme. However, the detector only require to sort out the K minimum outputs, and it does not need to sort out the whole correct-order output sequence. Thus, the simplified Batcher's Odd-Even Sorting Algorithm (SBOESA) [10] is used to reduce the number of comparators effectively, compared with the previous BOESA method. For example, in Fig. 3(b), when the detector is arranged to sort out 4 possible minimum outputs, the paths, marked by the red color,

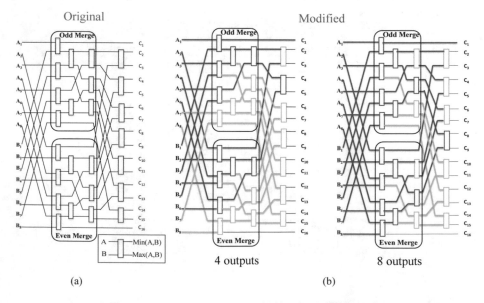

Fig. 3. (a) BOESA (b) Simplified BOESA (SBOESA)

Table 2. Comparison of different sorting schemes

Sorting algorithm	16 inputs	32 inputs	64 inputs
Parallel bubble sort	120	496	2016
BOESA	63	191	525
4 outputs (K = 4)			
Applied SBOESA	45	101	213
8 outputs (K = 8)			
Applied SBOESA	53	131	287

notes the required hardware part for sorting, and the paths, marked by the green dashed-lines, are not required because only 4 minimum outputs are needed instead of 16 outputs. In Fig. 3(a), the previous BOESA design needs 25 comparators. However, in Fig. 3(b), the applied SBOESA scheme only needs 11 comparators. Thus, more than half of comparators are decreased by the applied SBOESA sorting scheme. Table 2 shows the comparison of different sorting schemes. For 4 minimum outputs, in Table 2, the applied sorting architecture only needs 45 comparators for sorting the random-order inputs. To sort the K minimum outputs, the used SBOESA scheme requires a smaller number of comparators compared with the previous BOESA method.

3   Performance Simulation and Architecture Design of the Proposed K-Best Detector

The traditional K-Best algorithm uses a fixed K value. The K mother nodes are fixedly selected on the screening survival path of each layer, and these nodes continue to extend the child nodes. After QR decomposition, the expected values between the diagonal elements show that the channel gain increases with the number of detection layers. Thus, the K-Best algorithm can apply different K values to achieve the approximate the performance of using a fixed K value in each layer. Except that the first layer K value is equal to the total number of constellation points, the K values in layers can be designed with a decreasing order. Because the proposed hardware is a pipelined architecture, and the proposed design can use different combinations of hardware designs in each layer. The hardware design of the proposed 4 × 4 MIMO system in each layer are described as follows.

Since the proposed hardware is designed for 4 × 4 MIMO systems, the K values have 8 levels that can be determined, and the highest modulation mode in our system is 64QAM. In the system simulation, the used K value is up to 8. To achieve the suitable performance, the candidate parent nodes can be reserved as many as possible in the first few layers, and then the higher K value is selected. In the later layers, due to the preliminary screening in previous layers, the system reduces the K value to gradually find out the best path. By the simulation, we combines different K values to compare which combination is closer to the performance by using a fixed value in each layer. Figure 4 shows the comparison of variable K values in layers for 4 × 4 MIMO at 64QAM. The performance with the combination {88884441} is better than that with the combination {88844221}. However, the hardware complexity with the combination {88884441} is higher than that with the combination {88844221}. Besides, the operational speed with the combination {88884441} is less than that with the combination {88844221}. Therefore, the combination of K values with {88844221} is applied for the proposed hardware design. After the suitable combination of K values is decided, the hardware in different detection layers can be designed to achieve the hardware optimization.

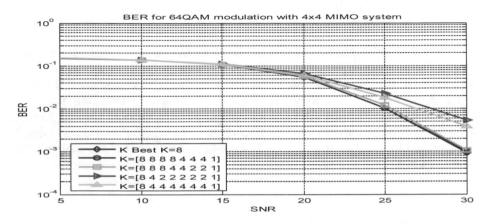

Fig. 4. Comparison of different variable K values for 4 × 4 MIMO at 64QAM

The 8th layer is the first layer of the K-Best detection. In this layer, the PED of all parent nodes are calculated, and it does not require the pre-screening enumeration and the sub-node sorting. The first layer is the last detection layer, and the K-Best detection ends up with the output results, which are closest to the transmitter signals. Then the system compares a path with the smallest PED. In the 7th to 4th layers, the K values are set at 8 and 4, respectively. Since the K value determines the number of child nodes to be sorted, and the data sequence of more ordered nodes is needed, and then the applied SBOESA method provides better performance. When the number of sorted nodes is less, the efficiency of parallel sorting is greater than that of the applied SBOESA method. Thus, the system adopts the SBOESA method in the 7th-5th layers, and uses the parallel sorting method in the 4th-2th layers. In the 2th layer, the K value is set at 2, and the PED in this layer is calculated by counting the two nodes listed in the Pre-Detection (PD), and the system adopts the parallel sorting scheme.

By the variable K-value design, the hardware complexity of the proposed MIMO detector is decreased. Table 3 lists the variable K-value design in eight layers for the proposed hardware design. By the variable K-value scheme, the total number of searching nodes is reduced efficiently, and the applied pre-screening and SBOESA schemes drops the hardware cost further. To raise the throughput, the pipelined and parallel architecture is utilized to implement the proposed K-Best detector. The proposed MIMO detector has four functional modules, which include Selection Constellation Point (SCP), Pre-Detection (PD), Partial Euclidean Distance (PED), and Sorting

(ST) modules. Figure 5 shows the architecture of the proposed K-Best detector. In each detection layer, the four functional module are replicated and processed by parallel and pipelined computations.

Table 3. Variable K-value scheme for the proposed design

	K value	Pre-screening (No. of enumerations)	Sorting method	No. of computed nodes
Layer 8	8	N.A.	N.A.	8
Layer 7	8	4	Simplified Batcher's OESA sorting	32
Layer 6	8	4		32
Layer 5	4	4		32
Layer 4	4	4	Parallel sorting	16
Layer 3	2	2		8
Layer 2	2	2		4
Layer 1	1	1	Direct comparison	2

Before hardware implementation, the designer must do a fixed-point simulation analysis. Figure 6 illustrates the system simulation and performance comparison between the floating-point and fixed-point models at the 16/64 QAM modes. It is necessary to use several bits to evaluate the hardware design to get the system performance, which is close to the original floating-point model. By system simulations via Gaussian channel with white noise, the fixed-point system model gets the quantized results of each module to determine the bit widths of each module. Table 4 shows the used bit-width information in the modules of the proposed hardware design.

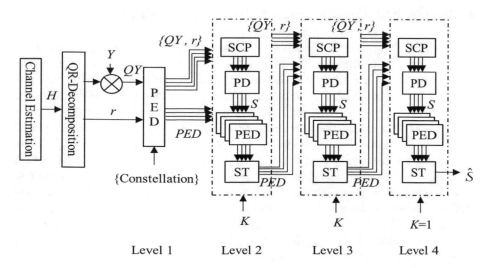

Fig. 5. The designed K-Best architecture is similar to that in [10]

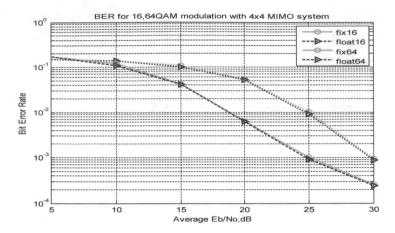

Fig. 6. System simulations between the floating-point and fixed-point models at the 16/64 QAM modes

Table 4. Bit-width information for the proposed hardware design

	Word length/Integer/Fraction
$\tilde{y}$	16 bits/6 bits/9 bits
R	13 bits/3 bits/9 bits
PD	14 bits/4 bits/9 bits
PED	14 bits/4 bits/9 bits
ST	14 bits/4 bits/9 bits

4   Results and Comparisons of VLSI Implementation

At first, the floating-point MATLAB model is used to verify the system performance at the algorithm level. Next, the fixed-point MATLAB model is followed to meet the BER performance. Then, the fixed-point MATLAB model is converted to the Verilog HDL hardware model. To implement the proposed hardware, the cell-based design flow with TSMC 90-nm CMOS technology is used for functional simulation, gate-level timing simulation, and hardware implementation. At the 64QAM mode, the proposed 4×4 MIMO detector requires 345 gate counts, and the design operates up to 184 MHz operational frequency. By applying the pre-screening enumeration and fast sorting schemes at 64QAM, the throughput of the proposed 4×4 K-Best detector achieves up to 4.4 Gbps (i.e. 4 symbols/cycle $\times$ 6 bits/symbol $\times$ 184 MHz). Besides, the normalized hardware efficiency (NHE) also reaches 12.8 Mbps/K gates. Table 5 lists the comparison of VLSI results among different K-Best detector designs. Compared with the previous K-Best detector designs [7, 9, 11], the proposed detector provides larger throughputs and performs better hardware efficiency.

Table 5. Comparison of VLSI results among different K-Best detector designs.

MIMO detectors	[9]	[7]	[11]	Proposed
Antenna	2×2 to 4×4	4×4	8×8	4×4
Modulation	QPSK to 64QAM	64QAM	64QAM	64QAM
Algorithm	Early-pruned K-Best	DKB	Modified K-Best	Modified K-Best
K Value	10	10	8	≈ 8
Model (Real/Complex)	Real	Real	Complex	Real
Output type	Hard	Hard	Hard	Hard
Technology	0.13 μm	0.13 μm	45 nm	90 nm
Gate count (KG)	491	114	63.75	345
Max. clock rate (MHz)	138	282	181.8	184
Max. throughput (Mbps)	1100	675	1090.8	4416
Power (mW)	127	135	782	189
NHE (Mbps/KG)	3.2	8.6	8.55	12.8

$$Throughput = \frac{2M \log_2^N}{t_{clk}} \times \text{maximum clock rate},\ NHE = \frac{\text{Throughput} \cdot (\frac{\text{Process}}{90})}{\text{Gate count}}\ [13].$$

5 Conclusion

To raise the throughput with the cost-effective architecture, in the paper, the efficient pre-screening and fast sorting schemes are used for the proposed K-Best detector. Firstly, the pre-screening based enumeration decreases almost half number of leaf nodes for searching, and the searching number of leaf nodes is reduced by the applied pre-screening based scheme. Secondly, the fast sorting method reduces the hardware complexity in the sorting process. By TSMC 90 nm CMOS process, the proposed MIMO detector is implemented for VLSI design. The throughput of the proposed 4 × 4 K-Best detector achieves up to 4.4 Gbps at 64QAM mode. Compared with previous K-Best hardware designs, the proposed design provides larger throughputs and performs superior hardware efficiency.

In the paper, for 4 × 4 MIMO at 64QAM, assuming K value is 8 in each layer, and there are 456 nodes that need to be visited. However, only 232 nodes need to be visited through the applied pre-processing scheme, and approximately 49% reduction in node visiting is equivalent to reducing nearly half of the PED calculations. It effectively reduces the hardware complexity of the K-Best algorithm. Besides, compared with the original Batcher's sorting design, the applied SBOESA sorting method reduces the average number of required comparators by 45%. The next-generation multi-antenna communication systems must provide a high-speed data transmission rate, and high throughput is a very important issue. In future works, the proposed hardware for MIMO detections will be extended to support massive multiple-antenna MIMO systems.

Acknowledgments.. This work was supported by the Ministry of Science and Technology, Taiwan (R.O.C.). The authors thank the National Chip Implementation Center in Taiwan for EDA supports.

References

1. Cho, Y.S., Kim, J., Yang, W.Y., Kang, C.G.: MIMO-OFDM Wireless Communications with MATLAB. John Wiley & Sons Inc, Singapore (2010)
2. Hassibi, B., Vikalo, H.: On the sphere decoding algorithm I. Expected Complexity. IEEE Trans. Signal Process. **53**(8), 2806–2818 (2005)
3. Agrell, E., Eriksson, T., Vardy, A., Zeger, K.: Closet point search in lattices. IEEE Trans. Inform. Theory **48**(8), 2201–2214 (2002)
4. Wong, K., Tsui, C., Cheng, R., Mow, W.: A VLSI architecture of a K-best lattice decoding algorithm for MIMO channels. Int. Symp. Circ. Syst. **3**, 273–276 (2002)
5. Liao, C.H., Wang, T.P., Chiueh, T.D.: A 74.8mW soft-output detector IC for 8 × 8 spatial-multiplexing MIMO communications. IEEE J. Solid-State Circ. **45**(2), 411–421 (2010)
6. Yazdi, S., Kwasniewski, T., Yazdi, S., Kwasniewski, T.: Configurable K-best MIMO detector architecture. In: 3rd International Symposium on Communications, Control and Signal Processing, pp. 1565–1569 (2008)
7. Shabany, M., Gulak, P.G.: A 0.13 μm CMOS 655Mbs/4 × 4 64-QAM K-Best MIMO detector. In: IEEE International Solid-State Circuits Conference, pp. 256–257 (2009)
8. Mondal, S., Eltawil, A., Shen, C.A., Salama, K.N.: Design and implementation of a sort-free K-best sphere decoder. IEEE Trans. VLSI Syst. **18**(10), 1497–1501 (2010)
9. Liu, L., Ye, F., Ma, X., Zhang, T., Ren, J.: A 1.1-Gb/s 115-pJ/bit configurable MIMO detector using 0.13-μm CMOS technology. IEEE Trans. Circ. Syst. II **57**(9), 701–705 (2010)
10. He, J.J., Fan, C.P.: Design and VLSI implementation of novel pre-screening and simplified sorting based k-best detection for MIMO systems. In: The 2015 International Symposium on VLSI Design, Automation and Test, Hsinchu, Taiwan (2015)
11. Rahman, M., Choi, G.S.: Hardware architecture of complex K-best MIMO decoder. Int. J. Comput. Sci. Secur. (IJCSS) **10**(1), 1–13 (2016)
12. Burg, A., Borgmann, M., Wenk, M., Zellweger, M., Fichtner, W., Bolcskei, H.: VLSI implementation of MIMO detection using the sphere decoding algorithm. IEEE J. Solid-State Circ. **40**, 1566–1577 (2005)
13. Studer, C., Benkeser, C., Belfanti, S., Huang, Q.: A 390 Mb/s 3.57 mm2 3GPP-LTE turbo decoder ASIC in 0.13 μm CMOS, In: IEEE Solid-State Circuits Conference, pp. 274–275 (2010)

Meal Information Recognition Based on Smart Tableware Using Multiple Instance Learning

Liyang Zhang$^{(\boxtimes)}$, Kohei Kaiya, Hiroyuki Suzuki, and Akio Koyama

Departmnet of Informatics, Graduate School of Science and Engineering,
Yamagata University, 4-3-16 Jonan, Yonezawa-shi, Yamagata 992-8510, Japan
{ttf04338, tsn82821}@st.yamagata-u.ac.jp,
{shiroyuki, akoyama}@yz.yamagata-u.ac.jp

Abstract. In recent years, people are paying more and more attention to the management of healthy meals because of the influence of lifestyle-related diseases, and some diet management systems are trying to help people lead a healthy life. Moreover, some studies have found that the proper meal habits can play a role in preventing disease to some extent. This paper introduces the smart tableware consisting of an acceleration sensor and a pressure sensor, which is used to obtain information about the sequence and content of meals. In addition, a method of analyzing and processing meal information through Multiple Instance Learning (MIL) is proposed to help people prevent diseases that are affected by lifestyle habits. At the same time, the acquisition process of MIL dataset is introduced and Support Vector Machine (SVM) is used. The performance evaluation results show that good results can be achieved by using MIL.

1 Introduction

Lifestyle disease is defined as a disease associated with the way a person or group of people lives. Lifestyle diseases include atherosclerosis, heart disease and stroke, obesity and type 2 diabetes, and diseases related to smoking, alcohol and drug abuse [1]. The world's obese population has almost tripled since 1975. In 2016, 39% of adults aged 18 and over were overweight and 13% were obese [2]. Overweight and obesity are major risk factors for a number of chronic diseases, including diabetes, cardiovascular diseases and cancer [3]. Diabetes caused 1.6 million deaths worldwide in 2016, compared with less than 1 million in 2000 [4]. Cardiovascular diseases kill 17.9 million people every year, accounting for 31% of global deaths. These diseases are caused by smoking, unhealthy diets, lack of physical activity and harmful use of alcohol, which in turn are manifested by raised blood pressure, elevated blood glucose, overweight and obesity [5].

Information systems related to people's behavioral habits are becoming popular in recent years. These systems collect important life log information through sensors and other devices to guide people to live a healthy life. Some systems that use wearable sensors have been proposed to capture people's behavior in life. There has system that wearable sensors are placed below the outer ear to detect the characteristic jaw motion during the chewing and to detect the frequency of chewing [6, 7]. In order to monitor the intake behavior automatically and accurately, at the same time, identify the food

© Springer Nature Switzerland AG 2020
L. Barolli et al. (Eds.): NBiS-2019, AISC 1036, pp. 189–199, 2020.
https://doi.org/10.1007/978-3-030-29029-0_18

intake of active users, Farooq et al. proposed a new wearable device that includes an accelerometer, a piezoelectric strain sensor and a data acquisition module to detect food intake even when the user is active or talking [8]. A watch-like configuration of sensors was proposed to continuously track wrist motion throughout the day and automatically detect the time of eating. Dong et al. found a correlation between eating activity and wrist motion, and tracked the linear and rotational motion of the wrist by wearing a watch-like configuration of accelerometers and gyroscopes [9].

Meanwhile, there are systems that support personal dietary life by recording meal content and using image recognition. FoodLog is a photo-based multimedia system that allows users to record their food intake by taking meal photos [10, 11]. PlateMate is also a system that allows users to take photos of meal [12]. These systems can record meal content or calories while helping people manage their daily health and monitor their progress on dietary goals. However, it is time-consuming and inaccurate to perform food logging by manually entering meal information, taking photos or self-reporting. Moreover, there is a general problem with these systems that it is not easy to obtain information such as the sequence during the meal. Studies have shown that intake of vegetables, whey protein and olive oil before intake of carbohydrates can improve postprandial glucose excursions in type 2 diabetes [13]. Therefore, in order to be healthy, we need to pay attention to the sequence of meal, and a reasonable sequence of meal can control or prevent diseases such as diabetes to a certain extent.

In our previous studies, we implemented the meal information collection system that collects meal information such as meal sequence, meal time and meal contents automatically by using the IoT wireless tag attached to tableware [14]. In order to obtain more information about the status of smart tableware, we used multiple sensors to collect meal information [15]. We hope that through such research, we can provide reasonable meal advice to help people prevent lifestyle-related diseases.

This paper introduces the process of using smart tableware equipped with an acceleration sensor and a pressure sensor to obtain meal information in the process of meal. At the same time, a method of analyzing and processing data using MIL is proposed to predict the sequence and time of meal of smart tableware. The process of using MIL is introduced in detail and examples are given to illustrate it. Finally, the experimental results are given and the performance evaluation is carried out.

The structure of this paper is as follows. In Sect. 2, we introduce the proposed method in detail. We discuss the performance evaluation in Sect. 3. Finally, some conclusions are given in Sect. 4.

2   Methodology

Existing studies have found that meal sequence and meal time affect the blood glucose levels [13, 16]. In order to collect information such as meal sequence and meal time, we use smart tableware composed of an acceleration sensor and a pressure sensor to collect meal information.

Here, we judge the meal content according to the sensor ID, and collect the movement status and weight information of the tableware according to the data obtained by sensors. However, the data obtained cannot be used directly. We need to extract useful feature information, such as meal sequence and meal time. Then, machine learning algorithms are used to determine whether food is eaten in a certain period of time. However, in our experiments, we found that unambiguously labeled positive and negative instances are not easy to obtain, we propose using MIL to solve it and introduce it in detail.

2.1 Smart Tableware

We use an acceleration sensor and a pressure sensor to capture the information of smart tableware, as shown in Fig. 1. For the acceleration sensor, "IoT wireless tag TWE-Lite-2525A" manufactured by Mono Wireless Inc. is used to obtain acceleration information. For the pressure sensor, "pressure sensor AS-FS" manufactured by AsakuasGiken Co., LTD. is used to obtain pressure information. We set up smart tableware by embedding acceleration sensor in tableware and placing tableware on the pressure sensor, as shown in Fig. 2. These sensors wirelessly transmit data related to the status of tableware.

Fig. 1. An acceleration sensor and a pressure sensor.

So far, we have considered using 3 types of information: sensor ID, acceleration information and pressure information. Since each tableware is equipped with a sensor of a different sensor ID, one tableware corresponds to one sensor ID, and here we assume that only one kind of food is contained in one tableware. If we assign food to each tableware in advance, the food corresponds to a sensor ID naturally. Through the movement of tableware obtained by acceleration sensor and the change of weight of tableware obtained by pressure sensor, we can grasp the state of the tableware, and combined with the sensor ID, the information about meal time and meal sequence can be judged.

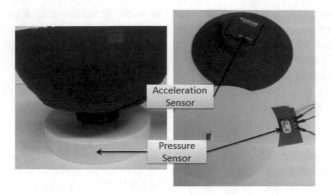

Fig. 2. The structure of the smart tableware.

2.2   Data Conversion

It is not easy to use the information obtained by sensors directly. We need to analyze them and convert data into what we need. We use information such as sensor ID, acceleration information and pressure information currently. According to the food assigned in advance, the sensor ID will correspond to the tableware. The information obtained from the acceleration sensor includes "Timestamps", "Tag ID", "Action Mode" and so on (Fig. 3). "Action mode" means that the object moves or remains stationary. The acceleration information shows the process of tableware movement. Information such as "Timestamps", "Tag ID", and "Voltage Value" can be obtained from the pressure sensor, as shown in Fig. 4. The change of "Voltage Value" reflects the change of object weight. The pressure information shows the change in the weight of tableware. Through the combination of acceleration information and pressure information, the state of the entire tableware can be grasped. The meal time and sequence are obtained based on the information.

;585;00000000;159;146;1015ee5;3050;0008;0000;1367;0723;X;-004;0003;-103;
Timestamps   Tag ID   Action Mode

Fig. 3. Log example of "IoT wireless tag TWE-Lite-2525A".

;588;00000000;195;108;10f1780;3050;1391;9084;1514;1391;S;
Timestamps   Tag ID   Voltage Value

Fig. 4. Log example of "Pressure sensor AS-FS".

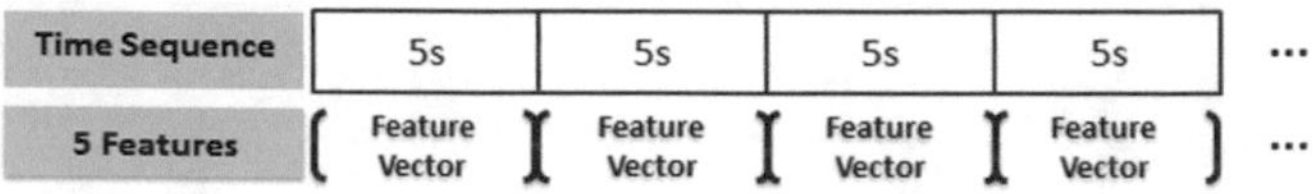

Time Sequence	5s	5s	5s	5s	...
5 Features	Feature Vector	Feature Vector	Feature Vector	Feature Vector	...

Fig. 5. An example of feature extraction based on time sequence.

Here, based on the experience gained from the experiment, we use 5 features and perform feature extraction every 5 s in time sequence according to the information obtained from sensors (Fig. 5). These 5 features include 2 features extracted from acceleration information and 3 features extracted from pressure information. The 2 acceleration features are "the detection count above the threshold" and "the continuous number above the threshold", respectively. These acceleration features are used to judge whether the tableware is moving and whether it is in continuous motion. The 3 pressure features are "pressure difference (voltage difference)", "continuous number above pressure difference average" and "deviation rate of moving average", respectively. These pressure features show the process and trends of food weight changes. By using these features, combined with MIL we can classify "eat" or "not eat".

2.3 MIL

The framework of MIL was proposed by Dietterich et al. [17] to solve drug activity predictions initially. Since then, MIL have been developed and applied in many fields. There are many studies based on MIL, which describes the real world object as a set of instances. In MIL, the training set consists of a number of bags, and each bag contains a number of instances. The label is associated with a bag, and the instances have no labels. Based on the description of MIL, each positive bag contains at least one positive instance. That is to say, the positive bag may contain negative instances, besides positive instances. Furthermore, all instances in negative bags are negative. The aim of MIL is to train a MIL classifier on the labeled bags and use the classifier to predict unknown bags. Note that although the training bag is labeled, the label of instance is unknown.

The Reason for Using MIL. There are many situations where it is difficult to distinguish between positive and negative categories in reality. In our research, when using chopsticks to pick up food from a tableware, the weight of food obtained by the smart tableware is changing. At this time, the food is only picked up through the chopsticks, but the food may not be delivered to the mouth. When we move the tableware toward the mouth, the acceleration sensor has already sensed the movement of the tableware, but the food may not have been delivered into mouth, and so on.

In the process of labeling, we found that it is not easy to define strictly whether an action is related to "eat" or "not eat". As shown in Fig. 6, we found that during the meal, the following 3 actions may be involved: (I) start picking up food from tableware with chopsticks; (II) delivering food to the mouth; (III) chewing. In these 3 actions, we are sure that the food is eaten in (III), which can be labeled "eat", but what about (I) and (II)? Without (I) and (II), food is not easy to enter the mouth.

Fig. 6. An example of actions during meals.

Therefore, some forms of weak supervision such as MIL are quite available. MIL considers a special form of weak supervision, in which the training set consists of a set of positive bags and a set of negative bags. If the bag contains at least one instance related to the subject of interest, the bag is classified as positive, and if the bag does not contain any instance related to the subject of interest, the bag is classified as negative. Based on the description of the MIL, we put the instance related to the subject of interest "eat" into the positive bag, and put the instance unrelated to "eat" into the negative bag, as shown in Fig. 7. We will introduce the specific process next.

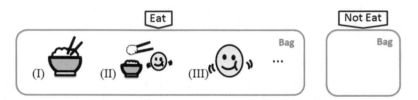

Fig. 7. An example of bags and instances.

Form the MIL Dataset. In order to describe the process of eating, we put such a process into a positive bag that is, from the beginning of a bite of food intake to the end of chewing, and label the bag as "eat". The food intake cycle [18] is generally defined as initially picking up the food from the tableware by manipulating a utensil like chopsticks or a fork, then continuing to move up to the mouth with the hand so that the food can enter the mouth, and finally moving the utensil away from the mouth with the hand. For a negative bag, it consists of an instance without any eating action.

Figure 8 shows an example of a positive bag and a negative bag. For example, eating a bite of rice involves the process of using chopsticks to take food from tableware to put it in the mouth until the end of chewing or to start a new bite of food intake. In the bag labeled "eat", we can see that the 1st period contains the process of getting rice from tableware, the 2nd period includes the process of delivering food to the mouth and starting to chew, and the 3rd and 4th periods are chewing. This is just an example of a positive bag. We treat it as a negative bag for the process of no food intake or chewing.

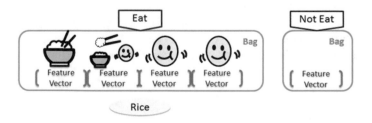

Fig. 8. An example of a positive bag and a negative bag.

Assume that the 40-s meal process is shown in Fig. 9. According to the classification of "eat" or "not eat", it is judged that there is no food is eaten in the 5th period, and rice is eaten in the 1st to 4th periods, miso soup is eaten between the 6th and 8th periods. In this case, the meal time of rice is 20 s, and the meal time of miso soup is 15 s. The sequence of the meal can be considered as eating rice first and then eating miso soup. In this way, the meal sequence and meal time can be obtained, which is information during the meal.

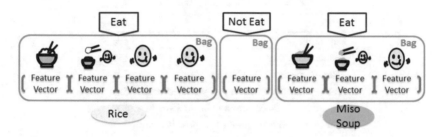

Fig. 9. An example of meal sequence and meal time.

SVM Algorithms for MIL. SVM is a supervised learning algorithm, while MIL considers a special form of weak supervision. In traditional supervised learning, an object is represented by an instance (i.e., a feature vector) and associated with a label, while in MIL, an object is described by a number of instances and associated with a label [19]. Therefore, the MIL data set needs to be adjusted so that SVM can be used.

Here, we use 2 SVM algorithms for MIL: NSK (Normalized Set Kernel) and SIL (Single Instance Learning) [20]. A bag is represented by the sum of all instances in it, in the description of NSK, normalized by its 1 or 2-norm, and then trained by SVM. In order to solve MIL with SIL, we need to apply bag label to all instances in it and convert the MIL data set into the standard supervised representation, then SVM is used.

3 Evaluation

We use smart tableware for the meal experiments, and assume that one food corresponds to a smart tableware. In order to facilitate labeling, we recorded the meal process with a video camera. For the data obtained from the smart tableware, the different foods are distinguished based on the sensor ID, features are extracted and MIL is used.

Here, we use F-score to evaluate performance, as shown in Eq. (3). The true positive Tp means the number of bags correctly classified as "eat", $Tp + Fp$ is the number of bags predicted to be "eat", $Tp + Fn$ describes the number of bags actually classified as "eat".

$$\text{Precision} = Tp/(Tp + Fp) \tag{1}$$

$$\text{Recall} = Tp/(Tp + Fn) \tag{2}$$

$$\text{F-score} = 2 \cdot \text{Precision} \cdot \text{Recall}/(\text{Precision} + \text{Recall}) \tag{3}$$

Table 1.   Details of the data set

Attribute	Value
Bag number	406
Instance number	1125
Average instance number in bag	2.8
Average instance number in positive bag	5.2
Maximum instance number in positive bag	13
Minimum instance number in positive bag	2
Average instance number in negative bag	1

To facilitate performance evaluation, we conducted a total of 10 meal experiments, performed data conversion and used MIL as described in the paper. For information on the data set, see Table 1. We can see that through the 10 meal experiments, we get a total of 406 bags and 1125 instances. Each positive bag contains an average of 5.2 instances and each bag contains an average of 2.8 instances. The maximum number of instances in a positive bag is 13 and the minimum is only 2. In the experiment, we used 10-fold cross-validation and used SVM (SMO) [21] training classifier. The 2 SVM algorithms for MIL are used, and results of F-score are compared. Figures 10 and 11 show the F-score of NSK and SIL, respectively, as well as the results of precision and recall (Table 2).

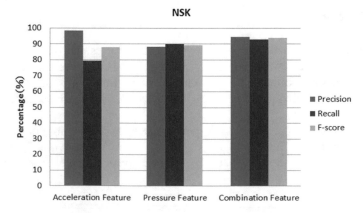

Fig. 10. The percentage results of NSK.

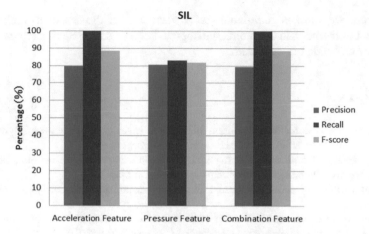

Fig. 11. The percentage results of SIL.

Table 2. Results of SVM algorithms for MIL based on combination feature

	NSK	SIL
Precision	94.6%	79.5%
Recall	93.0%	99.8%
F-score	93.8%	88.5%

According to the experimental results, we can see that the F-score of combination feature of NSK obtains 93.8%, while SIL does not reach 90%. Compared with SIL, NSK achieves better experimental results. This shows that in order to describe meal information, it is better to use the sum of all instances to represent the bag than give the bag label to all instances in it. For the acceleration feature, it can be seen that the F-score of NSK and SIL are similar, while the pressure feature is different obviously. In the pressure feature of NSK, not only F-score, but also precision and recall approximate 90%, while SIL is higher than 80% slightly. We find that the pressure feature play a better role in NSK than SIL.

4 Conclusion

This research aims to help people manage their meals effectively. This paper introduces the composition of smart tableware and the process of collecting meal information. At the same time, a new method using MIL is proposed, and the feasibility of method is proved by experiments. In the experiment, 2 SVM algorithms for MIL are compared, and NSK achieves better experimental result in F-score.

In the future, we will consider using more features to describe the state of tableware, not just the acceleration information and pressure information that has been used.

In addition, we consider improving the performance of the sensor to make it more sensitive. Furthermore, in the future, we hope to achieve independent tableware, that is, the sensor embedded in tableware.

References

1. MedicineNet.com: Medical definition of lifestyle disease. https://www.medicinenet.com/script/main/art.asp?articlekey=38316. Accessed 8 May 2019
2. World Health Organization: Obesity and overweight. https://www.who.int/en/news-room/fact-sheets/detail/obesity-and-overweight. Accessed 8 May 2019
3. World Health Organization: Obesity. https://www.who.int/topics/obesity/en/. Accessed 8 May 2019
4. World Health Organization: The top 10 causes of death. https://www.who.int/news-room/fact-sheets/detail/the-top-10-causes-of-death. Accessed 8 May 2019
5. World Health Organization: Cardiovascular disease. https://www.who.int/cardiovascular_diseases/world-heart-day/en/. Accessed 8 May 2019
6. Sazonov, E.S., Fontana, J.M.: A sensor system for automatic detection of food intake through non-invasive monitoring of chewing. IEEE Sens. J. **12**(5), 1340–1348 (2012)
7. Wang, S., Zhou, G., Hu, L., Chen, Z., Chen, Y.: CARE: chewing activity recognition using noninvasive single axis accelerometer. In: Proceedings of the 2015 ACM International Joint Conference on Pervasive and Ubiquitous Computing and Proceedings of the 2015 ACM International Symposium on Wearable Computers, Osaka, Japan, pp. 109–112 (2015)
8. Farooq, M., Sazonov, E.: A novel wearable device for food intake and physical activity recognition. Sensors (Basel) **16**(7), 1067 (2016)
9. Dong, Y., Scisco, J., Wilson, M., Muth, E., Hoover, A.: Detecting periods of eating during free-living by tracking wrist motion. IEEE J. Biomed. Health Inform. **18**(4), 1253–1260 (2014)
10. Aizawa, K., Ogawa, M.: FoodLog: multimedia tool for healthcare applications. IEEE Multimedia **22**(2), 4–8 (2015)
11. Foo.log Inc: FoodLog. http://www.foo-log.co.jp. Accessed 8 May 2019
12. Noronha, J., Hysen, E., Zhang, H., Gajos, K.Z.: PlateMate: crowdsourcing nutrition analysis from food photographs. In: Proceedings of the 24th Annual ACM Symposium on User Interface Software and Technology, UIST 2011, pp. 1–12 (2011)
13. Kuwata, H., Iwasaki, M., et al.: Meal sequence and glucose excursion, gastric emptying and incretin secretion in type 2 diabetes: a randomized, controlled crossover, exploratory trial. Diabetologia **59**(3), 453–461 (2016)
14. Kaiya, K., Koyama, A.: Design and implementation of meal information collection system using IoT wireless tags. In: Proceedings of 10th International Conference on Complex, Intelligent, and Software Intensive Systems (CISIS 2016), CISIS 2016.66, pp. 503–508 (2016)
15. Kaiya, K., Suzuki, H., Koyama, A.: Meal Information Collection System Using Smart Tableware. The Institute of Electronics, Information and Communication Engineers, MVE2017-20, pp. 31–36 (2017)
16. Medical Xpress: Chewing habits determine blood sugar levels after a carbohydrate-rich meal. https://medicalxpress.com/news/2016-06-habits-blood-sugar-carbohydrate-rich-meal.html. Accessed 8 May 2019
17. Dieterich, T.G., Lathrop, R.H., Lozano-Perez, T.: Solving the multiple instance problem with axis-parallel rectangles. Artif. Intell. **89**(1–2), 31–71 (1997)

18. Kyritsis, K., Tatli, C.L., Diou, C., Delopoulos, A.: Automated analysis of in meal eating behavior using a commercial wristband IMU sensor. In: Proceedings of 39th Annual International Conference of the IEEE Engineering in Medicine and Biology Society (EMBC), pp. 2843–2846 (2017)
19. Zhou, Z.-H., Zhang, M.-L., Huang, S.-J., Li, Y.-F.: MIML: a framework for learning with ambiguous objects. CoRR abs/0808.3231 (2008)
20. Bunescu, R.C., Mooney, R.J.: Multiple instance learning for sparse positive bags. In: Proceedings of the 24th International Conference on Machine Learning. Corvallis, OR, pp. 105–112 (2007)
21. Hall, M., Frank, E., Holmes, G., Pfahringer, B., Reutemann, P., Witten, I.H.: The WEKA data mining software: an update. SIGKDD Explor. **11**(1), 10–18 (2009)

From Ivory Tower to Democratization and Industrialization: A Landscape View of Real-World Adaptation of Artificial Intelligence

Toshihiko Yamakami[✉]

ACCESS, Tokyo, Japan
Toshihiko.Yamakami@access-company.com

Abstract. Deep learning is a hot research topic during 2010s. The rapid advances and data and computing-intensive characteristics of deep learning isolated ordinary people from the technology. However, as the technology advances, there are some visible signs of socialization in deep learning. The author presents the concept of democratization and industrialization of deep learning. Then, the author describes this trend as a new landscape view of AI technology. Finally, the author describes a 3-stage model of interaction between social community and technology in the era of digital transformation.

1 Introduction

Rises in machine learning such as deep learning witness a turning point where a new paradigm of intelligence integration increases its visibility in artificial intelligence. In this paper, the author deals with deep learning as a symbol token for socialization in artificial intelligence.

AI was considered to be something mysterious in the past. It is partly because there was not much accountability of complicated AI (Artificial Intelligence). After the impressive advances in deep learning, the value creation by massive data, heavy computing resources and intelligent learning management has been widely recognized. However, deep learning programming is still isolated from traditional coding. It is because there is no transparent method to fill the last mile gap between a chaotic status and a final learned status. Modular and step-by-step code practices are generally lacking in the deep learning process. The final step to reach the stable and comprehensive solution is almost invisible in the most part of the deep learning coding.

Elastic computing that deal with the flexible computing resource allocation enables a continuous and automatic integration of data gathering, learning and deployment of learned models.

Thanks to the rapid advances of best practices, tools and computing environments, the distance between the ivory tower of deep learning and the everyday world has become small.

© Springer Nature Switzerland AG 2020
L. Barolli et al. (Eds.): NBiS-2019, AISC 1036, pp. 200–211, 2020.
https://doi.org/10.1007/978-3-030-29029-0_19

There are some visible signs of socialization of artificial intelligence. One is democratization. The border between common people and deep learning experts come to shrink compared to the early days. Jeremy Howard spends extensive efforts in fast.ai [3] to make neural networks uncool again. He wants to provide non-experts with the deep learning tool belt to deal with every real world problems.

The other is industrialization. The border between researchers and practitioners come to shrink in some aspects. Considering the amount of data and computing, it is an incredible shrink in some senses. These socialization enables the utilization of real world data and provides a step toward so-called digital transformation.

Digital transformation is a sort of hype. However, considering the advances of communications, computation, data analytics, and deep learning, it is recognized as an inevitable evolution path as a matter of time.

It is important to understand the landscape of socialization of artificial intelligence. It is also necessary to capture the patterns of interaction between technology and societies.

The author coins a 3-stage model of interaction patterns between them and describes its implication for the future of the computing and societies.

2 Background

2.1 Purpose of Research

The aim of this research is to describe a transition model of artificial intelligence in the context of socialization in computing.

2.2 Related Work

Research on relationship of technology and societies consists of (a) democratization, (b) industrialization, and, (c) interaction of technology and societies.

First, in regards to democratization of technology, Fischer et al. discussed democracy in the networked society [citeFischer:2018:TNM:3209281.3209284. Sundar et al. discussed voting ICT adoption in India [6]. Tanenbaum et al. discussed DIY as a democratic technology [7]. Wanyiri discussed the Foodni approach in the development in the developing world [8].

Second, in regard to industrialization, Kagermann et al. discussed Industry 4.0 as a new industrial revolution. Wohlin presented top 10 challenges of software engineering research with industry [9]. Polycarpou et al. discussed engagement in industry-academia partnership [4].

Third, in regard to interaction of technology and societies, Carr discusses technology impacts on human future [2]. Rainey raised the issue of machine citizenship [5].

The originality of this paper lies in its identification of democratization and industrialization of deep learning in the viewpoint of transitive relationship of technology and societies.

2.3   Terms

The author defines the terms on the social encounter of technology as depicted in Table 1.

Table 1. Definitions

Term	Description
Democratization	Democratization of technology is commoditization of technology. With advances of tools and support environments, users can make use of technology without deep knowledge of implementation and deployment
Industrialization	Industrialization of technology is generalization and automation. With advances of knowledge, toolskits, and support environments, each application of technology can be managed automatically by non-human computing entities
Symbiosis	Symbiosis of technology is a departure of technology as a tool. Some technology deeply penetrates the everyday world and derives a qualitative transformation of societies. Technology becomes an inseparable part of a newly transformed society

It may be strange that a computer engineer talks about industrialization. In the last decade, we have witnessed DepOps, which integrates development and operation. Before that, software development and operation were separated and a bug-fixed software was manually and painfully brought to the deployment stage in a labor-intensive manner. In software development, deployment, monitoring and reliability management, there are still many labor-intensive aspects in software engineering in the real world. In that sense, labor-dependence and implicit knowledge dependence prevents real automation of software operation in many aspects.

3   Method

The author performs the following steps:

- Observe the current trend in deep learning,
- Describe factors that drive democratization and industrialization of deep learning,
- Describe a 3-stage model of interaction of technology and societies.

4   Observation

4.1   Progress of Deep Learning

A stage model of AI evolution is depicted in Fig. 1.

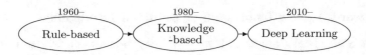

Fig. 1. This is a 3-stage model of AI evolution.

Artificial intelligence (AI) was born as early as computers in the 20th century. When a calculating machine was born, people came to recognize the concept of thinking machine. In early days, it was difficult to coin a computing hardware and software to deal with a high-level capability to match those of human beings. In the first stage, people tried to articulate a rule-base system with a concept that many rules could be accumulated to a high-level inference system. Rules were difficult to describe. And the relationships among rules presented larger challenges.

In 1980s, people made another attempt to deal with a high-level inference system by articulating a pile of knowledge. Rules are procedures. Knowledge seemed to be more static and easier to handle. This attempt did not produce the promised deliverables.

In 2010s, people made the third attempt to make so-called deep learning. The basic concept is to emulate neural networks which consist of human brains. The concept dated back to the 20th century. In the incubation phase, there lacked the resources, data and computing power, to drive deep learning. In 2010s, wide adoption of social network services and camera-equipped smartphones enabled a large amount of image data available for research. Also, the cloud computing enabled storing and processing of such data. Advances of GPU also helped deep learning from the resource side. Breakthroughs such as back propagation also pushed forward deep learning.

Factors that drive the deep learning are depicted in Table 2.

Table 2. Multiple factors that drive the deep learning

Factor	Description
Data availability	Cloud technology and smartphone technology enables large data sets of photos and videos to enable deep learning
Elastic computing	Scaling in computing is enabled by elastic computing in clouds
Data augmentation	In order to obtain a large set of learning data, data augmentation techniques have evolved
Prevention of overfitting	A variety of techniques to suppress overfitting is available

4.2   Towards a New Landscape of Artificial Intelligence

Basic theoretical backgrounds of deep neural networks were provided until 1980s, but the factors to enable deep neural networks were not available until 2010s. When the time is ripe, multiple factors to remove the past limits drove the eye-opening advances of deep learning.

Network integration and intelligence integration are compared in Table 3.

Table 3. Network Integration and Intelligence Integration are compared.

Shift	Pre-state	Drivers	Needs for integration
Network integration	Multiple communication protocols were exploited. No single framework or architecture dominated the world	TCP/IP, SMTP, HTTP dominated and standard protocols spread widely	Advances and evolution of versions of standard protocols made configurations and adaptations major challenges in the private network management
Intelligence integration	Ad hoc learnings and models were explored and developed as a case-base	Standard frameworks and learned models. Software-based tunings of meta-parameters of deep learning	IoT-based life-cycle integration and open learning frameworks made a shift from case-based learning to software-based tuning of deep learning

In 1980s, there was a shift toward open computing. This enabled a wide range of openly available network components. However, his put large pains in the integration part.

Driving factors of integration are depicted in Table 4.

Network integration was a big challenge during 1990s. In 2010s, thanks to elastic computing, we have witnessed automatic integration in many computing areas including deep learning. One example is Amazon Web Services.

Transitions of Amazon Web Services are depicted in Fig. 2. Starting from virtualization, the two movements: (a) verticalization and (b) softwarization of operation are in progress. At the end, in Amazon SageMaker, we witness the integration of these two trends in machine learning (mainly in deep learning, but it includes other machine learning components.).

This is one of the trends of intelligence integration, a migration trend of Artificial Intelligence into the real world. Softwarization enabled by cloud services enables this kind of migration.

Table 4. Multiple driving factors of integration

Item	Description
Standardization of components	When each component is proprietary, each component is vertically integrated by specific vendors. Configuration and dependency resolution are embedded in the total construction
Maturity of modules	When each component is mature, work focus shifts from manufacturing to use
Complexity of configuration and Dependency	When each module is standalone and isolated, integration work does not evolve. Heavy dependency resolution leads to the heavy integration work in configuration and dependency resolution
Growth	When the use of each technology is stable, large work is not necessary at integration. Strong growth and widespreadness drives the work focus from one-time deployment to continuous integration of added facilities and devices

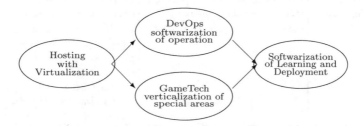

Fig. 2. Amazon Web Services make transitions with DevOps and verticalization to lead to an integrated learning platform.

4.3   Democratization

Technology is always advancing. The points that the author comes to think about democratization are depicted in Table 5.

It should be noted that some trends are driven by so-called GAFA, IT giants intention of dominating the key technology such as machine learning. Colaboratory provides empathy to Google in the community of machine learning, which may benefit Google intention of collecting more talents in this area. It is also noted that democratization of deep learning does not directly negatively impact Google's cash cow, advertising business.

The abundance of cash-flow in advertising business enables Google's generous contributions to the deep learning research community. Democratization requires affluence in most cases.

Table 5. Multiple factors that drive democratization of deep learning

Factor	Pre-democratization	Democratization
Know-how sharing	There were many fine-tuned parameter tuning in deep learning. They were difficult to share, and the sharing environment was poor	OSS-based tools (e.g. Jupyter Notebook) are used to share and replicate the result. Jupyter Notebook is a browser-based environment, so it is easy to use independent of underlying computing environments
Data sharing	The reliable data for deep learning were not abundant	Many data are shared with open access. Competition like Kaggle provides further open access to massive data eligible for large-scale data analytics
Reasonable environment	Computing-intensive problem solving needs expensive resources to replicate or try	Elastic computing environment lowers the bar for deep learning entry. Colaboratory (by Google) provides a free test environment for deep learning with GPU support, which also uses Jupyter Notebook
Free courseware	AI coursewares were difficult to pursue without any easy replicable environment and tools	Free coursewares (e.g. fast.ai) provides easy knowledge bases for deep learning. The high-level and well-maintained tools enable easy learning with replicable source code
Transfer learning	Many problems required a large size of training data which were difficult to manage for many players, especially entry level	Transfer learning enables small data and small processing of models. For example, image processing does not need a full scratch neural network build-up. Competition-winning learned models (e.g. Resnet-50) can be used to a new neural network by replacing the final layer with the problem-specific layer

4.4   Industrialization

The factors to derive industrialization are presented in Table 6.

Table 6. There are multiple factors that drive industrialization of deep learning.

Factor	Pre-industrialization	Industrialization
Seamless integration of business problem solving	The phases of business problem solving were separately supported by a range of different tools	SageMaker [1] provides a seamless and continuous support for business problem solving with machine learning
Easy deployment	Deployment required special professional services and customized design	Deployment is integrated in a seamless manner with other components of web services
Scalable deployment	It required specialized tuning and custom enhancement when a scalability was required	Harmonization of elastic web services enable easy scalable deployment without heavy burdens of operation

As previously described, IT was labor- and implicit knowledge intensive world of integration of complicated technology components. Data collection of IoT made this messy integration more challenging in the past. Amazon Web Services provide mitigation of automation of machine learning with elastic computing and softwarization of operation.

There are 6 stages in business problem solving using machine learning as depicted in Fig. 3.

Components in SageMaker provides support of this machine learning problem-solving chain. This accelerates industrialization of deep learning and penetration of technology into everyday lives and businesses.

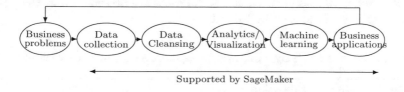

Supported by SageMaker

Fig. 3. There are 6 stages in business problem solving using machine learning.

5   Interaction of Technology and Societies

SageMaker provides integration of the everyday world and machine learning in some senses. SageMaker is just an example of such a trend. Elastic computing and softwarization of operation enable intensive everyday world machine learning, which can have a heavy impact in the human society.

The author proposes a 3-stage model of interaction between technology and societies as depicted in Fig. 4.

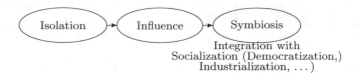

Fig. 4. Major technologies take 3 steps into everyday's life.

The detailed description of each step is described in Table 7.

Table 7. Detailed description of 3 stages

Stage	Description
Isolation	Technology is a tool. In that sense, it can be replaceable with anything to fill the purpose
Influence	The functions of technology starts to impact the society in some senses
Symbiosis	The function of technology brings fundamental qualitative change in societies. Societies require the technology in a deep-core level

Usually, technology starts as a method. As far as it fills the purpose, a method can be interchangeable. At some time, technology influence reaches a certain level, it causes a fundamental change in society. People may think the machine learning as just another tool. However, there is a possibility that it causes fundamental transformation of societies. Some cases are presented that the possible fundamental transformation in Table 8.

Table 8. Fundamental changes to cause a deep-core level transformation of a society

Item	Description
Economy	Machine learning can convert some parts of economy into sharing economy
Trust	Machine learning can be a indispensable part of social trust
Safety and defensive power	Safety and defensive power may depend on the IoT and machine learning

AI starts from ivory towers to catch the God-like wisdom using machines and algorithms. The rapid progress of deep learning, one of AI research domains, moves forward democratization and industrialization where ordinary people or no human intervention is feasible to apply the state-of-art techniques to everyday problem solving. This third stage of symbiosis is a new aspect of today's technology largely due to the emergence of IoT, AI, and cloud-based softwarization.

6   Discussion

6.1   Advantages of the Proposed Method

The author proposes new concepts of democratization and industrialization of deep learning. The hype of deep learning has driven the advances of models, modules, tool chains and underlying infrastructures.

The progress enables new waves of technological interaction with the real world. One is democratization. Advanced tools enable the problem solving by the non-deep learning experts. The other is automation and integration. Advanced tools enable softwarization of operation and automatic scaling with coordination of underlying infrastructures and tools.

The author refers it to socialization of technology to imply a new relationship of human societies. This is an attempt to capture the emerging new landscapes of interaction of technology and societies.

The progress of deep learning is considered a human identity risk so called "The Singularity." The advances of deep learning enables revisiting the Singularity where artificial intelligence surpasses human intelligence.

Emerging new trends of computing, such as massive open online courses (MOOCs), elastic computing, support tools, and softwarization of operation provide new opportunities for new type of interactions between technology and societies.

The author proposes a 3-stage model of interaction of technology and societies. It is crucial to understand the technology as the driving factor of transformation of societies. At the same time, with advances of penetrating invisible technologies, it is necessary to understand the new type of technology interaction such as symbiosis. Data analytics including IoT, and machine learning are such examples.

It is important to provide a building block to understand the new types of interaction that impacts everyday lives and businesses. The proposed model provides a new framework to revisit how we can manage interaction between technology and societies. The proposed views and models can be a first step to consider technology-symbiosis societies in the future.

6.2   Limitations

This paper is descriptive and qualitative. The quantitative measures to identify each transition are not discussed. The quantitative analysis of detailed case studies are missing.

The relationship between the proposed models and observable real world metrics are not defined in this paper.

Each driving factor is not quantitatively validated.

There are no verifiable case studies in this paper.

The methods to identify and assess the outcome of the proposed models are not presented in this research.

The quantitative measures to identify the influence and outcome of the proposed models are not explored in this paper.

Conditions of symbiosis are still exploratory and not exhaustive. Case studies on symbiosis between technology and societies remain future studies.

7   Conclusion

Artificial intelligence has a long history of research. In the most of 20th century, its deliverables did not match the hype and expectation. With advances in research, data and computing environment, deep learning engender human-compatible results in some areas. It brings not just achievement and also some socialization: democratization and industrialization.

When a technology is born, it is isolated from everyday world. It is in an ivory tower and can be comprehensible to a small number of experts. When it penetrates the everyday life, it encounters the real world and exhibits a wide range of various interaction patterns.

Deep learning is one of unique programming skills. Usually, a programming consists of modular building blocks. As each block is assembled to one piece, it gradually reaches a state of completion. In deep learning, it does not go that way in many cases. Until the very near to the final moment, the way to reach the right learned models are generally unknown. However, even in this isolation, we come to recognize the so-called socialization of deep learning, such as democratization and industrialization. This socialization is an outcome of advances of deep learning technology.

This is a stepping stone to understand the new kind interaction with technology. In some cases, technology is not just a tool, but it comes to penetrate into the real world and transform the everyday life. IoT, data analytics, machine learning, and softwarization will drive penetration of technology into deep inside of societies. The model toward symbiosis is an early attempt to capture such shifts of perspectives of technological interaction.

References

1. Amazon Web Services, Inc.: Amazon SageMaker (2019). https://aws.amazon.com/sagemaker//
2. Carr, N.: Closing keynote: computers, automation and the human future. In: Proceedings of the 2017 CHI Conference Extended Abstracts on Human Factors in Computing Systems, p. 4. CHI EA 2017, ACM, New York (2017)
3. fast.ai: fast.ai Making neural nets uncool again (2019). https://www.fast.ai/

4. Polycarpou, I., Andreou, P., Laxer, C., Kurkovsky, S.: Academic-industry collaborations: effective measures for successful engagement. In: Proceedings of the 2017 ACM Conference on Innovation and Technology in Computer Science Education, pp. 250–251. ITiCSE 2017, ACM, New York (2017)
5. Rainey, S.: Friends, robots, citizens? SIGCAS Comput. Soc. **45**(3), 225–233 (2016)
6. Sundar, S.S., Sreenivasan, A.: In machines we trust: Do interactivity and recordability undermine democratic technologies? In: Proceedings of the Seventh International Conference on Information and Communication Technologies and Development. ICTD 2015, pp. 60:1–60:4. ACM, New York (2015)
7. Tanenbaum, J.G., Williams, A.M., Desjardins, A., Tanenbaum, K.: Democratizing technology: pleasure, utility and expressiveness in DIY and maker practice. In: Proceedings of the SIGCHI Conference on Human Factors in Computing Systems. CHI 2013 pp. 2603–2612. ACM, New York (2013)
8. Wanyiri, J.: Foondi workshops: democratizing technology in Africa. Interactions **22**(4), 67–69 (2015)
9. Wohlin, C.: Empirical software engineering research with industry: top 10 challenges. In: Proceedings of the 1st International Workshop on Conducting Empirical Studies in Industry. CESI 2014, pp. 43–46. IEEE Press, Piscataway (2013)

Data Exportation Framework for IoT Simulation Based Devices

Yahya Al-Hadhrami[1]([⊠]), Nasser Al-Hadhrami[2], and Farookh Khadeer Hussain[1]

[1] University of Technology Sydney, Broadway, Ultimo, NSW 2007, Australia
yahya.s.al-hadhrami@student.uts.edu.au, farookh.hussain@uts.edu.au
[2] Information Technology Authority, Al-Athaiba, Seeb, Sultanate of Oman
nasser.alhadhrami@ita.gov.om

Abstract. Internet of things (IoT) is part of everyday life nowadays. Millions of devices are connected to the internet to collect and share data. Although IoT devices are evolving quickly in the consumer market where smart devices and sensors are becoming one of the main components of many households, IoT sensors and actuators are also heavily used in the industry where thousands of devices are used to collect and share data for different purposes. A need for an IoT simulation tool is necessary for development purposes and testing before deployments. One of the widely used tools among IoT researchers is the open-source tool Cooja simulator. Cooja has limitations—one is the lack of a way to export collected data as a data set for further processing. Therefore, this study introduces an extension tool to present and export the data into different forms.

Keywords: IoT · Security · Data mining

1 Introduction

IoT devices have gained a significant amount of attention in the last few years. Such an increase can be attributed to many factors; the main one is the decrease in CPU and memory cost. Now, almost anything can be transformed into an IoT device as long as it has the ability to connect to the internet. Therefore, a wide spectrum of applications can be integrated into the IoT infrastructure, which can be anything from military-based solutions to every day consumer applications, such as devices found in smart homes. The rapid increase in these devices did not come without its constraints and limitations, such as limited resources in battery and CPU power. Therefore, designing any application should consider these constraints to maintain a good life span for the IoT device. Another significant aspect to take into consideration is protocol design and implementation. Many studies have focused on performance testing and application development for MANET and ad hoc protocols, but there are limited studies on IoT networks. [1]. Therefore, the 6LoWPAN [2] and RPL protocols [3] were developed to address the limitation in existing protocols.

© Springer Nature Switzerland AG 2020
L. Barolli et al. (Eds.): NBiS-2019, AISC 1036, pp. 212–222, 2020.
https://doi.org/10.1007/978-3-030-29029-0_20

In this study, we are focusing on the Cooja simulator that emulates the Contiki operating system [4], which is embedded in operating system targeting devices with limited power and memory usage. Luckily, the developers of Contiki OS and Cooja have already implemented the system with a lightweight TCP/IP stack(uIP) and a duty cycling mechanism which helps save power consumption on nodes [5]. Although Cooja has a built-in extension for data collection and representation called Collect View the tool lacks the ability to export the collated data for further analysis, which is the case for most research. The need for a tool to export the collected data is essential since it allows the researchers to analyse the data and further studies the nodes and network behaviour. We propose a new framework that extends the current Collect View plugin for Cooja. Our tool is capable of exporting the data into CSV file or to a real-time MySQL database, which will simplify the process for researchers and developers in term of data collection and manipulation. The rest of this paper will explore first the related work in Sect. 2. Then in Sect. 4, background is detailed about the problem. In Sect. 4, we introduce the framework used to solve these issues of data exportation in Cooja. In Sect. 5, evaluation of the framework is discussed. Finally, we conclude this paper in Sect. 6.

2   Related Work

There are a few IoT network simulators available for researchers like OMNET++ [6], TOSSIM [7], NS3 [8], Cooja [4]. Due to its simplicity and extendibility, Cooja is by far the most widely used IoT simulator by researchers. Many tools have been developed to extend Cooja's functionality. One of these plugins is the Collect View plugin used to collect sensor data and display them in a user-friendly presentation. However, it lacks a data exportation tool to export the collected data for further processing. Another tool used to display sensor data in Cooja is the tool proposed by [9] where it tries to monitor different parameters of the sensor network and display them in a simple GUI interface. The problem with Cooja Trace is that it consumes CPU and memory power of the host computer. Furthermore, it does not provide any convenient method to export the collected data. Another tool has bee proposed by [10] that is used to display power levels of running nodes. However, this tool is limited to display the power-related parameters, such as the remaining power and current power consumption. Unfortunately, there has not been enough work done when it comes to exporting sensor data outside of Cooja simulator. Most studies focused on building tools addressing sensor optimization but not on data exportation and visualization.

3   Background

Before explaining the proposed framework in this paper, you have to understand the different components of the Cooja simulator. In this section, we explain the different components involved in this research.

3.1   Contiki OS

The Contiki OS is an event-driven system explicitly designed for low resources devices that have limited processing, memory, and bandwidth power. Although the system is event-driven based, it can support multithreading and multitasking. The Contiki OSs has already built in IPv6 stack [11] by following the RFC4294 [12] standards and the integration with 6LoWPAN protocol for reduced IPv6 header overhead.

3.2   Cooja

The Cooja simulator was developed to emulate the IoT based operating systems. This is a Java-based simulator that uses Java Native Interface (JNI) to interface between Java and the emulated OS, which is in our example of the Contiki OS. A full sensor emulation is achieved through the use of MSPSim [13], a Java-based instruction level emulator. That allows full emulation of a different platform networking stack, such as the Tmote Sky sensor. One of the main advantages of Cooja is its extendibility where developers and researches can extend the capabilities of the software through interfaces and plugins. One of the built-in plugins that extend Cooja's capability is the Collect View discussed below (Fig. 1):

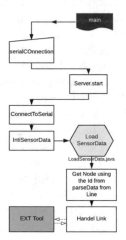

Fig. 1. Workflow of the collect-view

3.3   Collect-View

Collect View is a built-in plugin that comes with Cooja simulator, but it also can be executed as a stand-alone application for data collection. Collect View extends Cooja's capability by providing a way to collect and visualize data by altering the node firmware to send a set of sensor and network related information periodically. The sink node transmits all the acquired data in the network to Collect View using a special data handler. Unfortunately, this plugin lacks the ability to export data as a separate file for further processing. This is a gap this paper tries to fulfill through proposing a new framework. The Collect View consists mainly of two parts explained below:

3.4   Incoming Data Module

As we can see from the diagram, the Collect View module classes can be divided into two categories. The first category is the connection handling classes, where the module allows the user to choose between different connection mediums, such as UDP, TCP server, or through serial ports. These classes work as an interface between the data collection tool and the sink node. The sink node is responsible for collecting the data from nodes connected to the same RPL network and forwards them to the Collect View interface based on a predefined interval.

3.5   Data-Handling Models

To handle the incoming data, this model passes the stream of data coming from the sink node to the server where data is parsed and returned to its readable format using the SensorData.java and SensorInfo.java classes. The streamed data is then displayed in the Graphical User Interface (GUI), which contains a set of graphical JPanel used to display and visualize the information. Unfortunately, with all these functionalities, the Collect View does not provide a mechanism to store the collected data as a separate file for further data analysis. This where the DEF tool comes into play, where it allows you to monitor the data live and export them as needed. DEF utilises the available functionality of the Collect View plugin by passing the already established server and extracting all nodes data in the network. Each node contains a sensor data aggregator object that has a set of sensors information and network-related data.

4   Data Exportation Framework

As explained earlier, Cooja is a popular platform for IoT and Wireless Sensor Network (WSN) simulation and is widely used by the research community, but it lacks a tool for exporting collected data for in-depth analysis. Acquiring sensor data and network-related data in a simple textual format will allow the researcher to integrate this data into other platforms and analyse as needed. DEF is designed to integrate as an extension for the available Collect View plugin in Cooja. DEF simplifies the process by gathering and exporting the collected data into a CSV format file or MySQL-connected database. Such a tool will allow researchers and developers to focus more on their task rather than trying to

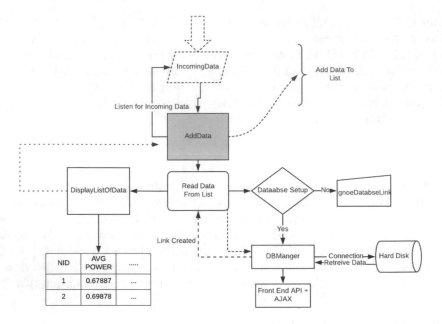

Fig. 2. Workflow of the DEF Tool

figure out how to export data to the correct format. To develop DEF the platform, we reverse engineered the Collect View model, profoundly modifying the
existing code to run DEF tool during run time. The DEF framework consists of
four main modules: the Data Handler Module (DHM), Data Visualization Module (DVM), Database Management Module (DMM), and API Module (APIM).

4.1   Data Handler Module DHM

Figure 2 shows all the different components of the DEF tool. The DHM module
extracts the collected data for each node and stores them as an ArrayList to be
embedded later in the table. To avoid blockage in the main thread, we separated
the data-gathering process into two threads, one for GUI and the other for
data handling and representation. This module main task is to handle, store,
and update the incoming sensor data. All the nodes in the network and their
information are stored and updated into an ArrayList, which later is embedded
into the data DVM. To address all of this, we designed Algorithm 1. It retrieves
a list of all nodes in the network and extracts the relevant sensor data as data
aggregator class. If the retrieved node does not have any data, the process of
adding sensor data to the sensor list is ignored. The received sensor list is passed
in real-time to the DVM module.

Algorithm 1. DEF Handel Incoming sensor Data

Input: A set of Sensor Nodes Data $A = \{a_1, a_2, \ldots, a_n\}$
Output: Aggregated Sesror Data

1 $SensorDataList \leftarrow a_1$
2 **for** $i < a_n$ **to** n **do**
3   **if** $a_i! = null$ **then**
4   add a_i to the $SensorDataList$
5   Update $SensorDataList$
6   **else**
7   Printout "Data is null"

8 **return** $SensorDataList$

4.2   Data Visualization Module DVM

The DVM is the unit responsible for displaying the incoming data in a presentable format. This module allows the network administrator to monitor nodes performance and export the data for further processing. In order to display the sensor data, we built our own custom table model by extending the Abstract Table Model class provided by Java to properly present the data in the correct form. This module was designed with user usability and convenience in mind, where the user has the ability to select the parameter of node information to display, a function not available in the original Collect View plugin. Such flexibility in display sensor information allows the researcher to focus on specific sensor parameters. Furthermore, the module allows the user to export the data as CSV from all nodes or each node selected. Figure 3 shows an example of the DEF display data, with emphasis on user convenience.

4.3   Database Management Module DMM

This module works as the database interface that handles the process of reading and writing into a database. We have chosen CSV and MySQL interfaces since most programming languages and tools have libraries that support both. The live data presented in the table can simultaneously update the table and MySQL database. To achieve these functionalities, the database manager class had to be developed. This main class function is to take user configuration and seamlessly create and update the table in the database. The Java JDBC driver [14] was utilized to achieve the connection between the DEF and the MySQL database. This will allow developers to extend the capability of the collected data outside the Collect View interface and monitor the behaviour of the node. One example is to utilise the Mysql [15] database to connect to a web interface that continuously monitors the nodes and their status without compromising the node power since all data is processed outside the IoT network. Moreover, since all the database processing is handled outside the resources of the IoT network, this widens the spectrum of applications that can use the extracted data.

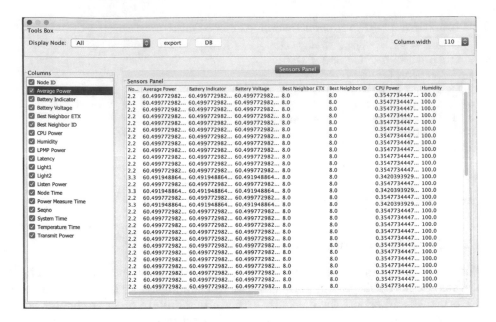

Fig. 3. Screenshot of the DEF Tool

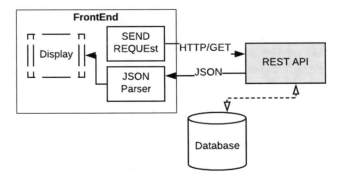

Fig. 4. EXT API

4.4 API Module APIM

To simplify the process, this study proposes a simple REST API to fetch the data from the database and convert it to a data stream that can be utilised in any third-party platform. This module is used to further enhance the productivity of the developer by abstracting all the database connection and data retrieval into simple API, ensuring a good integration with third-party applications. The API has two main classes described below:

– **Database() Class:** The database class used to establish a connection to the database containing all of the nodes records.

- **GetData() Class:** The class responsible for retrieving node-specific data from the database. Additionally, this class is responsible for converting the retrieved data into JSON format to allow third-party applications to utilise the retrieved data easily.

```
1  [{"NodeID":"2","avgPower":"60.4598"},{"NodeID":"3","avgPower"
     :"60.4319"},{"NodeID":"4","avgPower":"60.4939"}]
```

Fig. 5. Example of the JSON output

In the example workflow shown in Fig. 4, a request method is used to send an HTTP request to the REST API. The API initiated a database connection to retrieve the data and send it back to the front-end. The data retrieved is then encoded as JSON format using the GetData class. Figure 5 shows an example of the JSON output retrieved by the DEF API. It shows the sensors average power reading alongside the associated node ID. This JSON data structure can be utilised by any application or front-end that supports parsing JSON data, which is supported by most programming languages.

5   Evaluation

In this section, a detailed evaluation of the DEF framework is introduced. This section divided into two parts. First, simulation design where the design of the IoT network is discussed. The second part discusses the evaluation of the framework by monitoring CPU and memory usage.

5.1   Simulation Design

To evaluate the DEF tool architecture, we implemented a test bed use case scenario with Cooja emulated IoT devices. The example consists of 22 nodes scattered randomly across the working area. Figure 6 shows the distribution of the IoT nodes. All the client nodes implement the rpl-collect.c example available in the examples folder in Contiki OS. The sink uses the udp.sink.c. To emulate a real-world scenario where noise level can be introduced in the environment, a distributor node is placed within the network area to make signal noise in the network. Moreover, to observe how the DEF tool will perform in an attack scenario, attacker nodes are placed at the edge of the network affecting two neighbour nodes as shown in Fig. 4. The attacker is carrying a blackhole attack by advertising better rank to the sink node. To justify the use of the DEF tool, the evaluation scenario was executed twice with the same settings and run time except one run was without the DEF. This was done to compare CPU and memory consumption in both runs (Table 1).

Table 1. Workstation setup

CPU	Intel(R) i7-3740QM @ 2.70 GHz
Memory	8 GB RAM
OP	Linux contiki 4.13.0-21-generic

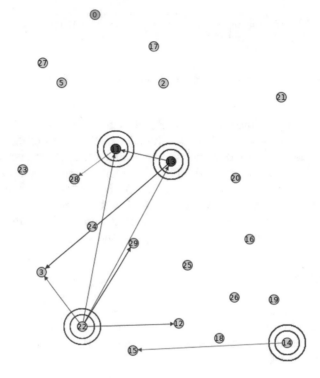

Fig. 6. IoT simulation network

5.2   Simulation Result and Discussion

We focused on the power and memory consumption of the host computer running the DEF tool, since the tool does not have any direct effect on the network simulation itself. Our focus is to evaluate how the tool evaluates in a simulation scenario explicitly talking about two parameters: memory and CPU benchmark with and without the DEF. To evaluate the performance of each scenario, we utilised the "top" command line in Linux to run every ten seconds to have a better evaluation of the current CPU and memory for each scenario.

Table 2 shows the final result of the evaluations. We ran the simulation six times in total—three for each scenario—with the DEF tool and without it. We can observe from the table that the longer the simulation time, the less CPU load. This can be explained because of the caching system in the CPU. Another observation is that there is almost no change in real-time when the DEF tool is

Table 2.

Scenario	AvgCPU %	AvgMemory %	SimTime (Hrs)	Realtime (S)
No DEF 1	117.56	4.4	1	49.71
No DEF 2	109.40	3.46	2	90.35
No DEF 3	103.00	6.5	3	115.44
DEF 1	118.00	5.87	1	59.71
DEF 2	110.45	4.92	2	91.15
DEF 3	107.67	6.97	3	117.22

running. The result shows a very minimal increase in CPU and memory usage with the DEF running in the background. This indicates that, although we are running a heavy task of data collection and representation, the DEF tool has minimal CPU and memory overhead.

6 Conclusion

In this study, the Cooja simulator and its limitation have been explored. To address the data exportation limitation, the DEF tool was presented in this paper. We have shown that the DEF tool simplifies the process of collecting and exporting sensor data from the Cooja simulator. Moreover, the result shows the DEF tool can perform for a longer period without increasing the CPU and memory resources.

References

1. Aschenbruck, N., Bauer, J., Bieling, J., Bothe, A., Schwamborn, M.: Let's move: adding arbitrary mobility to WSN testbeds. In: 2012 21st International Conference on Computer Communications and Networks (ICCCN), pp. 1–7. IEEE (2012)
2. Mulligan, G.: The 6LoWPAN architecture. In: Proceedings of the 4th Workshop on Embedded Networked Sensors, pp. 78–82. ACM (2007)
3. Winter, T., Thubert, P., Brandt, A., Hui, J., Kelsey, R., Levis, P., Pister, K., Struik, R., Vasseur, J.-P., Alexander, R.: RPL: IPv6 routing protocol for low-power and Lossy networks. Technical report (2012)
4. Eriksson, J., Österlind, F., Finne, N., Dunkels, A., Tsiftes, N., Voigt, T.: Accurate network-scale power profiling for sensor network simulators. In: European Conference on Wireless Sensor Networks, pp. 312–326. Springer (2009)
5. Österlind, F., Dunkels, A., Eriksson, J., Finne, N., Voigt, T.: Cross-level sensor network simulation with COOJA. In: First IEEE International Workshop on Practical Issues in Building Sensor Network Applications, SenseApp 2006 (2006)
6. Varga, A.: Omnet++. In: Modeling and tools for network simulation, pp. 35–59. Springer (2010)
7. Levis, P., Lee, N.: TOSSIM: a simulator for TinyOS networks. UC Berkeley, 24 September 2003

8. Henderson, T.R., Lacage, M., Riley, G.F., Dowell, C., Kopena, J.: Network simulations with the ns-3 simulator. SIGCOMM Demonstr. **14**(14), 527 (2008)
9. Strübe, M., Lukas, F., Kapitza, R.: Demo abstract: CoojaTrace, extensive profiling for WSNs. In: Poster and Demo Proceedings of the 9th European Conference on Wireless Sensor Networks, EWSN 2012, pp. 64–65. Citeseer (2012)
10. Österlind, F., Eriksson, J., Dunkels, A.: Cooja timeline: a power visualizer for sensor network simulation. In: Proceedings of the 8th ACM Conference on Embedded Networked Sensor Systems, pp. 385–386. ACM (2010)
11. Durvy, M., Abeillé, J., Wetterwald, P., O'Flynn, C., Leverett, B., Gnoske, E., Vidales, M., Mulligan, G., Tsiftco, N., Finne, N., et al.: Poster abstract: making sensor networks IPv6 ready. In: 6th ACM Conference on Networked Embedded Sensor Systems (2008)
12. Loughney, J.: RFC 4294: IPv6 node requirements (2006)
13. Eriksson, J., Dunkels, A., Finne, N., Osterlind, F., Voigt, T.: MSPsim–an extensible simulator for MSP430-equipped sensor boards. In: Proceedings of the European Conference on Wireless Sensor Networks (EWSN), Poster/Demo session, vol. 118 (2007)
14. Reese, G.: Database Programming with JDBC and JAVA. O'Reilly Media Inc., Sebastopol (2000)
15. DuBois, P., Michael Foreword By-Widenius: MySQL. New Riders Publishing, Thousand Oaks (1999)

Earthquake and Tsunami Workflow Leveraging the Modern HPC/Cloud Environment in the LEXIS Project

Thierry Goubier[1][✉], Andrea Ajmar[3], Carmine D'Amico[5], Paul Dubrulle[1],
Susanna Grita[3], Stephane Louise[1], Jan Martinovič[2], Tomáš Martinovič[2],
Natalja Rakowsky[4], Paolo Savio[5], Danijel Schorlemmer[6], Alberto Scionti[5],
and Olivier Terzo[5]

[1] CEA, List, PC 172, 91191 Gif-sur-Yvette Cedex, France
{thierry.goubier,paul.dubrulle,stephane.louise}@cea.fr
[2] IT4Innovations, VSB – Technical University of Ostrava,
17. listopadu 2172/15, 708 00 Ostrava-Poruba, Czech Republic
{jan.martinovic,tomas.martinovic}@vsb.cz
[3] ITHACA, via P. C. Boggio 61, 10138 Turin, Italy
{andrea.ajmar,susanna.grita}@ithaca.polito.it
[4] Alfred Wegener Institute, Helmholtz Centre for Polar and Marine Research,
Am Handelshafen 12, Bremerhaven, Germany
natalja.rakowsky@awi.de
[5] LINKS Foundation, via P. C. Boggio 61, 10138 Turin, Italy
{carmine.damico,paolo.savio,alberto.scionti,
olivier.terzo}@linksfoundation.com
[6] GFZ German Research Centre for Geosciences, Telegrafenberg,
14473 Potsdam, Germany
ds@gfz-potsdam.de

Abstract. Accurate and rapid earthquake loss assessments and tsunami early warnings are critical in modern society to allow for appropriate and timely emergency response decisions. In the LEXIS project, we seek to enhance the workflow of rapid loss assessments and emergency decision support systems by leveraging an orchestrated heterogeneous environment combining high-performance computing resources and Cloud infrastructure. The workflow consists of three main applications: firstly, after an earthquake occurs, its shaking distribution (ShakeMap) is computed based on the OpenQuake code. Secondly, if a tsunami may have been triggered by the earthquake, tsunami simulations (first a fast and coarse and later a high-resolution and computationally intensive analysis) are performed based on the TsunAWI simulation code that allows for an early warning in potentially affected areas. Finally, based on the previous results, a loss assessment based on a dynamic exposure model using open data such as OpenStreetMap is performed. To consolidate the workflow and ensure respect of the time constraints, we are developing an extension of a time-constrained dataflow model of computation, layered above and below the workflow management tools of both the high-performance computing resources and the Cloud infrastructure.

© Springer Nature Switzerland AG 2020
L. Barolli et al. (Eds.): NBiS-2019, AISC 1036, pp. 223–234, 2020.
https://doi.org/10.1007/978-3-030-29029-0_21

This model of computation is also used to express tasks in the workflow at the right granularity to benefit from the data management optimisation facilities of the LEXIS project. This paper describes the workflow, the associated computations and the model of computation within the LEXIS platform.

1   Introduction

Simulations in High Performance Computing (HPC) have long been capable (and used) to simulate natural disasters, from earthquakes, tsunamis, floods, forest fires to pest invasions and more. Additionally, geographic information systems have been used to map the extent of natural disasters and compute their effects on people and buildings, infrastructure and economy, by adding geo-referenced information onto the natural disaster simulation results.

Natural disasters are, however, complex phenomena to model and simulate. Local and small-scale irregularities, such as varying soil and sub-soil conditions or the shape of a bay, can have a huge impact on the shaking of an earthquake or the run-up height or inundation of a tsunami, respectively. To handle that, a regular increase in computational power and precision of the representation of the environment is used, which leads to the main difficulty of this domain: the time needed to simulate is incompatible with the time frame necessary for rapid and effective use of the results.

Multiple pragmatic concepts have been used to compensate for this. The tsunami early warning system for Indonesia, InaTEWS, employs the Tsunami Observation And Simulation Terminal (TOAST)[1] decision support system, connected to both a database of high-quality scenarios precomputed with TsunAWI and to the high-speed online algorithm easyWave with simpler model physics and coarse resolution: this allows for rapid tsunami warning announcements within minutes after an earthquake. While TsunAWI scenarios include inundation, e.g. for planning evacuation routes, they are based on predefined earthquake sources. In contrast, the real-time simulation easyWave starts with an earthquake source determined from the actual measurements and thus better assesses the actual situation, but, as a trade-off, easyWave can only deliver the wave height at the coast without inundation.

Disaster simulation results do not only provide the important early warnings but also valuable information and guidance for short-term emergency response (e.g. personnel to mobilize, shelter to be provided) and long-term disaster recovery actions (e.g. future supplies to order, disruptions to expect, disaster relief funding). Results from simulations have to be available at the right time for those decisions to be taken with the best available knowledge, because on-the-ground information will only become available after actions had to be taken to mitigate the disaster.

---

[1] https://www.gempa.de/products/toast/.

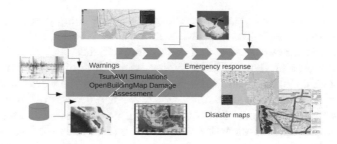

Fig. 1. A high level view of the earthquake and tsunami pilot flow.

To bridge the gap between detailed disaster simulations and rapid emergency decision support and to ensure that emergency response systems (ERS) have the best available information when needed, LEXIS combines knowledge and actors from different domains: natural disaster scientists and code optimization specialists. To achieve this, we plan to combine an earthquake damage assessment model and a tsunami simulation code under a single model of computation, linked to a satellite-based emergency mapping process, as shown in Fig. 1. This system is deployed on the LEXIS platform [1], crossing boundaries between High Performance Computing resources and the Cloud, and linking in data management acceleration technologies. We will detail the objectives, followed by a description of the workflow and its different components.

2   Objectives

In the LEXIS project, we are focused on innovating and accelerating the computation process of the ERS. The objectives are to leverage the LEXIS platform and create a new workflow for emergency response systems based on this platform. The main points of the LEXIS platform are federated HPC and Cloud resources, enhanced by the burst buffer technology.

This is comprised of three pillars:

- Infrastructure,
- Orchestration,
- Data management.

In practice, the LEXIS orchestration tool will distribute the computational resources, based on the available computational resources and data locality, to optimize the computational workflow and provide timely results, taking into account the time constraints expressed in the emergency workflow. Additionally, the orchestrator will know and allocate which parts of the computation should be sent to the HPC clusters and which should be sent to the Cloud infrastructure. From the point of the emergency response system, multiple TsunAWI simulations should be computed by the HPC resources, leveraging the burst buffer technology, while the data post-processing should be computed in Cloud, sharing the data from the simulation. Such workload distribution should provide the best efficiency in resource usage and data management.

From the point of view of the workflow itself, the objectives are the ability to express time constraints and time-based decisions in the workflow. Time constraints on the production of results, will drive the orchestration allocation of tasks to resources to ensure the required computations are done on-time. And time-based decisions, will allow the workflow to choose the best results at a point in time, and to manipulate tasks to ensure the most efficient use of the available computing resources given the time available. We expect the following two hypotheses to hold true in the context of this (and similar) workflows:

- Precise simulations will be too costly to prove that they will always end on-time, both for the complexity of those simulations, and for the necessary short time margin to ensure a worst case run ends before the deadlines, so wasting a huge amount of computing resources if the orchestration does not accommodate for simulations that may be unable to provide results before a deadline.
- Compute resources available to the workflow will be dynamic, with a reserved amount possibly guaranteed by contract between an infrastructure provider and civil authorities, and additional resources added on-line during the course of the emergency, mandating an elastic orchestration of tasks.

This is reflected in how the LEXIS technologies will be supporting this workflow, and how the workflow orchestration will be implemented. Knowledge of the compute characteristics of the various tasks of the workflow will also play a significant role, by allowing the manipulation of their parameters to ensure the best fit for the available computation and time resources.

3   The Earthquake and Tsunami Workflow

The earthquake and tsunami workflow is a combination of the OpenQuake code computing the shaking distribution (ShakeMap) of an earthquake, the tsunami simulation code TsunAWI, the dynamic exposure model, the loss assessment code, the satellite-based emergency mapping (SEM) process, and the HeScade time constrained dataflow model of computation, this deployed on the LEXIS platform.

The workflow is both event-triggered and permanently online and requires: provision and update of the best available estimates and results at all time, and provision of the best results so far at specific points of time after the event (those precise points will be defined in the course of the project by interacting with emergency response agencies).

The permanently online component of the workflow is the dynamic exposure model based on OpenStreetMap, which updates itself constantly from the crowd-sourced OpenStreetMap database in real-time, so as to benefit from up-to-date information.

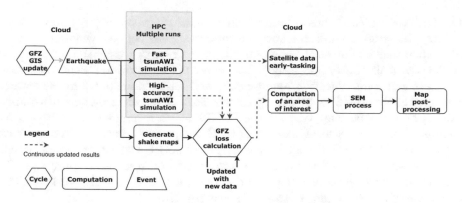

Fig. 2. ERS computational workflow on LEXIS infrastructure

All other parts are triggered upon reception of an earthquake event as shown in Fig. 2. Computations of the shaking distribution and the fast tsunami simulations are started immediately; more precise simulations are launched once more detailed information on the event are available. Fast simulations should provide results for emergency warning systems; more accurate results will be provided by the precise simulations (which would require more computation resources and time). The results define areas affected by shaking or inundation and are propagated to the remaining flow.

Upon reception of the earthquake ShakeMap, triggering an extraction of the relevant geographical subset of the world from the exposure model which is subsequently used for computing the loss assessment. This loss assessment will be continuously updated when the tsunami simulations provide results about inundated areas, with an eventual extension of the geographical area affected. Outputs are a geographical representation of the loss assessment and summaries.

The Satellite-based emergency mapping process is triggered by the availability of areas of interest. Such areas are determined by the earthquake event and its ShakeMap and by additional factors: the amount of tentative damage expected at that location, additional disaster information (extent of inundation or presence of inundation), the importance of settlements at that position. The area of interest is updated through the results of the TsunAWI simulation and the damage/loss assessment; fast simulations allow for a shorter time to determine the early areas of interest and better results allow for continuous refinement of the areas of interest. Damage assessment and inundation results are also integrated into the SEM products.

4   The Workflow Components

4.1   The HeScade Model of Computation

We selected a model of computation to allow for taking into account the Real-Time constraints of the ERS. To that end, we define a *Model of Computation*

(MoC) based on two types of actors: compute actors (tasks or jobs) and control actors, the latter orchestrate how the former should behave, when they start and when they can or must terminate (depending on the associated constraints within the goals of the application). The model of computation also encompasses the data and communication requirements and interactions between the actors, and can be described as belonging to the class of dataflow models of computation, and will be mapped over the LEXIS platform unified orchestrator, as well as within the platform Cloud and HPC components' task runtimes such as HyperLoom [3] or COMPS [2]. To this end, the LEXIS platform orchestrator will be responsible to install the software components on reserved resources, while the runtime tasks scheduler (e.g., HyperLoom or COMPS) will coordinate the execution of compute and control tasks accordingly.

The MoC defines an application as a K set of compute agents, C as a set of control agents, e as a set of communication edges in $K \times K \cup K \times C$ (*i.e.*, either between compute agents or between a compute agent and a control agent), d as a set of control edges in $C \times K$ *i.e.*, exclusively between a control agent and a compute agent. The edges are connected to so-called data ports and control ports of the agents so that:

- Any compute agent possesses at most one control port,
- No control agent can possess a control port,
- Any port is associated with a number of tokens (either data tokens or control tokens) that correspond to a predefined amount of data transfer which usually is an atomic production (for output port) or intake (for input ports),
- Any compute agent can have an arbitrary number of input or output data ports and each port is associated with a fixed series of numbers in $\mathbb{N}^k$ which defines the number of expected data tokens for each firing of the agent. When the last number of tokens from the series is reached, the series is restarted from the start, hence for any agent, the number of tokens consumed or produced on each port is a constant number throughout the super-period that includes all the agents. This ensures SDF [9] and CSDF [6] equivalence.

Contrary to the usual behaviour of processes in SDF or CSDF, agents in our MoC data input ports are not necessarily blocking when the number of expected data-token is not reached. The occurrence of a control token can "awaken" an actor so that depending on the planned behaviour, it can discard some of the input ports and only use the available data-token to run its processing and produce data-tokens on the output ports. Without control ports, compute agents must wait for all the input ports to reach the number of data-tokens specified by the intake behaviour of the agent, as in the SDF MoC.

The advantage of an SDF/CSDF equivalency is that there exists mathematical proof of when an application described this way is self-consistent and will avoid deadlock while running in finite memory. The extension of control ports allows providing real-time constraints to a system in which time-boundedness (for SDF/CSDF) is not guaranteed.

Therefore, by modelling the application with such a MoC, we can ensure that: first, the system can run without deadlocks and in finite memory provided it is well designed (which is a simple mathematical verification) and second, the correct use of control agent to enforce latency deadlines to the processes being fired by the orchestrator (see [4,10]). With at least one simulation with a parameter set allowing to meet the deadline with the initial resources reserved for the workflow, the availability of the outcome of at least this simulation is guaranteed at the deadline. At the specified deadline, the best available result is identified and can be forwarded to the next step in the workflow.

For the orchestrator engine managing the resources of the platform, this *a priori* knowledge of the availability of results makes the task of allocating resources easier. The resources reserved in case of an alert are used to start the simulation with guaranteed results at the deadline. Unused reserved resources can be used to start additional simulations. Also, considering that additional resources can become available after an alert, it is interesting to introduce a work-stealing approach that is driven by the data flow dependencies, as in [7]. In any case, the communication resources for the selected result can be reserved independently.

4.2   TsunAWI

TsunAWI was developed in the framework of the German-Indonesian Tsunami Early Warning System (GITEWS, 2005–2011, funded by the German Federal Ministry of Education and Research, [12]) and simulates all stages of a tsunami from the origin, the propagation in the ocean to the arrival at the coast and the inundation on land [11]. It solves the non-linear shallow water equations (SWE) on unstructured triangular finite elements that allow covering the coastal areas with a high resolution, while the long tsunami waves in the deep ocean can be represented by a coarse mesh, thus saving computation time and memory. The quality of the triangulation is crucial for the model results. We usually employ a mesh generator based on Triangle [14] with additional smoothing steps as described in [5]. Starting from a model domain defined within a topography/bathymetry data set, the mesh generator builds a mesh based on refinement rules depending on the water depth, the gradient of the bathymetry, and user defined criteria like minimum and maximum resolution and optional foci on regions of interest.

The scenarios for the Indonesia tsunami early warning system InaTEWS have a resolution of 12 km in the deep ocean, 150 m at the coast, and 50 m in project regions and around tide gauges. In hindcasts of real events, the resolution of 50 m is sufficient to simulate realistic waveforms at most tide gauge locations. For inundation studies, however, the horizontal resolution should be as fine as 20 m, as shown in [8]. Accordingly, the setup for LEXIS consists of a regional mesh with 20 m resolution in the city of Padang, 150 m elsewhere at the coast, and 5 km in the ocean, see Fig. 3 (Fig. 4).

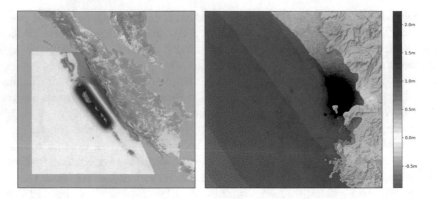

Fig. 3. Mesh with an initial condition for the simulation of tsunami inundation in the city of Padang. Map tiles by Stamen Design, under CC BY 3.0. Data by Open-StreetMap, under ODbL

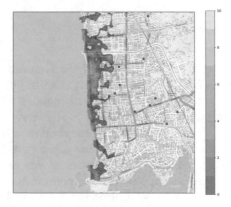

Fig. 4. Inundation simulation for Padang Map tiles by OpenStreetMap

4.3   Global Dynamic Exposure

Detailed understanding of local risk factors regarding natural catastrophes requires an in-depth characterization of the local exposure. In LEXIS we aim at using an exposure model [13] providing exposure and vulnerability indicators for all building footprint present in OpenStreetMap (OSM). OSM is the rich and constantly growing geographical database, that contains more than 5 billion geographical nodes, more than 1/3 of a billion building footprints (growing by more than 100,000 per day), and a plethora of information about school, hospital, and other critical facilities.

To ensure timely delivery of exposure and vulnerability indicators for an area affected by an earthquake or a tsunami, all indicators need to be precomputed and kept up-to-date. This requires the assessment of indicators anytime a building is added, changed, or deleted from OSM or a change is affecting the indicators of a building. This near-realtime processing is computationally demanding because in the order of 1 million buildings globally need to be (re-)assessed daily. With this approach, we increase the resolution of existing exposure models from aggregated exposure information to building-by-building vulnerability.

The impact of the LEXIS project on the exposure model and loss assessment based on the exposure model will be twofold:

- The process is data and compute intensive, and the project platform will be used to accelerate and allow the process to scale. Key elements are the need to keep the database up-to-date all the time so as not to have to base a damage estimate on outdated data, and the ability to extract the impacted area from the global database without delay upon reception of an event.
- In emergencies and for disaster response, having the possibility to ensure that results are delivered to the point where they are needed, and so benefit from acceleration and orchestration technologies ensuring urgent tasks are deployed timely and on the right resources.

4.4   Integration into an Satellite-Based Emergency Mapping Workflow

The ultimate goal of an SEM mechanism is to improve disaster relief effectiveness and thus to help reduce suffering and fatalities before, during and after a disaster event occurs [15]. Specifically dealing with the response phase immediately after a disaster, which typically lasts from several days to a few weeks, time is the major constraint: this is where automatic models and procedures potentially provide the greatest benefits. The Copernicus Emergency Management Service (CEMS), the European Commission SEM mechanism, is composed of an on-demand mapping component providing rapid maps for emergency response and risk & recovery maps for prevention and planning. Two are the specific CEMS phases that can be significantly improved by the exploitation of models and algorithms integrated into the LEXIS platform: (a) early-tasking of satellite acquisitions, exploiting the output of the TsunAWI model and (b) automatically generate a first estimate product (FEP), combining TsunAWI inundation outputs with building exposure data.

Most of earth observation satellite platform acquire images on request and have specific cut-off time to program an acquisition: missing this cut-off time would lead to a missing opportunity for data acquisition and to at least a 24 h delay. TsunAWI estimated wave height on the coast, crossed with some globally available exposure datasets, such as population places or population distribution, can be used to define and prioritize a set of Areas of Interest (AoI) polygons, to be submitted to the satellite data provider in the shortest time frame possible.

A First Estimate Product (FEP) provides a very fast yet rough assessment of the most affected locations. Coastal area inundation extent and maximum water height can be integrated with available exposure data at a single building level, in order to quickly generate first damage estimates, while waiting for the results of the final analysis which is more time consuming. In big disasters, like tsunamis,

FEP allows understanding the situation quickly and supports identifying focus areas for further image tasking and analysis.

Figure 6 displays the benefit of the LEXIS workflow, compared with a standard one triggered by a user activation request: in the reproduced scenario, thanks to the early-tasking, the first post-event image is expected to be delivered 1 day in advance. Furthermore, the FEP product is generated 1.5 days earlier, thanks to the availability of the TsunAWI inundation and the building exposure data (Fig. 5).

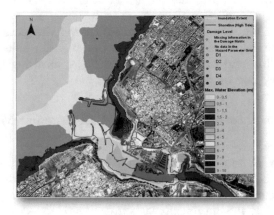

Fig. 5. Example of a FEP based on inundation extent and maximum water height combined with building exposure

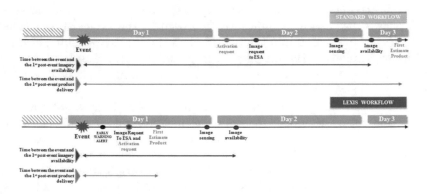

Fig. 6. Comparison, in terms of product timeliness, of a standard workflow and one based on LEXIS workflow

5   Conclusion

We have shown how, in the LEXIS project, an earthquake and tsunami workflow will be built, leveraging an orchestrated heterogeneous environment combining high-performance-computing and Cloud resources.

The tsunami simulation code TsunAWI will be used to provide both fast online simulations and accurate compute intensive simulations after an earthquake event. The dynamic exposure model, kept up-to-date with the latest information from the OpenStreetMap database, will be used to compute damage and loss assessments based on the earthquake ShakeMap and the tsunami inundations. Finally a satellite-based emergency mapping process will produce early, accurate maps of the affected areas. A time constrained dataflow model of computation will be used to ensure that the orchestration of the workflow will respect the time constraints and optimize for the computing resources available.

This will be deployed over the three key technologies of the LEXIS project, that are the infrastructure, the orchestration and the data management. Infrastructure services will be ensured by the LRZ and IT4Innovations centres, and heterogeneous acceleration capabilities will be employed to accelerate key points of the workflow.

Acknowledgements. This work was supported by the LEXIS project - the European Union's Horizon 2020 research and innovation programme under grant agreement No. 825532.

References

1. Terzo, O., Walter E., Levrier, M., Hachinger, S., Magarielli, D., Goubier, T., Louise, S., Parodi, A., Murphy, S., D'Amico, C., Ciccia, S., Danovaro, E., Lagasio, M., Donnat, F., Golasowski, M., Quintino, T., Hawkes, J., Martinovic, T., Riha, L., Slaninova, K., Serra, S., Peveri, R., Scionti, A., Martinovic J.: HPC, cloud and big-data convergent architectures: the lexis approach. In: Conference on Complex, Intelligent, and Software Intensive Systems (CISIS) (2019)
2. Badia, R.M., Conejero, J., Diaz, C., Ejarque, J., Lezzi, D., Lordan, F., Ramon-Cortes, C., Sirvent, R.: Comp superscalar, an interoperable programming framework. SoftwareX **3**, 32–36 (2015)
3. Cima, V., Böhm, S., Martinovič, J., Dvorský, J., Janurová, K., Vander Aa, T., Ashby, T.J., Chupakhin, V.: HyperLoom: a platform for defining and executing scientific pipelines in distributed environments. In: Proceedings of the 9th Workshop and 7th Workshop on Parallel Programming and RunTime Management Techniques for Manycore Architectures and Design Tools and Architectures for Multicore Embedded Computing Platforms, PARMA-DITAM 2018, pp. 1–6. ACM, New York (2018)
4. Do, X., Louise, S., Cohen, A.: Transaction parameterized dataflow: a model for context-dependent streaming applications. In: 2016 Design, Automation & Test in Europe Conference & Exhibition, DATE 2016, Dresden, Germany, 14–18 March 2016, pp. 960–965 (2016)
5. Frey, W.H., Field, D.A.: Mesh relaxation: a new technique for improving triangulations. Int. J. Numer. Methods Eng. **31**(6), 1121–1133 (1991)

6. Lauwereins, R., Bilsen, G., Engels, M., Peperstraete, J.A.: Cyclo-static data flow. IEEE Trans. Signal Process. **44**(2), 397–408 (1996)
7. Gautier, T., Besseron, X., Pigeon, L.: KAAPI: a thread scheduling runtime system for data flow computations on cluster of multi-processors. In: Proceedings of the 2007 International Workshop on Parallel Symbolic Computation, PASCO 2007, pp. 15–23. ACM, New York (2007)
8. Griffin, J., Latief, H., Kongko, W., Harig, S., Horspool, N., Hanung, R., Rojali, A., Maher, N., Fuchs, A., Hossen, J., Upi, S., Dewanto, S.E., Rakowsky, N., Cummins, P.: An evaluation of onshore digital elevation models for modeling tsunami inundation zones. Front. Earth Sci. **3**, 32 (2015)
9. Lee, E.A., Messerschmitt, D.G.: Synchronous data flow. Proc. IEEE **75**(9), 1235–1245 (1987)
10. Louise, S., Dubrulle, P., Goubier, T.: A model of computation for real-time applications on embedded manycores. In: 2014 IEEE 8th International Symposium on Embedded Multicore/Manycore SoCs (MCSoc), pp. 333–340, September 2014
11. Rakowsky, N., Androsov, A., Fuchs, A., Harig, S., Immerz, A., Danilov, S., Hiller, W., Schröter, J.: Operational tsunami modelling with TsunAWI - recent developments and applications. Nat. Hazards Earth Syst. Sci. **13**, 1629–1642 (2013)
12. Rudloff, A., Lauterjung, J., Münch, U. (eds.): The GITEWS Project (German-Indonesian Tsunami Early Warning System). NHESS - Special Issues. Copernicus Publications, Göttingen (2009)
13. Schorlemmer, D., Beutin, T., Hirata, N., Wyss, M., Cotton, F., Prehn, K.: Global dynamic exposure and the OpenBuildingMap - communicating risk and involving communities. In: EGU General Assembly Conference Abstracts, volume 20 of EGU General Assembly Conference Abstracts, p. 12871, April 2018
14. Shewchuk, J.R.: Triangle: engineering a 2D quality mesh generator and delaunay triangulator. In: Lin, M.C., Manocha, D. (eds.) Applied Computational Geometry: Towards Geometric Engineering, volume 1148 of Lecture Notes in Computer Science, pp. 203–222. Springer, Heidelberg (1996). From the First ACM Workshop on Applied Computational Geometry
15. Voigt, S., Giulio-Tonolo, F., Lyons, J., Kučera, J., Jones, B., Schneiderhan, T., Platzeck, G., Kaku, K., Hazarika, M.K., Czaran, L., Li, S., Pedersen, W., James, G.K., Proy, C., Muthike, D.M., Bequignon, J., Guha-Sapir, D.: Global trends in satellite-based emergency mapping. Science **353**(6296), 247–252 (2016)

Alternative Paths Reordering Using Probabilistic Time-Dependent Routing

Martin Golasowski[✉], Jakub Beránek, Martin Šurkovský, Lukáš Rapant, Daniela Szturcová, Jan Martinovič, and Kateřina Slaninová

IT4Innovations, VSB - Technical University of Ostrava,
17. listopadu 2172/15, 708 00 Ostrava-Poruba, Czech Republic
{martin.golasowski,jakub.beranek,martin.surkovsky,lukas.rapant,
daniela.szturcova,jan.martinovic,katerina.slaninova}@vsb.cz

Abstract. In this paper we propose an innovative routing algorithm which takes into account stochastic properties of the road segments using the Probabilistic Time-Dependent Routing (PTDR). It can provide optimal routes for vehicles driving in a smart city based on a global view of the road network. We have implemented the algorithm in a distributed on-line service which can leverage heterogeneous resources such as Cloud or High Performance Computing (HPC) in order to serve a large number of clients simultaneously and efficiently. A preliminary experimental results using a custom traffic simulator are presented.

1 Introduction

Cities around the world observe the ever-increasing level of traffic in their streets. Evolution of personal computing devices such as smart phones and spread of IoT sensor networks is enabling innovative approaches to traffic optimization. We have developed a server-side navigation service which provides optimal paths based on the current global view of the road network. The global view of the network is produced periodically by a fusion of different data sources such as traffic monitoring, weather forecast, floating car data (FCD), etc. The service can be easily used from a smartphone app or an in-car navigation system through simple HTTP API. It is also designed to be deployed on heterogeneous Cloud/High Performance Computing (HPC) infrastructure for dynamic scaling and efficient handling of thousands of cars driving in a Smart City.

This paper focuses mainly on description and implementation of routing algorithm implemented in the routing service. This algorithm can be seen as an extension of k-alternative routing with routes reordering and also it takes into account stochastic properties of traffic network. The algorithm can be roughly divided into three parts. The first part is a static routing which provides several alternative paths between origin and destination. The second part is the Probabilistic Time-Dependent Routing (PTDR) which estimates travel time distribution for each path. The third part then computes weights for each of the paths and determines the optimal path for a given time of departure. This routing

© Springer Nature Switzerland AG 2020
L. Barolli et al. (Eds.): NBiS-2019, AISC 1036, pp. 235–246, 2020.
https://doi.org/10.1007/978-3-030-29029-0_22

algorithm pipeline is thoroughly described in Sect. 2. Section 3 briefly describe implementation of the proposed algorithm in the routing service. Experimental results obtained with a custom traffic simulator are presented in Sect. 4.

1.1 Related Work

Computation of the travel time distribution on a path belongs to a well known class of problems related to optimal routing in stochastic networks. The problem can be seen as an attempt to maximise the probability of arrival within a certain time budget. This category is commonly referred to as the Shortest path with on-time arrival reliability or SPOTAR. In this formulation, the optimal path is selected from a set of known paths [6]. A pseudo-polynomial heuristic for SPOTAR is presented in [8]. In case of normally distributed travel times on the edges, a poly-logarithmic algorithm exists [9]. From the more current works, we have chosen article [2]. It aims at solving the routing problem of several cars from more general from the graph theory perspective. Another direction of current research are several articles concerning the use of Ant colony optimization (ACO) together with some protocol for vehicular communication to perform the routing from the more global perspective of several cars. Example of this approach can be found for example in [7].

In the context of the traffic optimization, our algorithm can be seen as an approximation of the SPOTAR problem, where paths are determined *a priori* and their stochastic properties are determined afterwards. This approach allows us to create an algorithm which can determine the path on-line, compared to the exact algorithms which cannot be used in an on-line service due to their complexity.

2   Advanced Routing Pipeline

Traffic in a city can be affected by a lot of various factors. Intensity and direction of the traffic varies rapidly over time, can be affected by holidays, festivals or simply by a work schedule. Accidents and repairs can block major arteries of the city, which can increase load across a very wide area. Therefore, traffic can have stochastic properties which have to be taken into account when determining an *optimal* path.

The notion of optimality gets more complicated when additional information like travel time distribution or traffic load prediction are available for the paths obtained from the static routing algorithm. This additional information makes the problem of finding the true optimal path through the traffic network almost unfeasible. This knowledge leads to a need to develop some form of heuristic algorithm which approaches the optimal solution as close as possible. Also this additional information creates kind of an uncertainty in what does the optimality mean. For somebody, optimal path would be the one which very reliably takes him to the destination in some reasonable time, while someone other would prefer a path that can be faster, but with some potential to backfire and take much longer time.

Given these facts, the problem of optimality must be extended with more complicated criteria. For example approach from the later example can be called a *risky* path and can be described as the path with the smallest mean duration without taking account some less probable outcomes. On the opposite, the former user can prefer the *safe* path which is determined by high (90%) percentile of the estimated travel time distribution instead of its mean. This may lead to a path that is not usually faster but has very stable duration. We call these specific routing approaches as strategies.

Those are just a few examples of more complicated criteria of path optimality. Availability of other data sources like prediction of load on the segments for certain times of day complicates the path selection process even further. Therefore, in our algorithm we make use of the process called reordering which defines a relationship between the different metrics provided for the initial k-paths. This approach is inspired by the vector model results reordering from information retrieval domain [4].

The pipeline is represented by a pseudocode in Algorithm 1. Input of the pipeline consists of origin and destination points and time of planned departure. Individual algorithms of the pipeline then use different representations of the current global view and have their own sets of parameters. Details of the algorithms are explained in the subsections below. Output of the pipeline is a list of paths between the two points sorted according to the selected optimality criteria.

Algorithm 1. Routing algorithm with PTDR and reordering

---

function GetRoutes(origin, destination, departureTime, configuration, speedProfiles)
    routes ← GetAltRoutes(origin, destination, configuration)
    drivingTimes ← PTDR(routes, departureTime, configuration, speedProfiles)
    bestRoute ← Reorder(routes, drivingTimes, configuration, speedProfiles)
    return *orderedRoutes*
end function

2.1   k-Alternative Paths

Initial set of k-alternative paths in our approach is determined by the Plateau method [1]. It is a variant of point-to-point routing heuristic operating on a static graph. It works by performing two passes of a basic Dijkstra algorithm between two points, while one pass is forward (origin to destination) and one pass is backwards.

Two subgraphs are constructed and the algorithm then searches for an intersection of these two subgraphs which forms the plateau. k shortest paths whose nodes appear in the plateau are returned as a result of the algorithm. Main input of this part is a graph representation of the road network, where some of the segment metadata (optimal speed, road closure) can be provided by the global view.

2.2 Probabilistic Time-Dependent Routing (PTDR)

The next step in the routing algorithm pipeline is used to estimate a distribution of travel time along the k provided paths. The PTDR part uses a Monte Carlo simulation to sample a *probabilistic* speed profiles assigned to the individual segments of the paths. The profiles are essentially an empirical probability distribution of a travel speed on the segment with fixed amount of support points.

The support points correspond to a notion of *Level of Service* (LoS) where the point with the highest speed corresponds to the most optimistic situation with no congestion at all and the point with the lowest speed corresponds to a complete traffic-jam on the segment. Each profile is valid only within a given time interval. At this point, we use only fixed size time interval for all profiles, variable size interval may be implemented in the future. The data used to derive these profiles may not be available for all time intervals or all segments, in this case an *optimal* speed determined from geographic properties of the segment is used as value of the current sample in the simulation. More details of the PTDR algorithm[1] are in [3,13]. A synthetic set of the probabilistic speed profiles for selected road segments in Prague is available on Zenodo [11].

Output of this step is a sample of travel times for each path whose size is given by a configuration. Estimated numeric characteristics of the travel time distribution are then used to describe the stochastic properties of the individual paths.

Number of samples affects precision of the output and run-time of this phase. A trade-off between the precision and run-time can be made using an on-line autotuner such as mArgot. We have successfully integrated the autotuner in this phase within the ANTAREX H2020 project [12].

2.3 Path Reordering

The path reordering is the most important part of the proposed routing pipeline as it weights the incoming paths according to a given criteria of optimality. Input of this phase consists of k-paths and result of the PTDR. The clients of the routing service can also provide an itinerary, which consists of current path the client is going to take at the moment.

Path R_n is composed of an ordered set of segments $s_i, i \in \langle 1, |r_n| \rangle$. Weight w_n is computed for each of k paths as a linear combination of selected metrics $\sigma = \{\sigma_1, ..., \sigma_k\}$

$$w_n = \sum_{i=1}^{k} \alpha_i \cdot \sigma_i,$$

where $\alpha_i \in\ <0; 1>$, $\sum_{i=1}^{k} \alpha_i = 1$ is a weight factor of individual metric σ_i which can be used to adjust influence of this metric.

---

[1] PTDR Source code - https://github.com/It4innovations/PTDR.

There are many metrics that can be utilized for the reordering. For example, we have utilized the following metrics:

- Dynamic and static properties of individual road segments (i.e. traffic intensity, road quality,...)
- Parameters of estimated travel time distribution (mean, deviation, quantiles)
- Difference from the current itinerary (set of waypoints).

Given these weights and metrics, we can find

$$R_{opt} = \arg\min_{n \in \{1,...,k\}} w_n,$$

where R_{opt} represents the best of k paths given its dynamic and static properties (other than its static speed or length utilized in the first part of the algorithm). In our case, R_{opt} is chosen as the path with the minimal weight. Please note, that this equation can be transformed to the task of finding the maximum weight w_n. This choice only depends on construction of individual metrics and given our implemented metrics, we have utilized the minimization of weights. This algorithm can also easily be modified to return a number of the best routes from the reordering. This property allows the user to choose slightly less optimal route if he prefers.

3   High Performance Implementation

In a recent article [5] we have presented a simplified version of a distributed system that serves routing requests in real time. That version was used to test the scalability in an HPC environment and its capability to route tens of thousands of vehicles. From this perspective, the used routing algorithm was insignificant. In order to use more sophisticated algorithms in the architecture, such as PTDR with reordering, we needed to update the architecture in a way that the algorithms have a possibility to share data[2] globally.

The updated architecture is depicted in Fig. 1. The main change is the addition of *key-value storage* (KV storage) that represents a global view of "the world". It composes of all the possible data that can influence that global view. From the logical point of view, the storage can be viewed as a stack of layers where each layer keeps some information about the route segments, e.g. speed, number of vehicles, weather conditions, and so forth. This is illustrated in Fig. 2 where each layer keeps different sort of information.

Workers using a particular routing algorithm can store the data they need to share into a specific layer of the KV storage. However, it takes a certain amount of time until the data becomes available for other workers because they can only access the data from their local *cache storage*. The separation of the global view and the local cache storage introduces additional latency, however it also increases the scalability of the entire architecture. The routing algorithms

---

[2] Data computed by a routing algorithm.

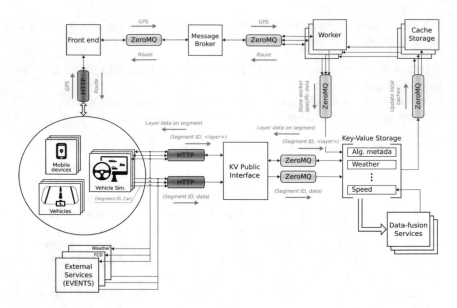

Fig. 1. Architecture of the routing system

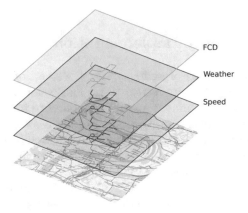

Fig. 2. Visualization of various layers that can be stored in the key-value storage.

have to be able to cope with the fact that their view of the virtual world may be slightly out of date. Nevertheless, this is a necessary condition anyway, since data from real-world vehicles is injected into the system with a certain latency of its own.

Another way to influence the global view is via *external services*. There is a public HTTP interface that allows any external service both to query the KV storage and modify it. In this way the global view can be informed about the current state of the world. For example, data from sensors about actual traffic, number of vehicles on the road or weather conditions.

The last way to modify the global view is via *data-fusion services*. These can modify the speed on segments based on data in other layers, e.g., slow down the speed on segments during heavy rain. Data-fusion services can be implemented as external services as well. Nevertheless, they are supposed to modify speed on segments based on data that is already present in the storage. Therefore, it is better for them to work with the KV storage directly for performance reasons.

The public routing interface was enhanced with the possibility to specify a strategy for computing the route. As outlined in the architecture, the real-world devices communicate via HTTP only with the front end. They send a request with the origin and destination and receive a computed route. The information about current traffic, number of vehicles on roads, and so forth, are pushed from external services asynchronously.

There is one specific component – *vehicle simulator* – that communicates with both the routing service and KV storage. It is used to add virtual vehicles to the global view and simulate heavy load of the system. Hence, there is no data from real sensors about these virtual vehicles, the simulator has to supply them itself. The original implementation contained a *segment speed simulator*. Its functionality is covered by the KV storage in the updated version – a data-fusion service modifies the speed on segments based on the number of vehicles (including the virtual ones) on the segments.

4   Experiments

We have performed a set of experiments using a traffic simulator which is used to simulate cars which drive in a city according to paths provided by the routing service [10]. Purpose of the experiment is to verify, that travel time can be lower for a defined group of cars driving in a city. Baseline of the experiment is simulation which uses just a basic static routing algorithm (Dijkstra) without the global view. We will run the same simulation scenario again with the full routing pipeline and observe how long the simulated cars spent driving.

The simulation scenario consists of cars, events, capacity of the virtual traffic network and routing service settings. Further description of the simulation parameters can be found in [10]. The first scenario is placed in Prague and is based on creating artificial traffic jams on certain bridges and observing behaviour of the traffic spilling on other bridges unaffected by the jam.

Origin (red) and destination (blue) points of this scenario are shown Fig. 3. The first three origin points are placed in an area behind the river, close to a bridge which is part of the fastest path to the destination. The fourth origin point is placed on the other side of the river close to the destination point to further affect traffic in this area by generating additional load.

The scenario runs 2,000 cars evenly distributed between the four origin points. The cars depart in 2 s. intervals simultaneously from the origin points from the beginning of the simulation. Each car sends a request for new route every 30 s.

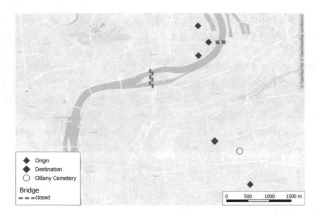

Fig. 3. Prague bridges scenario, four red points are the origin locations, blue point is the destination near the Olšany cemetery in Prague

Table 1 contain sequences of snapshots of virtual road network segments affected by the simulation with the actual speed corresponding to the given time step. Recording of the simulation progress is also available online[3].

Color of the segment shows current speed on the segment. Lower speed on the segment means high traffic level and vice versa. The left column of the table shows the results from the simulation executed using simple pipeline, with a basic routing algorithm. The pictures obtained for different simulation time steps show that the behaviour of the simulation is correct and when the capacity of the road is full, the system prefers less congested roads. This is visible especially on the last picture, when traffic increases in the smaller streets around the points of origin. This resulting traffic congestion in the network is used as a baseline for later optimization.

The next step after running the scenario with the simple routing is to repeat its execution several times; each time reducing speed on a different bridge in order to simulate traffic jam events in the road network. Average speeds gathered from the virtual road network are then combined together in speed profiles and probabilistic speed profiles in order to reflect differences in the traffic behaviour induced by the events. The scenario is executed with the generated probabilistic speed profiles using the full routing pipeline, including the PTDR and reordering phases. The expected result is that the mean trip time and total time of simulation will be shorter compared to the original run with the simple routing algorithm. Result of this run is visible in the right column of the Table 1.

It is clearly visible that segments affected in the simple scenario (left) for the same time step had faster speeds for the optimized run (right). Moreover, thanks to the optimization, a smaller area has been affected by the traffic, compared to the non-optimized situation. The following section provides a detailed summary of the trip time statistics for both runs and its comparison.

---

[3] https://youtu.be/OBXgGI7w_EA.

Table 1. Simulation progress

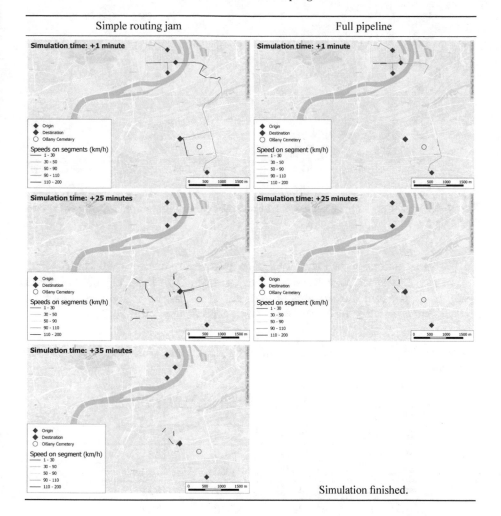

4.1 Driving Time Statistics

Table 2 contains numeric characteristics of car trip times in the two simulation runs. It is clear that mean trip time in the optimized run is 13.18% better than the un-optimized run. The highest trip time in the simulation has been reduced by 22.32%. Another metric of interest is the total simulation run time computed as time spent between departure of the first car and arrival of the last car. In this case, the total run time is 32.22% shorter in the optimized run. This experiment successfully validates the routing service which uses the full pipeline with PTDR and reordering phases. The PTDR uses the probabilistic speed profiles provided by the data collection and the processing phase.

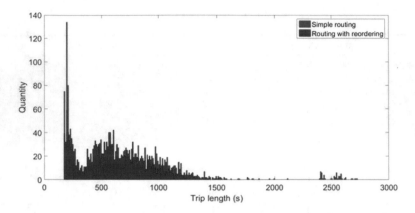

Fig. 4. Histogram of driving times with simulation runs with simple routing (blue) and reordering (red)

The *blue* histogram in Fig. 4 was generated from trip times for the run with simple routing. It shows a large number of outliers in the right side of x-axis meaning that significant amount of cars got stuck in the traffic jam and spent several times higher time on the road compared to the optimized run in *red*.

Table 2. Driving time statistics from simulation runs with simple routing and reordering

	Simple [s]	Full pipeline [s]	Difference [%]
Minimum	173.0	171.0	1.16
1st quartile	399.0	393.5	1.38
Median	671.0	584.0	12.98
Mean	705.6	612.6	13.18
Std. deviation	455.3	300.7	33.96
3rd quartile	948.0	800.0	15.61
Maximum	2,729.0	2,120.0	22.32
Total run time [s]	3,408.0	2,310.0	32.22

5   Conclusion and Future Work

We have proposed a novel routing algorithm which takes into account estimation of stochastic properties of the road segments and weights a set of alternative routes between two points. The algorithm has been implemented in an on-line distributed routing service deployed on a HPC system.

We have verified function of the algorithm with a custom traffic simulator which simulates the individual clients driving around the city along paths provided by the service. We managed to obtain a reduction of mean travel time by

13.18% on a simulated traffic scenario, compared to run which used only basic routing algorithm without any global view on the traffic network.

As the current implementation assumes that all drivers in the city drive along routes provided by our service, we would like to determine the minimal proportion of the traffic that is required to use the service in order to optimize the traffic in a given city in the future. Also we plan to experiment with more complicated routing strategies which can reflect various driver types.

Acknowledgements. This work was supported by The Ministry of Education, Youth and Sports from the National Programme of Sustainability (NPS II) project 'IT4Innovations excellence in science - LQ1602', by the IT4Innovations infrastructure which is supported from the Large Infrastructures for Research, Experimental Development and Innovations project 'IT4Innovations National Supercomputing Center – LM2015070', and partially by the SGC grant No. SP2019/108 'Extension of HPC platforms for executing scientific pipelines', VŠB - Technical University of Ostrava, Czech Republic.

References

1. Bader, R., Dees, J., Geisberger, R., Sanders, P.: Alternative route graphs in road networks. In: International Conference on Theory and Practice of Algorithms in (Computer) Systems, pp. 21–32. Springer (2011)
2. Blumer, S., Eichelberger, M., Wattenhofer, R.: Efficient traffic routing with progress guarantees. In: 2018 IEEE 30th International Conference on Tools with Artificial Intelligence (ICTAI), pp. 953–957 (2018). https://doi.org/10.1109/ICTAI.2018.00147
3. Golasowski, M., Tomis, R., Martinovič, J., Slaninová, K., Rapant, L.: Performance evaluation of probabilistic time-dependent travel time computation. In: IFIP International Conference on Computer Information Systems and Industrial Management, pp. 377–388. Springer, Cham (2016)
4. Martinovič, J., Snášel, V., Dvorský, J., Dráždilová, P.: Search in documents based on topical development. In: Advances in Intelligent Web Mastering-2, pp. 155–166. Springer (2010)
5. Martinovič, J., Golasowski, M., Slaninová, K., Beránek, J., Šurkovský, M., Rapant, L., Szturcová, D., Cmar, R.: A distributed environment for traffic navigation systems. In: The 13th International Conference on Complex, Intelligent, and Software Intensive Systems, CISIS 2019 (2019, accepted)
6. Miller-Hooks, E., Mahmassani, H.: Path comparisons for a priori and time-adaptive decisions in stochastic, time-varying networks. Eur. J. Oper. Res. **146**(1), 67–82 (2003)
7. Mouhcine, E., Mansouri, K., Mohamed, Y.: Intelligent Vehicle Routing System Using VANET Strategy Combined with a Distributed Ant Colony Optimization: Methods and Protocols, pp. 230–237 (2019)
8. Nie, Y.M., Wu, X.: Shortest path problem considering on-time arrival probability. Transp. Res. Part B Methodol. **43**(6), 597–613 (2009)
9. Nikolova, E., Kelner, J., Brand, M., Mitzenmacher, M.: Stochastic shortest paths via quasi-convex maximization. Algorithms-ESA **2006**, 552–563 (2006)
10. Ptošek, V., Ševčík, J., Martinovič, J., Slaninová, K., Rapant, L., Cmar, R.: Real time traffic simulator for self-adaptive navigation system validation (2018)

11. Rapant, L., Golasowski, M., Martinovič, J., Slaninová, K.: Simulated probabilistic speed profiles for selected routes in Prague (2018). https://doi.org/10.5281/zenodo. 2275647
12. Silvano, C., Agosta, G., Bartolini, A., Beccari, A.R., Benini, L., Besnard, L., Bispo, J., Cmar, R., Cardoso, J.M., Cavazzoni, C., et al.: Antarex: a dsl-based approach to adaptively optimizing and enforcing extra-functional properties in high performance computing. In: 2018 21st Euromicro Conference on Digital System Design (DSD), pp. 600–607. IEEE (2018)
13. Tomis, R., Rapant, L., Martinovič, J., Slaninová, K., Vondrák, I.: Probabilistic time-dependent travel time computation using Monte Carlo simulation. In: Kozubek, T., Blaheta, R., Šístek, J., Rozložník, M., Čermák, M. (eds.) High Performance Computing in Science and Engineering, pp. 161–170. Springer, Cham (2016)

A Proposal of Recommendation Function for Solving Element Fill-in-Blank Problem in Java Programming Learning Assistant System

Nobuo Funabiki[1(✉)], Shinpei Matsumoto[1], Su Sandy Wint[1],
Minoru Kuribayashi[1], and Wen-Chun Kao[2]

[1] Department of Electrical and Communication Engineering, Okayama University,
Okayama, Japan
funabiki@okayama-u.ac.jp
[2] Department of Electrical Engineering, National Taiwan Normal University,
Taipei, Taiwan
jungkao@ntnu.edu.tw

Abstract. To assist *Java programming* educations, we have developed a
Web-based *Java Programming Learning Assistant System (JPLAS)* that
offers several types of programming exercises to cover different levels
of students. Among them, the *element fill-in-blank problem (EFP)* was
designed for novice students to study grammar and basic programming
skills through filling in the blank elements in a Java source code. In
general, an EFP instance can be generated by selecting an appropriate
code in a textbook or a Web site, and applying the *blank element selection
algorithm* to it. Since EFP is expected to involve the Java grammar,
plenty of EFP instances have been prepared for JPLAS, where a student
may select an EFP instance matching his/her level. In this paper, we
propose a *recommendation function* of selecting a proper EFP instance
to be solved next by considering the *difficulty levels* and *grammar topics*.
We verify the effectiveness of our proposal through applications to 48
students in the Java programing class in Okayama University.

1 Introduction

Nowadays, the object-oriented programming language *Java* has been extensively
used in a variety of application systems in societies and industries due to its high
reliability, portability, and scalability. Moreover, Java was selected as the most
popular object-oriented programming language in 2015 [1]. Hence, the strong
demand in advancing Java programming educations has emerged from industries. A typical Java programming education in a school consists of grammar
instructions with textbooks in classes and programming exercises with computer
operations.

© Springer Nature Switzerland AG 2020
L. Barolli et al. (Eds.): NBiS-2019, AISC 1036, pp. 247–257, 2020.
https://doi.org/10.1007/978-3-030-29029-0_23

To assist Java programming educations, we have studied a Web-based *Java Programming Learning Assistant System (JPLAS)* [2,3], which offers several types of programming exercises, starting from code reading and grammar study until practical code writing using object-oriented programming concepts. In each type, an answer from the learner will be marked automatically on the server. Currently, JPLAS provides the *element fill-in-blank problem* [4], the *code completion problem* [5], the *value trace problem* [6], the *statement element fill-in-blank problem* [7], and the *code writing problem* [8], to support self-studies of Java programming.

The *element fill-in-blank problem (EFP)* is intended for a student who learns the Java grammar and basic programming skills through code reading. In an EFP instance, a student is required to fill in the blanks in the given Java code with several blank elements called the *problem code*. The code should be of high-quality, most worth for *code reading*. To solve an EFP instance, a student needs to carefully read the problem code and understand the structure, the algorithm/logic, and the semantics. Subsequently, by exercising knowledge of grammar and applying syntax rules to each statement, it is presumed that the student will fill in the blanks correctly.

At present, a new EFP instance is generated by selecting an appropriate Java source code in a textbook or a Web site, and applying the *blank element selection algorithm* [4] to the selected code. Since EFP instances should cover the whole Java grammar, a great number of EFP instances have been prepared for JPLAS. As a result, each student is allowed to choose EFP instances which suit him/her.

In this paper, we propose a *recommendation function* of selecting a proper EFP instance by considering the current answering result of the student in JPLAS. This function considers both the *difficulty levels* and the *grammar topics* that this student has solved successfully in the past, and then, it will selects a new EFP instance with more advanced level and different grammar topics. The *difficulty level* is defined to represent the difficulty in solving an EFP instance, by analyzing solution results of students in Okayama University.

As the evaluation, we implemented the recommendation function in the JPLAS server, and applied it to 48 students in the Java programing class in Okayama University. Eventually, we not only confirmed the effectiveness of the proposal, but identified several problems to be improved in future works.

The rest of this paper is organized as follows: Sect. 2 reviews our preliminary works to this paper. Section 3 presents the recommendation function. Section 4 shows the evaluation results of the proposal. Finally, Sect. 5 concludes this paper with future works.

2   Element Fill-in-Blank Problem in JPLAS

In this section, we review the element fill-in-blank problem in JPLAS.

2.1   Software Platform for JPLAS

In the JPLAS server, we adopt *Linux* for the operating system, *Tomcat* for the Web application server, *JSP/Java* for application programs, and *MySQL* for the database as shown in Fig. 1.

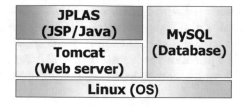

Fig. 1. JPLAS server platform.

2.2   Overview of Element Fill-in-Blank Problem

In the *element fill-in-blank problem (EFP)*, an *element* represents the least unit of a code which contains a reserved word, an identifier, a control symbol, and an operator. A *reserved word* signifies a fixed sequence of characters defined in the grammar to represent a specific function. An *identifier* is a sequence of characters defined in the code by the author to represent a variable, a class, or a method. A *control symbol* indicates other grammar elements such as "." (dot), ":" (colon), ";" (semicolon), "(,)" (bracket), "{, }" (curly bracket). An *operator* is used in a conditional expression to describe a condition to determine a logic in a code, such as "<" and "&&"..

2.3   Generation of Element Fill-in-Blank Problem

On the whole, a new EFP instance will be generated through the four steps: (1) to select a source code that covers the grammar topics to be studied, (2) to transform the code in (1) into a sequence of lexical units or elements and classify the type of each element by applying *JFlex* and *jay*, (3) to select the blank elements that have grammatically correct and unique answers by applying the *blank element selection algorithm* to the results in (2), and (4) to upload the generated EFP instance in (3) to the JPLAS server.

3   Proposal of Recommendation Function

In this section, we present the recommendation function of selecting the EFP instance to be solved next. It is noted that by changing the keyword list and assigning the weights properly, this proposal can be applied to other programming languages than Java.

3.1   Difficulty Level of EFP Instance

First, we present how to evaluate the *difficulty level* of an EFP instance.

3.1.1   Keywords for Grammar Topics

In this paper, the difficulty level of an EFP instance is calculated by taking the sum of the weights associated with the keywords that appear in the source code. A *keyword list* consists of the reserved words for Java, the identifiers, and the patterns in the regular expressions that should be mastered by a student in each grammar topic. Table 1 provides the keyword list with the grammar topics and their given weights. The grammar topics are referred from the JPLAS workbook in [9].

Table 1. Keyword list.

Weight	Grammar topic	Keywords
1	variable	var, stdin
1	operator	++, assignment operator
1	conditional statement	if, while, switch, relational operators, logical operators
1	for, while, do-while, break, continue	for, while, do-while, break, continue
1	arrays	length, 2D array, declare array
2	package, file	package, import
2	exception	try-catch, try-catch-finally, throws
3	class: field, method, member	field, method
3	class: overload, constructor, this	private, public, overload, constructor, this
3	class: library, String, class method	static, String, StringBuffer, Integer
3	class: inheritance, superclass, override	extends, super, protected, override
3	interface	interface, implements

3.1.2   Keyword Weight

The weights in Table 1 are selected based on the solving results of 33 students in the 2017 Java programming course at Okayama University. The average number of answer submissions to the server among all the students is regarded as an excellent index to represent the difficulty of the EFP instance, because most of the instances were correctly solved by the students after submitting the answers repeatedly. Then, the weights for the keywords were selected so that the obtained difficulty level has the highest correlation with the average number of answer submissions. Table 2 compares the correlation coefficients among the proposal, the number of statements in the code, and the number of blanks in the instance. This is, the proposal appears to provide a higher correlation coefficient than the others.

Table 2. Comparison of correlation coefficient.

Index	Correlation coefficient
Proposal	0.73
# of statements	0.68
# of blanks	0.43

3.1.3   Difficulty Level Calculation

First, each keyword in Table 1 is extracted from the source code. Next, the sum of the weights associated with these keywords is calculated, which becomes the difficulty level of this instance. For example, the following code has three keywords, *while*, *relational operator ("<")*, and ++. Thus, the difficulty level is $3 (= 1 + 1 + 1)$.

Listing 1: Example code for difficulty level.

```
 1 public class WhileSample {
 2    public static void main(String[] args) {
 3       int i = 0;
 4       while (i < 3) {
 5          System.out.println(i);
 6          i ++;
 7       }
 8       System.out.println("end");
 9    }
10 }
```

3.2   Solution Flow by Recommendation Function

The EFP instance solution flow by a student using the recommendation function is described here.

3.2.1   Recommendation of Next EFP Instance

The EFP instance to be solved next by the student is selected by applying the *instance recommendation algorithm* that will be presented below. This algorithm detects a proper EFP instance based on the correct answer rate and the number of answer submissions when this student solved the previous EFP exercise. When a student logs in JPLAS and starts solving EFP instances, the default initial instance stored in the server will be selected without exception. However, the initial instance should reflect the previous solution results from the student, which will be explored in future works.

3.2.2   Solution of Recommended EFP Instance

The student solves the selected EFP instance from the interface in Fig. 2. This interface shows the ID, the difficulty level, and the keywords of the instance. The student needs to fill in the blank forms with ideal elements. Then, when "Report" is clicked, the answers will be sent to the server to examine the correctness. If the student gives up searching for correct answers of specific blanks, "Give up" should be clicked. Then, the correct answers to the blanks appear on the forms so that the student will have knowledge of them. When all the forms are correct or "Give up" is clicked, "Report" is disabled and "Recommendation Question" is enabled. Then, by clicking "Recommendation Question", a new EFP instance will appear in the interface to be solved.

3.3   Instance Recommendation Algorithm

The *instance recommendation algorithm* identifies a proper EFP instance from the correct answer rate and the number of answer submissions at the previous EFP instance.

3.3.1   Update of Difficulty Level

The difficulty level L is updated by $L + \Delta L$. This change of the difficulty level ΔL is calculated depending on the correct answer rate $x(\%)$ in the previous EFP instance solution. When all the blanks were correctly solved ($x = 100$), ΔL is positive. Here, ΔL is changed by the number of submissions, since submitting answers repeatedly implies that the student may not understand the previous instance thoroughly. When "Give up" was clicked and x is smaller than the given threshold $M(\%)$ ($x < M$), ΔL is negative. Otherwise, ΔL is zero. The procedure to calculate ΔL is given as follows:

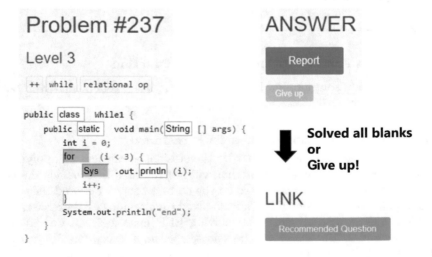

Fig. 2. EFP instance solving interface.

- If $x = 100$ and $y \leq N$,
$$\Delta L = D. \tag{1}$$

- If $x = 100$ and $y > N$,
$$\Delta L = \frac{D}{y - (N-1)}. \tag{2}$$

- If $x < M$,
$$\Delta L = D(\frac{x}{M} - 1). \tag{3}$$

- Otherwise,
$$\Delta L = 0. \tag{4}$$

In this algorithm, the difficulty level L is increased by ΔL when the correct answer rate x is 100%. This level change ΔL is controlled by the number of submissions y, because fewer submissions represent that the student has solved the previous EFP instance without difficulty. On the other hand, L is decreased when x is smaller than M, and ΔL is controlled by x. If x is small, the student does not understand the previous instance and should transfer to easier instances.

Table 3. Definitions of variables and parameters.

L	Current difficulty level	Parameter in paper
ΔL	Difficulty level change	
x	Correct answer rate (%)	
y	Number of submissions	
D	Maximum difficulty level change	8
N	Submission times parameter	3
M	Maximum correct answer rate (%)	80

3.3.2   Extraction of Candidate Instances
All the unsolved EFP instances whose difficulty levels exist between $L - 2$ and $L + 2$ are extracted.

3.3.3   Selection of Next EFP Instance
If the correct answer rate x is 100%, the EFP instance containing the largest number of elements that were failed through five or more answer submissions at solving the previous instance is selected from the extracted ones. Otherwise, the EFP instance containing the largest number of unsolved elements in solving the previous instance is selected.

4   Evaluation

In this section, we evaluate the proposed recommendation function.

4.1   Evaluation Setup

To evaluate the proposal, we implemented the function and prepared EFP instances whose difficulty levels are distributed from 1 to 31. Then, we asked 48 third-year students taking the Java programming course at Okayama University to solve them by using the function which starts from the specified instance whose difficulty level is 1 on JPLAS for 120 min during the class. Before this experiment, these students have studied the fundamentals of Java programming. When a student solved the instance at the highest level, he/she stopped solving a new instance, and was requested to answer the two questions in the questionnaire with five grades.

4.2   Questionnaire Result

Figure 3 shows the questionnaire result. On average, 60% students replied that they could recognize the weak points in Java programming and repeat to solve the relevant instances. For example, the solution record of such a student indicates that EFP instances using *void* and *StringBuffer* class were repeatedly selected because he often failed to solve them. Thus, the proposed function is effective in suggesting weak points in Java programming in general.

Still, around 20% students expressed that it was challenging for them to do so. When the solution record of one student is analyzed, the instances related to *BufferedReader* class and *Interface* were repeatedly selected. However, since he made numerous mistakes in other elements that were not considered in the recommendation function, he was unable to recognize his weak points clearly.

Most of the students who completed this experiment with the least number of EFP instances claimed that the increase of the difficulty level was too fast for them, and they were eager to solve more instances. To be specific, the five instances solved by them covered merely 50% of the keywords in Table 1. In

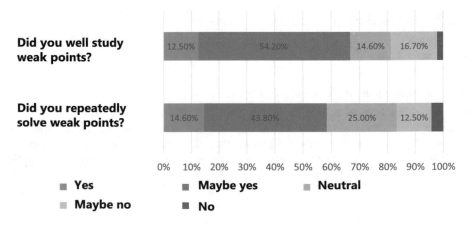

Fig. 3. Questionnaire results.

future works, we will improve the algorithm so that all the keywords may be covered by the assigned instances to each student even if he/she can solve the instances correctly with a few submissions. Then, we can avoid missing to detect the demanding grammar topics for the student.

4.3 Difficulty Level Change Results

Figure 4 shows the changes of the difficulty levels of the recommended instances by three levels of students. The high-level student completed this experiment by solving only five instances, where the difficulty level increases constantly. The middle-level student repeated instances several times at the level eight that are related to *void* and *StringBuffer*. It seems that this student did not understand the exercises entirely. After solving the instances, he completed the experiment. The low-level student saturated the level before reaching 20, and could not complete it. To encourage such a student to continue the study, it is important to develop a hint function, which will be explored in future works.

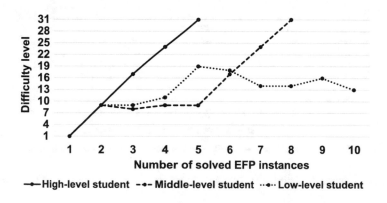

Fig. 4. Changes of difficulty levels of recommended instances.

4.4 Difficulty Level Change Results for Different Parameters

Then, we investigate the changes of the difficulty levels when the values of D, N, and M are different from Table 3. Figures 5, 6, 7 the results for (D, N, M) = (2, 3, 80), (3, 3, 75), and (5, 2, 80), where only 10 students solved them. It is found that a *high-level* reached the highest difficulty level in the shortest way, a *middle-level student* also reached the highest level but stayed at a certain level for a while, and a *low-level student* could not reach there. Actually, this student decreased the level after he/she could not solve EFP instances at the middle level, when a smaller value is used for D. Thus, the quick increase and the slow decrease of the level can be a good choice both for high-level and low-level students.

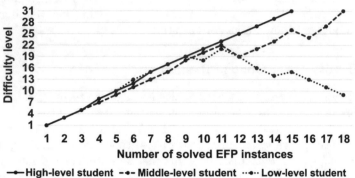

Fig. 5. Changes of difficulty levels of recommended instances for (2, 3, 80).

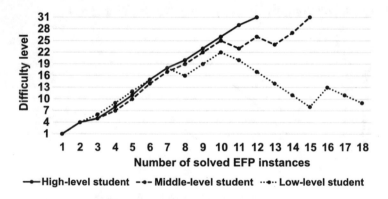

Fig. 6. Changes of difficulty levels of recommended instances for (3, 3, 75).

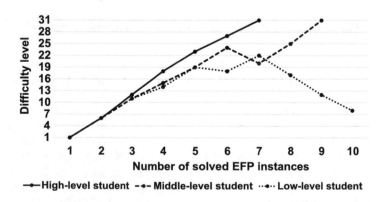

Fig. 7. Changes of difficulty levels of recommended instances for (5, 2, 80).

5   Conclusion

In this paper, we proposed the *recommendation function* of selecting the EFP instance to be solved next by considering the *difficulty levels* and *grammar topics*. We verified the effectiveness through applications to 48 students in the Java programing class in Okayama University. In the future, we will select the initial instance by reflecting the previous solution results provided by the student, improve the algorithm so that all the keywords will be involved in the assigned instances, examine existing recommendation techniques [11], implement a hint function to encourage a student to continue the study, and apply JPLAS with the proposal to Java programming courses.

References

1. Cass, S.: The 2015 Top Ten Programming Languages. http://spectrum.ieee.org/computing/software/the-2015-top-ten-programming-languages/?utm_so
2. Ao, S., et al. (ed.): IAENG Transactions on Engineering Sciences - Special Issue for the International Association of Engineers Conferences 2016, vol II, pp. 517–530. World Sci. Pub. (2018)
3. Ishihara, N., Funabiki, N., Kuribayashi, M., Kao, W.-C.: A software architecture for Java programming learning assistant system. Int. J. Comput. Softw. Eng. **2**(1), 116 (2017)
4. Funabiki, N., Tana, Zaw, K.K., Ishihara, N., Kao, W.-C.: A graph-based blank element selection algorithm for fill-in-blank problems in Java programming learning assistant system. IAENG Int. J. Comput. Sci. **44**(2), 247–260 (2017)
5. Kyaw, H.H.S., Aung, S.T., Thant, H.A., Funabiki, N.: A proposal of code completion problem for Java programming learning assistant system. In: Proceedings of CISIS, pp. 855–864 (2018)
6. Zaw, K.K., Funabiki, N., Kao, W.-C.: A proposal of value trace problem for algorithm code reading in Java programming learning assistant system. Inf. Eng. Express. **1**(3), 9–18 (2015)
7. Ishihara, N., Funabiki, N., Kao, W.-C.: A proposal of statement fill-in-blank problem using program dependence graph in Java programming learning assistant system. Inf. Eng. Express. **1**(3), 19–28 (2015)
8. Funabiki, N., Matsushima, Y., Nakanishi, T., Amano, N.: A Java programming learning assistant system using test-driven development method. Int. J. Comput. Sci. **40**(1), 38–46 (2013)
9. Tana, Funabiki, N., Zaw, K.K., Ishihara, N., Matsumoto, S., Kao, W.-C.: A fill-in-blank problem workbook for Java programming learning assistant system. Int. J. Web Inform. Syst. **13**(2), 140–154 (2017)
10. JUnit. http://junit.org/
11. Lu, J., Wu, D., Mao, M., Wang, W., Zhang, G.: Recommender system application developments: a survey. Decis. Support Syst. **74**, 12–32 (2015)

Web-Based Interactive 3D Educational Material Development Framework and Its Authoring Functionalities

Daiki Hirayama[1], Wei Shi[2], and Yoshihiro Okada[1,2]([⊠])

[1] Graduate School of Information Science and Electrical Engineering,
Kyushu University, Fukuoka, Japan
okada@inf.kyushu-u.ac.jp
[2] Innovation Center for Educational Resources (ICER), Kyushu University
Library, Kyushu University, Fukuoka, Japan

Abstract. Recent advances of web technologies besides computer hardware technologies have made web-based applications popular. Similarly, web-based educational materials have become popular because learners can study at anytime, anywhere using such materials on web browsers in their mobile devices like smartphones. This paper treats the development framework for such web-based educational materials. To increase the learning motivation of students, it is necessary to provide the students with attractive educational materials using state-of-the-art ICT like 3D CG. The authors have already proposed a web-based interactive 3D educational material development framework and several educational materials as its examples. However, the proposed framework has not had enough authoring functionalities. In this paper, the authors added several authoring functionalities to the proposed framework in order to make it possible to develop educational materials for standard users who do not have programming knowledge and skills like teachers. This paper introduces the added authoring functionalities to justify the usefulness of the proposed framework.

Keywords: e-Learning · 3D graphics · Educational materials · Development framework · Authoring environment

1 Introduction

In this paper, we treat a web-based interactive 3D educational material development framework. This research is one of the activities of Innovation Center for Educational Resources (ICER) [1] of Kyushu University Library, Kyushu University, Japan. For the efficient education, we have to provide attractive educational materials realized using state-of-the-art ICT like 3D CG, because current students, whose are sometimes called video game generation, are used to operate such contents [2, 3]. However, it is difficult for ordinary teachers to develop such contents because they are requested to have technological knowledge and programming skills which they do not have. Therefore, teachers need some tools for easily developing such attractive educational materials. For supporting such kind of users, we have already proposed a framework dedicated for the development of web-based interactive 3D educational materials [4, 5].

© Springer Nature Switzerland AG 2020
L. Barolli et al. (Eds.): NBiS-2019, AISC 1036, pp. 258–269, 2020.
https://doi.org/10.1007/978-3-030-29029-0_24

The proposed framework supports various types of media data like texts, audios, videos and 3D geometry data with their animation data, which are necessitated in attractive educational materials. In some cases like teaching a certain ceremony taken in a certain place at a certain era in the history study, teachers need story-based materials. So, our proposed framework also supports such story information. As case studies, we have already developed two educational materials using the framework. They are used for teaching certain ancient manners of the two ceremonies called 'Kanso' and 'Jimoku' taken in the ancient imperial court called 'Gosyo' in Japanese history study. These are web-based interactive 3D educational materials so that they can work on various platforms such as iPhone, iPad, and Android tablets besides standard desktop PCs. This is significant for BYOD (Bring Your Own Device) classes and Mobile Learning. Furthermore, to enable the development of more and more attractive educational materials using Virtual Reality (VR)/Augmented Reality (AR), we also introduced functionalities for that to the framework [6, 7].

The purpose of this research is to provide ordinary teachers with the framework that can enable the teachers to develop attractive educational materials. However, the current version of the proposed framework does not have enough authoring functionalities so that this time we added several authoring functionalities to the framework. In this paper, we introduce the added authoring functionalities to justify the usefulness of the proposed framework.

The remainder of this paper is organized as follows: Sect. 2 describes related work. We briefly explain the proposed framework, its design and functional components in Sect. 3, and introduce the two actual educational materials of Japanese history study. In Sect. 4, we explain newly introduced authoring functionalities. Finally, we conclude the paper and discuss about our future work in Sect. 5.

2 Related Work

There have been many development systems and tools for 3D contents so far. Some of them are commercial 3D CG software such as 3D Studio Max, Maya, and so on. However, these products can be used only for creating 3D CG images or 3D CG animation movies. Usually, they cannot be used for creating interactive contents. As a development system for 3D interactive applications, there is *IntelligentBox*, a constructive visual software development system for 3D graphics applications [8]. This system seems useful because there have been many applications actually developed using it so far. However, it cannot be used for creating web-based contents. Although there is the web-version of *IntelligentBox* [9], it cannot be used for creating story-based contents. With Webble world [10, 11], it is possible to create web-based interactive contents through simple operations for authoring of course, and it is possible to render 3D graphics assets as one interactive content. However, it does not have full functionalities of *IntelligentBox*.

There are some electronic publication formats like ePub, EduPub, iBooks and their authoring tools. Of course, these contents are used as e-learning materials. However, basically, these are not available on the web and do not support 3D Graphics except iBooks. iBooks supports rendering functionality of a 3D scene and control interfaces of its viewpoint. However, story-based contents cannot be created using it.

From the above situation, for creating web-based interactive 3D educational contents, we have to use any dedicated toolkit systems. The most popular one is Unity, one of the game engines [12], that supports creating web contents. The use of Unity requires programming knowledge and skills of the operations for it. Therefore, it is impossible to use Unity for standard end-users like teachers. As a result, in this paper, we extended our previous framework [4–7] for easily developing materials by adding newly authoring functionalities explained in the following sections.

3   Web-Based Interactive 3D Educational Material Development Framework

Our first targets of the framework were contents of Japanese History study. In general, a history consists of several stories. So, the framework should support creating a story. Each story is realized with several 3D scenes consisting of several architecture objects like buildings and furnitures, and several moving characters like humans who have their own shape model and animation data. Firstly, such 3D assets data should be prepared. Next, contents creators have to define a story for the content as one Java-Script file called 'Story Definition File'.

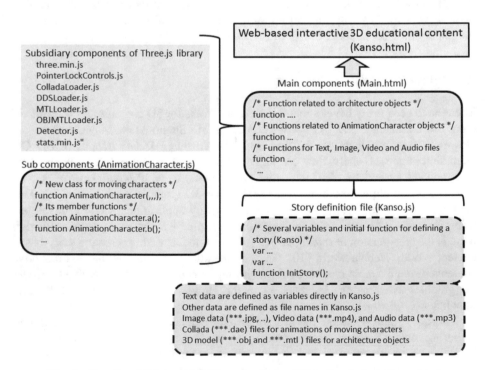

Fig. 1. Functional (Main and Sub) components of Framework, (ex) Kanso.html.

Figure 1 shows functional components of our proposed framework consisting of main components (Main.html) and sub-components (AnimationCharacter.js). The main components include functions related to architecture objects and functions related to AnimationCharacter objects represented as moving characters. The sub-components include the constructors of new AnimationCharacter class and its member functions. Besides main and sub-components, our framework uses Three.js [13], one of the WebGL based 3D Graphics libraries as a subsidiary component. When developing a story-based interactive 3D educational content with the proposed framework, a teacher has to prepare one story definition file and 3D assets for it. See the papers [4, 5] for more details.

'Kanso' and 'Jimoku' of Japanese History Study

As case studies, we have been developing two types of contents with the proposed framework. One is for learning certain events and ceremonies that were taken in the Imperial court called 'Gosyo', including ancient manners of the Emperor called 'Tennou' and cabinet members called 'Daijin' and 'Daiben' and so on as recorded in Japanese history documents. We have already finished the story called 'Kanso' and have been developing the story called 'Jimoku'.

'Kanso' is one of the ceremonies that were taken in 'Gosyo'. 'Kanso' includes various types of manners of 'Tennou' and 'Daijin' etc. In Japanese history study, students have to learn such manners by reading old documents about 'Kanso'. However, it is very difficult to understand the manners from the old documents. Using this interactive 3D educational material of 'Kanso' shown in Fig. 2, it becomes easy to understand.

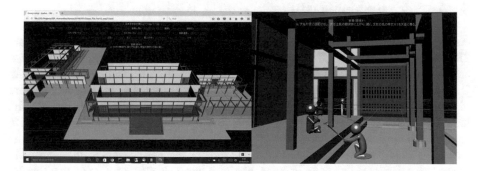

Fig. 2. Screen shots of 'Kanso' content

As shown in the left part of Fig. 2, there are three houses as architecture objects. The center one is 'Shishinden', the left one is 'Seiryoden', and the right one is 'Jinnoza'. These architecture objects are prepared as three 'obj' format files with their material files. The right part of Fig. 2 shows a scene at 'Jinnoza'. As moving characters of 'Kanso', there are seven characters, i.e., 'Tennou', 'Daijin', 'Daiben', etc. Each moving character is implemented as an instance of AnimationCharacter class in JavaScript. AnimationCharacter includes several animation data prepared as 'dae' format files. For this content, we prepared around 100 dae format files. To reduce the cost to prepare these data files, we employ 3D human pictograms for the shape data of moving characters. Since their colours are different, it is possible to recognize each character by the colour. For example, the colour of 'Tennou' is yellow, 'Daijin' is red and so on. There are seven polylines in a different colour shown in the right part of Fig. 2. Each is the moving path of the corresponding moving-character. Its colour is the same as the pictogram. At the upper and center position of the scene, texts data are displayed in white colour. These texts explain the behaviours of the moving characters. By reading the texts, the students can understand the manners in 'Gosyo' more deeply with both reading the texts and watching 3D CG animations (Fig. 3).

Fig. 3. Screen shots of 'Jimoku' content

4   Authoring Functionalities

As the first stage, creators of educational materials using the framework have to prepare 3D assets data those are 3D models of architecture objects like building, and also animation data for each animation character. After that, in 'Story definition file (Kanso. js)', the creator need to define the location, orientation and size of each architecture object, and to define the initial position and size of each animation character. Finally, the creator need to define a story that means the time-sequence of animation data. In spite of defining these by writing a program in 'Story definition file', by using the authoring functionalities introduced here, the creator can perform authoring intuitively as shown in Table 1. The main components introduced as authoring functionalities are a scene editing table and an animation editing table.

The following subsections, we explain the details of editing a scene, editing animation characters and saving/loading functionalities.

4.1   Editing a Scene

Figure 4 shows one of the screen images of the framework. As shown in the left upper part and the right part, there are several buttons, respectively used for editing a scene, editing animation characters, saving an editing information as a file and loading the

Table 1. Procedures for authoring of educational materials

---

1.   Preparing 3D assets data
3D models of architecture objects and animation data of animation characters should be stored in a certain folder of the proposed framework.
2.   Editing a scene
Users open the scene editing table to add architecture objects, and to visually adjust numerical values of their locations, orientations and sizes through watching the changes in the displayed 3D scene.
3.   Editing animation characters
Users open the animation editing table to add animation characters by entering their names.
  3.1 Setting animation files
  User open the dialog by pressing the cell corresponding to a target animation character. In the dialog, users can select and add an animation file name of the target animation character.
  3.2 Adjusting animations
  Users can adjust numerical values of the location, orientation and size of the target animation character in this table by looking at its preview animation. The initial values of the location, orientation and size are determined as the final values of its previous animation.
  3.3 Setting task numbers of each step
  Each animation is divided into several tasks as units of playback. Each task has a task number as an integer [0 or larger than 0]. Each task consists of several sequential steps. Then, the sequential steps that have the same task number are played atomically. The animation editing table displays task numbers on each step that can be edited by pressing one of them and by entering an integer number.
4.   Editing text explanations
The framework is possible to display an explanatory text of its corresponding animation. Users can edit the text on the animation editing table.
5.   Saving and Loading
When users complete the editing of a scene, animation characters and text explanations, they can save the editing information as a named data into the web server of the framework, and can load it from the server when needed.

---

editing information from the file. After pressing the button for editing a scene, the scene editing table shown in Fig. 5 will appear.

Scene Editing Table
This table has multiple buttons and shows architecture objects' names. By editing the contents of this table, it is possible to add new objects to the current scene or to remove already existing objects from it. In addition, by pressing the edit button prepared for each architecture object, a dialog for editing its attributes values shown in Fig. 6 will appear. The displayed numbers are the values of the location, orientation and scale of each architecture object according to its x-axis, y-axis, and z-axis. Each value can be changed by clicking '+' or '−' button or directly inputting a number. Whenever the

value is changed, it will be applied to the corresponding architecture object and the 3D scene will be updated simultaneously so that users can modify the values interactively by looking at the updated 3D scene.

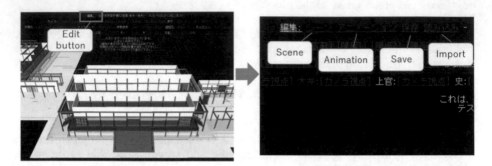

Fig. 4. Buttons for editing a scene, editing animation characters, saving an editing information and loading an editing information.

△ ▽	Scene Edit[Max Size:10]	[Add] [Close]
0	jinnoza_s3.obj	[Edit] [Delete]
1	mityodai.obj	[Edit] [Delete]
2	seiryoden_s3.obj	[Edit] [Delete]
3	shishinden_s3.obj	[Edit] [Delete]

Fig. 5. Screen image of the scene editing table.

jinnoza_s3.obj		[Close]	
Position[STEP]	[-]	1	[+]
Position[X]	[-]	135	[+]
Position[Y]	[-]	1.34	[+]
Position[Z]	[-]	-130	[+]
Rotation[STEP]	[-]	1	[+]
Rotation[X]	[-]	0	[+]
Rotation[Y]	[-]	90.00000000000001	[+]
Rotation[Z]	[-]	0	[+]
Scale[STEP]	[-]	1	[+]
Scale[X]	[-]	0.5	[+]
Scale[Y]	[-]	0.5	[+]
Scale[Z]	[-]	0.5	[+]

Fig. 6. Dialog for editing architecture object's attributes.

4.2   Editing Animation Characters

After pressing the button for editing animation characters shown in the left of Fig. 4, the animation editing table shown in Fig. 7 will appear. The left side of this table is shown in Fig. 8, there are the names of animation characters and a set of buttons that allow you to add a new animation character or delete one of the already existing animation characters. The animation information of each animation character is displayed on the right side as shown in Fig. 9. Each cell in the orange frame corresponds to a certain animation character and its step, and if the cell is assigned an animation file, the file name is displayed on the cell.

By pressing one of the cells, a dialog shown in Fig. 10 will appear, in which it is possible to edit the detailed information of the animation. The position, rotation and size information can be edited by the same operation in the same editing table. As well, it is possible to select an animation file. Since the animation is previewed whenever these operations are performed, users can edit the contents interactively.

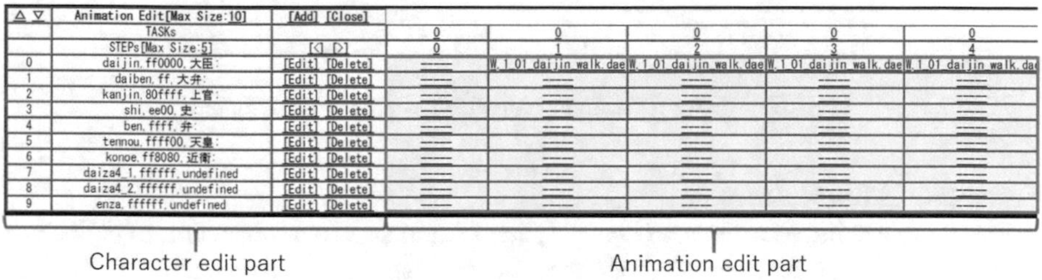

Fig. 7. Screen image of the animation editing table.

△ ▽	Animation Edit [Max Size:10]	[Add] [Close]
	TASKs	
	STEPs [Max Size:5]	[◁ ▷]
0	daijin, ff0000, 大臣:	[Edit] [Delete]
1	daiben, ff, 大弁:	[Edit] [Delete]
2	kanjin, 80ffff, 上官:	[Edit] [Delete]
3	shi, ee00, 史:	[Edit] [Delete]
4	ben, ffff, 弁:	[Edit] [Delete]
5	tennou, ffff00, 天皇:	[Edit] [Delete]
6	konoe, ff8080, 近衛:	[Edit] [Delete]
7	daiza4_1, ffffff, undefined	[Edit] [Delete]
8	daiza4_2, ffffff, undefined	[Edit] [Delete]
9	enza, ffffff, undefined	[Edit] [Delete]

Fig. 8. Character edit part of the animation editing table.

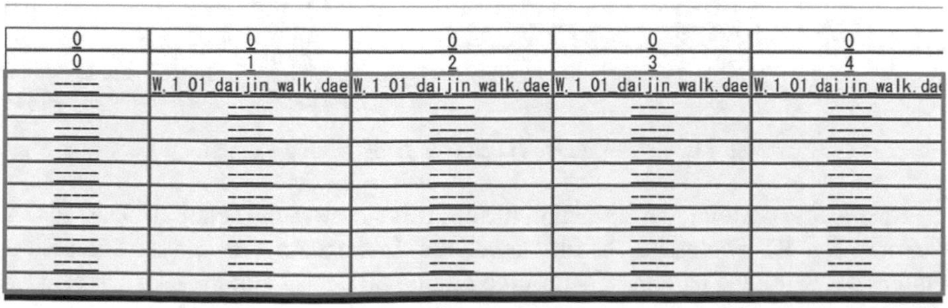

Fig. 9. Animation edit part of the animation editing table.

daijin		[Close]	
Position[STEP]	[-]	1	[+]
Position[X]	[-]	215	[+]
Position[Y]	[-]	0	[+]
Position[Z]	[-]	-140	[+]
Rotation[STEP]	[-]	1	[+]
Rotation[X]	[-]	0	[+]
Rotation[Y]	[-]	90.0000000000001	[+]
Rotation[Z]	[-]	0	[+]
Scale[STEP]	[-]	1	[+]
Scale[X]	[-]	5	[+]
Scale[Y]	[-]	5	[+]
Scale[Z]	[-]	5	[+]
File:		1_01_daijin_walk.dae	
Action type:		walk	

Fig. 10. Dialog for editing animation character's attributes.

Animation Editing Example

Using the following four figures (Figs. 11, 12, 13 and 14), this subsection explains how animation editing is carried out about 'step 14' and 'step 15' of a character "daijin". By pressing the cell of 'step 14' from the cells corresponding to the character "daijin" in the animation editing table shown in the left upper part of Fig. 11, the dialog shown in the right lower part will appear. Then, in this dialog, the user changed the action type into "Action".

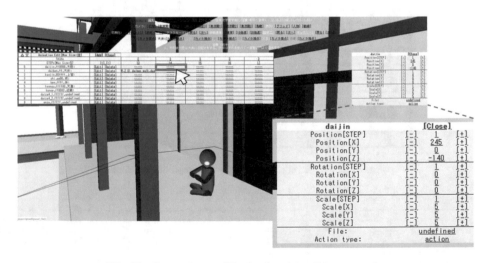

Fig. 11. Screen image (1) of animation editing example.

Next is for specifying an animation file corresponding to the action of 'step 14'. As shown in Fig. 12, by pressing the file "undefined" in the dialog, the file selection dialog window shown in the left upper part of the figure will appear and the user selected one of the animation files.

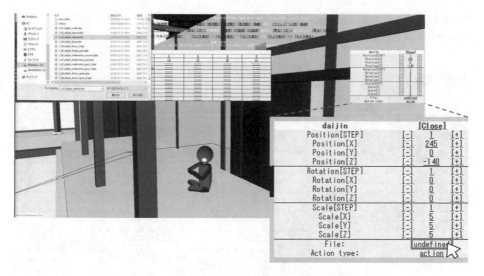

Fig. 12. Screen image (2) of animation editing example.

Similarly, by pressing the cell of 'step 15' from the cells corresponding to the character "daijin", the dialog will appear. Then, the action type can be change into "walk" as shown in Fig. 13. Also, the user adjusted the character position through looking at the displayed character.

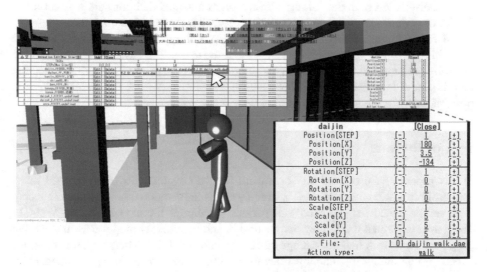

Fig. 13. Screen image (3) of animation editing example.

'Step 14' and 'step 15' are included in the same 'task 3' as shown in Fig. 14 so that these two steps will be performed atomically. Next, for specifying 'task 4', by pressing the next cell of 'task number' cells in the animation editing table, the input dialog window will appear, and the user entered an integer number '4' for this 'task number'.

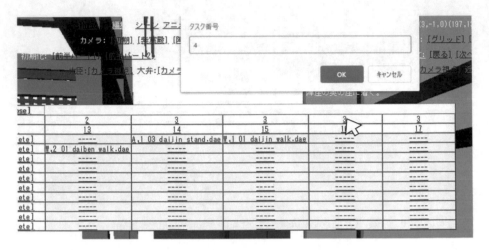

Fig. 14. Screen image (4) of animation editing example.

4.3 Saving/Loading Functionalities

Once the editing of a scene, animation characters and text explanations is completed, the framework can save the editing information as a named data into the web server of the framework. Users can load it from the server later. By pressing the button shown in Fig. 4 for saving an editing information, a dialog will appear for entering a name and the editing information can be saved by the entered name. The saved data can be edited again by pressing the import button and by entering its data name. As well, since the data is stored in a data storage directory on the web server of the framework, it is possible to share the edited information with other users.

5   Conclusions

In this paper, we introduced the framework for the development of web-based inter-active 3D educational materials. The framework is implemented using the most recent Web technologies, i.e., HTML5 and various JavaScript libraries including WebGL. This paper explained the framework, its design and functional components. Indeed, the proposed framework had not enough authoring functionalities. Therefore, this time, we added several authoring functionalities to the framework for enabling standard users who do not have programming knowledge and skills like ordinary teachers. So, in this paper, we mainly proposed the newly added authoring functionalities.

As future work, we will try to ask teachers to use the proposed framework for actually creating web-based interactive 3D educational materials and we will consult the teachers for evaluating development efficiency of the framework. Furthermore, we will create several educational materials with the framework and ask students to learn using the materials and we will consult the students for evaluating educational efficiency of the materials. From these evaluations, we will clarify the usefulness of our proposed framework.

Acknowledgements. We wish to thank Mr. Hiroaki Tokunaga because some parts of the authoring functionalities introduced in this paper were proposed in his master thesis. This research was partially supported by JSPS KAKENHI Grant No. JP17H00773.

References

1. ICER, 08 June 2018. http://www.icer.kyushu-u.ac.jp/en
2. Sugimura, R., et al.: Mobile game for learning bacteriology. In: Proceedings of IADIS 10th International Conference on Mobile Learning 2014, pp. 285–289 (2014)
3. Sugimura, R., et al.: Serious games for education and their effectiveness for higher education medical students and for junior high school students. In: Proceedings of 4th International Conference on Advanced in Information System, E-Education and Development, ICAI-SEED 2015, pp. 36–45 (2015)
4. Okada, Y., Nakazono, S., Kaneko, K.: Framework for development of web-based interactive 3D educational contents. In: 10th International Technology, Education and Development Conference, pp. 2656–2663 (2016)
5. Okada, Y., Kaneko, K., Tanizawa, A.: Interactive educational contents development framework based on linked open data technology. In: 9th Annual International Conference of Education, Research and Innovation, pp. 5066–5075 (2016)
6. Okada, Y., Kaneko, K., Tanizawa, A.: Interactive educational contents development framework and its extension for web-based VR/AR applications. In: Proceedings of the GameOn 2017, Eurosis, 6–8 September 2017, pp. 75–79 (2017). ISBN 978-90-77381-99-1
7. Ma, C., Srishti, K., Shi, W., Okada, Y., Bose, R.: E-learning material development framework supporting VR/AR based on linked data for IoT security education. In: Proceedings of 6th International Conference on Emerging Internet, Data & Web Technologies, EIDWT 2018, Tirana/Albania, 15–17 March 2018, pp. 479–491 (2018). ISBN 978-3-319-75928-9
8. Okada, Y., Tanaka, Y.: IntelligentBox: a constructive visual software development system for interactive 3D graphic applications. In: Proceedings of Computer Animation 1995, pp. 114–125. IEEE CS Press (1995)
9. Okada, Y.: Web version of IntelligentBox (WebIB) and its integration with Webble world, Vol. 372 of the series, Communications in Computer and Information Science, pp. 11–20. Springer (2013)
10. Tanaka, Y.: Meme Media and Meme Market Architectures: Knowledge Media for Editing, Distributing, and Managing Intellectual Resources. Wiley, New York (2003)
11. Webble World, 08 June 2018. https://github.com/truemrwalker/wblwrld3
12. Unity, 08 June 2018. https://unity3d.com/jp
13. Three.js, 08 June 2018. https://threejs.org/

Data Relation Analysis Focusing on Plural Data Transition for Detecting Attacks on Vehicular Network

Jun Yajima[1,2(✉)], Takayuki Hasebe[2], and Takao Okubo[1]

[1] Institute of Information Security, 2-14-1, Tsuruya-cho, Kanagawa-ku,
Yokohama, Kanagawa 221-0835, Japan
dgs194101@iisec.ac.jp
[2] Fujitsu Laboratories Ltd., 4-1-1, Kamikodanaka, Nakahara-ku,
Kawasaki 211-8588, Japan
jyajima@fujitsu.com

Abstract. On the Controller Area Network (CAN), there are three types of messages, namely periodic messages, event based periodic messages, and non-periodic messages. On the attack detection, there are many high accuracy methods for the periodic messages. For the event based periodic messages and the non-periodic messages, detection methods utilizing the relation of plural data in some messages are effective. In such methods, the relations between plural data are investigated from some messages that were collected beforehand. And, obtained relation information is used as attributes of statistic detection. In this paper, a new attack detection method utilizing the number of occurrences of specific values and changes of values in messages is proposed. And the derivation algorithm that derives relation information efficiently is also proposed. By using the derivation algorithm, we found 582 relations for the detection method.

1 Introduction

Recent cars are electronically controlled by in-vehicle networks. Some examples of it are running, curving, stopping, window-opening, window-closing, and etc. The typical one of the in-vehicle networks is Controller Area Network (CAN) [2]. It is pointed out that CAN is weak against cyber-attacks. By injecting malicious messages to CAN, network nodes called Electronic Control Unit (ECU) connected to the CAN might do operation not intended. Actually, in 2015, a remote control attack against actual vehicles was proposed [1]. It caused first recall by security reasons.

We categorize CAN messages to three types by transmission specification designed by car manufacturer or ECU maker.

- Periodic messages: All messages are transmitted periodically.
- Event based periodic messages: All messages are transmitted periodically in normal situation. However, when an event is occurred, a message is transmitted in not periodic.
- Non-periodic messages: All messages are transmitted non-periodically.

L. Barolli et al. (Eds.): NBiS-2019, AISC 1036, pp. 270–280, 2020.
https://doi.org/10.1007/978-3-030-29029-0_25

As one of the security measures for cyber-attacks, many attack detection methods on CAN were proposed [3–8]. In [3,4], detection methods that can detect attacks in high accuracy for the periodic messages were proposed. However, there are the event based periodic messages and the non-periodic messages. So, attack detection methods for the event based periodic messages and the non-periodic messages are needed. In [5–8], detection methods that can detect attacks for the event based periodic messages and the non-periodic messages were proposed.

In this paper, we introduce a detection method which focuses on data changing to another value and data becoming specific value in messages. In this method, information of CAN data relations is needed. In this paper, we also propose a relation analysis method which can be used for the introduced detection method and some existing methods. Finally, we show the effectiveness of proposed method by describing results of computer experiment.

This paper is structured as follows. In the Sect. 2, CAN is introduced. Existing attack detection methods are described in the Sect. 3. In the Sect. 4, we introduce an attack detection method. And, in the Sect. 5, typical data relation analysis method is introduced. In the Sect. 6, we propose a new data relation analysis method. The computer experiment and results are shown in the Sect. 7. Finally, we summarize conclusion and future works in the Sect. 8.

2 CAN

2.1 Overview

CAN is a network protocol used in control systems, for example, the in-vehicle networks in cars. CAN was developed by BOSCH and standardized as ISO11898 [2]. ECUs that use CAN communicate with each other by using the voltage difference between two communication lines. For in-vehicle networks, the network generally composes the bus topology. CAN is a broadcast communication protocol, so data transmitted from any ECU reach all ECUs connected with the same bus. Each bit of a CAN message is either dominant (0) or recessive (1). The dominant part of the message is given priority when messages are transmitted to communication lines by two or more ECUs at the same time. This mechanism is called Carrier Sense Multiple Access with Collision Avoidance (CSMA/CA). CAN has four data format types (data-frame, remote-frame, error-frame, and overload-frame). This paper focuses on the data-frame, so the details are explained as follows.

2.2 Data-Frame

Data-frame contains standard format and extended format. There are an 11-bit ID field in the standard format and a 29-bit ID field in the extended format. Although this paper only mentions the standard format, the discussion is also applicable for the extended format. The ID in the ID field indicates the meaning of the message and is used for communication arbitration. The communication arbitration is a mechanism whereby only the high priority message is processed when two or more ECUs transmit the data frame at the same time. Because the dominant is given priority over the recessive, the message where the dominant appears first in the most significant bit (MSB) of the ID field is processed first in all transmitted frames. When the dominant appears in the MSB of the ID field simultaneously, priority is similarly determined by the continuing bit. The transmission node transmits the transmission message and receives the message from the CAN simultaneously. The transmission node stops transmitting the transmission message if the ID of the received message has higher priority than the transmitted message. In general, IDs are operated that the message with the same ID is not used when transmission ECUs are different. In addition, the messages that are received by each ECU are decided beforehand based on the ID, and messages with differing IDs are ignored by each ECU. Data-frame has data field for storing data. The data field can have 64-bit data at maximum. In general, the length of data is decided by each ID.

2.3   Message Periodicity

On the CAN data-frame, many messages are transmitted at the designated transmission cycle determined for each ID. There are various cycle intervals, which range from about ten milliseconds to ten seconds. We call them periodic messages. Additionally, there are event based periodic messages and non-periodic messages. On the event based periodic messages, all messages are transmitted periodically in normal situation. However, when an event is occurred, a message is not transmitted in periodic. On the non-periodic messages, all messages are transmitted non-periodically. All CAN messages belong to these three patterns by each CAN ID.

3   Attack Detection on CAN

3.1   Attack Detection for Periodic Messages

Many of attack detection methods for periodic messages utilize reception time and reception cycle of each message. Some of examples of these detection methods are cumulative sum detection [3] and delayed decision cycle detection [4]. These methods can detect attacks in high accuracy for the periodic messages.

3.2   Attack Detection for Event Based Periodic Messages and Non-Periodic Messages

In general, the number of the event based periodic messages and the non-periodic messages are less than the periodic messages. However, to achieve secure vehicles, it is desirable to be able to detect attacks for these messages. On attack detection for these messages, it seems to be able to detect attacks utilizing the relation of each data-frame of message [5–8]. In [5,6], prediction methods of each data type (for example, constant, sensor, and counter) in data-frame in CAN message from many receptions of single message are proposed. After the prediction, attack detection using the data type is performed. For example, situations that data in the constant field is changed when message is received are detected as an attack. In [7,8], correlation value or pattern-matching for plural messages is used for attack detection.

In this paper, in order to detect attacks, we also focus on plural messages. We introduce a detection method that can detect attacks for not only event based periodic messages but also non-periodic messages. This method focuses on data changing to another value or becoming specific value in messages. And, we propose data relation analysis method for this detection method. We think that this method can be also applied to existing methods.

4   Attack Detection Focusing on Data Transition of Plural Messages

In this section, an attack detection method focusing on data transition in plural messages is described. In Fig. 1(a), a CAN log example of event based periodic messages is shown. On this log, Fig. 1(b) and (c) show the log for each ID. In this example, on both ID = 0x100 and 0x200, messages are received about each 0.3 s when no event are occurred. However, when an event is occurred, the message interval is not about 0.3 s. For example, on ID = 0x100, the interval is about 0.1005 s between the message reception time 1.100004 s and 1.200503 s. And at the same timing, the 1st byte of message becomes 0x80 (top bit becomes '1'). Similarly, on ID = 0x200, the interval is about 0.1005 s between the message reception time 1.200003 s and 1.300498 s. And at the same timing, the 4th byte of message changes from 0xC5 to 0x45. Such trends can be seen from logs obtained from real-vehicle. Against such situation (legitimate ECUs continue transmitting the legitimate messages), we assume an attacker tries to send a malicious message of ID = 0x200 at any timing. Now, we consider the detection of the attack. If attack detector has an attack detection algorithm based on the property that the interval time is not deviated when received message is not changed from the previous one, the detector can detect the attack when the 4th byte of malicious message is same as the previous legitimate message. However, if the attacker knows the property about the message, namely, change the 4th byte of a malicious message to another data of previous legitimate data can deviate the interval timing of messages, the detector cannot detect the attack because the detector cannot identify the legitimate messages and the malicious one. Figure 2 shows an example of such attack. In this situation, if the detector knows the following property, the detection possibility can be improved.

Property. The number of 1st byte of message which ID is 0x100 become specific value is same to the number of 4th byte of message which ID is 0x200 changes to another value from the previous one.

(a)CAN log			(b)ID=0x100			(c)ID=0x200		
ID	Time	Data(HEX)	ID	Time	Data(HEX)	ID	Time	Data(HEX)
100	0.500001	00 01 12 13	100	0.500001	00 01 12 13	200	0.600005	00 F2 E3 C5
200	0.600005	00 F2 E3 C5	100	0.799997	00 01 12 13	200	0.900002	00 F2 E3 C5
100	0.799997	00 01 12 13	100	1.100004	00 01 12 13	200	1.200003	00 F2 E3 C5
200	0.900002	00 F2 E3 C5	100	1.200503	**80** 01 12 13	200	1.300498	00 F2 E3 **45**
100	1.100004	00 01 12 13	100	1.500505	00 01 12 13	200	1.600502	00 F2 E3 45
200	1.200003	00 F2 E3 C5						
100	1.200503	**80** 01 12 13						
200	1.300498	00 F2 E3 **45**						
100	1.500505	00 01 12 13						
200	1.600502	00 F2 E3 45						

Fig. 1. An example of event based periodic messages

ID	Time	Data(HEX)	
200	0.600005	00 F2 E3 C5	
200	0.900002	00 F2 E3 C5	
200	1.200003	00 F2 E3 C5	
200	1.300498	00 F2 E3 45	
200	**1.350004**	**00 F2 E3 91**	Attacked!
200	1.600502	00 F2 E3 45	

Fig. 2. An example of attacked situation that attack detection algorithm cannot detect attack

When the detector uses this property, the detector can detect the attack because the number of becoming specific values (=1) in messages of ID = 0x100 is not equal to the number of changes (=3) in messages of ID = 0x200. When the detector tries to detect attacks using this property, the detector must know the position of the specific/changed value, and the number of the specific/changed values before attacks are injected. Therefore, the detection strategy becomes the following two phases.

Strategy of detection

1. Extraction Phase: Derive the bit positions of specific/changed values and the number of the specific/changed values for each message ID.
2. Detection Phase: Anomaly detection using the derived property.

In this paper, we focus on the phase 1. And the above discussion can be achieved to non-periodic messages.

5   Data Relation Analysis

On the data relation analysis for attack detection, we focus on plural messages that their CAN-IDs are different each other. We evaluate the number of becoming specific value and changing another value at each position of obtained CAN data. In the extraction phase described in the strategy of detection in Sect. 4, on the data relation analysis of obtained CAN log, the number of the becoming specific value is derived for each message whose CAN-ID is different. Additionally, for the attack detection, the timing of becoming a specific value or changing to another value must be evaluated. It seems that the data relation analysis can be achieved by using the following procedure. In this paper, data is defined as one of the data which is positioned at possible position and its length is possible length in a message.

Algorithm 1

1. For each data, evaluate the deviating-timing from the normal transmission interval (or transmission timing for non-periodic messages.)
2. Among the deviating timing of 1, derive the number of becoming a specific value or changing to another value.
3. Derive all data-pairs that deviating-timing evaluated in 1 is same between two data, and number derived in 2 is same between two data.

In the phase 1 of the above algorithm, for all data, the deviating-timing from normal transmission interval is evaluated. However, the evaluation is difficult. The reason is that in order to derive the deviating data, for each message, each message reception interval is evaluated by using a permitted boundary. Namely, permitted boundaries are used for each data. In general, much number of evaluation using permitted boundary cause rough results of deriving deviating-timing, Therefore, a derived data-pair from the phase 3 are the result of the evaluation twice by using the permitted boundaries. So, the above algorithm has following problem.

Problem of the Algorithm 1

1. The evaluation result of the deviating-timing becomes inaccurate by 2-times evaluations that use the permitted boundary.

In the next section, we propose an algorithm that solved this problem.

6   Proposed Method

6.1   Overview

In order to solve the problem of the Algorithm 1, we reduce the number of evaluations that use the permitted boundary. The essence of the problem of the Algorithm 1 is to check whether to deviate from normal interval in each message. Now, we propose the following algorithm. In this algorithm, we judge reception intervals of a message is nearly equal to that of another message by using one permitted boundary. The timing deviation can be evaluated more accurately by reducing the number of evaluations which use the permitted boundary.

Algorithm 2 (proposed method)

1. For each data, derive the number of becoming a specific value or changing to another value.

2. For each data, derive message reception intervals.
3. Derive all data-pairs that number of 1 is same between two data, and intervals of 2 is same between two data (use permitted boundary).

6.2   On Processing in Each Phase of the Algorithm 2

6.2.1   On the Phase 1 in Algorithm 2

The length of the payload of each message can be derived from observing CAN. Generally, plural data are stored in one payload. However, we don't know how many data is stored in one payload and where one payload is divided in each data. Though we derive both the number of becoming a specific value and the number of changing to another value, at first, we show the method of counting the number of changing. Because it is simpler than counting the number of becoming a specific value.

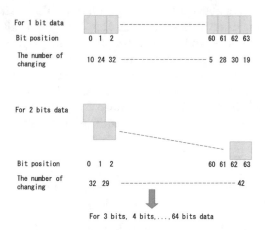

Fig. 3. Overview of counting

   We prepare counters for all positions which have possibility that each data is stored. For example, because the maximum length of CAN message is 64 bits, there are 64 position patterns (the 0th bit, the 1st bit, ..., the 63rd bit from the top of payload) for 1 bit data. For 2-bit data, there are 63 position patterns (the 0th–1st bits, the 1st–2nd bits, ..., the 62nd–63rd bits from the top of payload). Similarly, there are 62 position patterns for 3-bit data, there are 61 position patterns for 4-bit data, ..., there is 1 position for 64-bit data. Finally, There are $64 + 63 + 62 + \ldots + 1 = 2080$ patterns, and 2080 counters are prepared for each position pattern. After counters preparation, we count the number of changing to another value at each data length at each position. The number of changing is counted at each data length at each position about a message which has target CAN-ID while reading a log of CAN from the top of it. The overview of this counting is shown in Fig. 3.

Next, we show the method of counting the number of becoming a specific value. For counting it, we prepare counters at each length, position, and value. The pattern of values is derived by data length. For example, when data length is 2 bits, there are 4 patterns namely 00, 01, 10, and 11. For 2-bit data, patterns of positions is 63. Therefore, $4 \times 63 = 252$ counters are needed for 2-bit data. However, for example, there is 2^{64} patterns for 64-bit data, all counters cannot be prepared for long data. To solve this problem, we prepare counters corresponding to the values that appears in the CAN log. Adopting this method, we can reduce large memory space in a PC. An example of counting the number of changing to another value and becoming a specific value is shown in Fig. 4. In order to understand clearly, this example show the counting manner of the 4-bit data though 4-bit data is not allowed in CAN specification. In Fig. 4, on the counter of the number of becoming a specific value, we don't prepare counters for patterns whose value is '0'.

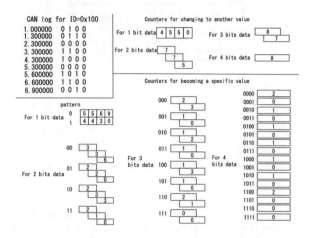

Fig. 4. An example of counting

6.2.2   On the Phase 2 in Algorithm 2
In this phase, message intervals are derived for each data. On counters for changing to another value, intervals of data are derived. The interval indicates the timing of changing the data-value to another value. Similarly, on counters for becoming a specific value, intervals that indicate the timing of becoming a specific value are derived.

6.2.3   On the Phase 3 in Algorithm 2
In this phase, data relations are derived from the results of phase 1 and phase 2. At first, data whose number of changing to another value, or becoming specific value is the same are collected from the results of phase 1. And a set S_i is made by the data whose number is i. Next, by using the results of the phase 2,

data whose timing of first changing to another value or becoming specific value is the same are collected from each element of S_i. And among the collected data, data whose changing/becoming intervals are also same are collected. And, the collected data are composed to $S_{i,j}, S_{i,j+1}, \ldots$ by each number and timing. Finally, output $S_{i,j}, S_{i,j+1}, \ldots$ as related data.

7 Experiment

We have confirmed the proposed data relation analysis method by computer experiment.

Table 1. The result of the experiment

Log length	About 315 s
The variation of ID	122 patterns
Total messages	376873
Extracted relations	582

Table 2. One example of relation (the number of changing/becoming is 17)

ID-X	$(2, 8, 166), (8, 8, 128), (59, 5, -1)$
ID-Y	$(1, 8, 85), (8, 8, 128), (59, 5, -1)$

ID	Time	Data(HEX)	ID	Time	Data(HEX)
X	57.043154	29 00 00 00 00 00 28 4C	Y	57.054185	2A 00 00 00 00 00 28 4C
X	**57.943272**	29 **80** 00 00 00 00 28 4D	Y	**57.954177**	2A **80** 00 00 00 00 28 4D
X	67.942846	29 00 00 00 00 00 28 4D	Y	67.953193	2A 00 00 00 00 00 28 4D
X	77.942821	29 00 00 00 00 00 28 4D	Y	77.954915	2A 00 00 00 00 00 28 4D
X	**79.122813**	29 **80** 00 00 00 00 28 4E	Y	**79.133920**	2A **80** 00 00 00 00 28 4E
X	89.122655	29 00 00 00 00 00 28 4E	Y	89.132772	2A 00 00 00 00 00 28 4E
X	**92.002538**	29 **80** 00 00 00 00 28 4F	Y	**92.012815**	2A **80** 00 00 00 00 28 4F
X	**100.682696**	29 **80** 00 00 00 00 28 50	Y	**100.694782**	2A **80** 00 00 00 00 28 50

Fig. 5. A part of log on ID-X and ID-Y

7.1 Preparation of Data

We prepared CAN data from an actual car. We achieved the proposed method to the CAN data. We extracted related data whose number of changing or becoming is the same, and also whose characteristic of deviation of its period is the same. In this experiment, we set the targets whose number of changing and becoming to 2 at minimum and to 64 at maximum.

7.2   Result

We found 582 relations from the CAN data. The result is shown in Table 1. An example of found relation is shown in Table 2. In Table 2, The triplet (a, b, c) indicates $(bitposition, datalength, datatype)$. $datatype$ means that $datalength$-bit data is at position $bitposition$, and its value become $datatype$ when message intervals are deviated. When $datatype$ is "−1", it means the data changes to another data when message intervals are deviated. On the Table 2, for messages of ID-X and ID-Y, the 8-bit length data at bit position 8 become 128 (0x80) when message intervals are deviated. Figure 5 shows the part of CAN data. From this log, we can confirm that the characteristic of data $(8, 8, 128)$ on message of ID-X is the same as that of ID-Y. From found results, we show the number of relations in each number of changing/becoming in Table 3.

Table 3. Found relations

Num.[a]	Rel.[b]	Num.	Rel.	Num.	Rel.	Num.	Rel.	Num.	Rel.	Num.	Rel.	Num.	Rel.	Num.	Rel.
2	315	10	5	18	1	26	0	34	0	42	0	50	0	58	0
3	126	11	6	19	0	27	0	35	0	43	0	51	0	59	0
4	28	12	5	20	4	28	1	36	0	44	0	52	0	60	0
5	12	13	3	21	3	29	1	37	0	45	0	53	0	61	0
6	33	14	0	22	0	30	0	38	0	46	0	54	0	62	0
7	14	15	4	23	1	31	0	39	0	47	0	55	0	63	3
8	9	16	1	24	1	32	0	40	0	48	0	56	0	64	0
9	3	17	1	25	1	33	0	41	1	49	0	57	0		

[a] The number of changing/becoming
[b] Found relations

7.3   Consideration

From Table 3, we found many relations whose relations are low. We can also see that low number of changing/becoming have many relations. In this subsection, we will discuss the reason of it and how we should do. We think the reason is that the low number of changing/becoming derives low number of checking timing about the timing-deviation from its period at the phase 3 in Algorithm 2. So, even if the relation is thin between set $S_{i,j1}$ and $S_{i,j2}$, the algorithm judges that $S_{i,j1}$ and $S_{i,j2}$ have same characteristics by the checking of low number. Therefore, in the detection phase, relations that have low number of changing/becoming should be excluded. Deriving the threshold of this is as a future work.

8   Conclusion and Future Work

We introduced the detection method for in-vehicle networks. It can be applied to not only the periodic messages but also the event based periodic messages and the non-periodic messages. In this paper, we proposed an efficient data relation analysis method that can be used for the detection method. We confirmed the effectiveness of this method by a computer experiment. We found many relations that can be used for the attack detection. Deriving the threshold discussed in Sect. 7.3 is a future work. And actual attack detection using relations is marked as also future work.

References

1. Miller, C., Valasek, C.: Remote exploitation of an unaltered passenger vehicle. Black Hat 2015 (2015)
2. ISO11898: Road vehicles – Controller area network (CAN) (2003)
3. Yajima, J., Abe, Y., Hasebe, T.: Proposal of anomaly detection method "cumulative sum detection" for in-vehicle networks. In: escar Asia 2018 (2018)
4. Otsuka, S., Ishigooka, T., Oishi, Y.: CAN Security: Cost-Effective Intrusion Detection for Real-Time Control Systems. SAE Technical Paper 2014-01-0340 (2014)
5. Markovitz, M., Wool, A.: Field classification, modeling and anomaly detection in unknown CAN bus networks. In: escar Europe 2015 (2015)
6. Kishikawa, T., Maeda, M., Tsurumi, J., Haga, T., Takahashi, R., Sasaki, T., Anzai, J., Matsushima, H.: A generic CAN message field extraction method to construct anomaly detection systems for in-vehicle networks. In: 2017 Symposium on Cryptography and Information Security, SCIS 2017 (2017)
7. Hamada, Y., Yoshida, K., Adachi, N., Kamiguchi, S., Ueda, H., Miyashita, Y., Isoyama, Y., Hata, Y.: Intrusion detection for acyclic messages in in-vehicle network: a proposal. In: Computer Security Symposium 2018, CSS 2018 (2018)
8. Iehira, K., Kanamori, K., Inoue, H., Ishida, K.: Extraction of correlation between in-vehicle sensor information using pattern matching for automatic generation of anomaly detection rules. In: 2018 Symposium on Cryptography and Information Security, SCIS 2018 (2018)

Gait-Based Authentication for Smart Locks Using Accelerometers in Two Devices

Kazuki Watanabe[1], Makoto Nagatomo[1], Kentaro Aburada[2],
Naonobu Okazaki[2], and Mirang Park[1(✉)]

[1] Kanagawa Institute of Technology,
1030 Shimo-Ogino, Atsugi, Kanagawa 243-0292, Japan
mirang@nw.kanagawa-it.ac.jp
[2] University of Miyazaki,
1-1 Gakuen-Kibanadai-Nishi, Miyazaki, Miyazaki 889-2192, Japan

Abstract. Smart locks can be opened and closed electronically. Fingerprint or face authentication is inconvenient for smart locks because it requires the user to stop for several seconds in front of the door and remove certain accessories (e.g., gloves, sunglasses). This study proposes a user authentication method based on gait features. Conventional gait-based authentication methods have low identification accuracy. The proposed gait-based authentication method uses accelerometers in a smartphone and a wearable device (i.e., smartwatch). We extracted 31 features from the acquired acceleration data and calculated identification accuracy for various machine-learning algorithms. The highest accuracy was 95.3%, obtained using random forest. We found that the maximum interval, minimum interval, and minimum value had the highest contributions to identification accuracy, and variance, median, and standard deviation had the lowest contributions.

1 Introduction

It has been predicted that the number of Internet of Things (IoT) products, which are devices that can communicate and interact with other devices over the Internet, will reach about 40 billion by 2020 [1]. Furthermore, wearable devices are increasing year by year, and are expected to increase to about 453 million in 2022 [2]. Therefore, it is predicting that many people will wear wearable devices. Various smart lock products, which can be electronically locked and unlocked, have been released, including Qrio [3], August [4], Kwikset [5], those smart locks using some authentication (e.g., PIN, password, fingerprint, face authentication). However, authentication for smart locks has several problems. For instance, passwords can be guessed by an attacker and require to remember the password, which is memory load for a user. Biometric authentication, which uses human physical and behavioral features, has been developed. Furthermore, fingerprint authentication requires the user to stop for several seconds in front

© Springer Nature Switzerland AG 2020
L. Barolli et al. (Eds.): NBiS-2019, AISC 1036, pp. 281–291, 2020.
https://doi.org/10.1007/978-3-030-29029-0_26

of the door and cannot be used with gloves. Face authentication requires the user to stop for several seconds in front of the door and cannot be used with sunglasses or hats. In this paper, we propose an authentication method based on gait features using acceleration from a smartphone and a wearable device (e.g., smartwatch). This method, assuming a smart lock, authentication starts at a place far from the smart lock and ends at a place immediate the smart lock. Hence, this method is possible to automatically authentication, and the low burden for the user.

Previous studies have proposed gait-based authentication using smartphone accelerometers [6–8]. These methods require the smartphone to point in a specific inclination during authentication and have low identification accuracy. One study proposed a gait authentication method based on many wearable sensors [9]. Although identification accuracy was higher than that obtained using a single sensor, this method requires the user to wear many devices, making it inconvenient.

In this paper, we propose a system model of gait-based authentication based on a smartphone and a wearable device (i.e., smartwatch) for smart locks. We calculate composite acceleration because authentication without being conscious of devices inclination. Furthermore, we extract 31 features from the composite acceleration data and calculate accuracy for various machine-learning algorithms. As a result, the highest accuracy was 95.3%, obtained using random forest, which experiments are conducted with a smartphone and a smartwatch to evaluate the proposed method.

The rest of this paper is organized as follows. Section 2 describes related work on gait-based authentication. Section 3 presents the proposed system model for gait-based authentication based on two devices and shows the processing of acceleration data used for authentication and feature extraction. Section 4 describes the implementation and experiments. Section 5 shows the accuracy of the proposed method and the features that contribute most to identification accuracy. Section 6 gives the conclusions and ideas for future work.

2   Related Work

2.1   Gait Authentication Using Machine Learning

Hou et al. [6] proposed a gait-based authentication method that uses x-, y-, and z-axis accelerations acquired from a smartphone accelerometer. This method classifies users using machine-learning based on the extracted acceleration features (e.g., average value, standard deviation, absolute standard deviation, average composite acceleration, time between peaks, bottle distribution). The false acceptance rate was 0.6% and the false rejection rate was 8.7% when using Decision Tree J48 as the classification algorithm; they were 0.3% and 3.8%, respectively, when using a neural network. The average values of x-, y-, and z-axis accelerations, respectively, contributed to identification accuracy, whereas the average combined acceleration and the time between peaks did not. Furthermore, the accuracy was calculated using 56 classification algorithms in Weka, a collection of machine-learning algorithms for data mining tasks. It was found that the accuracy greatly depends on the classification algorithm. However, this method does not consider the inclination of the device.

Konno et al. [7] proposed a gait-based authentication method that uses two sensors, namely the accelerometer and gyroscope in a smartphone. This method calculates the distance between the sensor data and reference data. The user is identified by a classifier based on the distance. It is necessary to authenticate in the same inclination of the device as at registration because uses accelerometers and gyroscope.

Iwamoto et al. [8] estimated gait state (stationary, walking, running) and smart device position (front pocket, back pocket, chest pocket, watch device screen, shake arm) and user. Experiments were conducted on five test subjects. Accuracies of 99.7%, 99.4%, and 97.0% were obtained for gait estimation, device position estimation, and user identification, respectively. From the result of the smart device position estimation, it was found that wearing sensors at multiple position on the body improved accuracy.

2.2 Gait Authentication with Multiple Wearable Sensors

Mondal et al. [9] attached eight rotation sensors on subjects' bodies (right shoulder, left shoulder, right arm, left arm, right hip, left hip, right foot, left foot) and asked the subjects to walk. Using machine-learning, this method was able to discriminate users with an accuracy of 100%. However, since many sensors must be worn, this method is inconvenient.

In the present study, we achieve high-accuracy gait-based authentication without considering the inclination of the device by placing a wearable device (smartwatch) and a smartphone at different positions on the body. This method is practical because many people have such devices.

3 Proposal Method

3.1 System Model

The proposed method performs smart lock authentication using acceleration data obtained from two devices, namely a smartphone and a wearable device. Figure 1 shows the system model. The proposed system model consists of a smartphone, a wearable device equipped with an accelerometer, and a smart lock. The acceleration data are used to authenticate the user. The smartphone and wearable device use proximity detection based on the iBeacon protocol (based on Bluetooth). iBeacon uses three levels of proximity, namely immediate (less than about 2 cm), near (about 2 cm to about 1 m), and far (about 1 m to about 50 m). The proposed method uses the near and far levels. The authentication procedure is as follows:

(1) The two devices detect the iBeacon at the far level.
(2) The two devices start acceleration measurement.

(3) The two devices detect the iBeacon at the near level.
(4) The two devices stop the acceleration measurement.
(5) The two devices transmit the acceleration data to the smart lock.
(6) The smart lock performs authentication using the obtained acceleration data.

The smart lock stores reference user gait data in advance. In the authentication phase, user authentication is performed using the reference data. In step (6), the smart lock processes the acceleration data and extracts features. The times at which acceleration data are acquired from the two devices are expected to be different due to differences in communication delay and proximity detection. Therefore, features that are not influenced by acquisition time, such as the average value and the maximum value, are extracted. Using the extracted features, the user is judged by the classifier and authentication is performed. A supervised machine-learning algorithm is used as the classifier.

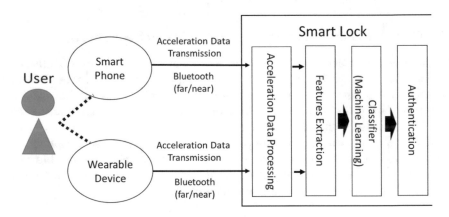

Fig. 1. Proposed system model

3.2   Acceleration Data Processing

We process the acceleration data to facilitate feature extraction. Gait-based authentication is performed without using the inclination of the device. This is achieved by using the composite acceleration. We represent the combined acceleration r_i^1, r_i^2 of the smartphone and wearable device as follows. The obtained x-, y-, and z-axis acceleration data are represented as x_i^1, y_i^1, z_i^1 for the smartphone and x_i^2, y_i^2, z_i^2 for the wearable device.

$$r_i^1 = \sqrt{(x_i^1)^2 + (y_i^1)^2 + (z_i^1)^2} \tag{1}$$

$$r_i^2 = \sqrt{(x_i^2)^2 + (y_i^2)^2 + (z_i^2)^2} \tag{2}$$

In addition, d_1 and d_2 are shown below for the set of composite acceleration data acquired from the two devices. The times at which the i-th data point was acquired (with respect to the start of measurement) are t_i^1 and t_i^2 for the smartphone and wearable device, respectively. n_1 and n_2 are the numbers of data points acquired during measurement.

$$d_1 = \left\{ (t_i^1, r_i^1) | i \in \{1, ..., n_1\} \right\} \tag{3}$$

$$d_2 = \left\{ (t_i^2, r_i^2) | i \in \{1, ..., n_2\} \right\} \tag{4}$$

Since the numbers of acceleration data points for the two devices are different due to differences in performance and measurement time, the number of data points are adjusted. For example, when $n_1 > n_2$, the set d_1', d_2' of data when combining the sample numbers are as shown below.

$$d_1' = \left\{ (t_i'^1, r_i'^1) | i \in \{1, ..., n_2\} \right\} \tag{5}$$

$$d_2' = \left\{ (t_i'^2, r_i'^2) | i \in \{1, ..., n_2\} \right\} \tag{6}$$

Here, $t_i'^1$, $r_i'^1$ are calculated as follows:

$$t_i'^1 = t^1_{\underset{j \in \{1, ..., n_1\}}{\arg\min} (|t_j^1 - t_i^2|)} \tag{7}$$

$$r_i'^1 = r^1_{\underset{j \in \{1, ..., n_1\}}{\arg\min} (|t_j^1 - t_i^2|)} \tag{8}$$

Since the variation of acceleration data is small in the stationary state and large in the walking state, a section of the acceleration data with a high standard

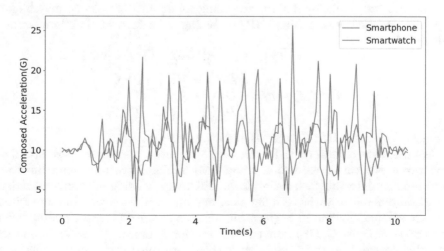

Fig. 2. Composite acceleration after acceleration processing

deviation is extracted as the walking state. Noise removal is then applied to the acceleration data using a low-pass filter. Figure 2 shows the waveforms obtained after processing the acceleration data from the two devices (smartphone in the right pocket and smartwatch worn on the left wrist).

3.3 Feature Extraction

To identify the user, we extracted the following features from the composed acceleration data: average value, median value, standard deviation, variance, maximum value, minimum value, maximum value interval, minimum value interval, cycle, and similarity. We also used the differences between these features obtained from the smartphone and wearable device as features. For similarity, the acceleration data from the two devices were shifted along the time axis and the highest similarity value was taken. To calculate similarity, we used three template matching methods, namely sum of absolute difference (SAD), sum of squared difference (SSD), and normalized cross correlation (NCC). Equations (9) to (11) show the calculation of these three similarities, respectively. Here, the number of samples for comparing the degree of similarity decreases if you shift it the number of samples. Therefore, the number of shifts is $k \in \{1, ..., \frac{2}{3}n\}$

$$SAD = \min_{k \in \{1, ..., \frac{2}{3}n\}} \frac{1}{n-k} \sum_{i=1}^{n-k} |r_i'^1 - r_{i+k}'^2| \tag{9}$$

$$SSD = \min_{k \in \{1, ..., \frac{2}{3}n\}} \frac{1}{n-k} \sum_{i=1}^{n-k} (r_i'^1 - r_{i+k}'^2)^2 \tag{10}$$

$$NCC = \max_{k \in \{1, ..., \frac{2}{3}n\}} \frac{\sqrt{\sum_{i=1}^{n-k} r_i'^1 r_{i+k}'^2}}{\sqrt{\sum_{i=1}^{n-k} (r_i'^1)^2} \sqrt{\sum_{i=1}^{n-k} (r_{i+k}'^2)^2}} \tag{11}$$

We also use similarity determined using minimum cost elastic matching, a type of dynamic programming (DP) matching, as a feature. This similarity is calculated as:

$$\text{Cost: } C = |r_i'^1 - r_j'^2|$$

$$DP(i,j) = \min \begin{cases} DP(i-1,j-1) + C \\ DP(i-1,j) + C \\ DP(i,j-1) + C \end{cases} \tag{12}$$

For $i, j \in \{1, ..., n\}$, $DP(n, n)$ is taken as the features.

We calculate a total of 31 features (two means, two medians, two standard deviations, two variances, two maxima, two minima, two maximum intervals, two minimum intervals, two cycles, mean difference, median difference, standard deviation difference, variance difference, maximum difference, minimum difference, maximum interval difference, minimum interval difference, period difference, SAD, SSD, NCC, DP) from the acceleration data from the smartphone and wearable device. User authentication is then performed using machine-learning classification.

4   Implementation and Experiments

To acquire acceleration data from a device, we developed a Java program. The smartphone was a Sony Xperia XZs and the wearable device was a Sony Smart-Watch 3. A server was used instead of a smart lock. A program developed in Python was used to receive acceleration data and calculate the accuracy of machine-learning classification. Bluetooth was used to transmit and receive data. Acceleration measurement was conducted by transmitting an acceleration measurement command and a measurement end command to each device. For machine-learning, we used the Python module scikit-learn [10]. We calculated the accuracy of support vector machine (SVM), naïve Bayes, random forest, and neural network classifiers.

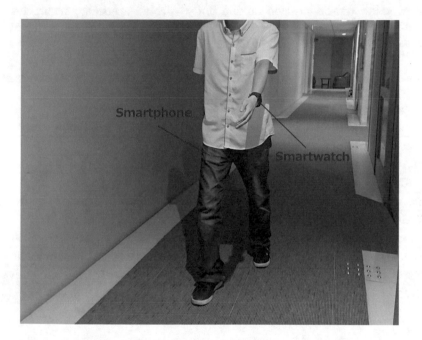

Fig. 3. Experiment conditions

We conducted experiments to confirm the effectiveness of the proposed gait-based authentication method. As shown in Fig. 3, subjects put the smartphone in their right pocket and wore the smartwatch on their left wrist. Each subject walked 20 times in a corridor for about 10 m. The subjects were 15 students from Kanagawa Institute of Technology. The experiment procedure was as follows:

(1) The two devices started acceleration measurement.
(2) The subject stopped for 5 s.
(3) The subject walked 10 m in the corridor.

(4) The subject stopped for 5 s.
(5) The two devices ended acceleration measurement.

Steps (2) and (4) were used to facilitate the extraction of the walking state in step (3). The user was identified using the acceleration data. We evaluated the identification accuracy.

5   Evaluation

We calculated accuracy based on 300 acceleration data points acquired in the above experiment. We used half the data as training data, and the remaining data as test data. Accuracy was calculated as the correctly classified data points divided by the total number of test data points, given as a percentage. To determine whether sensor location on the body affected accuracy, accuracies were measured when using only the smartphone in the right pocket and only the wearable device worn on the left wrist.

Table 1. Identification accuracy obtained using only smartphone for various classifiers

Machine learning			
SVM	NaiveBayes	RandomForest	NeuralNetwork
Features			
Mean	Mean	Mean	Mean
Median	Median	Median	Median
Standard deviation	Variance	Maximum value	Standard deviation
Variance	Maximum value	Minimum value	Variance
Maximum value	Minimum value	Maximum interval	Maximum value
Maximum interval	Maximum interval	Minimum interval	Minimum value
Minimum interval	Minimum interval		Maximum interval
Cycle			Minimum interval
			Cycle
Accuracy			
57.0%	71.3%	76.7%	64.7%

When using only the smartphone, the differences and similarity were not used as features. We extracted the mean, median, standard deviation, variance, maximum interval, minimum interval, and cycle period as features. Table 1 shows the accuracies obtained using only the smartphone in the right pocket for various classifiers and the corresponding features. The highest accuracy was 76.7%, obtained using random forest.

We also calculated accuracies obtained using only the smartwatch on the left wrist. Table 2 shows the results and the corresponding features. The highest accuracy was 76.0%, obtained using random forest.

Table 2. Identification accuracy obtained using only smartwatch for various classifiers

Machine learning			
SVM	NaiveBayes	RandomForest	NeuralNetwork
Features			
Median	Mean	Mean	Mean
Standard deviation	Median	Median	Standard deviation
Maximum value	Standard deviation	Variance	Variance
Minimum value	Maximum value	Maximum value	Maximum value
Maximum interval	Minimum value	Minimum value	Minimum value
Minimum interval	Maximum interval	Maximum interval	Maximum interval
Cycle	Minimum interval	Minimum interval	Minimum interval
		Cycle	
Accuracy			
47.3%	69.3%	76.0%	51.3%

Table 3. Identification accuracy obtained using both smartphone and smartwatch for various classifiers

Machine learning			
SVM	NaiveBayes	RandomForest	NeuralNetwork
Features			
Mean	Standard deviation	Median	Mean
Median	Minimum value	Variance	Median
Standard deviation	NCC	Minimum value	Standard deviation
Variance	DP	SSD	Variance
Maximum value	Maximum interval	NCC	Maximum value
Minimum value	Minimum interval	Maximum interval	Minimum value
SAD		Minimum interval	SSD
NCC		Cycle	NCC
Maximum interval			Minimum interval
Minimum interval			
Accuracy			
86.7%	90.0%	95.3%	83.3%

We also evaluated the proposed method using both the smartphone and wearable device. The features were divided into 13 groups (mean, median, standard deviation, variance, maximum value, minimum value, maximum interval, minimum interval, period, SAD, SSD, NCC, DP) to reduce calculation time. For instance, the mean group included the mean of acceleration data from the smartphone, that from the wearable device, and the difference between the means

Table 4. Most frequently used features

Machine learning	
NaiveBayes	RandomForest
Number of frequent	
Minimum value (2)	Maximum interval (363)
NCC (2)	Minimum value (342)
Maximum interval (2)	Average (332)
Minimum interval (2)	Minimum interval (317)
Average (1)	NCC (281)
Standard deviation (1)	SSD (197)
Maximum value (1)	DP (197)
SSD (1)	Cycle (194)
DP (1)	Maximum interval (184)
Median (0)	SAD (184)
Variance (0)	Standard deviation (181)
SAD (0)	Median (176)
Cycle (0)	Variance (173)

obtained for the two devices. Accuracy was calculated by combining the feature groups. Table 3 shows the accuracies obtained using the two devices and the corresponding features. The highest accuracy was 95.3%, obtained using random forest.

Tables 1, 2, and 3 indicate that using two devices gives better accuracy compared with that obtained using one device. Table 4 shows the features used in machine-learning that exceeded an accuracy of 90% in order of most frequent. The maximum interval, minimum interval, and minimum value had the highest contributions to identification accuracy, and variance, median, and standard deviation had the lowest contributions.

6   Conclusion

This study proposed a system model of gait-based authentication based on two devices (a smartphone and a wearable device). The system extracts 31 features from acceleration data obtained during walking. We carried out experiments with 15 subjects walking in a corridor to confirm the effectiveness of the proposed system. The subjects put the smartphone in their right pocket and the smartwatch on their left wrist. We calculated the identification accuracy for various machine-learning algorithms. The highest accuracy was 95.3%, obtained using random forest. The maximum interval, minimum value, and minimum interval had the highest contributions to identification accuracy, and variance, median, and standard deviation had the lowest contributions.

In future work, we plan to calculate another features that improve the accuracy, and evaluate attack against this method. In addition, other types of wearable device, such as glasses, will be evaluated.

Acknowledgements. This work was supported by JSPS KAKENHI Grant Numbers JP17H01736, JP17K00139.

References

1. Ministry of Internal Affairs and Communications: 2018 White Paper on Information and Communication in Japan (2018). http://www.soumu.go.jp/johotsusintokei/whitepaper/ja/h29/pdf/n3300000.pdf
2. Gartner: Gartner Says Worldwide Wearable Device Sales to Grow 26 Percent in 2019, 07 June 2019. https://www.gartner.com/en/newsroom/press-releases/2018-11-29-gartner-says-worldwide-wearable-device-sales-to-grow-
3. Qrio: Qrio Smart Lock, 07 May 2019. https://qrio.me/smartlock
4. August: August Smart Lock, 07 May 2019. https://august.com
5. Kwikset: Door Locks Door Hardware Smart Locks & Smart key Technology, 16 April 2019. https://www.kwikset.com
6. Hou, R., Watanabe, Y.: A Study on authentication at the time of the walk of using the acceleration sensor of smartphone. In: Computer Security Symposium 2013, pp. 21–23 (2013). (in Japanese)
7. Konno, S., Nakamura, Y., Shiraishi, Y., Takahashi, O.: Improvement of gait-based authentication by using multiple wearable sensors. IPSJ J. **57**(1), 109–122 (2016). (in Japanese)
8. Iwamoto, T., Sugimori, D., Matsumoto, M.: A study of identification of pedestrian by using 3-axis accelerometer. IPSJ J. **55**(2), 734–749 (2014). (in Japanese)
9. Mondal, S., Nandy, A., Chakraborty, P., et al.: Gait based personal identification system using rotation sensor. J. Emerg. Trends Comput. Inf. Sci. **3**(3), 395–402 (2012)
10. Scikit-learn: scikit-learn machine learning in Python Scikit-learn 0.19.1 documentation, 07 May 2019. http://scikit-learn.org/stable/index.html

Automatic Vulnerability Identification and Security Installation with Type Checking for Source Code

Shun Hinatsu[1]([✉]), Koichi Shimizu[1], Takeshi Ueda[1], Benoît Boyer[2], and David Mentré[2]

[1] Mitsubishi Electric Corporation,
5-1-1 Ofuna, Kamakura, Kanagawa 247-8051, Japan
Hinatsu.Shun@bc.MitsubishiElectric.co.jp
[2] Mitsubishi Electric R&D Centre Europe,
1 allée de Beaulieu CS 10806, 35708 Rennes Cedex 7, France
B.Boyer@fr.merce.mee.com

Abstract. Cyber security has been an important issue for control systems, however, there may be a shortage of security experts in the near future. Most of security engineering methods focus only at high-level such as architectures, and do not consider the availability of the control system. In addition, it is often difficult use them without security expertise. In this paper, the author proposed a novel method for solving this problem. The method automatically identifies vulnerabilities in a source code using a formal and static method (type-checking), and fixes them by installing security functions accordingly. Since this approach is entirely automated, it can be seamlessly used by regular engineers, i.e. without security expertise. The authors implemented the method and experimentally investigated its effectiveness on a simple example including the characteristics typical of communicating control systems.

1 Introduction

It has been very important to investigate security engineering for control systems against cyber attacks, because more and more systems are getting connected to outside networks [14]. However, some study warns of the shortage of security experts [6]. Hence, there may be insufficient staffs who can investigate security engineering for control systems in the near future.

However, most of security engineering methods focus on security at high-level such as architectures, not at low-level such as source code. In addition, they do not consider the availability of the control system. Furthermore, it is often difficult to use them without security expertise.

In this paper, we propose a novel security engineering method to solve the problem above. It automatically identifies vulnerabilities for each variable in a source code of a system by utilizing formal method. Then it automatically inserts

© Springer Nature Switzerland AG 2020
L. Barolli et al. (Eds.): NBiS-2019, AISC 1036, pp. 292–304, 2020.
https://doi.org/10.1007/978-3-030-29029-0_27

security functions in the code based on the identification result and the value of each variable. Since the method is entirely automated, even no security experts can secure control system software at source code level, considering availability of control systems.

2 Related Work

If designers or developers investigate security engineering, roughly speaking, they identify (analyze) vulnerabilities which may lead to cyber attacks, and install (implement) security functions in it to reduce or prevent them. We present some of security engineering methods for "vulnerability identification" and "security installation" and their limitations here, and then motivate the direction of the proposed approach.

2.1 Vulnerability Identification

Microsoft develops "Threat Modeling Tool" [7] which enables the user to design systems and analyze threats which may occur in them. However, it focuses only high-level system configuration such as network connection between a client and a server. Therefore, it cannot be used when implementing the system at low-level such as source code-level.

Some tools enable the user to detect vulnerabilities in a system illustrated in block-shape models [12,13]. However, it is desirable that the user has security knowledge and experience, if they implement security function to the system based on the result. Generally speaking, the integration of security into a system by hand is difficult [5]. Therefore, there is often a gap between vulnerability identification and security installation.

2.2 Security Installation

Some tools extend UML (Unified Modeling Language) and enable the user to design security system such as role-based access control [3,5]. However, they focus only high-level design such as architecture, and don't output source codes automatically. There may be a gap between design by them and implementation. Therefore, it is difficult to use them for implementing security at low-level such as source code-level which describes a data flow.

The method by Livshits et al. [4] executes static analysis for detecting input points (source) to output points (sink) in a source code, and outputs candidate insertion place of sanitizer (security function) for a web application. However, it does not consider a value of each variable. If it is applied to a control system, it may add security functions to some not important variable, which may lower the performance (availability) of the system. However, availability is the most important for control systems in information security triad (CIA triad; Confidentiality, Integrity, and Availability).

2.3   Direction of Our Approach

Generally, the existing security engineering methods have these limitations:

- They focus on security at high-level such as network or architecture. They cannot be used for security at low-level such as source codes. There is often a gap between security design and implementation.
- They do not consider the performance (availability) of the system. They may apply countermeasures for not important assets or variables. If there are several assets, their values in terms of CIA triad are different from each other. For example, password for web shopping should be protected in terms of confidentiality, while measured data integrity is important in IoT (Internet of Things) [2].
- Users should have security knowledge and experience. It is often difficult for regular engineers who do not have security expertise to use them.

We propose a novel method for security engineering, which has features below:

- It has two processes at low-level (source code-level)
    1. Vulnerability identification:
       analyze a existing source code and behaviour of each variable to find vulnerabilities
    2. Security installation:
       add security functions in the source code by using the result of vulnerability identification
- Security installation adds appropriate security function based on the value of each asset (i.e. does not add for not important asset), which leads to satisfying least functionality (defined in technical control system component requirements IEC 62443-4-2 CM-7) and considering availability
- Both of the two processes can operate automatically, hence even engineers without security expertise can use the method.

3   Method

In this section, we explain how to identify vulnerability and install security in a source code by our proposal method. Figure 1 describes the overview of the method. First, the method analyzes an input source code by using a formal and static method in "Vulnerability identification", and shows the vulnerable points as "Vulnerability information". Second, it adds security functions in the input source code in "Security installation". Finally, it outputs a source code which has security functions.

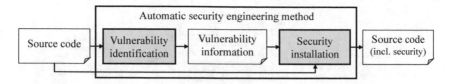

Fig. 1. Overview of the proposal method (some inputs are omitted)

3.1 Vulnerability Identification

3.1.1 Overview

Figure 2 describes the overview of vulnerability identification. It consists of two modules: Generation of vulnerability specification and Type checker. This identification needs a variable list and a function list in addition to a source code as inputs, and outputs the information about functions and variables which may have vulnerability as "Vulnerability information".

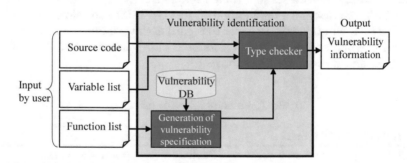

Fig. 2. Overview of vulnerability identification

In this part, we use a taint-analysis-like method for finding vulnerability of each variable. As illustrated in Fig. 3, taint analysis labels input data from untrusted sources by adding "taint (tag)", then tracks how the tainted attribute propagates, and checks whether tainted data is used in unsafe ways by checking the tag [10]. Our method also focuses on input (source) and output (sink) of each variable. Input from the outside may include unsafe data, hence we annotate a type (taint) for each variable to regard it as unsafe. If a variable is processed by security function (sanitizer) before output to the outside (or some critical process), the taint is cleansed and the variable can be regarded as a trusted value (the upper of Fig. 3). Otherwise, it remains unsafe (the lower of Fig. 3). We use formal method for checking taints (unsafe or trusted).

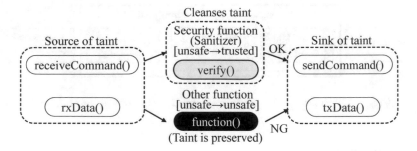

Fig. 3. Overview of taint analysis [15]

3.1.2   Generation of Vulnerability Specification

Before analysis (as in Fig. 3), general information on vulnerabilities and the target source code are needed. For coverage of vulnerabilities, we used CWE (Common Weakness Enumeration), which is a community-developed list of common software security weaknesses [8]. Table 1 shows a part of vulnerability database (DB), which we created for the method. It has CWE ID and information of sanitizer ("Detail"), which is used in order to regard each variable as unsafe or trusted by checking whether the sanitizer exists or not. We investigated the content of each CWE, and decided sanitizer for it.

Table 1. Vulnerability DB (excerpt)

CWE ID	CWE remark	Countermeasure
200	Information exposure	Sanitizer: encryption, decryption
494	Without integrity check	Sanitizer: MAC generation, verification

Extracting function names of source/sink/sanitizer in the source code ("Function list" in Fig. 2), e.g. `encrypt()` for encryption, we can embody Table 1 to create vulnerability specification (Table 2) for the input source code.

Table 2. Example of vulnerability specification

CWE ID	Countermeasure
200	Source: `receiveCommand()`, Sink: `sendCommand()`, Sanitizer: `encrypt()`, `decrypt()`
494	Source: `receiveCommand()`, Sink: `sendCommand()`, Sanitizer: `generateMAC()`, `verify()`

3.1.3   Type Checker

In vulnerability identification, we use type checking for each variable (as in Fig. 3), because it is a lightweight formal method, and it can be applied to security theory [1]. In the method, we annotate a type unsafe or trusted for each variable when it is declared or used as an argument. Then, by checking the change of the type, we check whether the security function is applied on the targeted variable or not before its content is shared outside (or some critical process). Existence or nonexistence of the security function may be factor of the possibility of the vulnerability in the source code.

Listing 1 is a sample of unsafe function unsafe_read(). It has no annotation on default function output, which means "Returned value is unsafe".

Listing 1. Sample of unsafe function

```
int unsafe_read(void nothing);
```

Listing 2 is a sample of input C source code which has vulnerability. critical_process() is defined as a function which requires a trusted parameter and returns a trusted value. If a variable which has unsafe is not processed by the sanitizer, it remains unsafe, and is not considered a secured variable.

Listing 2. Sample of C source code which has vulnerability (excerpt)

```
int __attribute__((trusted)) critical_process(
                        int __attribute__((trusted)) input);

int main() {
    while (1) {
        int tmp = unsafe_read();

        int error = critical_process(tmp);
```

Listing 3 is a sample of security function sanitize(), which is defined as a function which returns a trusted value, regardless of which the input is unsafe or trusted. If the variable which has unsafe is processed by sanitize(), it becomes trusted, and is considered a secured variable.

Listing 3. Sample of security function

```
int __attribute__((trusted)) sanitize(int input);
```

Listing 4 is a sample of input C source code which has no vulnerability. There is sanitize() between unsafe_read() and critical_process(), while there is not sanitize() in Listing 2. Therefore, Listing 4 is considered to be secured.

Listing 4. Sample of C source code which has no vulnerability (excerpt)

```
int __attribute__((trusted)) critical_process(
                         int __attribute__((trusted)) input);

int main() {
    while (1) {
        int tmp = unsafe_read();
        int safe = sanitize(tmp);
        int error = critical_process(safe);
```

By checking a type unsafe or trusted of each variable, we can obtain information about functions and variables which may have vulnerability. This module outputs the information mentioned above as a result ("vulnerability information"), which is used to add appropriate functions to each variable in security installation.

3.2 Security Installation

3.2.1 Overview

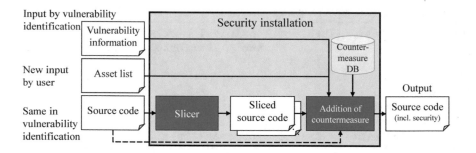

Fig. 4. Overview of security installation

Figure 4 describes the overview of security installation. It has two modules; Slicer and Addition of countermeasure. This installation needs an asset list in addition to the vulnerability information and the source code as inputs, and outputs a source code which has security functions.

3.2.2 Slicer

The proposal method slices the input source code into a small code (sliced source code) for each variable to get its data flow including functions which has the variable as the argument. Then countermeasures (security functions) are added for each variables in the next module. This slicer uses AST (Abstract Syntax Tree), which is the parsing result of C source code. It has the information such as the line number where the variable exists in the code. We chose Clang AST [16] which provides the line number of variable declaration and assignment. If the slicer is applied to Listing 2, it can output Listing 5 by choosing the statements related to tmp in **int** main().

Listing 5. Sample of sliced source code

```
int tmp = unsafe_read();
int error = critical_process(tmp);
```

3.2.3   Addition of Countermeasure

At high-level, e.g. hardware-level, there are many countermeasures [9]. In order to treat countermeasures at low-level (source code-level), the method uses countermeasure database (DB) for each element of CIA triad. They have security function calls and their insertion places which uses other function's position as a reference. Table 3 is a sample of countermeasure DB, which describes that Sanitizer `sanitize()` is added after function `unsafe_read()` in terms of integrity (I).

Table 3. Sample of countermeasure DB

Name	Function	Insertion place (reference)	CIA
Sanitizer	`sanitize()`	after `unsafe_read()`	I

Before adding functions in Table 3, this module decides which variable should have countermeasure based on vulnerability information (output of vulnerability identification), which provides information about functions and variables which may have vulnerability. Then it decides security functions for each variable based on the asset value in "asset list" (Table 4). The value is the grade of each asset (variable) for each element of CIA triad, which is defined in advance. "**H** (High)" means important, and "L (Low)" means not important. By this process, we can add only necessary functions, which leads to considering the availability of the system.

Table 4. Sample of asset list

Asset	Variable	Confidentiality	Integrity	Availability
Temporal data	tmp	L	**H**	L

For example, if vulnerability information includes "variable `tmp` is unsafe, and function `critical_process()` requires `trusted` variable" in Listing 2, security function should exist for `tmp`. According to Table 4, `tmp` (Integrity: **H**) is important and needs security function for integrity. Therefore, sanitizer `sanitize()` in the DB for integrity (Table 3) can be placed after `unsafe_read()` (Listing 6).

Listing 6. Sample of security installation result (excerpt)

```
int tmp = unsafe_read();
sanitize(); //added by security installation
int error = critical_process(tmp);
```

Thus the proposal method can install security functions in source codes.

4   Experiment

We implemented vulnerability identification tool and security installation tool based on Sect. 3, and conducted an experiment in order to investigate the effectiveness of the proposal method. In the experiment, we input a C source code into the two tools, and checked the output from them.

4.1   Example

We assumed a simple communicating control system consisting of a HMI (Human-Machine Interface), a controller and a device (Fig. 5). The controller receives commands from the outside (HMI), and sends commands to the other outside (Device). We prototyped a C source code (Listing 7) in the controller, and assumed it as input for the proposal method. The code (Listing 7) receives commands (Number num and Message msg) from commandline in PC, and displays some results. Listing 8 (in a header file) shows the functions' input and output types.

Listing 7. Input C source code (excerpt)

```
void controller(void) {
    int num = receiveCommand(arg_v1);
    int msg = receiveCommand(arg_v2);

    sendCommand(num, msg);
```

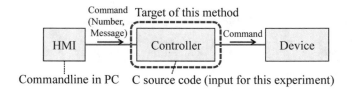

Fig. 5. Target system for this experiment

Listing 8. Functions in input C source code

```
int receiveCommand(char input);

int sendCommand(int __attribute__((trusted)) arg_num,
                int __attribute__((trusted)) arg_msg);
```

Table 5. Countermeasure DB

Name	Function	Insertion place (reference)	CIA
Encryption	`encrypt()`	before `sendCommand()`	C
Decryption	`decrypt()`	after `receiveCommand()`	C
MAC generation	`generateMAC()`	before `sendCommand()`	I
Verification	`verify()`	after `receiveCommand()`	I

In order to check the basic operation and the automaticity of the method, we focused on only encryption/decryption for confidentiality, and tamper detection (generation of MAC (message authentication code) and verification) for integrity. Confidential data is encrypted before sending, and decrypted after receiving. For checking its integrity, MAC are generated, and verified later. Therefore, 4 functions were defined in Countermeasure DB (Table 5).

Table 6 shows an of asset list for this experiment. Their values in CIA are different from each other.

Table 6. Asset list

Asset	Variable	Confidentiality	Integrity	(Availability)
Number	num	L	**H**	(L)
Message	msg	**H**	L	(L)

4.2 Result of Vulnerability Identification

We input the code (Listing 7) into the vulnerability identification tool, and got Listing 9 as output of the type checker. Listing 9 indicated that `sendCommand()` requires `trusted` variables, not `unsafe` ones such as num and msg. This result satisfies Listing 8, which shows that `receiveCommand()` outputs `unsafe` value and `sendCommand()` accepts only `trusted` values. Therefore, the tool automatically notified that both num and msg might have vulnerabilities. In security installation, countermeasures are added for them in the code based on this result.

Listing 9. Output of type checker (vulnerability information)

```
[type-check] controller.c:15: User Error:
  illegal call to function 'sendCommand':
  parameter #1 (arg_num) should be of type int trusted
               found expression of type int unsafe: num
[type-check] controller.c:15: User Error:
  illegal call to function 'sendCommand':
  parameter #2 (arg_msg) should be of type int trusted
               found expression of type int unsafe: msg
```

4.3   Result of Security Installation

We input the code (Listing 7) and the asset list (Table 6) into the security installation tool. Considering the results of vulnerability identification (Listing 9),
countermeasures were needed for both num and msg. Finally, we got Listing 10
as output of security installation.

Listing 10. Output C source code (excerpt)

```
void controller(void) {
    int num = receiveCommand(arg_v1);
verify();
    int msg = receiveCommand(arg_v2);
decrypt();

generateMAC();
encrypt();
    sendCommand(num, msg);
```

Listing 10 included these security functions;

- verify() after num = receiveCommand()
- decrypt() after msg = receiveCommand()
- generateMAC() and encrypt() before sendCommand(num, msg)

Considering the DB (Table 5) and the asset list (Table 6), num (Integrity: **H**)
needed generateMAC() which can be placed before sendCommand(). It also
needed verify() which can be placed after receiveCommand(). msg (Confidentiality: **H**) needed encrypt() which can be placed before sendCommand() It
also needed decrypt() which can be placed after receiveCommand(). Therefore, the tool succeeded in installation of security functions for each variable.

5   Discussion

We proposed the method to identify vulnerabilities and install security automatically at source code-level. Then we implemented each methods, and conducted
an experiment to investigate the effectiveness of them. In the experiment, we
input a sample source code and checked inputs and outputs of the two tools,
and both of them seemed to go well. Although we expressed one function call
for one security function, e.g. encrypt() for an encryption, "encryption" may
include several steps; e.g. initialization, encryption, and clean up [11]. Therefore,
we should integrate the tool-chain and apply it to more concrete and larger system as well as define the insertion place for each step of each function, depending
on target source codes.

Each element in CIA triad often has trade-off relationship with another [17].
Generally, availability is the most important for control systems, and often lowered by security functions for confidentiality or integrity. However, the proposal
method decides the suitable security function for each variable based on the

result of vulnerability identification and the value of each variable (asset), which leads to reduction of unnecessary functions. Therefore, the user can add only essential functions to the system, which leads to least functionality and considering the availability. Least functionality is a requirement defined in IEC 62443-4-2 CM-7 (Security for industrial automation and control systems - Part 4-2: Technical security requirements for IACS components). We need an evaluation of performance based on data such as each function's time cost and capacity used.

The security functions in Listing 10 have no arguments (`encrypt()` without `msg`). We have to include the arguments for each of them in our future works. However, there are no method to identify vulnerabilities and install security at source code-level with consideration of availability. Therefore, even function calls will be helpful information for no security experts. In addition, if the result of the security installation includes arguments for countermeasures, e.g. `encrypt(msg)`, we can also verify the correctness of the installation result by using the same type checker.

6   Conclusion

This paper proposed a novel security engineering method for control systems. The method automatically identifies vulnerability in a source code by formal method, and adds appropriate security function for each variable, considering the availability of the system. An experiment was conducted to investigate the effectiveness of the proposal method by applying it to a simple control system. As a result, the method succeeded in detecting the vulnerable points and adding the appropriate security functions in the source code automatically. This method can be expected to support regular engineers who do not have security expertise.

References

1. Evans, D., Larochelle, D.: Improving security using extensible lightweight static analysis. IEEE Softw. **19**(1), 42–51 (2002)
2. Fekade, B., Maksymyuk, T., Kyryk, M., Jo, M.: Probabilistic recovery of incomplete sensed data in IoT. IEEE Internet Things **5**(4), 2282–2292 (2017)
3. Jürjens, J.: UMLsec: extending UML for secure systems development. In: International Conference on the Unified Modeling Language, pp. 412–425. Springer (2002)
4. Livshits, B., Chong, S.: Towards fully automatic placement of security sanitizers and declassifiers. ACM SIGPLAN Not. **48**, 385–398 (2013)
5. Lodderstedt, T., Basin, D., Doser, J.: SecureUML: a UML-based modeling language for model-driven security. In: UML 2002 - The Unified Modeling Language, pp. 426–441 (2002)
6. McAfee: Hacking the skills shortage: a study of the international shortage in cybersecurity skills. https://www.mcafee.com/content/dam/enterprise/en-us/assets/reports/rp-hackingskills-shortage.pdf
7. Microsoft: Microsoft Threat Modeling Tool - Microsoft Docs. https://docs.microsoft.com/ja-jp/azure/security/azure-security-threat-modeling-tool

8. MITRE Corporation: CWE - Common Weakness Enumeration. http://cwe.mitre. org/index.html
9. Moro, N., Heydemann, K., Encrenaz, E., Robisson, B.: Formal verification of a software countermeasure against instruction skip attacks. J. Cryptogr. Eng. 4(3), 145–156 (2014)
10. Newsome, J., Song, D.X.: Dynamic taint analysis for automatic detection, analysis, and signature generation of exploits on commodity software. In: NDSS, vol. 5, pp. 3–4. Citeseer (2005)
11. OpenSSL Software Foundation: Cryptography and SSL/TLS - OpenSSL. https:// www.openssl.org/docs/manmaster/man3/EVP EncryptInit.html
12. Pedroza, G., Apvrille, L., Knorreck, D.: AVATAR: a SysML environment for the formal verification of safety and security properties. In: 2011 11th Annual International Conference on New Technologies of Distributed Systems, pp. 1–10. IEEE (2011)
13. Poirier, C., Kriaa, S., Pebay-Peyroula, F., Mraidha, C., Zille, V.: A tool for I & C system architecture design: the French connexion cluster. In: International Symposium on Future I & C for Nuclear Power Plants/International Symposium on Symbiotic Nuclear Power Systems 2014, pp. 1–7 (2014)
14. Sadeghi, A.-R., Wachsmann, C., Waidner, M.: Security and privacy challenges in industrial internet of things. In: 2015 52nd ACM/EDAC/IEEE Design Automation Conference (DAC), pp. 1–6. IEEE (2015)
15. Subbrao, M.: Are Static Application Security Testing (SAST) Tools Glorified Grep? https://www.synopsys.com/content/dam/synopsys/sig-assets/ebooks/are-sast-toolsglorified-grep.pdf
16. The Clang Team: Introduction to the Clang AST - Clang 9 documentation. https:// clang.llvm.org/docs/IntroductionToTheClangAST.html
17. Zeng, W., Chow, M.-Y.: Optimal tradeoff between performance and security in networked control systems based on coevolutionary algorithms. IEEE Trans. Ind. Electron. 59(7), 3016–3025 (2012)

Blockchain-Based Malware Detection Method Using Shared Signatures of Suspected Malware Files

Ryusei Fuji[1], Shotaro Usuzaki[1], Kentaro Aburada[1(✉)], Hisaaki Yamaba[1], Tetsuro Katayama[1], Mirang Park[2], Norio Shiratori[3], and Naonobu Okazaki[1]

[1] Department of Computer Science and Systems Engineering, University of Miyazaki, 1-1 Gakuen-Kibanadai-Nishi, Miyazaki 889-2192, Japan
aburada@cs.miyazaki-u.ac.jp
[2] Kanagawa Institute of Technology,
1030, Shimo-Ogino, Atsugi, Kanagawa 243-0203, Japan
[3] Chuo University, 1-13-27 Kasuga, Bunkyo-ku, Tokyo 112-8551, Japan

Abstract. Although rapid malware detection is very important, the detection is difficult due to the increase of new malware. In recent years, blockchain technology has attracted the attention of many people due to its four main characteristics of decentralization, persistency, anonymity, and auditability. In this paper, we propose a blockchain-based malware detection method that uses shared signatures of suspected malware files. The proposed method can share the signatures of suspected files between users, allowing them to rapidly respond to increasing malware threats. Further, it can improve the malware detection by utilizing signatures on the blockchain. In the evaluation experiment, we perform a more real simulation compared with our previous work to evaluate the detection accuracy. Compared with heuristic methods or behavior-based methods only, the proposed system which uses these methods plus signature-based method using shared signatures on the blockchain improved the false negative rate and the false positive rate.

1 Introduction

Malware causes serious damage to modern society, not just computers; therefore the detection is very important. An example of damage caused by malware infection is ransomware. If a computer is infected with ransomware, the data inside it is encrypted, and the user loses control of the computer. The attacker will request a ransom from the user in exchange for releasing the restriction on access to the computer or its data. In 2017, computers around the world were infected with malware called Wannacry, caused severe damage. Even now, more than one million computers have a high risk of the infection [1]. The amount of pecuniary damage from ransomware in 2017 is said to be USD 5 billion [2], and is expected to be increased in the future. In addition, malware targeting Internet of Things (IoT) equipment has been emerging. In 2016, denial of service

© Springer Nature Switzerland AG 2020
L. Barolli et al. (Eds.): NBiS-2019, AISC 1036, pp. 305–316, 2020.
https://doi.org/10.1007/978-3-030-29029-0_28

attacks of up to 1.5 Tbps have been executed by exploiting IoT devices infected with "Mirai" [3]. According to Security Report 2017/2018 [4] published by AV-TEST, in recent years the number of new malware programs observed is more than 100 million per year, which means that about 4 new malware programs are discovered per second. Clearly, malware has a serious adverse effect on modern society by impacting the Internet and computers. In order to reduce the damage caused by malware, we need to detect it quickly.

Malware detection techniques are roughly divided into three types: signature-based methods, behavior-based methods, and heuristic methods [5]. Signature-based methods are commonly used to detect malware. The signature is a sequence of bytes with features extracted from malware, and if the contents of an inspected file match one of these signatures, the file is determined to be malware. Signature-based methods have the advantage of reliably detecting known malware, but they have the disadvantage of not being able to detect unknown malware. Behavior-based methods perform their malware detection by actually executing the file under inspection and observing its behavior. These methods can detect unknown malware that cannot be detected by signature-based methods. However, these methods have the disadvantage of a high false positive rate (FPR), which is the rate of benign files being labeled as malicious files. Heuristic methods are techniques that detect malware using data mining and machine learning techniques. The features utilized in heuristics methods include API call sequences issued to the operating system and machine language instruction. The main advantages and disadvantages of heuristics methods are similar to those of the behavior-based detection methods.

Hashimoto et al. [6] provided information on malware that could not be detected by anti-virus software to the vendor to calculate the subsequent malware detection rates, and evaluated the anti-virus software. According to the study, the malware detection rates 30 days after providing the malware information were 40% at most. Therefore, it is difficult to detect malware quickly with only anti-virus software provided by the vendor. To solve such a problem, we investigated a system for sharing malware information using blockchain technology with the aim of sharing the signatures of suspected malware files [7].

In recent years, blockchain technology has attracted much attention due to its four main characteristics of decentralization, persistency, anonymity, and auditability [8]. The technology was proposed by Nakamoto [9] in 2008 to realize the Bitcoin network, which enables rapid transactions between users with a low cost and without mediation of a central authority. Nowadays the technology is the fundamental technology of various virtual currencies and expected to be used in various fields [10].

In our previous work [7], we investigated a system for sharing the signatures of suspected malware files using blockchain technology, and performed a simple evaluation. In this paper, we perform a more real simulation of the proposed method to evaluate the detection accuracy.

2   Related Work

There are a few related works in our study. Jingjing et al. [11] proposed a framework, called Consortium Blockchain for Malware Detection and Evidence Extraction (CB-MDEE), that detects and classifies malware for mobile devices. The CB-MDEE is composed of two blockchains, a public blockchain (PB) and a consortium blockchain (CB). Users belonging to the PB use a multi-feature model created from, for example, sensitive behavior graphs and installation packages, to detect and classify malware, and store the information on the PB for subsequent malware detection and classification. Members of malware detection organizations belonging to the CB use the information to create a fact base for updating the malware feature database. In evaluation experiments, the CB-MDEE achieved a classification accuracy of 94% for android malware.

Roman et al. [12] proposed a system to support cyber analysts by classifying and managing cyber incident reports using blockchain technology and a deep autoencoder neural network. When a cyber expert enters a cyber incident report into the system, the system classifies the report and returns past similar incident reports. Because the classification and management are executed automatically, the cyber expert can adopt suitable countermeasures quickly. In the evaluation, they used 5,850 training documents and 584 test documents to validate the effectiveness of their proposed system. For the "fulldisclosure" category, they achieved a true positive rate 0.991 and an FPR of 0.059.

This study is based on the assumption that users belonging to the blockchain have a different malware detection system using behavior-based methods or heuristic methods. Then, these detection results are saved as votes on the blockchain for later use. As a result, we can detect and eliminate malware by utilizing the results of our own malware detection system and votes of other users stored on the blockchain, which is different from the above studies.

3   Blockchain Technology

Blockchain technology was proposed as a fundamental technology for realizing Bitcoin in the paper published by Nakamoto [9] in 2008. Bitcoin was realized by combining several inventions to decentralize functions, such as currency issuance and mediation of transactions, which banks typically do. Due to the decentralized function, Bitcoin makes it possible to issue currency and create transactions among users without third-party institutions, such as banks.

The blockchain has four main characteristics: decentralization, persistency, anonymity, and auditability [8]. Based on the above characteristics, this study adopted blockchain technology, which enables rapid sharing of suspected malware signatures between users without the intervention of a central organization, such as an anti-virus vendor. It is possible to use these signatures to eliminate malware.

Blockchain is the basis for various platforms, such as Ethereum [13] and Hyperledger [14]. Ethereum is a platform for building blockchain-based decentralized applications (Dapps) in an open-source development environment.

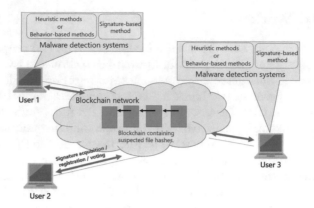

Fig. 1. Overview of proposed system

Dapps save and use certain information on a blockchain, and their use is expected to increase. We can develop various applications by executing a programmed contract called a smart contract on the Ethereum blockchain. We adopted Ethereum because it is already used as a blockchain platform in many Dapps, such as uPort [15].

4   Proposed System

In this section, first, we explain an overview of the proposed system and some assumptions in the system. Next, we describe records which are stored on the blockchain, and the elimination decision formula to calculate the maliciousness degree based on the records. Finally, we state a countermeasure against mass voting by malicious users.

Table 1. Definitions for each symbol

Symbols	Definitions
M_d	Maliciousness degree
M_t	Threshold for malicious degree
V_t	Threshold for total votes
V_b	The number of votes of 'benign'
V_m	The number of votes of 'malicious'
R_v	Voting confidence rate
R_s	Self confidence rate
D_r	Detection result of own malware detection system

4.1   Overview of Proposed System

An overview of the proposed system is shown in Fig. 1. The blockchain network is composed of users who want to share and obtain malware information. Here, it is assumed that each user's computer hosts a heuristic or behavior-based malware detection system and a signature-based system. Also, we suppose the malware detection system use different features or methods. The blockchain is used to store signatures (file hash values) and other information from suspected malware files.

When a user downloads an executable file, heuristic or behavior-based malware detection is executed first. If the downloaded executable file is judged as malware, the user sends the file hash value to the blockchain network as a suspected malware file identity. When another user downloads the same executable file, the user first checks whether the file hash value of the executable file is already registered as a suspected malware file identity on the blockchain. If the same file hash value exists on the blockchain, the user's heuristic or behavior-based malware detection system judges whether the file is malicious, and the result is sent as a vote (malicious or benign) to the blockchain network. Thereafter, based on the voting results on the blockchain and the results of its own malware assessment, the user's detection system decides whether to remove the suspect file.

A flowchart showing the process for each user is provided in Fig. 2, and symbols used in this paper are defined in Table 1.

4.2   Malware Detection Systems on User Computers

In this study, we assume that each user belonging to the blockchain network installs the following two malware detection systems on the computer:

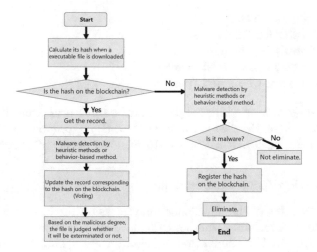

Fig. 2. Flowchart of malware detection performed by each user

- Malware detection system using heuristic or behavior-based methods
  This program is executed when the user downloads an executable file. In this
  study, it is assumed that each user detects malware using different features
  or methods.
- Malware detection system using signature-based methods.
  This system is responsible for investigating whether signatures already exist
  on the blockchain. Also, it calculates the degree of maliciousness and eliminates the downloaded file according to the result.

4.3   Detection of Suspected Files and Transmission of File Hash Values

When a user downloads an executable file, the heuristic or behavior-based malware detection is executed first. Next, the user's computer checks whether the
file hash value is already registered as a suspected malware file hash value on
the blockchain. If the file hash value does not exist on the blockchain, and if the
malware detection system determines that the downloaded file is malware, the
computer sends the file hash value to the blockchain network to share it and
then eliminates the file. When the file hash value exists on the blockchain, the
user's detection system sends out the result of its own malware analysis as a vote
("malicious" or "benign") to the blockchain network and then decides whether
to remove the file with the elimination decision formula (see Sect. 4.5).

4.4   Record Components

Here we describe data such as file hash values and the number of votes to be
stored on the blockchain. Data stored on the blockchain can be represented as
a record, and the record is represented by the following five elements:

- Suspected file hash value
- Number of votes for "malicious"
- Number of votes for "benign"
- Addresses of users who voted "malicious"
- Addresses of users who voted "benign".

The recording of user addresses prevents the same user from illegally voting
more than once. Here, the user address is not an IP address but the address used
on the blockchain, such as "0xca35b7d915458ef540ade6068dfe2f44e8fa733c".

4.5   Elimination Decision Formula

When the hash value of the downloaded file exists on the blockchain, the user's
detection system determines whether to remove the file based on the maliciousness degree given by the elimination decision formula. That is, when Eq. (1) is
satisfied, the file is not deleted, and when Eq. (2) is satisfied, the file is deleted.

$$M_d \leq M_t, \tag{1}$$
$$M_d > M_t,$$
$$0 \leq M_d \leq 1. \tag{2}$$

4.5.1   When $V_m + V_b \geq V_t$

The user's system uses only the voting result on the blockchain and calculates the maliciousness degree with Eq. (3).

$$M_d = \frac{V_m}{V_m + V_b}. \tag{3}$$

4.5.2   When $V_m + V_b < V_t$

The user's system calculates maliciousness degree with expression (4), the results of voting on the blockchain, and its own malware detection results by heuristic or behavior-based methods. Here, it is assumed that the malware detection system outputs 1 when the file is malware and 0 when it is benign. That is, $D_r \in \{0, 1\}$.

$$M_d = \frac{V_m}{V_m + V_b} \times R_v + D_r \times R_s. \tag{4}$$

where R_v and R_s are defined by the following expressions:

$$R_v = \frac{V_m + V_b}{V_t}, \tag{5}$$
$$R_s = 1 - R_v. \tag{6}$$

4.6   Countermeasure Against Mass Voting by Malicious Users

The records stored on the blockchain include the addresses of users who voted in order to prevent duplicate voting by the same address. However, since any user can generate an unlimited number of addresses, it is insufficient to use only the measures described above. Therefore, we tried to solve this problem by establishing a web server by trustworthy organizations and institutions to register voting addresses and limit the addresses eligible to vote.

First, users who wish to participate in the Ethereum blockchain network access the web server installed by trustworthy organizations and institutions and register an address to be used for voting. The web server accesses the Register smart contract on the Ethereum blockchain and registers the address. After that, the user joins the network and acquires, registers, and votes for signatures. The Vote smart contract, which is responsible for signature acquisition, registration, and voting, accesses the Register smart contract and checks whether the address exists in the Register smart contract. If the address exists, acquisition, registration, and voting of the signature are accepted; otherwise, these are rejected.

Here, the web server prohibits the registration of consecutive addresses from the same IP address and confirms the human by CAPTCHA. From the above, it is possible to prevent the registration of addresses by malicious users and bots.

5   Evaluation

In this section, we explain an evaluation experiment to verify the effectiveness of the proposed system. The purpose of the experiment is to investigate whether the accuracy of detecting and removing malware is improved by using the proposed system. We performed a more real simulation compared with our previous work [7] to evaluate the detection accuracy. To conduct the experiment, we created a prototype of the proposed system using a Python script without using the blockchain platform. An overview of the experiment is shown in Fig. 3. The users acquire, register and vote for the signatures via a database instead of the blockchain. The simulation continued until each user performed malware detection and removal for the predefined files regarded as malicious or benign.

In the evaluation experiment, it is assumed that no malicious user exists and each user possesses the malware detection systems described in Sect. 4.2. In addition, rather than implementing user-specific malware detection systems, we created pseudo malware detection systems that have FPR and FNR as parameters.

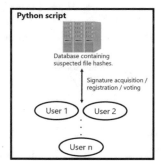

Fig. 3. Overview of experiment environment

Table 2. Parameter definitions for Eqs. (7) and (8)

Symbol	Definition
True Positive (TP)	The number of malware detected as malware correctly
True Negative (TN)	The number of benign files judged as benign files correctly
False Positive (FP)	The number of benign files detected as malware mistakenly
False Negative (FN)	The number of malware judged as benign files mistakenly

5.1   Detection Accuracy Indicators

We used the false negative rate (FNR) and FPR as evaluation indexes for malware detection and removal accuracy. FNR is the rate at which malware is mistakenly judged as a benign file, and FPR is the rate at which a benign file is erroneously detected as malware.

$$FNR = \frac{FN}{TP + FN}, \tag{7}$$

$$FPR = \frac{FP}{TN + FP}. \tag{8}$$

The definitions of the symbols in Eqs. (7) and (8) are given in Table 2.

5.2 Parameters

We conducted the experiment with the parameters shown in Table 3. The value for M_t was determined empirically, and the value for V_t was determined based on a statistical approach. In addition, the FPR and FNR of the pseudo malware detection systems were set with reference to the literature [16], where Windows malware was detected using machine instruction sequences. Specifically, malicious instruction extraction and malicious sequential pattern extraction (MSPE) were used to efficiently and effectively obtain malicious sequences by heuristic methods. In the evaluation experiment, the detection result with MSPE combined with all-nearest-neighbor was the best result, achieving a detection rate of 96.17% (FNR of 3.83%) and FPR of 6.13% They also experimented with combinations of other classifiers, and we set the FNR to 5% and FPR to 6% based on their experiment results.

Table 3. Parameters

Parameters	Values
The number of user addresses	5000
The number of file hash assumed as malicious	2000
The number of file hash assumed as benign	2000
Threshold for total votes (V_t)	385 or 664
Threshold for malicious degree (M_t)	0.5
False negative rate (FNR)	0.05
False positive rate (FPR)	0.06
Cover rate	0.05–1

In order to perform a more real simulation, we introduced cover rate, which means the percentage of the total file hash the user will inspect. For example, when cover rate is 1, each user performs the malware detection of heuristic methods or behavior-based methods for all the predefined files regarded as malicious or benign, and registers, votes, and obtains information from the blockchain as necessary.

5.3   Results and Discussion

In the experiment, we calculated FPR and FNR in the following two cases:

- Case A
  Heuristic methods or behavior-based methods only.
- Case B
  Heuristic methods or behavior-based methods plus signature-based method
  using signatures on the blockchain.

FNR and FPR in Case A are the values from Sect. 5.2. The experimental
results for FPR and FNR in Case B are shown in Figs. 4 and 5 respectively.
FNR and FPR are the average for all users.

From Figs. 4 and 5, we can see that as the cover rate was increased, FPR
and FNR were improved. In addition, as V_t increased, the effectiveness of the

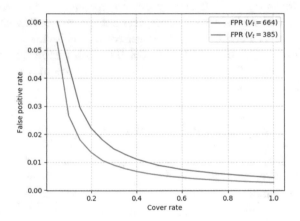

Fig. 4. Experimental results for FPR

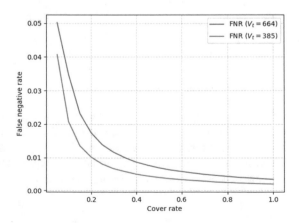

Fig. 5. Experimental results for FNR

proposed system decreased. However, if V_t is set to a very small value, the user makes the judgment using a small number of voting results on the blockchain, which may cause a false positive or false negative. In contrast, if V_t is too large, voting results on the blockchain can not be used effectively for malware detection. Therefore, to make the proposed system work effectively, it was revealed that we need to determine V_t carefully and get enough votes for the parameter.

6 Conclusions

In this paper, we propose a blockchain-based malware detection method that uses shared signatures of suspected malware files. The proposed system allows users to share the signatures of suspected files, and we can more rapidly respond to increasing malware without a centralized organization, such as an anti-virus vendor. In the evaluation experiment, we perform a more real simulation compared with our previous work to evaluate the detection accuracy. The evaluation experiment showed that the proposed system improved the FNR and the FPR. However, it was revealed that as the cover rate decreases, the effectiveness of the proposed system also decreases. Therefore, we need to determine V_t carefully and get enough votes for the parameter. For future work, it is necessary to set each parameter and precondition so as to more accurately reflect the real world. Further, we need to devise a more effective system for mass voting by malicious users.

Acknowledgements. This work was supported by the Japan Society for the Promotion of Science, KAKENHI Grant Numbers JP17H01736, JP17K00139, and JP18K11268.

References

1. Two years after WannaCry, a million computers remain at risk. https://techcrunch.com/2019/05/12/wannacry-two-years-on/. Accessed 17 May 2019
2. Sultan, H., et al.: A survey on ransomware: evolution, growth, and impact. Int. J. Adv. Res. Comput. Sci. **9**(2) (2018)
3. Barrera, D., Molloy, I., Huang, H.: IDIoT: Securing the Internet of Things like it's 1994. arXiv preprint arXiv:1712.03623 (2017)
4. AV-Test "Security report 2017/18". https://www.av-test.org/fileadmin/pdf/security_report/AV-TEST_Security_Report_2017-2018.pdf. Accessed 02 Dec 2018
5. Bazrafshan, Z., et al.: A survey on heuristic malware detection techniques. In: Information and Knowledge Technology (IKT) 2013 5th Conference, pp. 113–120 (2013)
6. Hashimoto, R., Yoshioka, K., Matsumoto, T.: Evaluation of anti-virus software based on the correspondence to non-detected malware. Distributed Processing System (DPS), pp. 1–8 (2012). (in Japanese)
7. Fuji, R., et al.: Investigation on sharing signatures of suspected malware files using blockchain technology. In: International Multi Conference of Engineers and Computer Scientists (IMECS), pp. 94–99 (2019)

8. Zheng, Z., et al.: An overview of blockchain technology: architecture, consensus, and future trends. In: IEEE 6th International Congress on Big Data, pp. 557–564 (2017)

9. Nakamoto, S.: Bitcoin: A peer-to-peer electronic cash system (2008)

10. Wüst, K., Gervais, A.: Do you need a Blockchain? In: 2018 Crypto Valley Conference on Blockchain Technology (CVCBT), pp. 45–54. IEEE (2018)

11. Gu, J., et al.: Consortium blockchain-based malware detection in mobile devices. IEEE Access **6**, 12118–12128 (2018)

12. Graf, R., King, R.: Neural network and blockchain based technique for cyber threat intelligence and situational awareness. In: 2018 10th International Conference on Cyber Conflict (CyCon). IEEE (2018)

13. Ethereum Project. https://www.ethereum.org/. Accessed 02 Dec 2018

14. Hyperledger - Open Source Blockchain Technologies. https://www.hyperledger.org/. Accessed 02 Dec 2018

15. uPort.me. https://www.uport.me/. Accessed 02 Dec 2018

16. Fan, Y., Ye, Y., Chen, L.: Malicious sequential pattern mining for automatic malware detection. Expert Syst. Appl. **52**, 16–25 (2016)

Draft Design of Li-Fi Based Acquisition Layer of DataLake Framework for IIoT and Smart Factory

ByungRae Cha[1], Sun Park[1], Byeong-Chun Shin[2],
and JongWon Kim[1(✉)]

[1] School of Electronical Engineering and Computer Science, GIST,
Gwangju, South Korea
{brcha, sunpark, jongwon}@gist.ac.kr
[2] Department of Mathematics, College of Natural Sciences,
Chonnam National University, Gwangju, South Korea
bcshin@jnu.ac.kr

Abstract. The next-generation digital revolution will have more changes in aspect of producers than consumers. In the traditional manufacturing environment, the starting point for the transition to smart factories is the collection, storage, and integration of data scattered across manufacturing environments. One of the solutions to solve clean environment problem of manufacturing industry is to apply Li-Fi to Acquisition Layer of DataLake framework. And we described draft design of data parsing stage of acquisition layer as pre-stage for TDA analysis in detail.

1 Introduction

The next-generation digital revolution will have more changes in aspect of producers than consumers. Especially the digitization of the manufacturing process will have tremendous effects in terms of efficiency and cost savings. The key points of the smart factory are speed, customization and cost. It focuses on the challenges of traditional manufacturing industries such as derivative products, niche products, option products, market volatility, and labor-intensive and costly fixed production. The most attractive point of the Smart Factory is expected remarkable productivity improvement and reduced manufacturing cost by 10–20% [1].

In the traditional manufacturing environment, the starting point for the transition to smart factories is the collection, storage, and integration of data scattered across manufacturing environments. In order to apply and utilize Industrial IoT smoothly for smart factory, it is very important to consider wireless communication transforming of wired communication for clean environment of manufacturing industry and resistance to noise for wireless communication. One of the solutions to solve these problems is to apply Li-Fi to acquisition layer of DataLake framework. And we described draft design of data parsing stage of acquisition layer as pre-stage for TDA analysis in detail.

© Springer Nature Switzerland AG 2020
L. Barolli et al. (Eds.): NBiS-2019, AISC 1036, pp. 317–324, 2020.
https://doi.org/10.1007/978-3-030-29029-0_29

2 Related Work

2.1 Data Lake

The history of data storage and utilization has long been developed and complemented. As the amount of data stored in the enterprise database increased, DB marketing was the first attempt to use the data. And DW (Data Warehouse) was born with the need to collect and manage useful data. In order to analyze more intelligently, a method of analyzing not only structured data but also unstructured data has been studied and developed. As a result, Big Data trends have emerged. Data Lake is not just about loading and analyzing data; it can integrate with existing DWs and provide integrated data as a service. The ultimate goal of Data Lake is to be central for all services. Figures 1 and 2 present Microsoft and Oracle's concept diagram of Data Lake.

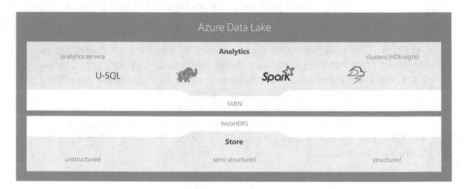

Fig. 1. Concept diagram of Azure Data Lake

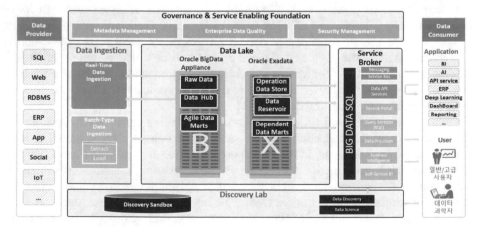

Fig. 2. Concept diagram of Oracle's Data Lake construction

2.2 Industrial IoT, Smart Factory, and Digital Twins

Industrial IoT (Internet of Things), a major trend in the global economy, is changing the business environment of various industries. Three approaches - creating new business models, promoting innovation, and switching people - have an important impact on the successful growth of production and growth businesses. In particular, IDC analyzed that the application and use of Industrial IoT will accelerate innovation in the areas of manufacturing products, services and operations. Moreover, the Smart Factory is defined by Deloitte Univ. as a flexible system that can self-op-timize performance across a broader network, self-adapt to and learn from new conditions in real or near-real time, and autonomously run entire production processes. The Smart Factory should have the five key characteristics of connectivity, optimization, transparency, proactivity, and agility in Fig. 3, and the shift from traditional supply chain to digital supply network at the same time must be done as shown in Fig. 4 [2]. And a Digital Twin can be defined by Deloitte Univ., fundamentally, as an evolving digital profile of the historical and current behavior of a physical object or process that helps optimize business performance in Fig. 5. The digital twin is based on massive, cumulative, real-time, real-world data measurements across an array of dimensions. These measurements can create an evolving profile of the object or process in the digital world that may provide important insights on system performance, leading to actions in the physical world such as a change in product design or manufacturing process [3].

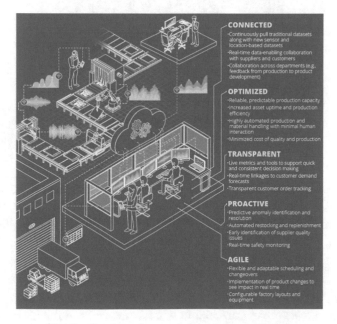

Fig. 3. Five key characteristics of a smart factory

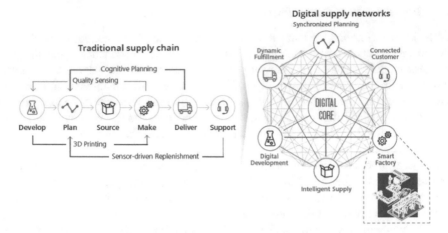

Fig. 4. Shift from traditional supply chain to digital supply network

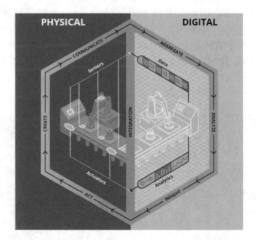

Fig. 5. Manufacturing process Digital Twin model

The key differences between smart factory and digital twin are IT (Information Technology)/OT (Operation Technology) and physics/cyber. The smart factory is an improvement of productivity through the combination of IT and OT, and Twin is a good service through the combination of physical system and cyber.

2.3   Li-Fi

Li-Fi (Light Fidelity), a concept of transmitting data using visible light, is attracting attention as a complementary material to wireless networks in the telecommunication industry. Li-Fi is possible to transmit a lot of data at a higher speed compared with Wi-Fi, and it is expected to be utilized in a specific field (inside of a plane, hospital, etc.) because it can be used without interference in an area sensitive to electromagnetic waves. In particular, the fundamental problem that data cannot be transmitted beyond the walls that light cannot penetrate is pointed out as a barrier to technology diffusion [4, 5].

3   Draft Design of Li-Fi Based Acquisition Layer of DataLake Framework

The Datalake framework is structured as shown in Fig. 6 [6, 7]. The Acquisition Layer of DataLake framework for Industrial IoT & Smart factory should collect, store and integrate data from various sensors or data sources of various sites or sources in the manufacturing field by various communication technologies.

In aspect of shifting to a smart factory in the traditional manufacturing sector, priority is given to ensuring smooth data collection and data-related security. The advantage of Li-Fi is that there is an unlimited frequency resource of visible light, which is more 10,000 times available than Wi-Fi. In addition, it is particularly advantageous in areas where security is important due to the nature of optical wavelengths that can't be hacked. And there is no interference with other frequencies and stable data transmission is possible.

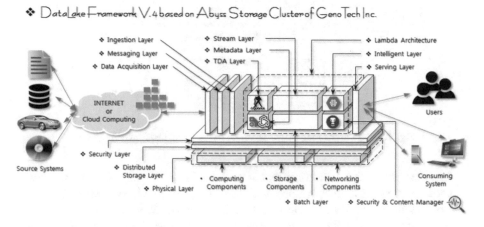

Fig. 6. Concept diagram of DataLake framework V.4

As shown in Fig. 7, the goal of acquisition layer of DataLake framework is to paint a picture of where and how information stored out there and what kind of attribute information-processing mechanisms for TDA (Topological Data Analysis [8]) analysis. In organization, data exists in various forms: Structured data, Semi-structure data, or Unstructured data. Some of the examples of structured data are relational database, XML/JSON data, messages across systems, and so on. Semi-structured data is also very prevalent from an organization perspective, particularly in the form of email, chats, documents and so on. And unstructured data also exists in the form of raw texts, images, videos, audio and so on. For all of these types of data, it may not be possible to always define a schema. Schemas [9] are very useful while translating data into meaningful information. While defining the schema of structured data would be very straightforward, a schema cannot be defined for semi-structured or unstructured data. One of the key roles expected from the acquisition layer is to be able to convert the data into messages that can be further processed in a DataLake framework; hence the acquisition layer is expected to be flexible to accommodate a variety of schema specifications. At the same time it must have a fast connect mechanism to seamlessly push all the translated data messages into the DataLake framework.

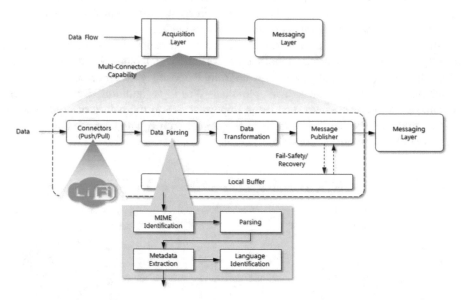

Fig. 7. Li-Fi and functions of acquisition layer of DataLake framework

The data acquisition layer may be composed of multi—connector components on the acquisition side and push the acquired data into a specific target destination. In the case of DataLake framework, the target destination would be the messaging layer. There are specific technology frameworks that enable low-latency acquisition of data from various types of source systems; for every data type, the acquisition connectors are generally required to be configured & implemented depending on the framework used. The data acquisition layer is expected to perform limited transformation on the data acquired as to minimize the latency. The transformation within the data acquisition layer should be performed only to convert the acquired data into a message/event that it can be posted to the messaging layer. In the event that messaging layer is not reachable, the data acquisition must also support the required fail-safe [10] and fail-over [11] mechanisms automatically. For this layer to fail-safe, it should be able to support local and persistent buffering of messages such that, if needed, the messages can be recovered from the local buffer as and when the message layer is available again. This component should also support fail-over and if one of the data acquisition processes fails, another seamlessly takes over. For this layer to support low-latency acquisition, it needs to be built on fast and scalable parsing and transformation components. As shown in the preceding Fig. 7, an acquisition layer's simplified component view comprises connectors, data parsers, data transformers, and a message publisher.

In data parsing component of acquisition layer, the steps required to construct data indexing. The architecture is composed by key components: parser framework, a MIME (Multipurpose Internet Mail Extensions [12]) detection mechanism, language detection. In pre-stage, connectors can be leverage that gathers the pointers to the available data on the network. First, after collecting the set of pointers to data (file, block and objects) of interest, we iterate over that set and then determine each data's media type by MIME detection. Second, once the data's media type is identified, a suitable parser can be selected, and then used by parser to provide both extracted textual content, as well as extracted metadata from the underlying data. The metadata can be used to provide additional information on a per-media-type basis. Lastly, language identification is a process that discerns what language the data is codified in. Indexing can use this information to decide whether a link to an associated translation service should be provided along with the original data.

4 Conclusions

The next-generation digital revolution will have more changes in aspect of producers than consumers. Especially the digitization of the manufacturing process will have tremendous effects in terms of efficiency and cost savings. In the traditional manufacturing environment, the starting point for the transition to smart factories is the collection, storage, and integration of data scattered across manufacturing environments. In order to apply and utilize Industrial IoT smoothly for smart factory, it is very important to consider wireless communication transforming of wired communication for clean environment of manufacturing industry and resistance to noise for wireless communication. One of the solutions to solve these problems is to apply Li-Fi to Acquisition Layer of DataLake framework.

Acknowledgments. This research was supported by Basic Science Research Program through the National Research Foundation of Korea (NRF) funded by the Ministry of Science, ICT and Future Planning (2017R1E1A1A03070059).

References

1. Berger, R.: The 4th Industrial Revolution - The Future Already. DASAN 3.0 (2017)
2. Burke, R., Mussomeli, A., Laaper, S., Hartigan, M., Sniderman, B.: The Smart Factory - Responsive, Adaptive, Connected Manufacturing. Deloitte University Press, Westlake (2017)
3. Parrott, A., Warshaw, L.: Industry 4.0 and the Digital Twin. Deloitte University Press, Westlake (2017)
4. Patel, H.: Survey on Li-Fi technology and its applications. Int. J. Inf. Sci. Tech. (IJIST) **6** (1/2) (2016)

5. Li-Fi. https://en.wikipedia.org/wiki/Li-Fi
6. Cha, B., Park, S., Shin, B.-C., Kim, J.: Draft design of datalake framework based on abyss storage cluster. Smart Media J. **7**(1), 9–15 (2018)
7. Cha, B., Park, S., Shin, B.-C., Kim, J.: Design and verification of connected data architecture concept employing DataLake framework over Abyss storage cluster. Smart Media J. **7**(3), 57–63 (2018)
8. TDA (Topological data Analysis). https://en.wikipedia.org/wiki/Topological_data_analysis
9. Schema. https://en.wikipedia.org/wiki/Database_schema
10. Fail-safety. https://en.wikipedia.org/wiki/Fail-safe
11. Fail-over. https://en.wikipedia.org/wiki/Failover
12. MIME (Multipurpose Internet Mail Extensions). https://en.wikipedia.org/wiki/MIME

A Novel Hybrid Recommendation System Integrating Content-Based and Rating Information

Tan Nghia Duong[(✉)], Viet Duc Than, Tuan Anh Vuong, Trong Hiep Tran, Quang Hieu Dang, Duc Minh Nguyen, and Hung Manh Pham

Hanoi University of Science and Technology, Hanoi, Vietnam
nghia.duongtan@hust.edu.vn

Abstract. Collaborative filtering (CF), the most efficient technique in recommendation systems, can be classified into two types: neighborhood-based model and latent factor model. Both are only based on the user-item interaction, or rating information, and do not take into account the item's content-based information which may contain valuable knowledge. In this work, we propose a hybrid content-based and neighborhood-based recommendation system which utilizes the genome tag associated with each movie in the MovieLens 20M dataset. Experiment results show that our proposed system not only achieves a comparable accuracy but also performs at least 2 times faster than the "pure" CF methods.

Keywords: Recommendation system · Collaborative filtering · Similarity measure · Neighborhood-based · Matrix factorization

1 Introduction

Thanks to rapid development of information technology and network, the volume of data has been sharply increasing over the last decade, which results in not only convenience but also information overload. People are living among millions or even billions of products and services that make them confused about which to buy or consume. Recommendation systems appear to help solve such problems. An effective recommendation system has to be capable of providing the most appropriate suggestions to its users in a timely manner. Therefore, recommendation system has become an active topic research and attracted increasing attention in information-retrieval and data-mining communities [1–5].

Traditionally, there are three main approaches to implement a recommendation system. The first approach is the content-based model where items, e.g. products or services, similar to the ones a user liked in the past are suggested to her. Here, the similarity degree between two items is calculated using the content-related features associated with each item. The second one is based on the *demographic* of the users, assumed that groups of users with the similar demographic profile should share the same interest in some specific items.

© Springer Nature Switzerland AG 2020
L. Barolli et al. (Eds.): NBiS-2019, AISC 1036, pp. 325–337, 2020.
https://doi.org/10.1007/978-3-030-29029-0_30

Both of these two approaches often provide suggestions promptly; however, in practice, the quality of such suggestions is not good enough due to the lack of metadata associated with item and personal information of user. Furthermore, even when information related to item and user is available, it is not possible to guarantee that it is correct and reliable [2,6].

Currently, the most popular and accurate approach to recommendation system is based on collaborative filtering (CF) technique [2,7], which leverages the user-item interaction data, represented as the user preference matrix containing either explicit feedback (also called *ratings*) or implicit feedback on the items given by users. There are two main methods to implement a CF system: neighborhood-based and matrix factorization. A typical example of the former is the item-oriented model which discovers items most similar to those a user bought/used and suggests them to her [8,9]. The latter, which is accepted as the-state-of-the-art technique, can explore the hidden factors under the preference matrix relating users to items. The typical matrix factorization models are **SVD** and **SVD++** [10,11]. Whilst the item-oriented models could capture local-level information and make reasonable recommendations promptly, their matrix factorization counterparts are capable of extracting global-level information embedded in the rating matrix in order to produce more accurate suggestions at the cost of time complexity. In addition to the accuracy and timing, another crucial factor is the ability to explain the reason of the suggestions which helps enhance the users' experience, encourage them to interact more actively with the system, fix wrong impressions and eventually improve the overall performance of the whole system. From this perspective, the neighborhood-based models often provide a better explanation for their recommendations than the matrix factorization ones [12,13].

In this paper we try to improve similarity measures between items and thus increase the prediction accuracy of the item-oriented CF models. The main contribution is a novel method to incorporate the content-based information associated with each item into the classic similarity degree formulas. Applying new similarity measures in the traditional item-oriented CF techniques constitutes a new hybrid content-based and neighborhood-based recommendation system which achieves an accuracy prediction equivalent to the state-of-the-art matrix factorization models but still reserves its own strengths such as better explanation and low time complexity.

The remainder of this paper is organized as follows. Section 2 summarizes the basic knowledge of neighborhood-based and matrix factorization CF models. Our previous work is briefly described in Sect. 3. Section 4 presents a detailed description of the proposed model. Experimental results are shown in Sect. 5. Finally, we conclude our work in Sect. 6.

2   Preliminaries

In this paper, u, v denote users and i, j denote items. The preference by user u for item i is denoted by r_{ui}, also known as the rating, where high values indicate strong preference. The notation $\hat{r}_{ui}$ denotes predicted ratings. The (u, i) pairs for which r_{ui} is known are stored in the set $\mathcal{K} = \{(u, i) | r_{ui} \text{ is known}\}$.

2.1 Baseline Estimates

In practice, some users may give higher ratings than others, and some items may have higher ratings than others. These effects must be taken into account by adjusting the data, which can be done by the *baseline estimates*. The average rating of the whole dataset is denoted by μ. A baseline estimate for an unknown rating r_{ui} is denoted by b_{ui} and accounts for the user and item effects:

$$b_{ui} = \mu + b_u + b_i \tag{1}$$

The parameters b_u and b_i correspond to the observed deviations of user u and item i, respectively, from the average. In order to estimate b_u and b_i one can solve the least squares problem:

$$\min_{b_*} \sum_{(u,i) \in \mathcal{K}} (r_{ui} - \mu - b_u - b_i)^2 + \lambda_1 \left(\sum_u b_u^2 + \sum_i b_i^2 \right) \tag{2}$$

Here, the first term, $\sum_{(u,i) \in \mathcal{K}} (r_{ui} - \mu - b_u - b_i)^2$, strives to find b_us and b_is that fit the given ratings. The regularizing term, $\lambda_1 \left(\sum_u b_u^2 + \sum_i b_i^2 \right)$, avoids overfitting by penalizing the magnitudes of the parameters [14].

2.2 Neighborhood-Based Models

Neighborhood-based model is a kind of CF system which is easy to implement and to explain the reason behind the suggestions. There are two ways to implement neighborhood-based CF models: (i) user-oriented (or user-user [15]) model which estimates unknown ratings of a user based on the ratings of its similar users in the dataset, and (ii) item-oriented (or item-item [9,16]) model which finds the most similar items to the item a user "liked" and recommends these items to her. Currently, the second way is more successful because of the following reasons [2].

- The ability to provide a rational explanation for recommendations.
- The size of the item-item similarity matrix is much more smaller than of the user-user one (in practical recommendation systems, the number of users are often much larger than of items).

Therefore, our concentration in this work is on item-oriented methods. Central to most item-oriented approaches is a similarity degree between items. The measure of the similarity degree between two items i and j can be computed by using cosine similarity function (**Cos**) as follows:

$$s_{ij}^{Cos} = \cos(x_i, x_j) = \frac{\sum_{u \in U_{ij}} r_{ui} r_{uj}}{\sqrt{\sum_{u \in U_i} r_{ui}^2} \sqrt{\sum_{u \in U_j} r_{uj}^2}} \tag{3}$$

where U_{ij} is the set of all users that rate both items i and j, U_i the set of all users that rate item i, and U_j the set of all users that rate item j.

Another commonly used measure of similarity degree in practice is Pearson correlation coefficients (**PCC**) defined in Eq. (4):

$$s_{ij}^{PCC} = \frac{\sum_{u \in U_{ij}} (r_{ui} - \mu_i) \cdot (r_{uj} - \mu_j)}{\sqrt{\sum_{u \in U_{ij}} (r_{ui} - \mu_i)^2} \cdot \sqrt{\sum_{u \in U_{ij}} (r_{uj} - \mu_j)^2}} \tag{4}$$

Recently, a modified version of Eq. (4) was proposed using baseline estimates b_{ui} instead of means μ_i, μ_j as follows:

$$\hat{\rho}_{ij} = \frac{\sum_{u \in U_{ij}} (r_{ui} - b_{ui}) \cdot (r_{uj} - b_{uj})}{\sqrt{\sum_{u \in U_{ij}} (r_{ui} - b_{ui})^2} \cdot \sqrt{\sum_{u \in U_{ij}} (r_{uj} - b_{uj})^2}} \tag{5}$$

Since the vast majority of ratings are unknown, it is expected that some items share only a small amount of common raters. As we can see, the computation of the **PCC** is based on only common users between two items. Accordingly, the reliability of the measure becomes more reliable when the amount of common users increases. A shrunk correlation coefficient which helps avoid overfitting when there are only few common users is integrated into Eq. (5) to create a new similarity measure named **PCCBaseline**:

$$s_{ij}^{PCCBaseline} = \frac{|U_{ij}| - 1}{|U_{ij}| - 1 + shrinkage} \cdot \hat{\rho}_{ij} \tag{6}$$

where $|U|$ is the number of common users between items i and j, and *shrinkage* is the shrunk correlation coefficient.

The final target is to estimate $\hat{r}_{ui}$ - the unobserved rating made by user u for item i. By computing the similarity degree between two items using aforementioned formulas, we can identify the k most similar items to i rated by user u. This set of k neighbors is denoted by $S^k(i, u)$. Then, the predicted value of r_{ui} can be computed as a weighted average of the ratings of similar items (named **kNNBasic** model):

$$\hat{r}_{ui}^{kNNBasic} = \frac{\sum_{j \in S^k(i,u)} s_{ij} r_{uj}}{\sum_{j \in S^k(i,u)} s_{ij}} \tag{7}$$

or as a weighted average of the ratings of the similar items while adjusting for user and item effects through the baseline estimates (named **kNNBaseline** model [14]):

$$\hat{r}_{ui}^{kNNBaseline} = b_{ui} + \frac{\sum_{j \in S^k(i;u)} s_{ij} (r_{uj} - b_{uj})}{\sum_{j \in S^k(i;u)} s_{ij}} \tag{8}$$

2.3   Latent Factor Models

Latent factor models form an alternative approach to neighborhood-based CF recommendation systems aiming to uncover latent features that explain the observed ratings. Currently, matrix factorization models have gained popularity thanks to their superior accuracy and flexible scalability proven during the Netflix Prize [17]. We start with the basic matrix factorization model, named **SVD**. Then another common model, named **SVD++**, which is based on the traditional **SVD** and integrates additional *implicit* information about user behavior in order to improve the prediction accuracy is briefly described.

2.3.1   SVD

Matrix factorization models map both users and items into a latent space of dimension f. A typical model associates each user u with a user-factors vector $p_u \in \mathbb{R}^f$, and each item i with an item-factors vector $q_i \in \mathbb{R}^f$. For a given item i, the elements of q_i measure the degree to which the item possesses those factors. For a given user u, the elements of p_u measure the degree of interest the user has in those factors. The prediction is done by taking an inner product $\hat{r}_{ui} = q_i^T p_u$. The more involved part is parameter estimation. Many of the recent works use explicit feedback information to learn the parameters while avoiding overfitting through an adequate regularized model, such as:

$$\min_{p_*, q_*} \sum_{r_{ui} \text{ is known}} \left(r_{ui} - q_i^T p_u\right)^2 + \lambda \left(||q_i||^2 + ||p_u||^2\right) \tag{9}$$

Here, the constant λ controls the extent of regularization. Minimization is typically performed by either stochastic gradient descent or alternating least squares [10,18]. Besides, an easy stochastic gradient descent optimization was popularized by Funk during Netflix Prize [17].

2.3.2   SVD++

Prediction accuracy is improved over the original SVD model by considering also implicit feedback which provides an additional indication of user preferences. Because of this reason, a second set of item factors is added, relating each item i to a factor vector $y_i \in \mathbb{R}^f$. The predicted rating is computed as follows:

$$\hat{r}_{ui} = q_i^T \left(p_u + |R(u)|^{-\frac{1}{2}} \sum_{j \in R(u)} y_j \right) \tag{10}$$

The set $R(u)$ contains the items rated by user u. Model parameters are determined by minimizing the associated regularized squared error function through stochastic gradient descent approach similar to **SVD** model [11].

3   Previous Work

Our previous paper [19] analyzed the distribution of the similarity scores in the traditional item-oriented CF systems. Specifically, the similarity scores are computed using aforementioned formulas (**Cos** and **PCC**) based on the rating information. Intensive experiments on the original MovieLens 20M dataset showed that the values of similarity degree between two arbitrary items are 97% distributed in the range of [0.85; 1] with a coefficient of variation of 4.83%. Such a small coefficient of variation makes it difficult to distinguish a pair of two similar items from a pair of two dissimilar ones using traditional functions. This badly affects the item-oriented models which utilize the similarity degree between two items to make useful recommendations.

Based on this observation, we proposed a novel similarity measure which could achieve a wider spectrum of the similarity degree by using the cubed version of the original functions, named **cubedCos** and **cubedPCC**, as follows.

$$s_{ij}^{cubedCos} = (s_{ij}^{Cos})^3 \tag{11}$$

$$s_{ij}^{cubedPCC} = (s_{ij}^{PCC})^3 \tag{12}$$

Experiments carried out on the original MovieLens 20M dataset proved that newly proposed measures totally outperform their original counterparts at both accuracy and timing indicators. Even when compared with the state-of-the-art matrix factorization model (**SVD**), the item-oriented CF system using our best similarity measure (**cubedPCC**) provides a 2.20% lower RMSE (Root Mean Squared Error) whilst working at least 2 times faster.

In this paper, a cubed version of **PCCBaseline**, named **cubedPCCBaseline**, is implemented along with **cubedCos** and **cubedPCC** for the purpose of comparison with the proposed similarity measure.

$$s_{ij}^{cubedPCCBaseline} = (s_{ij}^{PCCBaseline})^3 \tag{13}$$

4   Proposed System

4.1   Problems with the Traditional Similarity Measures

As described in Sect. 2.2, the similarity measure between two movies in the traditional item-oriented CF systems is calculated using the rating information. In more detail, two movies are considered to be "similar" if they are rated by the same users with the same ratings; otherwise, they are "dissimilar". In practice, there are some problems with this calculation.

Firstly, it is solely based on the interaction between users and movies. In most cases, users are not willing to provide their opinions (i.e. ratings) on movies so the rating matrix is highly sparse (up to 99.47% in the MovieLens 20M dataset). This often leads to the case that each movie may have a large number of ratings but be rated by only few common users. For instance, movie A has 100 ratings, movie B has 200 ratings; however there is only 2 users rating both A and B. In this situation, the calculation of the similarity degree between A and B in the traditional item-oriented CF algorithms is not reliable because it only considers a tiny piece of available information to evaluate the similarity between two movies and ignores most of the remaining useful information.

Secondly, in some cases, it is often observed that the films in a series tend to have analogous ratings by a specific user. In other word, if a user likes a film in a series and provides a high rating, she tends to like the others in that series and provides high ratings also. Clearly, the films in a series have a strong relationship to each other. Nevertheless, as previously discussed, users are often lazy at providing their ratings on movies they watched. Let's use the trilogy of "The Godfather" as an example: the first part - "The Godfather" - is rated 5 by user A, and the second part - "The Godfather Part II" - is rated 4.5 by user B. It means that user A really likes the first part and user B likes the second part a lot. Yet, user A didn't rate "The Godfather Part II" and user B didn't rate "The Godfather". A practical recommendation system is expected to suggest the second part to user A and the first part to user B based on the above observation: the films in a series are often liked/disliked by a specific user. In this case, evaluating the similarity between the first and the second part of "The Godfather" series using only rating information is not adequate: it ignores the fact that these two movies are in the same series and can lead to a conclusion that they are dissimilar because the similarity measure is very low.

Lastly, regarding to the timing, it can be easily seen that at each time calculating the similarity degree between two movies, it is necessary to find the common users who rated both movies. In practical systems, the number of users and items can reach millions; besides, new ratings constantly flow into the system requiring recomputing the similarity measures frequently. As a result, this calculation will consume a huge computational cost and timing.

4.2   Tag Genome in the MovieLens 20M Dataset

Different from the datasets previously released by GroupLens, the MovieLens 20M dataset includes a data structure, called Tag Genome, that contains tag relevance scores for movies. This structure is a *dense* matrix: each movie has a value for *every* tag in the genome. In more detail, the tag genome encodes how strongly movies exhibit particular properties represented by tags and is computed using a machine learning algorithm on user-contributed content such as tags, ratings, and textual reviews [20].

In the dataset, the number of tags for each movie is 1,128; the relevance between the movie and the corresponding tag is represented by a number in the range of [0; 1] named "genome score". In other word, each movie is characterized by a 1,128-element vector which reflects its content-based information. The heat map in Fig. 1 illustrates 10 movies with the genome scores associated with their first 10 tags where darker color indicates stronger relevance. As we can see, movie "GoldenEye" is strongly relevant to tags "007" and "007 (series)", and has almost no relationship to tags "18th century" or "19th century". It is clear that

"GoldenEye" is the 17th film in the James Bond 007 series which was released in 1995. Obviously, a genome score vector associated with each movie consists of fairly useful and reliable information which is contributed by numerous users, and can be used as a new way to describe the content-based characteristics of each movie. To our best knowledge, the information has not been effectively utilized by any other researchers. Therefore, we decide to take advantage of the genome score vectors to improve the performance of the current neighborhood-based collaborative filtering algorithms.

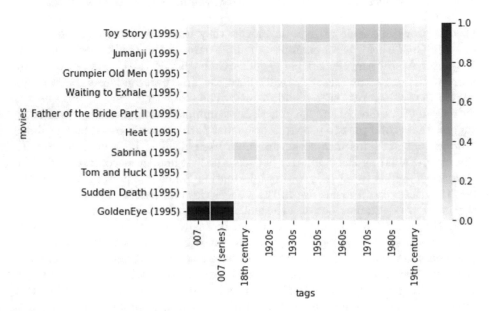

Fig. 1. Tag relevance scores for movies. The titles of movies are on the left of the map, the tags at the bottom, and the bar on the right depicting the relevance between 0 and 1.

4.3 Integrating Content-Based Information into Classic Neighborhood-Based CF Algorithms

To alleviate the drawbacks of the traditional similarity measures, we propose a new method to evaluate the analogy between two movies utilizing the content-based information. In particular, each movie is characterized by a 1,128-element genome score vector $\mathbf{g} = \{g_1, g_2, ..., g_{1128}\}$, and the similarity s_{ij} between movies i and j is now considered as the similarity $s_{\mathbf{g}_i,\mathbf{g}_j}$ between two corresponding vectors $\mathbf{g}_i$ and $\mathbf{g}_j$. In this paper, two different implementations originated from the common similarity formulas are used in order to calculate $s_{\mathbf{g}_i,\mathbf{g}_j}$ as follows.

$$s_{\mathbf{g}_i,\mathbf{g}_j}^{Cos_{genome}} = \frac{\sum_{k=1}^{G} g_{i.k} g_{j.k}}{\sqrt{\sum_{k=1}^{G} g_{i.k}^2}\sqrt{\sum_{k=1}^{G} g_{j.k}^2}} \tag{14}$$

$$s_{\mathbf{g}_i,\mathbf{g}_j}^{PCC_{\text{genome}}} = \frac{\sum_{k=1}^{G}(g_{i.k} - \bar{g}_i)(g_{j.k} - \bar{g}_j)}{\sqrt{\sum_{k=1}^{G}(g_{i.k} - \bar{g}_i)^2}\sqrt{\sum_{k=1}^{G}(g_{j.k} - \bar{g}_j)^2}} \qquad (15)$$

where $\bar{g}_i$ and $\bar{g}_j$ are the mean genome scores of vectors $\mathbf{g}_i$ and $\mathbf{g}_j$, respectively; and $G = 1128$ is the length of genome vectors.

Eventually, the process of estimating unknown ratings in common neighborhood-based CF systems is carried out as usual: $s_{\mathbf{g}_i,\mathbf{g}_j}^{Cos_{\text{genome}}}$ and $s_{\mathbf{g}_i,\mathbf{g}_j}^{PCC_{\text{genome}}}$ are substituted into Eqs. (7) and (8) instead of s_{ij} to make prediction for **kNNBasic** and **kNNBaseline** models, respectively. By integrating the content-related information in genome score vector associated with each movie into the traditional item-oriented algorithms, the proposed model can be regarded as a hybrid system between content-based and neighborhood-based methods. Exploiting both the content-based and user-movie interaction information is expected to improve the overall performance of the recommendation system.

5   Experimental Results

5.1   Dataset and Evaluation Criteria

In order to evaluate the performance of the models presented in this paper, the famous MovieLens 20M dataset is used as a benchmark. The dataset, released by GroupLens in 2016, originally contains 20,000,263 ratings and 465,564 tag applications across 27,278 movies created by 138,493 users (all selected users had rated at least 20 movies). The ratings are float values ranging from 0.5 to 5.0 with a step of 0.5. Different from the previously released datasets of GroupLens, this dataset includes a current copy of the Tag Genome which was computed using a machine learning algorithm on user-contributed content including tags, ratings, and textual reviews [20].

Because the proposed system makes use of the information in tag genome vectors, it is necessary to apply a preprocessing step into the original dataset. In more detail, we firstly drop out the movies which do not have tag genome. After that, only movies and users with at least 20 ratings are kept. The preprocessed dataset now consists of 19,799,049 ratings (approximately 98.99% compared with the original dataset) given by 138,493 users for 10,239 movies (Table 1).

Table 1. Summary of the original MovieLens 20M and the preprocessed dataset

	# Ratings	# Users	# Movies	Sparsity
Original dataset	20,000,263	138,493	27,278	99.47%
Preprocessed dataset	19,799,049	138,493	10,239	98.60%

After preprocessing step, the dataset is split into 2 distinct parts: 75% ratings of each movie are used as the training set and the 25% remaining ratings as the testing set. For the purpose of comparing the overall performance between models, two indicators are used: RMSE, a widely used metrics in the literature, for accuracy evaluation and Time [s] for timing evaluation. Here, RMSE is calculated using the following formula.

$$RMSE = \sqrt{\sum_{u,i \in \text{TESTSET}} (\hat{r}_{ui} - r_{ui})^2 / |\text{TESTSET}|} \tag{16}$$

where $|\text{TESTSET}|$ is the size of testing set, $\hat{r}_{ui}$ is the predicted rating estimated by the model, and r_{ui} is the actual rating made by user in the testing set. All experiments are carried out on a workstation consisting of an Intel®Xeon®Processor E5-2630 2.3 GHz, 12 GB RAM and no GPU.

5.2   Performance Evaluation

The performance of our proposed models are compared with the state-of-the-art recommendation systems including neighborhood-based models (**kNNBasic, kNNBaseline**) and matrix factorization models (**SVD, SVD++**). Furthermore, the best models described in our previous work (**cubedCos, cubedPCC, cubedPCCBaseline**) are also used as competitors. For neighborhood-based models, the number of neighbors is chosen as 40. For matrix factorization models, the number of latent factors is 50.

Table 2. Performance of the state-of-the-art neighborhood-based (**kNNBasic, kNNBaseline**) and matrix factorization (**SVD, SVD++**) models using rating information. Best models proposed in our previous work are also included for comparison.

Model		RMSE	Time [s]
kNNBasic (k = 40)	Cos	0.8636	2,143
	PCC	0.8304	2,493
	PCCBaseline	0.8128	2,531
kNNBaseline (k = 40)	Cos	0.8285	2,604
	PCC	0.8108	2,868
	PCCBaseline	0.8046	2,991
SVD (50 factors)		0.7922	18,220
SVD++ (50 factors)		**0.7894**	**144,324**
kNNBasic (k = 40)	cubedCos	0.8056	2,416
	cubedPCC	0.7924	2,792
	cubedPCCBaseline	0.7886	2,942
kNNBaseline (k = 40)	cubedCos	0.8012	2,845
	cubedPCC	0.7910	3,014
	cubedPCCBaseline	**0.7882**	**3,066**

As highlighted in Table 2, **kNNBaseline** model using **cubedPCCBaseline** to measure the similarity between two movies dominates other competitors at both performance indicators RMSE and Time [s]. Even with the state-of-the-art matrix factorization model **SVD++ (50 factors)**, our previous model works slightly better at accuracy (0.15% lower RMSE) and approximately 47 times faster at timing. It is noteworthy that all these models are using only user-movie interaction information to estimate unknown ratings.

With the same benchmark, the performance of our proposed hybrid recommendation systems in this paper is displayed in Table 3. As can be seen, when choosing the size of neighborhood as 40, **kNNBaseline** model with the similarity degree between two movies calculated by formula PCC_{genome} using genome score vectors provides rather good predictions. Its accuracy indicator RMSE is only 0.22% and 0.38% higher than the ones of **SVD++ (50 factors)** and **cubedPCCBasline**, respectively. However, at timing indicator, it can make predictions approximately 84.85 times and 1.8 times faster than **SVD++ (50 factors)** and **cubedPCCBaseline**, its counterpart using rating information, respectively.

Table 3. Performance of the proposed hybrid recommendation systems with $k = 40$ and $k = 10$.

Model		RMSE	Time [s]
kNNBasic (k = 40)	Cos_{genome}	0.8562	1,424
	PCC_{genome}	0.8268	1,653
kNNBaseline (k = 40)	Cos_{genome}	0.8202	1,521
	$\mathbf{PCC_{genome}}$	**0.7912**	**1,701**
kNNBasic (k = 10)	Cos_{genome}	0.8416	1,225
	PCC_{genome}	0.8266	1,352
kNNBaseline (k = 10)	Cos_{genome}	0.8037	1,360
	$\mathbf{PCC_{genome}}$	**0.7905**	**1,474**

Additional experiments are performed to find out the best value of the neighborhood size when applying our proposed system. This leads to a remarkable conclusion that reducing the size of the neighborhood does not negatively affect the accuracy of predictions whilst speeding up the whole system. After hyperparameter tuning, we find that the hybrid content-based and collaborative filtering system provides best performance with $k = 10$: its accuracy indicator RMSE is 0.7905, which is nearly 0.14% and 0.29% higher than the ones of **SVD++ (50 factors)** and **PCCBaseline** respectively, but costs only 1,474 [s], i.e performs about 98 times and 2 times faster than its competitors. This is a rational result because: (i) the unknown ratings are estimated utilizing all available information associated with each movie (instead of only rating information as in traditional CF systems), and (ii) evaluating the analogy between two fixed-length genome score vectors is a low computational task and can be simultaneously calculated for all vectors at the same time.

6   Conclusion

In this paper, a new method to calculate the similarity measure between two movies is described using the content-based information which is represented in the genome score vector of each movie. Applying the new similarity measure into the process of making predictions results into a hybrid content-based and neighborhood-based model which achieves an accuracy equivalent to the state-of-the-art item-oriented and matrix factorization models but performs at least 2 times faster.

Future work will focus on analyzing and optimizing the tag genome-based representation of each movie. Furthermore, a combination of the proposed model in this paper and the matrix factorization model may create a hybrid recommendation system with further improvement in accuracy.

Acknowledgement. This research is funded by Ministry of Science and Technology (MOST) under grant number 10/2018/ĐTCT-KC.01.14/16-20.

References

1. Christakopoulou, E., Karypis, G.: HOSLIM: higher-order sparse linear method for top-N recommender systems. In: Proceedings of the Pacific-Asia Conference on Knowledge Discovery and Data Mining. Springer (2014)
2. Ricci, F., Rokach, L., Shapira, B.: Recommender systems: introduction and challenges. In: Recommender Systems Handbook. Springer (2015)
3. Christakopoulou, E., Karypis, G.: Local item-item models for top-N recommendation. In: Proceedings of the 10th ACM Conference on Recommender Systems. ACM (2016)
4. He, X., Liao, L., Zhang, H., Nie, L., Hu, X., Chua, T.S.: NAIS: neural attentive item similarity model for recommendation. IEEE Trans. Knowl. Data Eng. **30**(12), 2354–2366 (2018)
5. Nguyen, H.P., Tran, Q.V., Miyoshi, T.: Video compression schemes using edge feature on wireless video sensor networks. Hindawi J. Electr. Comput. Eng. **2012**, 27 (2012)
6. Mooney, R.J., Roy, L.: Content-based book recommending using learning for text categorization. In: Proceedings of the 5th ACM Conference on Digital Libraries. ACM (2000)
7. Smith, B., Linden, G.: Two decades of recommender systems at Amazon.com. IEEE Internet Comput. **21**(3), 12–18 (2017)
8. Sarwar, B.M., Karypis, G., Konstan, J.A., Riedl, J.: Item-based collaborative filtering recommendation algorithms. In: Proceedings of the 10th International Conference on World Wide Web (2001)
9. Linden, G., Smith, B., York, J.: Amazon.com recommendations: item-to-item collaborative filtering. IEEE Internet Comput. **1**, 76–80 (2003)

10. Koren, Y., Bell, R.M., Volinsky, C.: Matrix factorization techniques for recommender systems. Computer **8**, 30–37 (2009)
11. Koren, Y.: Factorization meets the neighborhood: a multifaceted collaborative filtering model. In: Proceedings of the 14th ACM SIGKDD International Conference on Knowledge Discovery and Data Mining. ACM (2008)
12. Herlocker, J.L., Konstan, J.A., Riedl, J.: Explaining collaborative filtering recommendations. In: Proceedings of the 2000 ACM Conference on Computer Supported Cooperative Work. ACM (2000)
13. Tintarev, N., Masthoff, J.: A survey of explanations in recommender systems. In: The IEEE 23rd International Conference on Data Engineering Workshop, ICDE 2007. IEEE (2007)
14. Koren, Y.: Factor in the neighbors: scalable and accurate collaborative filtering. ACM Trans. Knowl. Discov. Data **4**, 1 (2010)
15. Herlocker, J.L., Konstan, J.A., Borchers, A., Riedl, J.: An algorithmic framework for performing collaborative filtering. In: ACM SIGIR Conference on Research and Development in Information Retrieval, SIGIR 1999 (1999)
16. Sarwar, B., Karypis, G., Konstan, J., Reidl, J.: Item-based collaborative filtering recommendation algorithms. In: The 10th International Conference on World Wide Web, WWW 2001 (2001)
17. Funk, S.: Netflix Update: Try This At Home (2006). http://sifter.org/~simon/journal20061211.html
18. Bell, R.M., Koren, Y.: Scalable collaborative filtering with jointly derived neighborhood interpolation weights. In: 7th IEEE International Conference on Data Mining, ICDM 2007. IEEE (2007)
19. Duong T.N., Than V.D., Tran T.H., Dang Q.H., Nguyen D.M., Pham H.M.: An effective similarity measure for neighborhood-based collaborative filtering. In: Proceedings of the 5th NAFOSTED Conference on Information and Computer Science. IEEE (2018)
20. Harper, F.M., Konstan, J.A.: The MovieLens datasets: history and context. ACM Trans. Interact. Intell. Syst. (TIIS) **5**, 19 (2016)

The 14th International Workshop on Network-Based Virtual Reality and Tele-Existence (INVITE-2019)

Proposal of a High-Presence Japanese Traditional Crafts Presentation System Integrated with Different Cultures

Yangzhicheng Lu[1(✉)], Tomoyuki Ishida[2], Akihiro Miyakwa[3],
Yoshitaka Shibata[4], and Hiromasa Habuchi[1]

[1] Ibaraki University, Hitachi, Ibaraki 316-8511, Japan
{18nm742y,hiromasa.habuchi.hiro}@vc.ibaraki.ac.jp
[2] Fukuoka Institute of Technology, Fukuoka, Fukuoka 811-0295, Japan
t-ishida@fit.ac.jp
[3] Nanao-city, Nanao, Ishikawa 926-8611, Japan
a-miyakawa@city.nanao.lg.jp
[4] Iwate Prefectural University, Takizawa, Iwate 020-0693, Japan
shibata@iwate-pu.ac.jp

Abstract. In recent years, the number of foreign visitors to Japan has increased year by year due to the global trend of Japanese culture. In such surroundings, not a few foreign visitors are interested in Japanese traditional culture and traditional crafts. Therefore, information dissemination for foreigners and development of overseas markets are required in the Japanese traditional craft industry. In this research, we implemented a high-presence Japanese traditional crafts presentation system with different cultures by using virtual reality technology with a head-mounted display. Our proposed system promotes the understanding of Japanese traditional culture to foreigners by fusing Japanese traditional crafts with foreign architectural styles. Furthermore, our research supports the dissemination of information and sales channels for Japanese traditional crafts.

1 Introduction

In recent years, the number of foreign tourists visiting Japan has increased year by year. Along with this, it is necessary for various industries involved in tourism to capture inbound demand and create new services and products. On the other hand, Japanese culture, such as animation, fashion and traditional crafts, has received high acclaim from overseas in recent years. These Cool Japan resources are diverse and have unique characteristics. For example, traditional crafts have carefully selected materials, advanced skills, and historical culture. Therefore, effective utilization of these Cool Japan resources is required for effective information dissemination based on the characteristics of the resources. On the other hand, the traditional craft industry has problems such as stagnant domestic demand and lack of information dissemination. Therefore, the Manufacturing Industries Bureau Ministry of Economy, Trade and Industry and the Association for the Promotion of Traditional Crafts Industries point

© Springer Nature Switzerland AG 2020
L. Barolli et al. (Eds.): NBiS-2019, AISC 1036, pp. 341–349, 2020.
https://doi.org/10.1007/978-3-030-29029-0_31

out the importance of promoting traditional crafts by using ICT technology, disseminating information and sales channels to domestic and foreign markets [1–3].

2 Our Previous Researches

In our previous research, we implemented a traditional Japanese craft presentation system using augmented reality (AR) technology and virtual reality (VR) technology to support the promotion of traditional craft products and the information dissemination of traditional cultures. The Japanese traditional craft presentation system using AR technology [4] superimposed virtual traditional crafts in the real world by using mobile terminals such as smartphones and tablets. By realizing this system, users can easily experience traditional crafts. However, because users browse virtual traditional crafts on mobile terminals, they cannot experience the texture, scale and taste of traditional crafts.

On the other hand, Japanese traditional craft presentation system using VR technology [5] realized the experience of the traditional craft with immersive feeling by using the head mounted display (HMD). Users can experience traditional crafts with visual sense close to real environment through this system. However, since this system operates VR space using a controller, it is difficult for the user to operate in VR space. Moreover, this system only provides Japanese room style, and information dissemination and develop sales channels on traditional crafts to foreigners visiting Japan are not considered.

3 Research Objective

In this research, we propose a high-presence Japanese traditional crafts presentation system that reproduces Japanese traditional crafts in a highly realistic virtual space by using VR technology. Users can experience the texture, scale and taste of Japanese traditional crafts by using this system. Also, by realizing the simulation experience of arranging Japanese traditional crafts in a room style in foreign countries, this system provides foreigners visiting Japan with an environment that integrates Japanese traditional culture and different cultures. This is expected to information dissemination and develop sales channels on traditional crafts to foreigners visiting Japan.

4 System Configuration

This system consists of three layers: VR Devices, Unity Application [6] and Network as shown in Fig. 1. The VR Devices of this system use the base station [7], data glove and Vive Pro HMD [8]. The base station acquires the position and direction information of the data glove and HMD, and transfers it to the Unity Application that controls the system. At the same time, operation information from the user is transferred to Unity Application through the data glove. Also, the VR space generated by the Unity Application is presented to the user as visual information through the HMD.

The Unity Application reflect the creation of VR space and user manipulation within VR space. In the Unity Application of this system, there is data storage in which traditional craft objects and house environment scenes are stored. The VR space is generated from objects stored in data storage and presented to the user through the HMD. The attitude and position of the user (avatar) in the VR space are controlled based on the position and direction information of the data glove and HMD. In addition, this system provides a space sharing function that allows multiple users to share and experience VR space through a network.

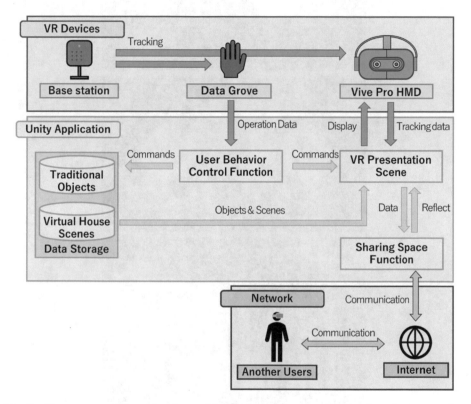

Fig. 1. The system configuration in our high-presence Japanese traditional crafts presentation system.

5   Prototype System

5.1   System UI

The system UI is provided to the user in an environment suitable for data glove because the system captures the user's operation using the data gloves. Also, system functions are provided by the user pressing the button on data glove. When the user presses the left "MENU" button, a menu panel for selecting system functions is displayed (Fig. 2).

The system functions provide users with room space changing function, traditional craft arrangement function, and space sharing function with remote users.

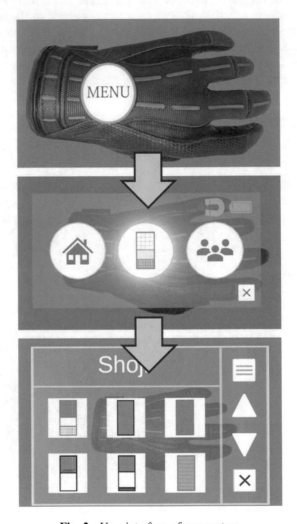

Fig. 2. User interface of our system.

5.2   Japanese Traditional Crafts

Japanese traditional crafts include types such as Tategu (fittings), Kimono (clothes), Somemono (dyed goods), Orimono (woven fabrics) and Shikki (lacquers). Furthermore, Tategu can be divided into types such as Ranma (transom), Itado (wooden door), Shoji (paper sliding door), Fusuma (sliding door), and Garasudo (glass door). In order to reproduce the texture and delicate making of traditional crafts, we introduced the CAD data of Tategu in Nanao City, Ishikawa Prefecture as virtual objects. However, the application of this system is constructed with Unity, and CAD data cannot be imported without data conversion. Therefore, we converted CAD data into ".fbx" format, which is a 3D model format that can be imported into Unity, and then imported it into the application as virtual objects (Fig. 3).

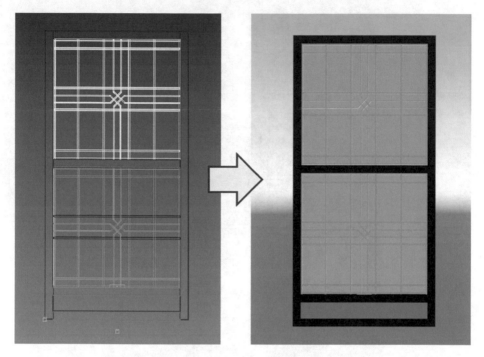

Fig. 3. Data conversion for importing virtual objects into this system.

5.3 Virtual House Scenes

The system introduces a room space designed by interior designers to the system in order to provide the user with a realistic VR experience. Some of these room spaces use 3D space provided by the interior design site "Justeasy [9]." The 3D space provided by this site is created with Autodesk 3ds max [10] and cannot be imported directly into Unity. Therefore, in order to import 3D data into this system, we converted the data format from ".max" format to ".fbx" format. As a result, this system provides the user with a highly realistic room space (Fig. 4).

5.4 Experience of High-Presence Japanese Traditional Crafts

This system uses the HMD to provide the user with the experience of Japanese traditional crafts in a high-presence VR space. The user can change the viewpoint or move in the VR space according to his/her movement. Since this research aimed at promoting Japanese traditional crafts to foreigners visiting Japan, we prepared room spaces of Chinese style, American style, and European style in addition to Japanese style. Therefore, users can experience the arrangement of Japanese traditional crafts in

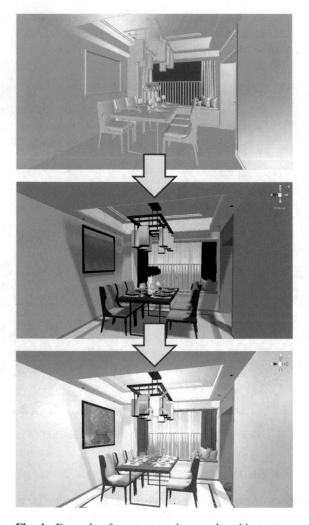

Fig. 4. Example of room space imported to this system.

various room styles. The user can also change the virtual room space and Japanese traditional crafts by selecting the button in the menu using the data glove (Fig. 5).

5.5   Interactive Operability Close to the Real Environment

We constructed an interactive operation method using data gloves to provide users with a highly realistic experience. By using the data glove, it is possible to capture the finger motion of the user, and the same motion as the real environment can be reproduced in the virtual space. In addition, the user's operation in the real environment is reflected in the virtual space by capturing the position information and direction information of the data glove by tracking from the base station. Furthermore, collision detection is set for

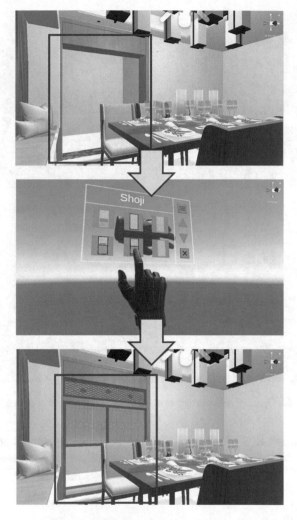

Fig. 5. Japanese traditional crafts arrangement experience in various room styles.

objects in the virtual space, and the user can operate Japanese traditional crafts such as Shoji and Fusuma according to the movement of the data glove. Therefore, the user can experience the function selection of the system and the interaction with the object according to the hand movement by wearing the data glove (Fig. 6).

5.6   Space Sharing Function

In this research, we constructed a space sharing function that allows many users at remote locations to experience the same virtual space. When using the space sharing function, the host user creates a virtual space, and performs network sharing processing for remote users to operate the virtual space. The remote users then performs an

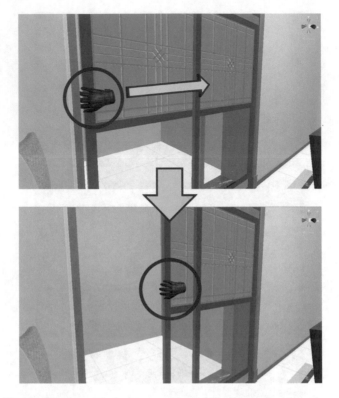

Fig. 6. Opening and closing operation of Shoji by using data glove.

operation to connect to the host user. Finally, the host user creates an avatar for the remote user to operate the virtual space, and gives the remote user the operation authority.

6   Conclusion

In this research, we implemented the high-presence Japanese traditional crafts presentation system integrated with different cultures. This system provides users with experience of texture, scale and taste of traditional crafts in a virtual environment, and support the dissemination of Japanese traditional culture. Furthermore, the provision of a virtual experience that integrates Japanese traditional crafts and different cultural room styles is expected to information dissemination and develop sales channels on traditional crafts to foreigners visiting Japan.

References

1. The Association for the Promotion of Traditional Crafts Industries: Current Situation. https:// kyokai.kougeihin.jp/current-situation/. Accessed 11 May 2019
2. The Association for the Promotion of Traditional Crafts Industries: 2018 Business Plan and Budget Statement. https://kougeihin.jp/system-manager/wp-content/uploads/20180405_jigyou.pdf. Accessed 11 May 2019
3. Regional Revitalization Study Group utilizing Cool Japan Resources for Tourism: Report by Regional Revitalization Study Group utilizing Cool Japan Resources for Tourism. https:// www.meti.go.jp/committee/kenkyukai/chiiki/cool_japan/pdf/report01_01_00.pdf. Accessed 11 May 2019
4. Iyobe, M., Ishida, T., Miyakawa, A., Shibata, Y.: Proposal of an AR traditional crafting mobile system using Kansei retrieval method. In: Proceedings of the 23rd International Symposium on Artificial Life and Robotics, pp. 588–591 (2018)
5. Ishida, T., Lu, Y., Miyakawa, A., Sugita, K., Shibata, Y.: Proposal of a virtual traditional crafting system using head mounted display. In: Proceedings of the 21th International Conference on Network-Based Information Systems, pp. 476–484 (2018)
6. Unity. https://unity.com/ja. Accessed 11 May 2019
7. Base station. https://enterprise.vive.com/jp/support/vive-pro/category_howto/base-stations. html. Accessed 11 May 2019
8. VIVE Pro. https://www.vive.com/jp/product/vive-pro/. Accessed 11 May 2019
9. Justeasy. https://www.justeasy.cn/. Accessed 11 May 2019
10. Autodesk 3DS MAX. https://www.autodesk.co.jp/products/3ds-max/overview. Accessed June 2019

Proposal of Interactive Information Sharing System Using Large Display for Disaster Management

Ryo Nakai[1(✉)], Tomoyuki Ishida[2], Noriki Uchida[2],
Yoshitaka Shibata[3], and Hiromasa Habuchi[1]

[1] Ibaraki University, Hitachi, Ibaraki 316-8511, Japan
{18nm728a,hiromasa.habuchi.hiro}@vc.ibaraki.ac.jp
[2] Fukuoka Institute of Technology, Fukuoka, Fukuoka 811-0295, Japan
{t-ishida,n-uchida}@fit.ac.jp
[3] Iwate Prefectural University, Takizawa, Iwate 020-0693, Japan
shibata@iwate-pu.ac.jp

Abstract. In this research, we implemented an interactive information sharing system for support decision making and information sharing in the disaster response headquarters. Our proposed system realizes interactive information sharing between a large display and mobile devices. In addition, this system realizes the facilitation of the discussion in the meeting by the playback function which can browse previous meeting contents on the mobile device and the reuse function which can reproduce and reuse the screen state of the previous meeting. This research realizes rapid disaster response by supporting the decision making of the disaster response headquarters at the time of large-scale natural disaster.

1 Introduction

When large-scale natural disaster occurs, such as the 1995 Great Hanshin Awaji Earthquake and the 2011 Great East Japan Earthquake, the disaster response headquarters is established in each local government. However, various problems of the disaster response headquarters have been pointed out, such as methods of information sharing, information gathering, information dissemination, and decision making. The Verification Report of the Correspondence of Kinugawa River Flood in Joso-City describes the following problems of the disaster response headquarters [1].

- It is difficult to organize information because the disaster response headquarters receives a large amount of information
- It is difficult to information sharing because the disaster response headquarters uses whiteboards for information sharing. This problem causes information gaps among local government officials

L. Barolli et al. (Eds.): NBiS-2019, AISC 1036, pp. 350–358, 2020.
https://doi.org/10.1007/978-3-030-29029-0_32

2   Research Objective

In this research, we propose the interactive information sharing system "HyperReM" that supports decision making and information sharing at the disaster response headquarters meeting. HyperReM has HyperReM-Master, which visualizes report items and common items on a large display, and HyperReM-Node, which shares information via a Web browser on a smartphone or tablet. In addition, the system implements the following two functions to facilitate meeting proceedings and discussions.

(1) A playback function that browses previous meeting contents.
(2) A reuse function that reproduces and reuses the screen state of previous meeting.

3   System Configuration

The system configuration of the HyperReM is shown in Fig. 1. HyperReM consists of HyperReM-Node, HyperReM-Master, and Content Management Server.

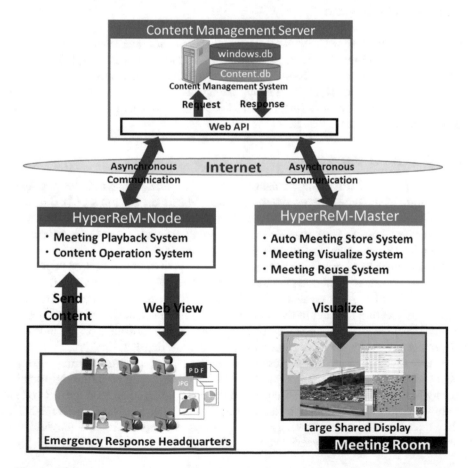

Fig. 1. The system configuration in our proposed interactive information sharing system.

- HyperReM-Master
  HyperReM-Master visualizes content (image, PDF, URL) transmitted from HyperReM-Node on a large display. In addition, HyperReM-Master continuously records the screen state (the current x-coordinate, y-coordinate, width, height of all contents) on the large display and saves it on Content Management Server. This provides the user with reuse function.
- HyperReM-Node
  HyperReM-Node transmits the information collected by the disaster response headquarters to HyperReM-Master via Content Management Server. HyperReM-Node provides users with playback function.
- Content Management Server
  Content Management Server supports exchange of information between HyperReM-Master and HyperReM-Node using WebSocket API. Also, Content Management Server continuously saves the screen state transmitted from HyperReM-Master.

4 Prototype System

HyperReM realizes HyperReM-Node as a web application that runs on smartphones and tablets. In addition, this realizes HyperReM-Master as a desktop application that runs on a PC. The user starts HyperReM-Master on a PC connected to the large display, and then connects to HyperReM-Node to advance a meeting. HyperReM-Node transmits information such as image, PDF, Web page, text data to HyperReM-Master via Content Management Server.

4.1 HyperReM-Master

We developed HyperReM-Master using Electron [2], an open source desktop application framework. When the user starts HyperReM-Master, the top screen (Fig. 2) is displayed. HyperReM-Master holds meeting information for each project. Therefore, when the user starts the meeting from the beginning, it is necessary to create a project by selecting "Create New Project" on the top screen.

On the other hand, when the user resumes a previous meeting, select "Open Project" on the top screen and open the project list screen (Fig. 3). The project list screen displays a list of projects created in the previous. When the user selects an arbitrary project, the meeting list held in the project is listed on the right side of the project list screen. Users can resume a previous meeting by selecting "Resume" at the bottom of the project list screen after selecting an item in the meeting list.

When the user resumes the previous meeting, the information sharing screen is displayed. Authentication information and QR code are displayed on the information sharing screen. When the user participates in the resumed meeting, read the QR code at the lower right of the information sharing screen and connect to the HyperReM-Node. After that, the information sharing screen displays the contents of the previous discussion transmitted from HyperReM-Node (Fig. 4).

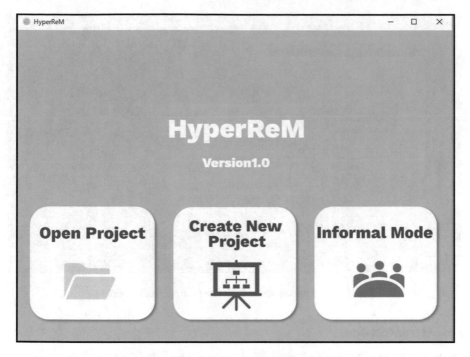

Fig. 2. Top screen of HyperReM-Master.

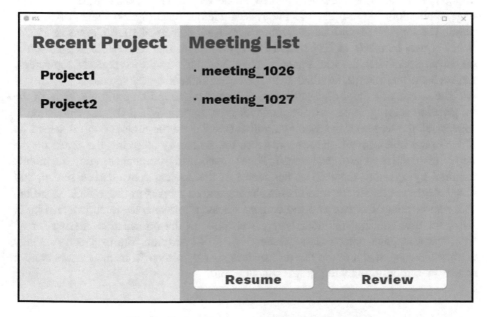

Fig. 3. Project list screen of HyperReM-Master.

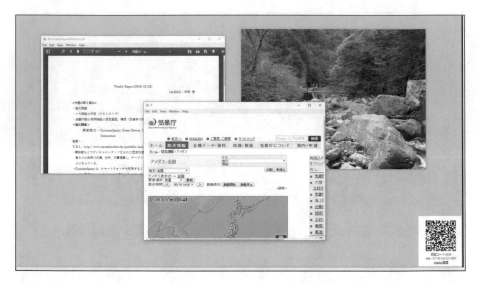

Fig. 4. Information sharing screen of HyperReM-Master.

4.2   HyperReM-Node

Users use HyperReM-Node to transmit content to HyperReM-Master. We developed HyperReM-Node as a Web application in consideration of simplicity and usage in various information devices. Also, since we developed HyperReM-Node as a single page application, the user can use the system without updating the screen of the web page. The top screen and authentication screen of HyperReM are shown in Fig. 5. When a user connects to HyperReM-Node web page, the HyperReM top screen is displayed first. When the user selects "Attend Meeting" on the HyperReM top screen, the screen of the mobile terminal switches to the authentication screen.

The user inputs the 4-digit authentication code displayed on the lower right of the information sharing screen on the authentication screen. When the authentication is successful, the screen of the mobile terminal switches to the remote control screen. In order to transmit content, the user switches the screen by selecting the menu on the upper left of the remote control screen. The user selects the content stored in the mobile terminal by selecting the icon at the center of the content transmission screen, and transmits the content to the information sharing screen by performing a flick operation. The remote control screen and the content transmission screen are shown in Fig. 6. Also, the user can transmit an arbitrary web page to the information sharing screen from the web page transmission screen (Fig. 7). In addition, the user can remotely control the content shared on the information sharing screen from the remote control screen by scrolling, pinching in and pinching out.

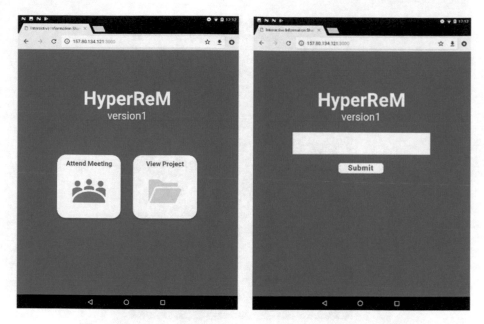

Fig. 5. Top screen and authentication screen of HyperReM-Node.

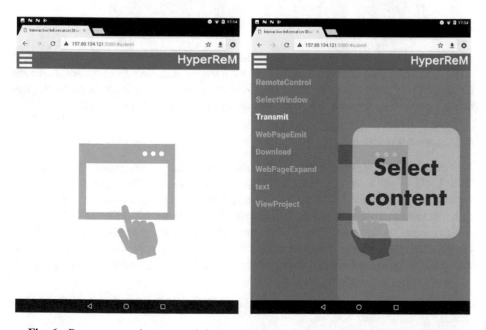

Fig. 6. Remote control screen and the content transmission screen of HyperReM-Node.

Fig. 7. Web page transmission screen of HyperReM-Node.

4.3 Playback Function

Users can view the screen states of previous and current meetings in chronological order (Fig. 8). The screen states of the meeting is acquired from windows.db on Content Management Server. The user can confirm the screen states in chronological order by scrolling the seek bar displayed on the playback screen.

4.4 Reuse Function

The project reuse function receives all screen states (x, y coordinates, width, height, text, URL of all contents) from windows.db of Content Management Server, and reproduces all contents on the project reuse screen. The user can view the screen states of the selected meeting in chronological order by operating the seek bar at the bottom of the project reuse screen (Fig. 9). By pressing the "Restart" button at the bottom of the project reuse screen, the information sharing screen is expanded on the mobile terminal. At the same time, all content is reproduced on the information sharing screen.

Fig. 8. Playback screen.

Fig. 9. Project reuse screen.

5   Conclusion

In this paper, we proposed an interactive information sharing system "HyperReM" for support the decision making and information sharing of the disaster response headquarters. HyperReM realizes interactive information sharing between a large display and mobile devices. Our proposed system consists of HyperReM-Node, HyperReM-Master, and Content Management Server. Also, this system has the playback function that allows participants to view previous meeting contents on a mobile terminal and the reuse function that can reproduce the screen states of the previous meeting.

References

1. Joso-City Flood-Control Measure Verification Committee: Verification Report of the Correspondence of Kinugawa River Flood in Joso-City (2015). http://www.city.joso.lg.jp/ikkrwebBrowse/material/files/group/6/kensyou_houkokusyo.pdf. Accessed May 2019
2. Github: Electron. https://electronjs.org/. Accessed May 2019

A TEFL Virtual Reality System for High-Presence Distance Learning

Steven H. Urueta$^{(\boxtimes)}$ and Tetsuro Ogi

Graduate School of System Design and Management, Keio University,
Yokohama, Japan
surueta@keio.jp, ogi@sdm.keio.ac.jp

Abstract. VR technologies can be effective for distance learning. However, currently available virtual reality systems for teaching English as a foreign language (TEFL) are limited: they are often a supplementary experience within a traditional classroom or do not offer a high level of student-teacher interaction. This research considers an HMD-based platform with two linked subsystems to address these issues. The first is a series of high-presence learning scenarios with roles played by the student(s) and teacher through avatars. The second is an interactive web portal within the VR environment that allows for learning, testing, and feedback. In order to assess the potential of this proposed system, student learning outcomes and ease of use are evaluated through preliminary testing, and this research also posits a more comprehensive system for managing longer-term courses.

1 Introduction

The field of educational technology has seen much advancement in virtual learning environments, including the use of online platforms, games, and various testing applications to improve educational outcomes [1–4]. All of these have been used with varying effects in distance learning [5, 6]. However, the use of consumer VR head-mounted displays (HMDs) is still in its growing stages, and their efficacy as a stand-alone teaching tool has been understudied for teaching English as a foreign language, also known as TEFL [7].

Though there is a plethora of new research examining the potential of VR as an educational aid, only a small portion has focused on the intersection of HMD-based systems and TEFL. Yet, there are many reasons why the two are complementary. To start, VR is useful for active learning [8], which has been suggested to be more effective for knowledge retention and testing performance than passive methods [9]. In addition, VR can be independent of the time and place in the physical world, so it can be used to provide virtual access to environments and situations that may ordinarily be inaccessible physically. This accessibility can be a good complement for task-based and scenario-based learning. Finally, distance learning technologies paired with VR may also provide a layer of safety, decreasing social phobias about speaking or interacting with others [10]. As many beginning English learners may be shy, this can facilitate productive communication.

© Springer Nature Switzerland AG 2020
L. Barolli et al. (Eds.): NBiS-2019, AISC 1036, pp. 359–368, 2020.
https://doi.org/10.1007/978-3-030-29029-0_33

There are a few more peculiarities that distinguish the current use of VR for TEFL. One of the most effective English-teaching methods is face-to-face instruction by a skilled, experienced teacher, who can provide students with real-time solving of pronunciation issues, grammar correction, and other tasks. This is a hurdle current software might have trouble jumping: such VR TEFL packages may rely on relatively passive learning [11], and even integrating software using adaptive learning can have its limits [12].

While those current packages can be quite useful, especially with scalability for large learning groups, this paper hopes to address the research gap on the use of a live, professional teacher. Many have expounded on the effectiveness of a live teacher using video chatting software for TEFL, but relatively little attention has been paid to the use of a high-presence VR environment and HMDs. Using an interdisciplinary approach, this research blends educational technology with system design to create a network-based system for a live instructor to teach and evaluate English skills through VR. This includes an integrated testing solution which students can operate without having to remove the headset or exit the virtual teaching environment.

This research seeks to answer three main questions. Firstly is, "Does this VR system have positive educational outcomes?" It will test if this VR-based teaching tool can be effective at transmitting English knowledge. Secondly is, "Is the VR system embraced by students?" It will assess if the students have trouble with system usability and find if they are interested in using VR for future learning opportunities. Third is, "Is the VR system reliable?" It will explore if the many features and components required for a standalone VR teaching system can be put together with the higher levels of reliability needed for educational settings.

Finally, this research also aims to be a stepping stone for a more elaborate system, one that can emulate the functionality of a traditional classroom and tailored for long-term TEFL use. Thus, areas for improvement will also be addressed.

The remainder of the paper is organized as follows. Section 2 provides an overview of the main subsystems, Sect. 3 covers the experimental design, Sect. 4 contains the results of the experiment, and Sect. 5 offers some future trajectories for this research and its conclusions.

2 System Architecture

As shown in Fig. 1, the left oval is the student-centered portion of the system. Using an HMD and powerful PC, the student can access a high-presence VR world through a chat program client, VRChat. This is paired with a web-based testing portal running through Oculus Dash, a virtual desktop display. On the right is the teacher-centered portion, in which the instructor can interact with the student using the HMD system. Once the lesson is over, the instructor can receive test data and conduct the lesson evaluation using a traditional desktop setup for increased workflow. In the middle stand the third-party servers that handle the VR scenario and testing management.

While the system network is separated into three sections, there are two subsystems that are the focal point of this research. These are the High-Presence Learning Scenario and the TEFL Web Portal. They appear on the student's VR display as two parts, as

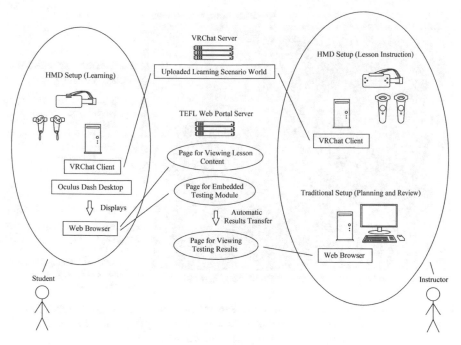

Fig. 1. An architectural diagram showing the major connections within the holistic VR system.

shown in Fig. 2. They are also relatively independent of one another, as the Web Portal can be used with other VR chat platforms (such as High Fidelity), and the Learning Scenario can be coupled with other web-based testing systems. They are explained in greater detail below.

2.1 The High-Presence Learning Scenario

The first subsystem of interest is the High-Presence Learning Scenario, which runs and is stored on the VRChat servers. VRChat is an online chat platform created with the Unity engine, and it is optimized for VR functionality. Features include voice chat using HMD microphones and object manipulation with HMD controllers. Using the Unity Editor alongside the VRChat SDK, a customized virtual learning scenario was created. The scenario takes place inside an office building during a mock career fair. The scenario is limited to one room, an interview office room. Here, the teacher leads the student by explaining business interview behavior, useful phrases, and guidelines for answering interview questions. Inside the room, various props and items fitting the scenario, such as books, manuals, and company products, have been inserted, and these items can also be manipulated using the HMD controllers. Blender was used to create and texture a portion of the models, and others were procured from content databases such as Sketchfab.

The use of a virtual office and movable objects allows for task-based learning [13]. As an example, one step of the scenario is for the teacher, roleplaying as a particular

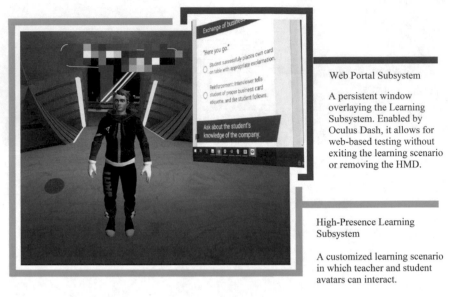

Web Portal Subsystem

A persistent window overlaying the Learning Subsystem. Enabled by Oculus Dash, it allows for web-based testing without exiting the learning scenario or removing the HMD.

High-Presence Learning Subsystem

A customized learning scenario in which teacher and student avatars can interact.

Fig. 2. An example of how the Web Portal Subsystem overlays the High-Presence Learning Subsystem. (Due to difficulties in simultaneously screen-capturing the Oculus Mirror Compositor and the Oculus Dash virtual desktop, a mockup was created).

company's representative, to ask, "Could you bring me the manual we give to new workers? It's on the furthermost table". This task allows testing multiple levels of comprehension. First is being able to discern the manual using context clues, as the manual is titled "Employee Guide" and features a cover with a relevant photo. The second is being able to move as if in physical space to the furthermost table, cementing the learning process.

An additional feature is that the teacher and student can use a virtual whiteboard. For this scenario the teacher uses it to emphasize important information, also asking the student to write or highlight certain points of interest.

2.2   The TEFL Web Portal

The second main subsystem is a web portal optimized for use with HMD controllers and displayed within the Learning Scenario. This has been accomplished through the recent introduction of Oculus Dash and its virtual desktop capabilities. While previous virtual desktop offerings had low integration with the HMD interface or had high levels of lag and stuttering, this new version is smoothly integrated and has a negligible performance impact.

The Web Portal subsystem is limited to the student during the teaching phase, and it is overlaid above the Learning Scenario as a window. The portal window is pegged not to the movement of the HMD but to the standing location of the student's avatar. Thus, the student can turn to the portal window when necessary and away from it when not.

Yet, the window will follow the student around rather than stay in a stationary point within the Learning Scenario world.

The web portal includes a few functions, the most important being a testing feature in which students can use the HMD controllers to scroll through multiple-choice questions and select the answers they wish. Future updates may have audio answer submissions that use speech-to-text technology to test pronunciation quality. This testing system uses Google Forms, which has a quiz-making function that can also automatically send data to a control panel that can be accessed by the teacher using a traditional desktop setup at a later time.

Special attention was made for the Web Portal's ease of use with HMD controllers, so the large buttons and simple menus (for example, no drop-downs, which failed in preliminary testing) require comparably little effort to navigate. Scrolling is also intuitively supported by Oculus Dash. Thus, after preliminary testing, an older system, which had one question at a time displayed and allowed going back and forward using buttons, was removed. The new system, with the entire quiz as a scrollable single page, was easier to use.

2.3   Hardware and Other Elements

As seen in Fig. 1 and mentioned earlier, the hardware can be separated into student and teacher setups. Most of the system components, both software and hardware, are easily available off-the-shelf. The student setup consists of an HMD (Oculus Rift, Consumer Version 1, including touch controllers) connected to a high-performance notebook (HP Omen 15, Gen. 8 Intel i7, Nvidia GTX 1070 Max-Q, 16 GB Ram). The teacher setup is an HMD (HTC Vive base setup, including controllers) paired with a high-performance desktop (Gen. 5 Intel i5, Nvidia GTX 980, 16 GB Ram). Of course, there are also the various servers of third parties (Google Forms, VRChat, etc.) and the VR platforms Oculus Home and Steam VR.

3   Research Methodology and Experiment Details

There are two main experiments. The first is an evaluation of student learning outcomes using data from the completed pre- and post-tests. The second is a usability study using data from student surveys. Eight students separately completed a full testing and survey session, each lasting around 45 min to an hour. Before each experiment, students were briefed on the system and its controls.

3.1   Testing Learning Outcomes

The learning outcomes evaluation was done through the pre- and post-testing method. Five questions were made following an example of the commonly-used "one-shot" pre- and post-test study [14]. As in the example by Karshmer and Bryan, the same quiz was used for pre-and post-testing, each question touching on one English learning concept as explained in the Learning Scenario. This includes vocabulary recall, context clues, and task completion. Students could select from four choices, as well as an "I don't know" to reduce the possibility of guessing influencing the results [15]. Students were also encouraged to not guess. Following Patton's guidelines [16], the experiment population was drawn from a selection of criteria, finding non-native English speakers with a self-assessed "intermediate" English level or higher on a set scale. In addition, some potential students with debilitating levels of VR sickness were winnowed out during the selection process. The testing was conducted entirely within the Learning Scenario while using the HMD.

3.2 Usability Study

The user interface was evaluated through a usability survey, which included multiple-choice questions about audio and video quality, lag, the learning experience, gained interest in using VR for education, and other qualities related to the research questions [17]. As VR has been shown to improve student motivation [18], confidence in being prepared for a similar situation in real life was also checked. These choices were given on a Likert scale. In addition, a free answer portion allowed students to respond about anything they liked or did not like about the experience that was not listed on the survey. During preliminary testing, testers were offered two different setups based on the different HMDs. The majority of preliminary testers preferred the virtual desktop function of the Oculus, so all students used it during the main testing phase. There were no other hardware or software preferences found with the preliminary testers.

4   Experiment Results

4.1   Learning Evaluation Results

It is clear that the system allows for the transfer of knowledge, though there are certainly some areas for improvement. While "I don't know" constituted the majority of answers in the pre-test, 52%, it decreased to 15% in the post-test. Correct answer rates improved from under 40% on average to over 70%. However, during the free answer portion of the survey, three students expressed that it was difficult to learn the lesson content because they were having to focus on manipulating the HMD interface. This is supported by past studies referencing the increased cognitive load of certain VR applications leading to a steeper learning curve [19]. Perhaps future, long-term experiments can lessen this problem after the first few lessons, given that people may become acclimated to VR and the HMD control system.

4.2   Usability Study Results

The majority of users had an overall positive experience with the system, as seen in Fig. 3. Of note is the high proportion of students who considered it an "interesting" experience and gained enthusiasm in using VR for English learning. The majority of students also responded positively when asked if they have more confidence in being prepared for a similar situation in the future.

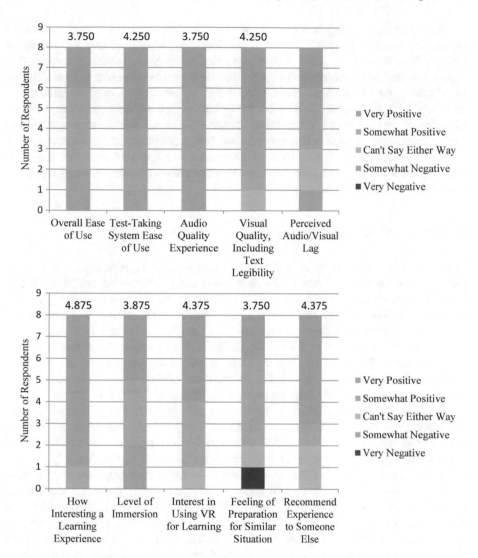

Fig. 3. Usability survey results from the eight fully-tested students. The number above each bar is the average of the Likert scale responses, with 5 points for "Very Positive" down to 1 point for "Very Negative".

However, three students complained of motion sickness to varying degrees. This has been dubbed "VR Sickness", and while it has been studied in learning situations, it seems there are currently no highly-reliable ways to eliminate it [20]. While some described the sensation as very mild, others said it was troublesome to the point of distraction. In addition, as mentioned earlier, two experiments had to be stopped in the preliminary stage because of extreme nausea. This may mean VR has some exclusionary qualities, as some may be left out during VR group activities or may have trouble joining a VR-exclusive course.

As a portion of testing was conducted at corporate offices, it was noticed that the experience was heavily dependent on network quality, which includes internet speed and a lack of firewalls or other blockages preventing access to necessary servers. One experiment had to be abandoned because of a network issue, and others were delayed. In addition, the system has a long setup time, an average of over 30 min, and a relatively high setup cost. This of course could be part of the reason for the slow adoption of desktop-based VR headsets in the educational field. VRChat is still an Alpha release, and there are often updates that change the program's content, as well as updates for Oculus Dash that lead to compatibility issues. After six of the eight experiments were completed, an update to VRChat changed some design elements for one of the virtual rooms used.

5 Future Work and Conclusion

There are three main areas in which this research aims to progress, starting with reconciling a similar VR system with an existing set of educational criteria or guidelines. Two examples would be the MEXT (Ministry of Education, Culture, Sports, Science, and Technology) educational guidelines for high schools in Japan or the Common European Framework of Reference for Languages. Second, there is a distinct lack of research on long-term VR use for TEFL, such as a standard half-year course or yearlong course. There is also little research on VR as the sole method of education, outside of the traditional classroom. The combination of these two qualities can lead to further steps towards a new goal, an accredited VR-based TEFL learning course.

Hardware will also be shifted to a homogenized system in which both the student and teacher use the same hardware and software setup, which will cut down on setup time. In addition, the use of the overlaid web portal will be expanded. While the current system has only the student using web portal contents and the teacher relying on a memorized lesson plan, the future portal aims to lower teacher prep time and increase usability by having a teacher control panel on the web portal. There, the teacher can access a lesson script for asking questions, as well as monitoring and controlling the flow of the lesson. Thus, many of the functions of a traditional classroom (writing, speaking, multimedia, testing) can be emulated. The lesson script can be modified to provide adaptive learning opportunities for students who have trouble progressing or understanding, choosing the next set of questions or activities based on student performance in the previous set. The expansion is also planned to have multiple learning scenarios connected by a single hub, in which students can use the "portal" feature of VRChat to travel among them. Last of all, a new research design, co-opting the plethora of previous research on CALL (Computer Assisted Language Learning), can be used, following the study of Virtual Reality Assisted Language Learning, or VRALL [21].

While there are many hurdles to large-scale educational adoption, including reliability issues, the targeted use of high-performance, high-presence VR scenarios can be useful for task-based language acquisition, increasing student interest and confidence, and providing alternative immersive learning methods with a high level of student-teacher interaction.

References

1. Bonner, E., Reinders, H.: Augmented and virtual reality in the language classroom: practical ideas. Teach. Engl. Technol. **18**(3), 33–53 (2018). IATEFL Poland
2. Mostafa, J.E., Mohsen, H.: Exploiting adventure video games for second language vocabulary recall: a mixed-methods study. Innov. Lang. Learn. Teach. **13**(1), 61–75 (2019). Taylor and Francis
3. Liu, K.: The MORPG-based learning system for multiple courses: a case study on computer science curriculum. Int. J. Distance Educ. Technol. **13**(1), 102–123 (2015). IGI Global
4. Chris, D.: Introduction to virtual reality in education. Themes Sci. Technol. Educ. **2**(1–2), 7–9 (2009). The Educational Approaches to Virtual Reality Technologies Laboratory, University of Ioannina
5. Thorsteinsson, G., Page, T., Lehtonen, M., Ha, J.G.: Innovation education enabled through a collaborative virtual reality learning environment. J. Educ. Technol. **3**(3), 10–22 (2006). I-Manager Journals
6. Kim, H., Ke, F.: OpenSim-supported virtual learning environment: transformative content representation, facilitation, and learning activities. J. Educ. Comput. Res. **54**(2), 147–172 (2016). SAGE Publications
7. Cooper, G., Park, H., Nasr, Z., Thong, L.P., Johnson, R.: Using virtual reality in the classroom: preservice teachers' perceptions of its use as a teaching and learning tool. Educ. Media Int. **56**(1), 1–13 (2019). Routledge
8. Reitz, L., Sohny, A., Lochmann, G.: Computer-assisted language learning: concepts, methodologies, tools, and applications. Int. J. Game-Based Learn. **6**(2), 46–61 (2019). IGI Global
9. Reitz, L., Sohny, A., Lochmann, G.: VR-Based Gamification of communication training and oral examination in a second language. Int. J. Game-Based Learn. **6**(2), 46–61 (2016). IGI Global
10. Anderson, P.L., Price, M., Edwards, S.M., Obasaju, M.A., Schmertz, S.K., Zimand, E., Calamaras, M.R.: Virtual reality exposure therapy for social anxiety disorder: a randomized controlled trial. J. Consult. Clin. Psychol. **81**(5), 751–760 (2013). American Psychological Association
11. Yildirim, G., Yildirim, S., Dolgunsoz, E.: The Effect of VR and traditional videos on learner retention and decision making. World J. Educ. Technol. Curr. Issues **11**(1), 21–29 (2019). SciencePark Research
12. Mirzaei, M.S., Zhang, Q., van der Struijk, S., Nishida, T.: Language learning through conversation envisioning in virtual reality: a sociocultural approach. In: EuroCALL, Jyväskylä, August 26 2018
13. Willis, J.: A Framework for Task-based Learning, 4th edn. Longman, Harlow (1996)
14. Bryan, J., Karshmer, E.: Assessment in the one-shot session: using pre- and post-tests to measure innovative instructional strategies among first-year students. Coll. Res. Libr. **74**(7), 574–586 (2013). Association of College and Research Libraries
15. Burton, R.: Quantifying the effects of chance in multiple choice and true/false tests: question selection and guessing of answers. Assess. Eval. High. Educ. **26**(1), 41–50 (2001). Routledge
16. Patton, M.Q.: Qualitative Research & Evaluation Methods, 4th edn, p. 238. SAGE Publications, Thousand Oaks (2002)
17. Rochlen, L.R., Levine, R., Tait, A.R.: First-person point-of-view-augmented reality for central line insertion training: a usability and feasibility study. Simul. Healthc. **12**(1), 57–62 (2001). Routledge

18. Li, S., Chen, Y., Wittinghill, D.M., Vorvoreanu, M.: A pilot study exploring augmented reality to increase motivation of chinese college students learning English. In: ASEE Annual Conference, Indianapolis, 15–18 June (2014)
19. Gorham, T., Jubaed, S., Sanyal, T., Starr, E.: Assessing the efficacy of VR for foreign language learning using multimodal learning analytics. In: Professional Development in CALL: A Selection of Papers, EuroCALL Teacher Education SIG, pp. 101–116 (2019)
20. Magaki, T., Vallance, M.: Measuring reduction methods for VR sickness in virtual environments. Int. J. Virtual Pers. Learn. Environ. 7(2), 27–43 (2017). IGI Global
21. Kaplan-Rakowski, R., Wojdynski, T.: Students' attitudes toward high-immersion virtual reality assisted language learning. In: EuroCALL, Jyväskylä, 22–25 August 2018

A Trial Development of 3D Statues Map with 3D Point Server

Hideo Miyachi[✉]

Department of Information Systems, Faculty of Informatics,
Tokyo City University, 3-1-1 Ushikubo-nishi, Tsuzuki-ku, Yokohama, Japan
miyachi@tcu.ac.jp

Abstract. I have been developing a data reduction system to 3D computer graphics data by converting from surface data to point data, and downsizing the point data. I have started to publish the system as Web service last year. It provides the data reduction services and the function of publishing the 3D point data in WebGL format. The advantage of the proposing data reduction method is an ability to control the data size regardless of the complexity and the volume of the original data. Using the advantage, it could contribute to make 3D CG sharing system even under an unstable communication line for smartphone. As a use case, I have developed a trial 3D Statues Map by using the Web service. It confirmed that the trial system operates with faster response than the system in AOBA DIGITAL MUSEUM in operation.

1 Introduction

The 3D geometric data that a computer should handle is getting bigger and more complex year by year [1, 2]. In recent years, 3D modeling by photo measurement has become available inexpensively, and anyone can easily acquire 3D geometric data. For example, from aerial photographs taken by drone, you can obtain 3D information of the city with a large amount of data. In order to share such a large amount of geometric data on the Internet, data reduction is required. To meet such needs, I have been proposing a new data reduction method which converts surface data to point data [3–5]. While the conventional data reduction method [6] aims to reduce the amount of data while maintaining the features of the original data, but the method I have proposed aims to reduce the amount of data while adapting to the resolution when the reduced data is viewed. Therefore, using the proposed method, it is possible to convert any large and complex data into data of the desired size. Of course, the proposed method has the disadvantage that it omits detailed information less than one pixel in the specified resolution. Therefore, this method should not be used to discuss in detail, but to share the appearance of data.

The concept of the proposed method is shown in Fig. 1. When 3D object, for example a teapot in Fig. 1, is shown in computer graphics, the final output is image that is represented in RGB values stored in an array, called frame buffer, with resolution size. Also, when the Z buffer method is used, the distance from a camera defined in 3D graphics is stored in an array called a Z buffer which has the same size as the frame buffer. The data combining both frame buffer and Z buffer forms 3D point data.

© Springer Nature Switzerland AG 2020
L. Barolli et al. (Eds.): NBiS-2019, AISC 1036, pp. 369–376, 2020.
https://doi.org/10.1007/978-3-030-29029-0_34

Gathering the 3D point data obtained by rendering from various directions such as left, right, front and rear and so on, it finally forms a 3D point cloud with density close to the resolution of the rendering window, which represents the approximate shape of the teapot.

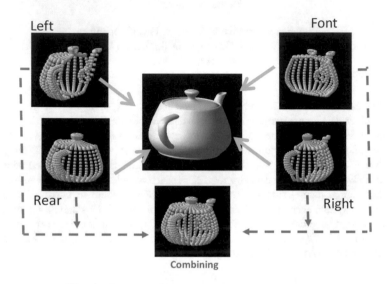

Fig. 1. Concept of proposed data reduction method.

This method has implemented as service program. Then it was published on the Internet on September in 2018 [7]. In this paper, I report a use case of the service. As an application example, 3D Statues Map system has been developed as prototype. The performance of the proposed method under a condition of a practical usage was measured with the system.

2   The Purpose of This Study

Aoba Ward is an administrative district located in the northernmost part of Yokohama City, which is famous as a port city in Japan, and has an area of 35.06 km^2 and a population of approximately 310,000 (as of October 1, 2018) [8]. Figure 2 shows the location of Aoba Ward in Japan. As one of the public relations activities of the ward, the homepage called AOBA DIGITAL ART MUSEUM is published and operated. Here, pictures, sculptures, cultural assets, etc. related to Aoba Ward have been introduced with digital images. The top page is shown in Fig. 3(a).

Our laboratory has been researching the advancement of the MUSEUM in cooperation with Aoba Ward Office. As the first step, I tried to make the 3D statues data. In December 2018, we took photographs of 11 statues and tried to create digital 3D statue data. As shown in Fig. 3(b), some of them have already published as 3D contents. Here, interactive 3D view embedded in the Web is provided by software POLY

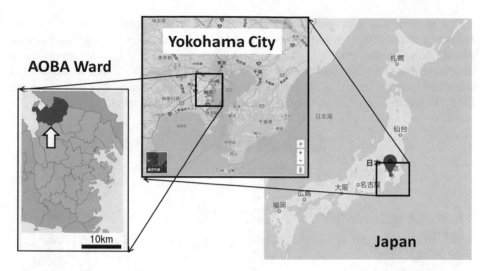

Fig. 2. Location of AOBA ward. It is a part of Yokohama City which is about 40 km south of Tokyo that is capital in Japan.

(a)   (b)

Fig. 3. Home Page of AOBA DIGITAL ART MUSEUM. (a) Top page, (b) 3D Statue's page. The right side of (b) is shown by POLY service managed by google.

supported by Google Inc. It takes about 10 s to show the 3D object after clicking "show 3D button". The purpose of this study is to make this response faster by using the proposed method.

3 The Development of 3D Statue Map

3.1 3D Digitization of Statues and Conversion to 3D Point Data

An overview of the proposed system is shown in Fig. 4. First, images of statues were photographed from the surroundings, and their images (① in Fig. 4) were converted into 3D data (② in Fig. 4). RECAP system provided by Autodesk was used in the 3D reconstruction. The 3D data was saved as OBJ format. At the stage, the information is stored in three files that consist of an OBJ file including geometric data, a JPEG file including texture data on the surface, and an MTL file connecting the two files. Second, the 3D data was converted into point data with color (③ in Fig. 4). A program I have developed for this study was used in the process. It extracts X, Y and Z coordinates values of each vertex and R, G and B color values corresponding to the vertex from OBJ file and JPG file respectively, and stored them in PLY format.

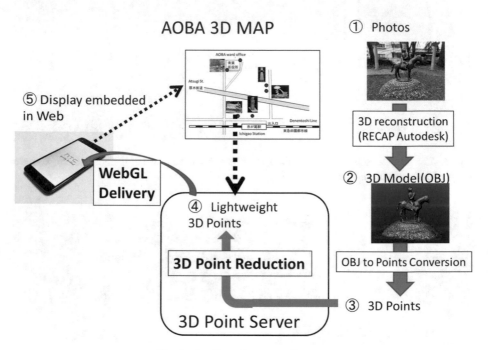

Fig. 4. Outline of proposed system.

3.2   The Work of the Proposed 3D Point Server

The point data with color generated in the previous section was converted into light-weight 3D points (④ in Fig. 4) by the proposed 3D Point Server. Here, the target resolution of the window in which the statues will be displayed set to 300×300 pixels. Eight kinds of statues' point data were generated. Some of them are shown in Fig. 5. The images are rendered from the data of 3D Model (OBJ) in Fig. 4.

The lightweight 3D points can be linked by a URL. Clicking on the button that each statue is located in AOBA 3D map, the URL corresponding to the lightweight data of the statue is called from the Web browser running in smartphone, 3D Point Server sends the point data with 3D viewer written in WebGL to the smartphone. The WebGL viewer embedded in Web browser (currently available only for Google Chrom) shows the 3D statue which can be rotated, scaled and translated by pointing operation (⑤ in Fig. 4). The WebGL viewer is shown in Fig. 6.

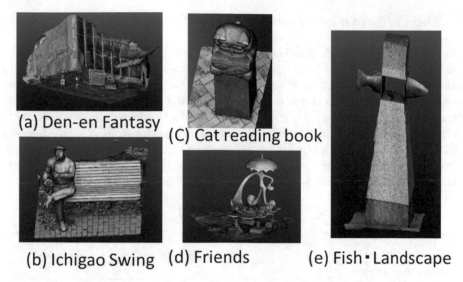

(a) Den-en Fantasy   (C) Cat reading book

(b) Ichigao Swing   (d) Friends   (e) Fish・Landscape

Fig. 5. Five statues presented in OBJ in my trial AOBA 3D Statues MAP (Stage ② in Fig. 4).

Fig. 6. Final display presented in points in PLY format (Stage ⑤ in Fig. 4).

4   The Performance Evaluation of the Proposed System

The reduction ratio to each statue data is listed in Table 1. The size of data indicates the file size stored in storage. Each row of the table presents the change of the data size along with the data conversion from ② to ④ in Fig. 4. The TITLE describes the name of statue and remarks. Columns A and B indicate the size of OBJ file and JPG file respectively, and column C indicates the sum of the two kinds of files. Column D indicates the size of point data with color after converting from column C. In the conversion, the connection data for forming polygon is removed, so the size is reduced. Column E indicates the size of reduced point data in ④ in Fig. 4. Column F and G indicates the reduction ratio from C to E and from D to E respectively.

Table 1. Reduction performance list by the proposed method.

No	TITLE		Original [KB]			Polygon with Color [KB]		Reduction Ration[%]	
	Japanese	Remarks	A	B	C	D	E	F	G
			geometry	texture	subtota (A + B)	Original	Deructed	E/C	E/D
			OBJ	JPG		PLY	PLY		
1	Neko	cat	104,920	26,742	131,662	19,408	5,666	4%	29%
2	Ichigao	music player	189,792	30,171	219,963	34,626	8,673	4%	25%
3	Sakana	fish	237,001	31,303	268,304	43,987	8,767	3%	20%
4	Denen	cuntyside	357,725	38,016	395,741	64,519	8,178	2%	13%
5	Friend	freiend	236,884	30,718	267,602	43,292	8,074	3%	19%
6	Kiroboshi	star	215,618	31,021	246,639	39,383	9,649	4%	25%
7	TOU	tower	61,214	24,940	86,154	11,712	7,285	8%	62%
8	Unicorn	unicorn	344,384	33,358	377,742	62,402	8,796	2%	14%

Table 2. Comparison of the number of points before and after reduction.

No	TITLE		Original	Deducted	Reduction Ration
	Japanese	Remarks			
1	Neko	cat	497,174	144,767	29%
2	Ichigao	music player	890,091	220,641	25%
3	Sakana	fish	1,102,824	196,917	18%
4	Denen	cuntyside	1,620,634	201,635	12%
5	Friend	freiend	1,101,533	204,825	19%
6	Kiroboshi	star	1,013,538	247,835	24%
7	TOU	tower	294,450	180,991	61%
8	Unicorn	unicorn	1,560,273	216,748	14%

The practical reduction ratios under this condition were 2 to 8% in conversion from OBJ format to lightweight point data, and the average was 3.9%. The reduction ratios under this condition were 13 to 62% in downsizing from D to E, and the average was 25.8%. Considering the feature of the proposed system, the coefficient of variation of C and D before applying reduction marked 40% and 43% respectively, but the coefficient of variation of E after reduction was reduced to 14%. It means that applying the proposed reduction method to various kinds and sizes of data could make the almost same size data. Therefore, the reduction effect for large data, e.g., No. 8 is high, and the reduction effect for small data, e.g., No. 7 is low.

To verify the effect of downsizing from D to E, the numbers of points at D and E are shown in Table 2. Since the data size in PLY format is approximately proportional to the number of points, the reduction rate of the number of points was almost the same as the reduction rate of file size.

At present, five kinds of data from No. 1 to No. 5 are published in the trial AOBA 3D Map on my Web site. Both point data before reduction and point data after reduction can be displayed on the site to experience the effect of the reduction. Using a smartphone connected via a 4G (MAX 100 Mbps) mobile line, I measured the time from the click of the map to the display completion. The average time for data before reduction was 15.4 s and the average time for data after reduction was 3.0 s. The response time has been reduced to 20%. The processing time per unit size for both indicated the same speed that was 22.6 Mbyte per seconds.

Currently, it takes about 10 s from button click to display completion on the official AOBA DIDIGAL MUSEUM, although this performance is just a test result in a day. However, it is suggested that the proposed system is effective for the publication of 3D statue information.

5 Conclusion

I have developed a data reduction system and have been publishing it as Web service. As a use case of the service, I have developed a trial version of 3D Statue Map. Using the system, I verified the practical performance of the proposed system by applying the reduction to a data set of 3D statues published at AOBA DIGITAL MUSEUM. As a result, the reduction ratio achieved 3.9% on average at the conversion from OBJ format data to lightweight point data. The reduction reduces the response time for viewing 3D data in 3D Map on a smartphone. In a test, using the reduced data by applying the proposed method, the response time was reduced to 20%. It suggested that the proposed system works effective for the large 3D geometric data publishing.

Acknowledgments. This research was supported by KAKENHI 17K00162.

References

1. Levoy, M., Pulli, K., Curless, B., Rusinkiewicz, S., Koller, D., Pereira, L., Ginzton, M., Anderson, S., Davis, J., Ginsberg, J., Shade, J., Fulk, D.: The digital michelangelo project: 3D scanning of large statues. In: Proceedings of ACM SIGGRAPH 2000, pp. 131–144, July 2000
2. Agarwal, S., Furukawa, Y., Snavely, N., Curless, B., Seitz, S.M., Szeliski, R.: Reconstructing rome. IEEE Comput. **43**, 40–47 (2010)
3. Miyachi, H.: Data reduction by points extraction in consideration of viewpoint positions of observer. Trans. VSJ **36**(8), 40–45 (2016). (in Japanese)
4. Miyachi, H.: Quality evaluation of 3D image represented by points. In: The 21st International Symposium on Artificial Life and Robotics, pp. 686–689 (2016)
5. Miyachi, H.: Quality evaluation of the data reduction method by point rendering. In: Proceeding of The Third International Symposium on Bio Complexity 2018 (ISBC 3rd 2018), pp. 986–990 (2018)
6. Heckbert, P.S., Garland, M.: Survey of polygonal surface simplification algorithms, Technical report CMU-CS, Carnegie Mellon University (1997)
7. Miyachi, H., Kuroki, I.: Server system of converting from surface model to point model. In: The 13th International Workshop on Network-based Virtual Reality and Tele-existence (INVITE-2018), vol. 22, pp. 467–475 (2018)
8. Naruhodo AOBA2018: Aoba ward office (2018). (in Japanese)

Driving Simulator System for Disaster Evacuation Guide Based on Road State Information Platform

Yoshitaka Shibata$^{(\boxtimes)}$ and Akira Sakuraba$^{(\boxtimes)}$

Iwate Prefectural University, Takizawa, Japan
{shibata,a_saku}@iwate-pu.ac.jp

Abstract. In this paper, we propose driving simulator system for disaster evacuation guide based on road state information platform to drill evacuation from disaster situation. The residents driving cars are conducted one by one from the current locations to the best evacuation places using evaluation guidance which is consisted of navigation system and voice guidance system through navigation road edge computing. The basic system functions and architecture of road state information platform are provided.

1 Introduction

On the large scale of tsunami disaster, it is very important for residents to escape from the coast area to safe evacuation places as fast as possible. According to the report of Cabinet Office [1] as shown in Fig. 1, more than 50% of residents who evaluated to the evaluation places used mobility in the East Japan Great Earthquake March 2011. However, among of one third of those residents encountered the traffic congestion and troubled evacuation activity. Basically, it is regulated that evaluation by mobility from tsunami is prohibited in Japan as shown in Fig. 2. However, although many people safely could complete their evacuations by walk, old people, patients and disable persons still could not escape by themselves without other's helps. For those reasons, safer and quicker evacuation system is indispensable those people.

To realize the system, the following evacuation problems have to be cleared. First, the problems of evacuation by mobility is how to avoid and eliminate the vehicle left unattended along the street in the occurrence of a disaster. The second is how to find and avoid the obstacles on the road before passing through. The third is how to avoid the evacuating persons along road by walk as indicated in Fig. 3.

To resolve those problem, we have to make the rules on each area depending on the geological condition, roads, residence areas and traffic condition. As indicated in Fig. 4 in flood areas, driving is not used and just should walk to the higher hill or mountain. When there is no evaluation places around the residents, wider road should be used by reducing the number of mobility by taking on the many people together as possible and evacuate to the wider evacuation place to go for away from the coast.

© Springer Nature Switzerland AG 2020
L. Barolli et al. (Eds.): NBiS-2019, AISC 1036, pp. 377–384, 2020.
https://doi.org/10.1007/978-3-030-29029-0_35

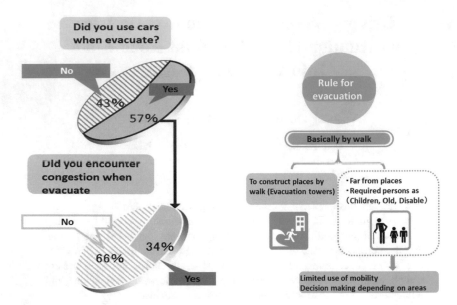

Fig. 1. Evacuation by car in Mar. 2011   **Fig. 2.** Rule of evacuation in local areas

Thus, before admitting system evacuation by mobility, deep study must made on each local area. So far there are simulation studies with safety of evacuation mobility for tsunami in the East Japan Great Earthquake as shown in the research papers [2–4].

In the research [2], simulation system was developed by considering the expression model of the individual activity, the network of evaluation routes and evacuation. In the research [3], they studied that the evaluation by walk was not enough to be in time to reach the specified safe evaluation for the actual town depending on the evacuation rule. In the research [4], they studied that traffic congestion occurred when the traffic density increasing the traffic density for large city.

In the research [5], they studied that more proper use method with mobility is required by considering the utilization of both the number of cars and parking space as political parameters.

In the research [6], they cleared the transmission and sharing temporal and spatial disaster information such in geological and as alarm and evacuation recommendation is very important. Through those researches, pre-studied of disaster evacuation route, driving rule on road, disaster information sharing function, evacuation drill in normal time are precisely considered.

As principle rule of evacuation, evacuation by walk is essential. If the evaluation places are far way, or considered persons (children, old persons, disable persons), limited use of mobility for those persons based on the role in emergence case as shown in Fig. 4. In order to smoothly and safely drive to evacuation place, the people should take evaluation drill as image the actual disaster evacuation.

In the followings, Road Information Plat Form is proposed in section two. Disaster Evacuation Guide System is explained in section three. Driving Simulator System to conduct the disaster evacuation place is shown in section five. Finally conclusion and future research in section six.

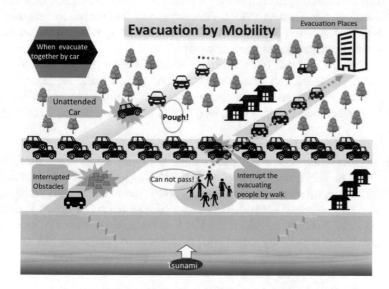

Fig. 3. Problems of evacuation by mobility

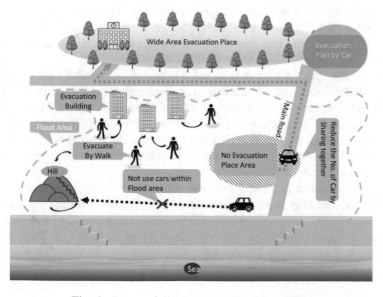

Fig. 4. Image of disaster evacuation by mobility

2 Road Information Platform

We introduce a new generation wide area road surface state information platform based on crowd sensing and V2X technologies as shown in Fig. 5 [7]. The wide area road surface state information platform mainly consists of multiple road side wireless nodes, namely Smart Relay Shelters (SRS), Gateways, and mobile nodes, namely Smart Mobile Box (SMB). Each SRS or SMB is furthermore organized by a sensor information part and communication network part.

The sensor information part includes various sensor devices such as semi-electrostatic field sensor, an acceleration sensor, gyro sensor, temperature sensor, humidity sensor, infrared sensor and sensor server. Using those sensor devices, various road surface states such as dry, rough, wet, snowy and icy roads can be quantitatively decided.

On the other hand, the communication network part integrates multiple wireless network devices with different N-wavelength (different frequency bands) wireless networks such as IEEE802.11n (2.4 GHz), IEEE802.11ac (5.6 GHz), IEEE802.11ah (920 MHz) and organizes a cognitive wireless node. The network node selects the best link of cognitive wireless network depending on the observed network quality by Software Defined Network (SDN). If none of link connection is existed, those sensing data are locally and temporally stored until approaches to another mobile node or road side node, and starts to transmit sensor data by DTN Protocol. Thus, data communication can be attained even though the network infrastructure is not existed in challenged network environment such as mountain areas or just after large scale disaster areas.

In our system, SRS and SMB organize a large scale information infrastructure without conventional wired network such as Internet. The SMB on the car collects various sensor data including acceleration, temperature, humidity and frozen sensor data as well as GPS data and carries and exchanges to other smart node as message ferry while moving from one end to another along the roads.

On the other hand, SRS not only collects and stores sensor data from its own sensors in its database server but exchanges the sensor data from SMB in vehicle nodes when it passes through the SRS in road side wireless node by V2X communication protocol. Therefore, both sensor data at SRS and SMB are periodically uploaded to cloud system through the Gateway and synchronized. Thus, SMB performs as mobile communication means even through the communication infrastructure is challenged environment or not prepared.

This network not only performs various road sensor data collection and transmission functions, but also performs Internet access network function to transmit the various data, such as sightseeing information, disaster prevention information and shopping and so on as ordinal public wide area network for residents. Therefore, many applications and services can be realized.

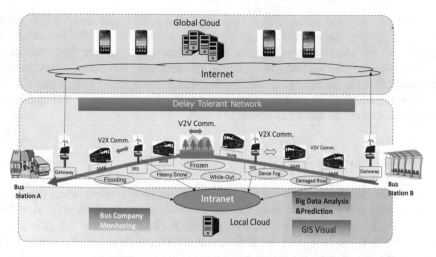

Fig. 5. Road state information platform

3 Disaster Evacuation Guide System

The Fig. 6 show a Disaster Evacuation Guide System using our proposed system for normal case and emergent case [8, 9]. In normal situation, many Japanese tourists and foreign tourists from different countries visit to the Points of Interest (POI) such as the historical places, national parks and mountains and spas where network information infrastructures are usually not well prepared in snow country in Japan. Even worse, it is difficult for foreigners to understand the information of Points POI written in Japanese. However, by introducing the proposed mobility information infrastructure, and using smart phones and tablet terminals, the tourists can ask and speak by their own voices and obtain information POI by their own languages, eventually being navigated to the objective places.

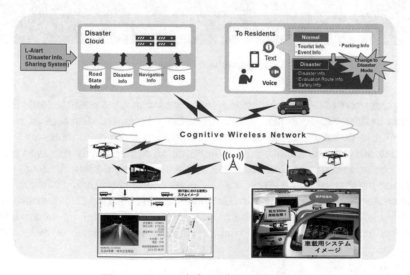

Fig. 6. Disaster information guide system

On the other hand, when disaster occurred, the push typed disaster information, evaluation information are automatically delivered to those tourists by their languages. Eventually, the tourist can safely evacuate to the proper shelter from the current location. Through the mobility information infrastructure, the disaster state information, resident safety information, required medicine, feeds and materials are also collected and transmitted by mobile nodes between the counter disaster headquarter and evacuation shelters as shown in Fig. 6.

4   Driving Simulator System

Figure 7 indicates the simulator system for disaster drill and is used for any people who needs to train the evacuation from normal time. In order to smoothly respond and evacuate on actual tsunami disaster, it is required for any persons to regularly practice antidisaster drill for any cases for different situations, different road conditions, different locations and not only Japanese but foreign people to obtain the knowledge and experiences in priori. For those reasons, driving simulator is used to realize the those any kinds of situations. Using driving similar system, antidisaster drill application can be realized based on virtual reality (VR) and augmented reality (AR) by changing many disaster situations. The disaster evacuation by mobility on past actual disasters, such as the East Japan Great Earthquake and Kumakoto Earth Earthquake can be experienced as if he or she were virtually driving the car from the disaster area to the evacuation places whole being navigated by the evacuation guide information system.

Fig. 7. Driving simulator for disaster evacuation

As parameters of driving simulator system, road width, the possible roads from the residents to the evacuation place, point of traffic signals, the distances of river and precipices can be set as various candidates of possible evacuation routes.

During the practice, various driving information can be timely observed and collected such as end to end time from the origin residence, driving distance, average speed, the number of brake action etc. By considering the driving information, the proper evacuation route candidate for individual can be selected. This practice can be proceeded with local government office to verify the effects for the area where the traffic road environment cannot be well developed. As verification of evacuation by mobility, the experimental drill using the driving simulator are executed by the experienced and non-experienced residents on the actual tsunami disaster by operating on the driving simulator. Through this experience, the proper Disaster Evacuation Guide System is verified not only for the residents in the experienced area but only non-experimented areas. To respond to the risk by Disaster Evacuation Guide System.

5 Conclusions and Future Works

On the large scale of tsunami disaster, it is very important for residents to escape from the coast area to evacuation places as fast as possible. The residents should drill evaluation action in ordinal time. In this paper, we propose driving simulator system for disaster evacuation guide based on road state information platform to drill evacuation from disaster situation. The residents driving cars are conducted one by one from the current locations to the best evacuation places using evaluation guidance which is consisted of navigation system and voice guidance system through navigation road edge computing. Once a disaster occurred such as tsunami, disaster cloud computing corrects the information with the disaster locations and scale, areas, road states and inform this information to the drivers and guides to escape to the nearest evaluation places. The basic system functions and architecture are provided.

Currently we are constructing a prototyped driving simulator system. As future works, we will combine this driving simulator system with road information platform and disaster evacuation guide system and evaluate the effects of our proposed system.

Acknowledgments. This research was supported by Strategic Information and Communications R&D Promotion Program (SCOPE) No. 181502003, Ministry of Internal Affairs and Communications, Japan.

References

1. A 2011 Report Cabinet Office of Disaster Evaluation Action against the East Japan Great Earthquake. http://www.bousai.go.jp/kaigirep/chousakai/tohokukyokun/7/pdf/1.pdf
2. Imamura, F., Suzuki, T., Taniguchi, M.: Development of a simulation method for the evacuation from the Tsunami and its application to Aonae, Okushiri Is., Hokkaido. Jpn. Soc. Nat. Disaster Sci. **20**(2), 183–195 (2001)
3. Kitamura, S., Suzuki, T., Imamura, F.: Evacuation simulation around Sendai Seaport. In: The 57th Annual Conference of Japan Civil Engineering, IV-302 (2002)
4. Suzuki, T., Imamura, F.: Simulation model of the evacuation from a tsunami in consideration of the resident consciousness and behavior. Jpn. Soc. Nat. Disaster Sci. **23–4**, 521–538 (2005)
5. Katada, T., Kuwasawa, N., Watanabe, H.: A consideration of the relation between evacuation by car and human damage. In: The 14th of Conference in Japan Disaster Information Society, pp. 152–155 (2012)
6. Katada, T.: Development of scenario simulator for evacuation measures and disaster education. Traffic Eng. **48–1**, 8–23 (2013)

7. Shibata, Y., Sato, G., Uchida, N.: A new generation wide area road surface state information platform based on crowd sensing and V2X technologies. In: The 21th International Conference on Network-Based Information Systems (NBiS2018). Lecture Notes on Data Engineering and Communications Technologies book series (LNDECT), vol. 22, pp. 300–308 (2018)
8. Ito, K., Hashimoto, K., Shibata, Y.: V2X communication system for sharing road alert information using cognitive network. In: The 8th International Conference on Awareness Science and Technology (2017)
9. Uchida, N., Ito, K., Hirakawa, G., Shibata, Y.: Evaluations of wireless V2X communication systems of for winter road surveillance systems. In: The 19th International Conference on Network-Based Information Systems, (NBiS2016), pp. 58–63. Technical University of Ostrava (2016)

The 13th International Workshop on Advanced Distributed and Parallel Network Applications (ADPNA-2019)

Estimation Method of Traffic Volume Using Big-Data

Kazuki Someya[1], Ryozo Kiyohara[2](✉), and Masashi Saito[3]

[1] Graduate School of Kanagawa Institute of Technology, Atsugi, Japan
[2] Kanagawa Institute of Technology, Atsugi, Japan
rkiyohara@gmail.com
[3] Kanazawa Institute of Technology, Nonoichi, Japan

Abstract. Traffic jams have recently become a significant problem in provincial cities that tend to have poor railway services in Japan. Therefore, the main means of transportation are public buses, taxies, and private vehicles. Moreover, traffic accidents and road construction sites frequently block traffic. It is therefore difficult to estimate the travelling time from the origin to destination in real-time. To estimate the travelling time, we must predict the behaviors of many vehicles that depend on an "origin to destination" (OD) traffic volume. In our previous study, we proposed an estimation method for OD traffic volume using two types of big data, a road traffic census and mobile spatial statistics. In this study, we evaluated our proposed method on various situations through a traffic simulation.

1 Introduction

In provincial cities, traffic jams have become one of the most significant problems for drivers. In the near future, autonomous vehicles will become available, and people who are unable to drive will have access to them.

In advanced cities, traffic jams can be avoided by using public transportation. However, public transportation seems to not be successful in provincial cities where large-scale road construction cannot be easily carried out because these cities cannot financially afford such construction. Thus, it would be beneficial for provincial cities if the behavior of a vehicle, such as the route a vehicle travels on, could be predicted.

To predict the behavior of a vehicle, the "origin to destination" (OD) information needs to be analyzed. Various methods exist for gathering OD information. One major method is an investigation of a person's travel information through a questionnaire survey or by interviewing the driver [1]. However, both types of personal travel data depend on human memory and have insufficient accuracy.

In this study, we propose a method for estimating the OD traffic volume using two types of big data, namely, a road traffic census (RTC) [2], in which the traffic volume is gathered from many different points along highways and main roads (e.g., national routes), and mobile spatial statistics (MSS) [3], in which cellular phones within a specific area are divided into a mesh.

© Springer Nature Switzerland AG 2020
L. Barolli et al. (Eds.): NBiS-2019, AISC 1036, pp. 387–395, 2020.
https://doi.org/10.1007/978-3-030-29029-0_36

The goal of the this study is to derive a regression equation. We then show that the traffic volume at any point can be predicted even during a traffic accident or during road construction.

2 Related Studies

Many studies have been conducted for predicting traffic volume. Kim et al. proposed the prediction of traffic using V2R communications [4] and proposed a toll gate system (e.g., ECT) as a V2R system. Although this method is limited to highways, it can predict the near-future traffic volume on a highway.

A traditional investigation into personal travel depends on human memory and therefore appears to be less accurate. A new method applying GPS and accelerometers to the vehicles has thus been proposed [5]. However, this method requires the installation of various applications, thus rendering it difficult to use without certain incentives.

A method for estimating the flow of people using the population in specific areas, which is same as MSS, has also been previously proposed [6].

NTT DOCOMO has been experimenting with a "near future people forecast" [7], which can predict movement several hours ahead by applying a real-time version of MSS and an online prediction technology for spatiotemporal variables. NTT DOCOMO is currently developing an artificial-intelligence taxi, which can predict the number of taxis that will be used within a 30-min period. Forecasting is a technique used for predicting the number of people in an area several hours in advance.

A smart access vehicle (SAV) has been proposed which involves a model of OD traffic using MSS and the number of people getting on and off. This technology predicts the number of people for each type of transportation [8]. However, it cannot predict the traffic volume on a road.

3 Two Types of Big Data

3.1 Mobile Spatial Statistics

MSS are open data provided by NTT DOCOMO, which periodically detect mobile phones within the area of each base station. The population during the day and night can be roughly predicted, and depending on the time, the number of people that commute and drive into the central city can be determined. The details of the data are shown in Table 1.

Table 1. Example of mobile spatial statistics.

City name	Date	Time	Area no.	Residence area no.	Age	Gender	No. of people
Kanazawa	19 Nov. 2019 Thu.	7:00	543665645	17201	20	1	7896
Kanazawa	19 Nov. 2019 Thu.	8:00	543665645	17201	20	2	8012
Nonoichi	06 Apr, 2016 Wed.	18:00	543654865	17203	30	1	1779
Nonoichi	06 Apr, 2016 Wed.	19:00	543654865	17203	40	1	1542
Hakusan	10 Feb. 2016 Wed	9:00	543614465	17210	50	2	823
Hakusan	10 Feb. 2016 Wed	10:00	543614465	17210	50	1	902

The population is the number of mobile phones. The data for the area are divided into meshes (200 m to 2 km). There are numerous base stations in urban areas such as Tokyo, and the areas are narrow. However, there are fewer base stations in provincial cities, and the areas are therefore wide.

These data have the following features for many different aspects.

- Erase personal information;
- Estimate the number of people based on the attributes of the terminals and the market share using such attributes; and
- Delete the information within a specific area with few terminals.

These data involve statistics for the distribution of mobile terminals. In this study, we use the mesh data in Kanazawa-City and Nonoichi-City, as shown in Fig. 1.

Fig. 1. Mesh number MSS in Kanazawa-City and Nonoichi-City

3.2   Road Traffic Census

A road traffic census (RTC) is made up of traffic volume data gathered every five years by the Japanese Ministry of Land, Infrastructure, and Transport (MLIT). The data on an RTC are gathered from highways and main roads (e.g., national routes) throughout Japan. Therefore, an RTC indicates the number of vehicles at specific points.

Table 2 shows an example and Fig. 2 shows a map of an RTC. These data do not cover all roads. It is therefore difficult to estimate the OD traffic volume from only an RTC. RTC data used in this study are from 2017, which is the most recent period of available data.

Table 2. Example of road traffic census (RTC)

Area number	Date	Direction	Kinds of vehicle	Traffic volume each time (number/time)					Traffic volume (day)
				7:00~	8:00~	9:00~	17:00~	18:00~	
17300080010	20151008	1	1	550	337	327	66	274	5968
17300080010	20151008	1	2	129	120	179	116	130	2274
17300080010	20151008	2	1	360	318	331	38	168	5777
17300080010	20151008	2	2	73	111	169	107	109	2376

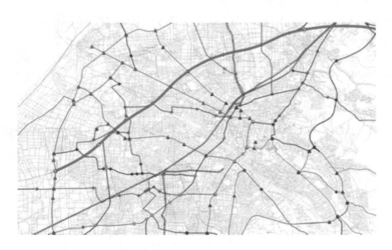

Fig. 2. Road traffic census survey area around Kanazawa station

4   Proposed Method

Both MSS and RTC are independent datasets and therefore have the following problems.

- An RTC is not observed on any specific roads and is made up of a few points of data.
- MSS cannot have the attributes of a terminal with walking pedestrians or driving vehicles.

- Neither types of data have an ID. Therefore, all data are independent, and the mobility of a specific vehicle cannot be analyzed.

Based on these points, it is difficult to predict the mobility of the vehicles using either data types.

We propose a new method that uses both MSS and RTC data. We targeted Kanazawa City and Nonoichi City in Ishikawa Prefecture. The number of meshes applied in MSS is 116, and RSS data are included in 85 of the meshes.

We analyzed the correlation between MSS and RTC. Because we are required to decide the representative value of population by time zone for one mesh of MSS, we used a method with less accumulated error. We used the median that was not edited because it has a low cumulative error. We split the area of RTC like MSS meshes. Each time zone median population is shown in Fig. 3 as a scatter plot. The X-axis denotes the MSS traffic volume by time zone, and the Y-axis is the RTC traffic volume by time zone. The whole regression equation is indicated by a solid line, and the regression equation for each city planning type is indicated by a dotted line. Green dots indicate residential areas, and orange dots indicate downtown areas. Equation (1) shows is the whole regression equation, and the coefficient of determination was 0.58.

$$y = -1.0 \times 10^{-6}x^{-2} + 0.69x + 1029.7 \tag{1}$$

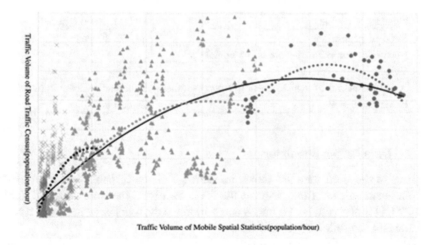

Fig. 3. Correlation between MSS and RTC

In this study, we propose a method for evaluating this equation. We define many types of vehicle behavior, traffic accidents, and road construction. We then simulate them using a traffic simulator. We can analyze the MSS and RTC on the traffic simulator and evaluate Eq. (1).

5   Simulation for Evaluation

5.1   Simulation Environment

We evaluated our equation using a traffic simulator called Scenargie [9], which has a multi-agent simulator function with a network simulator. In this evaluation, many vehicle agents drive on a road, and simulator's sensor counts the number of vehicle agents, similar to an RTC. We assume that each driver has a mobile device. Moreover, pedestrians are present on the road. We applied many different ratios of pedestrians.

Table 3 shows the details of the simulation environment. The term "people" indicates drivers and pedestrians. In this study, pedestrians are people who do not drive (including the people in buildings or houses).

The roads were imported from OpenStreetMap [10]. The MSS data comprise 81 mesh areas (2 km × 2 km).

We simulated nine patterns made up of the number of people (10,000, 20,000, and 30,000) and ratio of pedestrians (40%, 50%, and 60%). The OD of the vehicle is random, and the vehicle starts from one building and goes to another using the minimum distance. When the vehicle arrives at the destination, the vehicle disappears from the simulator. Therefore, the number of people is not always the same.

Table 3. Simulation environment

Number of people	10,000, 20,000, 30,000
Ratio of pedestrians	40, 50, 60 [%]
Maximum number of people in the vehicle	1
Velocity of the vehicle	60 [km]
Average interval of vehicle peering in the simulator	Exponential distribution
Experiment time	21,600 [s]

5.2   Log Data on the Simulator

The results of the simulation are shown in Tables 4, 5 and 6. Table 4 shows the MSS data from the simulator, Table 5 shows the RTC data from the simulator, and Table 6 shows the OD traffic volume. Figures 4 and 5 indicate the number of people in a certain timeframe and the following timeframes.

All data consist of the ID, type (driver or pedestrian), road, area, and time (s). In the simulator, we can identify the vehicle and thus can use the ID. The OD data in Table 6 show (with the ID) which road and what time the vehicle appeared on.

5.3   Evaluation

We derived the regression equations from the results of a simulation. Table 6 shows the coefficient of determination for each pattern. Table 7 shows the regression equations for each pattern. The average coefficient of determination is 0.83, and the coefficient of determination from real data is 0.58.

Table 4. MSS from simulation

ID	Type	Time	Road	x	y	Area
35986	Vehicle	3600		−1962.1	2633.79	440
13976	Pedestrian	7200		1660.41	3593.35	450
66209	Vehicle	10800	Route156	2343.66	1181.77	540

Table 5. RTC from simulation

ID	Type	Time	Road	x	y	Area
16226	Vehicle	3600	Route291	9142.75	6400.63	210
15742	Vehicle	7200	Route359	349.988	823.052	660
37261	Vehicle	10800	Route159	857.572	1980.01	550

Table 6. Example OD traffic volume

ID	Road1 (time1)	Road2 (time2)	Road3 (time3)	Road4 (time4)
10812	Route 159(3200)	Route 10(7101)	Route 157(10090)	Route 157(18044)
62928	Route 146(4500)	Route 10(8005)		
86092	Route 359(6455)	Route 159(12890)	Route157(17300)	

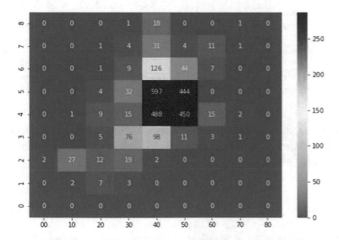

Fig. 4. Example of results (Number of peoples)

This shows that the value of coefficient of determination differs. Moreover, the differences among the nine patterns are not very large. The nine patterns are based on different parameters, namely, the number of people and their ratio. However, all patterns show the same trends, which appear to be similar to a real RTC and MSS.

We can evaluate them based on other aspects as well. During the simulation, we can gather the IDs in an RTC. The mobility of each vehicle can be analyzed using this

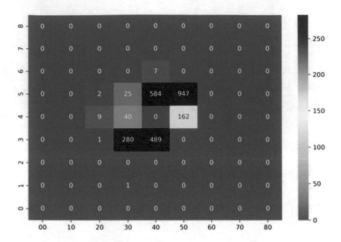

Fig. 5. Number of peoples next time of Fig. 4

Table 7. Regression equations from simulation

Riding rates	Number of people		
	10000	20000	30000
40	$y = -9.60 \times 10^{-3}x^2$ $+ 3.25x + 0.87$	$y = -5.60 \times 10^{-3}x^2$ $+ 3.44x + 0.12$	$y = -2.98 \times 10^{-3}x^2$ $+ 3.26x + 2.10$
50	$y = -8.38 \times 10^{-3}x^2$ $+ 3.36x + 0.77$	$y = -4.51 \times 10^{-3}x^2$ $+ 3.48x + 0.16$	$y = -1.85 \times 10^{-3}x^2$ $+ 3.25x + 3.04$
60	$y = -6.24 \times 10^{-3}x^2$ $+ 3.23x + 0.12$	$y = -3.06 \times 10^{-3}x^2$ $+ 3.22x + 2.25$	$y = -1.85 \times 10^{-3}x^2$ $+ 3.25x + 3.04$

ID. In this case, the average coefficient of determination is 0.48, and the central value is 0.49, which is extremely low. Therefore, it is difficult to predict the traffic volume from only an RTC using the IDs.

6   Conclusion

In this study, we propose an evaluation method of regression equations derived from RTC and MSS data. We show that if the vehicle ID information is added to the RTC and MSS, a better estimation accuracy can be achieved.

Acknowledgment. This work was supported by JSPS KAKENHI Grant No. 16K00433. Data on the Mobile Spatial Statistics in Kanazawa and Nonoichi were provided by NTT DoCoMo.

References

1. MILT: International visitor survey. http://www.mlit.go.jp/kankocho/en/siryou/toukei/syouhityousa.html
2. Uesaka, K., Monma, T., Matsumoto, S., Hashimoto, H., Mizuki, T.: The FY2010 Road Traffic Census Results of the General Traffic Volume Surveys (Overview). www.nilim.go.jp/english/annual/annual2012/84.pdf
3. Murase, A.: How mobile spatial statistics Began. NTT Docomo J. **14**(3), 1
4. Kim, J., Kurauchi, F., Uno, N., Hagihara, T., Daito, T.: Using electronic toll collection data to understand traffic demand. J. Intell. Transp. Syst. **18**(2), 190–203 (2014)
5. Wolf, J., Randall, G., Bachman, W.: Elimination of the travel diary: experiment to derive trip purpose from global positioning system travel data. Transp. Res. Rec. J. Transp. Res. Board **1768**(1), 125–134 (2001)
6. Akagi, Y., Nishimura, T., Kurashima, T., Toda, H.: A fast and accurate method for estimating people flow from spatiotemporal population data. In: Proceedings of the Twenty-Seventh International Joint Conference on Artificial Intelligence (IJCAI-18) (2018)
7. NTT DoCoMo: AI Bus. https://www.itsap-fukuoka.jp/demo/AI_Bus.html
8. Nakashima, H., Sano, S., Hirata, K., Shiraishi, Y., Matsubara, H., Kanamori, R., Koshiba, H., Noda, I.: One cycle of smart access vehicle service development
9. Takai, M., Martin, J., Kaneda, S., Maeno, T.: Scenargie as a network simulator and beyond. J. Inf. Process. **27**(1), 2–9
10. Open Street Map Japan. https://openstreetmap.jp

Experimental Evaluation of Publish/Subscribe-Based Spatio-Temporal Contents Management on Geo-Centric Information Platform

Kaoru Nagashima[1(✉)], Yuzo Taenaka[2], Akira Nagata[3], Katsuichi Nakamura[3], Hitomi Tamura[4], and Kazuya Tsukamoto[1]

[1] Kyushu Institute of Technology, Iizuka, Japan
nagashima@infonet.cse.kyutech.ac.jp, tsukamoto@cse.kyutech.ac.jp
[2] Nara Institute of Science and Technology, Ikoma, Japan
yuzo@is.naist.jp
[3] iD Corporation, Fukuoka, Japan
{a-nagata,k-nakamura}@intelligent-design.co.jp
[4] Fukuoka Institute of Technology, Fukuoka, Japan
h-tamura@fit.ac.jp

Abstract. Cross-domain data fusion is becoming a key driver to growth of the numerous and diverse applications in IoT era. Nevertheless, IoT data obtained by individual devices are blindly transmitted to cloud servers. We here focus on that the IoT data which are suitable for cross-domain data fusion, tend to be generated in the proximity, and thus propose a Geo-Centric Information Platform (GCIP) for the management of Spatio-Temporal Contents (STCs) generated through the cross-domain data fusion. GCIP enables to keep STCs near the users (at an edge server). In this paper, we practically examine the fundamental functions of the GCIP from two aspects: (1) Geo-location aware data collection and (2) Publish/Subscribe-based STC production. Furthermore, we implement a proof-of-concepts (PoC) of GCIP and conduct experiments on a real IPv6 network built on our campus network. In this experiment, we showed that multiple types of IoT data generated in the proximity can be collected on the edge server and then a STC can be produced by exploiting the collected IoT data. Moreover, we demonstrated that the Publish/Subscribe model has a potential to be effective for STC management.

1 Introduction

With the rapid growth of both sensor devices and wireless technologies, various type of things can be connected to the Internet, that is the IoT. In the IoT era, since the combination of various "things' data" could bring us new and undiscovered contents, the cross-domain (horizontal-domain) IoT data fusion attracts much attention. However, IoT data transmitted from the numerous IoT

L. Barolli et al. (Eds.): NBiS-2019, AISC 1036, pp. 396–405, 2020.
https://doi.org/10.1007/978-3-030-29029-0_37

devices are generally enforced to be gathered in cloud servers and be used for a specific IoT service, which is widely referred to as the vertical-domain.

In this paper, we focus on that some type of IoT data suitable for the cross-domain data fusion are generated in the geographical proximity. Along this line, we propose a new information platform, Geo-Centric Information Platform (GCIP) in which IoT data are kept in physically close edge servers (i.e., in the physical proximity) and then new contents can be produced as a result of analysis and processing of the data. Note that we define the produced contents, which are worth for users for the limited duration and at the limited location, as the Spatio-Temporal Contents (STCs).

First, we introduce a conceptual model for GCIP and then implement a proof-of-concept (PoC) with the fundamental functions of geo-location aware data collection and Publish/Subscribe-based STC production into a real environment. Next, we conduct the experiments to show the effectiveness of the GCIP by using IPv6 campus network. More specifically, we examine the feasibility of not only geo-location aware data collection but also Publish/Subscribe based STC production.

Rest of this paper is organized as follows, We first review the existing studies in Sect. 2. Then, the conceptual design of our proposed method is described in Sect. 3 and the demonstration environment is described in Sect. 4. We show and discuss the experimental results in Sect. 5, and finally Sect. 6 concludes this paper.

2 Related Work

In this section, we review the existing studies focusing on geo-location based network control and IoT data processing. In reference [1], MQTT has been extended to handle location information. Although all IoT devices are assumed to have its geo-location by using GPS, it is significantly difficult for IoT devices to load GPS because they are based on cheap, small, and low-powered design. As another example, the reference [2] discusses data processing on the premise of location information. Hence, if cloud server produces the contents from the collected IoT data, the server needs to be aware of location. As one of location-aware method, GCIP adopts network level approach. Thus, GCIP cloud collect IoT data in the geographical proximity by changing both of routing tables and identifier of the IoT devices.

In reference [3], various IoT data such as social data, media data, etc. are collected and analyzed to extract beneficial information with unique features in each region. However, there is a problem in scalability and performance because all of the IoT data is processed on a specific cloud server. The references [5] and [2] have mentioned several use cases in which various types of data are processed, i.e., cross-domain data fusion. However, since the processing are performed on the cloud server, the amount of traffic between the users and servers is significantly increased, thereby increasing not only the latency between the users and server but also packet losses. In contrast, GCIP produces STC on the edge server(s), which is relatively geographically closer to the users.

Reference [4] widely summarizes the existing studies focusing on the contents search method. However, none of existing studies tries to find the realtime/on-demand contents dynamically generated based on the collected IoT data. On the other hand, GCIP dynamically generates dynamic contents by considering the temporal and spatial characteristics.

From these points, we can say that GCIP provides new aspects of (1) geolocation-aware communication to IoT devices and (2) on-demand Spatio-Temporal content production based on the geographical proximity.

3   Geo-Centric Information Platform (GCIP)

This section first describes the GCIP design concepts. After that, we describe PoC design in terms of data collection and STC production, respectively.

3.1   Conceptual Design of GCIP

Majority of IoT devices are deployed and dedicated for specific services. That is, device and service are tightly associated. Moreover, each of network is individually managed by network operator without the consideration of the physical location. These situations make the cross-domain data fusion quite difficult. However, as stated in the introduction, cross-domain IoT data fusion is becoming a key driver to accelerate the production of new and beneficial services in the IoT era. To achieve cross-domain data fusion, we here focus on that the data suitable for the cross-domain data fusion tend to be generated in the geographical proximity. Thus, we introduce a concept of Geo-Centric Information Platform (GCIP), which allows us to collect, process the IoT data with consideration of geographic location where they generated. As a result, GCIP efficiently produces various types of STCs.

Figure 1 shows the conceptual design of GCIP. Procedures of GCIP consist from the following 4 steps. First, the data transmitted from IoT devices within some area are replicated at the intermediate router(s), irrespective of the network type (Step 1). Then, the replicated data are forwarded to a proximity edge server(s) having the analysis/process functions (Step 2). The server(s) generates STCs as a result of processing with the collected data (Step 3). Finally, the server sends the produced STC in response to users (Step 4). In this way, the proposed GCIP has a potential to be a fundamental infrastructure for the IoT era.

3.2   PoC Design for Geo-Location Aware IoT Data Collection

There are two requirements to collect data based on geographical location.

(i) Geo-location area for collecting IoT data can be identified by the intermediate routers.
(ii) Modification for geo-location aware communication has a backward compatibility.

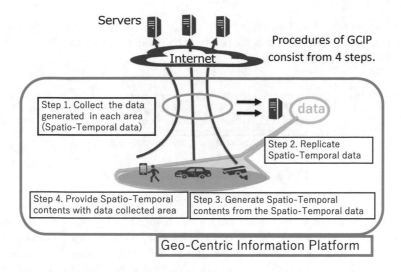

Fig. 1. Conceptual design of GCIP

We propose Physical Location-Aware Communication (PLAC) method that can satisfy the requirements described above. To achieve the first requirement, we design the hierarchical mesh-structured network topology, as shown in Fig. 2. The geographical space is divided into hierarchical meshes based on latitude and longitude lines, and each of which has a unique mesh code[1]. Size of the minimum mesh area is a square area of 39 m each side, by extending rule of Open-i area). As shown in Fig. 2, the length of the mesh code increases as the decrease in the size of geographical area decreases. Since a unique mesh code is allocated to each

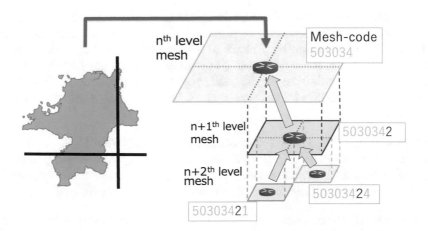

Fig. 2. Example of hierarchical mesh area and mesh code

---

[1] The mesh code is basically followed by the NTT Docomo open-i area [6].

of intermediate router, the router can identify its own belonging geographical area and handle all of IoT data containing the same mesh code. That is, the first requirement can be satisfied.

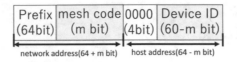

Fig. 3. Physical location-aware address

Next, to achieve PLAC with backward compatibility with the traditional Internet technology, we already proposed the geo-location aware address format [7]. As shown in Fig. 3, the mesh code with variable length depending on the mesh level is embedded with the IPv6 address. More specifically, it is located in the beginning of the lower 64 bits and used as a part of network address. Moreover, to maintain the backward compatibility with the Internet, we do not modify the part of network prefix (the upper 64 bits). From this, the extended network address (prefix + mesh code) can express the certain geographical area and the principle of the traditional IP routing process (i.e., longest match routing) can be used for PLAC, thanks to the backward compatibility. In this way, PLAC can be effective for IoT data collection at the routers on the mesh-structured network. Note that we hereinafter define intermediate routers on the mesh-structured network as "mesh router".

3.3   PoC Design for Publish/Subscribe-Based STC Production

In this section, we consider how various STCs, which are worth for users in the IoT area, are produced on the proposed GCIP. First, we need to consider that the STC is produced as a result of analyzing/processing to the collected IoT data; Therefore, some dedicated server(s) is necessity for IoT data analysis and processing, in addition to the mesh router. Furthermore, to flexibly and efficiently produce STCs on the server, the following three requirements should be satisfied.

Requirement 1 Type of IoT data should be uniquely identified.
Requirement 2 Collected IoT data should be flexibly stored and processed.
Requirement 3 To produce the STC with accurate geo-location information, the edge server needs to know where the IoT data was generated.

To satisfy these requirements, we focus on the Publish/subscribe communication model, which is a new concept for decoupling data collection and distribution (interactive communication is not mandatory). By employing the topic-based Publish/Subscribe model, requirement 1 can be achieved. Moreover, if the processing functions are implemented in the broker, requirement 2 also can be achieved. As for requirement 3, if the location information is included in the application data, it is possible. However, if the mesh router collecting IoT data from some geographical area serves as the publisher, we cannot touch the application data directly. Therefore, we need to achieve requirement 3, even without modification of the application data at all.

4   Demonstration Environment

In this section, we conduct experiments to show the effectiveness of the proposed GCIP from the view point of the following aspects.

- Can a mesh router on the mesh network topology collect various type of IoT data by using PLAC method?
- Can a mesh router collect various types of IoT data generated in the geographical proximity but transmitted from physical different network ?
- Can a mesh router (broker) produce the STCs by using the collected IoT data?

For the first experimentation, we used kyutech campus network. Figure 4 shows the mesh network topology. We placed 6 mesh routers, which are denoted as R1_1 , R1_2, and so on, for each mesh level. In contrast, 7 elements like D1_1, D1_2, and so on are IoT devices with various types of sensor, which are developed by Raspberry Pi. D1_1, D1_2 and D1_3 are located in the 10th level mesh network whose mesh router is R1_1. Similarly, D1_4, D1_5 and D1_6 are located in 10th level mesh network whose mesh router is R1_2. And, D1_7 are located in one of 10th level mesh network belonging to different 9th level mesh network, which belong to R1_3. These IoT devices periodically transmit sensor

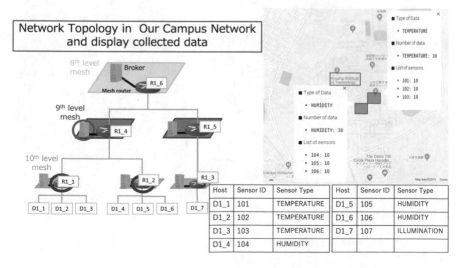

Fig. 4. Demonstration Environment for examining heterogeneous IoT data collection

data obtained from sensor to the cloud server on the Internet every 5 min. Since the transmission path is determined by the PLAC method, intermediate mesh routers including 9th level and 10th level mesh routers can collect the IoT data while associating with the geographical location. Then, the router replicates the collected data and transmits them to the broker. In this experiment, we confirm whether the R1_8 and R1_9 collect different types of IoT data by using PLAC method.

Next, we examine that PLAC method can collect the IoT data even when the data are transmitted through the different networks (those subnets/prefixes are different). Figure 5 shows the topology used for experimentation. We particularly focus on 10th level mesh network, which settings are the same on Network A, but we additionally assume Network B has a coverage in the same 10th level mesh area. R1_1 and R2_1 denote mesh routers for each of mesh network. In contrast, D1_1 to D1_3 and D2_1 to D2_2 denote the IoT devices with various types of sensor, which are developed by Raspberry Pi. IoT devices periodically (every 5 min) transmit data as in the same experimentation. In this experiment, we investigate that GCIP collects the IoT data generated in the same proximity but transmitted over different networks.

For the final experiment, we examine the feasibility on producing a STC using heterogeneous IoT data in a specific area. We here use the network topology of first experimentation again and employ the discomfort index [9] as the STC. The discomfort index (DI) is calculated as follows:

$$DI = T - (0.55 - 0.0055RH)(T - 14.5) \tag{1}$$

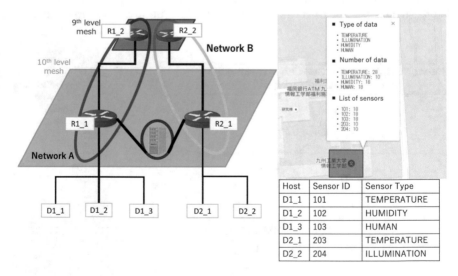

Fig. 5. Demonstration Environment for examining IoT data collection over multiple networks

where T indicates the mean value of air temperature in C and RH indicates the 5 min average relative humidity (Since the IoT data are received by the broker at the intervals of 5 min, the broker calculates the DI every 5 min based on the information of temperature and humidity. Note that the distribution method of the produced STC is out of scope in this paper. So, we use typical Web API (application) to demonstrate the detailed information of STC generated in the broker.

5 Result and Discussion

5.1 Experimental Results

The data collected by the routers R1_1 and R2_1 are shown on the right side of Fig. 4. We also illustrate the detailed information of the sensor number and the sensed values, which are collected by mesh routers, in the right map. Figure 4 shows that the mesh routers can collect the IoT data, irrespective of the data type because R1_1 can collected temperature data from D1_1, D1_2 and D1_3 and R1_2 can collected humidity data from D1_4, D1_5 and D1_6. Therefore, this result shows that various types of IoT data could be collected thanks to both the by PLAC address and its routing function.

Discomfort Index   It's 22 now in current area

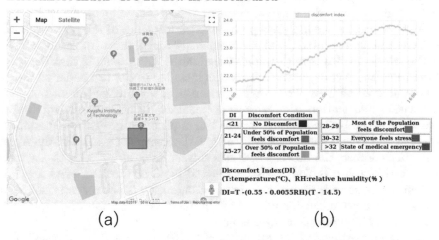

(a)   (b)

Fig. 6. Produced STC (left: spatial view, right: temporal view)

Figure 5 shows the results of second experiment. From Fig. 5, we can see that broker can collect data over different networks because R1_1 transmits data from D1_1, D1_2 and D1_3, while R2_1 transmit data from D2_1 and D2_2. As we can find from the sensor ID, we can see that the data passing through R1_1 and R2_1 is can be successfully collected. From this result, we can say that IoT data can be collected based on geographical proximity over different networks.

We describes the third experiment. Figure 6 shows content generated by the 9th mesh broker. This result can be viewed in the web application. Figure 6(a) shows a spatial area in where the produced STC is beneficial for the users. The broker (mesh server) inherently identifies the geo-location area by its own IPv6 address including mesh code. Moreover, Fig. 6(b) shows a time series variation of STC value. Since the collected data are stored in the broker (mesh server), we can naturally produce the time series variance of the STC.

In this way, the broker can identify the type of IoT data by the topic ID defined by introducing the topic-based Publish/Subscribe model (i.e., Apache Kafka [8]) (requirement 1). Furthermore, we found that the broker can store and process the IoT data collected by using PLAC method (requirement 2). Finally, as for requirement 3, the current implementation indicates the STC for 9th level mesh only. As a result, the broker naturally produces the STC with time series variance.

5.2   Discussion

First, we discuss the data collection performance of the proposed GCIP. The first and second experimentation show that the proposed PLAC mechanism can handle the IoT data based on the geographical location, thereby achieving the geo-location aware data collection by the router on the mesh network. More specifically, we also show that the IoT data can be naturally collected at mesh routers, independent of the access network.

Next, we discuss the IoT data processing performance of the proposed GCIP. We have already explained that requirement 1 and 2 are completely satisfied, but the requirement 3 is not completely satisfied. The reason is that the spatial information (mesh code) of the IoT data is included in the IP header as the part of the source IP address. The application layer of the broker cannot obtain the geo-location information of narrow area (such as 10the level mesh). However, since the Apache Kafka handles data at the application layer, the IP address is discarded and thus the mesh code of the IoT devices is not passed to the Apache Kafka. In such case, the broker illustrates the spatial area by using its own mesh code. Therefore, if a broker is always located in all of the 10th level mesh, the problem can be solved. However, since the assumption is not realistic, some kind of cross-layer mechanism for informing the mesh code in network layer to the Apache Kafka in application layer is needed.

Through experiments, the usefulness of GCIP was actually confirmed. First, GCIP can collect IoT data based on geography location by exploiting PLAC mechanism. When the cloud server tries to achieve the same thing, the geo-location information should be added to all of application packets (payload). That is, modification of application is mandatory. Second, GCIP naturally supports the store of the received IoT data on the mesh server. Therefore, the mesh server flexibly (e.g, temporally) process the stored data, thereby producing the STC effectively.

6 Conclusion

In this paper, we proposed the Geo-Centric Information Platform (GCIP) that can manage the Spatio-Temporal Contents (STCs) produced by the cross-domain data fusion. We practically examined the effectiveness of the GCIP in terms of (1) geo-location aware data collection and (2) Publish/Subscribe-based STC production through the experiments on campus IPv6 network. As a result, we confirmed the following outcomes.

- By using PLAC mechanism, IoT data transmitted from diverse devices can be collected at the edge router based on the geographical area independent of their access networks.
- STC can be successfully produced by exploiting the existing implementation (Apache Kafka) employing Publish/Subscribe model. Note that processing function should be developed on the implementation.

These results demonstrated the feasibility of the fundamental functions in the proposed GCIP. However, although the current implementation assumes that each of mesh network has one mesh router, it is difficult in the real environment. Therefore, we will extend the GCIP to adapt to the practical environment in the near future.

Acknowledgements. This work was supported by JSPS KAKENHI Grant Number JP18H03234 and the Commissioned Research of National Institute of Information and Communications Technology NICT).

References

1. Bryce, R., Srivastava, G.: The addition of geolocation to sensor networks. In: 13th International Conference on Software Technologies, pp. 762–768 (2018)
2. Yasumoto, K., Yamaguchi, H., Shigeno, H.: Survey of real-time processing technologies of IoT data streams. J. Inf. Process. **24**(2), 195–202 (2016)
3. Gonzales, E., Ong, B.T., Zettsu, K.: Searching inter-disciplinary scientific big data based on latent correlation analysis. In: Proceedings of Workshop on Big Data and Society (in Conjunction with IEEE BigData 2013), pp. 9–12, October 2013
4. Pattar, S., Buyya, R., Venugopal, K.R., Iyengar, S.S., Patnaik, L.M.: Searching for the IoT resources: fundamentals, requirements, comprehensive review, and future directions. IEEE Commun. Surv. Tutor. **20**, 2101–2132 (2018)
5. Al-Fuqaha, A., Guizani, M., Mohammadi, M., Aledhari, M., Ayyash, M.: Internet of Things: a survey on enabling technologies, protocols and applications. IEEE Commun. Surv. Tutor. **17**(4), 2347–2376 (2015)
6. NTT docomo: Open iArea guideline. https://www.nttdocomo.co.jp/binary/pdf/service/developer/make/content/iarea/domestic/open-iarea.pdf. (in Japanese)
7. Tamura, H.: Program for determining IP address on the basis of positional information, device and method. JP Patent 6074829, 20 January 2017
8. Apache Kafka. https://kafka.apache.org/
9. Talukdar, Md.S.J., Hossen, Md.S., Baten, A.: Trends of outdoor thermal discomfort in Mymensingh: an application of Thoms' discomfort index. J. Environ. Sci. Nat. Resour. **10**(2), 151–156 (2018)

DTN Sub-ferry Nodes Placement with Consideration for Battery Consumption

Kazunori Ueda[✉]

Kochi University of Technology, 185 Miyanokuchi, Tosayamada, Kami, Kochi, Japan
ueda.kazunori@kochi-tech.ac.jp

Abstract. When networking infrastructures are damaged due to a serious disaster such as earthquakes, communications on the Internet might be inconvenient. Many researches about Delay Tolerant Networking (DTN) have been progressed to solve the issue. The message ferrying is one of data transferring method in DTN. A message ferry is placed and moving around an area. Nodes in a network transfer data to the message ferry when it comes. If a message ferry cannot move around frequently, battery will be consumed due to waste data transfer and some opportunities for data transfer to the message ferry are lost. To solve this issue, we propose a sub message ferry that collect data when a message ferry does not stay in an area. A sub message ferry is chosen from nodes that have enough battery charge. Our proposed method has advantages about sustainable duration of mobile nodes in terms of battery charge by comparison with conventional message ferrying.

1 Introduction

Now a day, many people use mobile devices to communicate with family, friends, or other people. From this background, communications on the Internet might be inconvenient when networking infrastructures are damaged due to a serious disaster such as earthquakes. Delay Tolerant Networking (DTN) is attracted as a technology that can be used for such situations. Many researches about DTN have been progressed to solve the issue.

Basically, in networks based on DTN data are transferred with the epidemic routing [1]. Nodes that adopt the epidemic routing transfer data to all neighboring nodes. As a result, data are transferred in a wide area. This method enables nodes to transfer data toward a receiver node. However, many data transfers can be waste.

The message ferrying is one of data transferring method in DTN [2]. A message ferry is placed and moving around an area. Nodes in a network transfer data to the message ferry when it comes. If a message ferry cannot move around frequently, battery will be consumed due to waste data transfer and some opportunities for data transfer to the message ferry are lost.

To solve this issue, we propose a sub message ferry that collect data when a message ferry does not stay in an area. A sub message ferry is chosen from nodes

© Springer Nature Switzerland AG 2020
L. Barolli et al. (Eds.): NBiS-2019, AISC 1036, pp. 406–412, 2020.
https://doi.org/10.1007/978-3-030-29029-0_38

that have enough battery charge. Our proposed method has advantages about sustainable duration of mobile nodes in terms of battery charge by comparison with conventional message ferrying.

2   Delay Tolerant Networking

In poor environments due to serious disasters, disconnection or delay will occur frequently. Delay Tolerant Networking (DTN) is a networking method to achieve communication in such poor environments. Recent years, researches that integrate the DTN and mobile ad-hoc networking (MANET) have been progress and attracted. In DTN architecture, received data are stored in a storage and the data are transferred to other nodes when communication is available. Data are transferred hop by hop and the data reach at receiver nodes.

Nodes that adopt the epidemic routing transfer data to all neighboring nodes [1]. As a result, data are transferred in a wide area. This method enables nodes to transfer data toward a receiver node. However, many data transfers can be waste. Figure 1 shows data transferring model based on the epidemic routing method. Issues of the epidemic routing method are the same as reactive routing methods of mobile ad-hoc networking or peer/contents search of peer-to-peer networking model and applications. Since data transfer is performed on each node, total count of data transferring increases drastically if the number of nodes are large.

The message ferrying is one of data transferring method in DTN [2]. A message ferry is placed and moving around an area. Nodes in a network transfer data to the message ferry when it comes. If a message ferry cannot move around frequently, battery will be consumed due to waste data transfer and some opportunities for data transfer to the message ferry are lost. Figure 2 shows data transferring model based on the message ferrying method. One of issues of the message ferrying method is a period of moving around cycle. Message ferries are moving around multiple regions where some nodes stay and the nodes want to transfer data. It takes a long time to come to the same region again if there are many regions to be visited in networks.

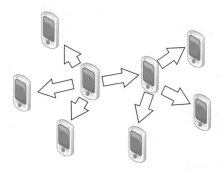

Fig. 1. Data transferring based on the epidemic routing method.

From issues mentioned above, it is needed to save networking resources or battery of nodes. Some researches have been progressed about saving consumption of networking resources or battery of nodes [3,4]. One of them is removing data that will not be transferred in order to avoid buffer consumption. Other of them is method that focuses on remaining battery.

3   Sub-message Ferrying Method

In this section, we explain details of the sub message ferrying method. The sub message ferrying method is proposed to extend sustainable time of networks and increase opportunities to communicate with a ferry node. Waste data transfer will be reduced owing to a sub message ferry. This method does not need to introduce a specific node to play a special role. Following subsections show considering environments and the details of proposed method.

3.1   Environments

Following environments are estimated under a serious disaster.

1. In a network, communications with outer networks are unavailable or very hard due to damage of network infrastructures such as communication facilities or networking cables.
2. Evacuees stay at evacuation centers and they have a smartphone to communicate with others and get information.
3. Evacuees move sometimes around the evacuation centers and communication time depends on battery of each smartphone without a mobile battery.
4. Communication between an evacuation center and other center or outer facilities are achieved with a message ferrying method on DTN.
5. A message ferry does not move around frequently due to lack of operation resources due to serious disaster damages.

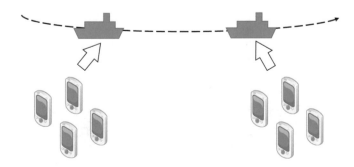

Fig. 2. Data transferring based on the message ferrying method.

3.2   Sub Message Ferry Method

Since frequency of moving around by a ferry node based on conventional message ferrying method is decreased, battery of each node in a network is consumed while the node waits for arriving the ferry node. To solve this issue, we introduce a sub-message ferry node into DTN. The sub-message ferry node collect data from other nodes in a network and transfer the data to a ferry node when the ferry node come to the network. This method can reduce battery consumption because the number of waste data transfer is decreased. Figure 3 shows a sub-message ferry node in a network is collecting data and sending the data to the message ferry node. A sub-ferry node is chosen by considering remaining battery of each node in a network. Nodes that have enough remaining battery based on a threshold are selected and sub-message ferry nodes do not transfer data to other message ferry nodes when the number of sub-message ferry node is more than one sub-ferry node.

4   Evaluations

Communication time of a network was used with comparison as performance evaluation of the sub-message ferry method. Time until remaining battery of all nodes would be 0 was regarded as communication time. Since it is difficult to compare remaining battery in an actual network environment, computer simulation was performed and results of the simulations was analyzed. In following subsections, simulation environments and results are shown.

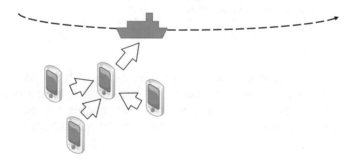

Fig. 3. Data transfer based on sub-message ferrying method.

4.1   Simulation Environments

As a network simulator focused on DTN, The Opportunistic Network Environment simulator (The ONE) was adopted [5]. The ONE enables researchers to simulate networking on DTN. The ONE supports various routing methods and can simulate battery consumption. Configuration file of The ONE includes places of nodes, moving model, or the number of nodes.

4.2   Simulation Conditions

Shape of simulation area was set to circle and radius of area was set to 100 [m] or 200 [m]. The number of nodes was set to 50 or 100, and the number of sub-message ferry node was set to 0, 1, or 3. Table 1 shows parameters about messages that were generated by mobile nodes. Messages were generated every 90 s. The size of message varied from 50 to 150 [KB] and Time To Live (TTL) was set to 300 s.

Table 1. Simulation conditions.

Parameter	Value
Area size	100, 200 [m]
The number of nodes	50, 100
Data generation period	90 [s]
Message size	50–150 [KB]
Time To Live	300 [s]

Table 2. Simulation conditions about battery consumption.

Parameter	Value
Initial remaining battery	3,000–6,000
Battery consumption of node search	22
Battery consumption of data transfer	12
Period between searches	15 [s]
Stopping time	0–600 [s]

Table 2 shows parameters about battery consumption. The initial amount of battery was set randomly between 3,000 and 6,000. Each node searched other node every 15 s. Moving speed of each node was set randomly between 0.5 to 1.5 [k/h] and stopping time was set randomly between 0 and 600 s. Simulation time was 3 h.

4.3   Simulation Results

This subsection shows results of simulations. In the simulations, communication time was measured. Tables 3, 4, and 5 show the results of simulations. When the sub-ferry node was adopted, communication time was longer than the case that no sub-ferry node existed.

Table 3. Communication time (radius: 100 m, nodes: 50).

The number of sub-ferry nodes	Communication time
0 (conventional)	3,783
1	4036
3	4080

Table 4. Communication time (radius: 200 m, nodes: 50).

The number of sub-ferry nodes	Communication time
0 (conventional)	3,831
1	4075
3	4090

Table 5. Communication time (radius: 100 m, nodes: 100).

The number of sub-ferry nodes	Communication time
0 (conventional)	3,829
1	3916
3	4028

5   Summary

When networking infrastructures are damaged due to a serious disaster such as earthquakes, communications on the Internet might be inconvenient. Many researches about Delay Tolerant Networking (DTN) have been progressed to solve the issue. The message ferrying is one of data transferring method in DTN. A message ferry is placed and moving around an area. Nodes in a network transfer data to the message ferry when it comes. If a message ferry cannot move around frequently, battery will be consumed due to waste data transfer and some opportunities for data transfer to the message ferry are lost. To solve this issue, we proposed a sub message ferry that collect data when a message ferry does not stay in an area. A sub message ferry was chosen from nodes that have enough battery charge. Our proposed method had advantages about sustainable duration of mobile nodes in terms of battery charge by comparison with conventional message ferrying.

Acknowledgement. The author would like to thank M. Nakabayashi for sharing his dataset with us.

References

1. Vahdat, A., Becker, D.: Epidemic routing for paritially connectedad hoc networks. Duke Technical report CS2000-06 (2000)

2. Ammar, M., Zhao, W., Zegura, E.: A message ferrying approach for data delivery in sparse mobile ad hoc networks. In: Proceedings of the MobiHoc 2004, pp. 187–198 (2004)
3. Kurose, J., Zhanga, X., Negliab, G., Towsley, D.: Perfomance modeling of epidemic routing. Comput. Netw. **51**, 2867–2891 (2007)
4. Haas, Z., Small, T.: A new networking model for biological applications of ad hoc sensor networks. IEEE/ACM Trans. Netw. **14**(1), 27–40 (2006)
5. Keranen, A., Ott, J., Karkkainen, T.: The one simulator for DTN protocol evalution. In: Proceedings of the SIMUTools 2009 (2009)

Concept Proposal of Multi-layer Defense Security Countermeasures Based on Dynamic Reconfiguration Multi-perimeter Lines

Shigeaki Tanimoto[1]([⊠]), Yuuki Takahashi[1], Ayaka Takeishi[1],
Sonam Wangyal[1], Tenzin Dechen[1], Hiroyuki Sato[2],
and Atsushi Kanai[3]

[1] Faculty of Social Systems Science, Chiba Institute of Technology,
Chiba, Japan
shigeaki.tanimoto@it-chiba.ac.jp,
y_takahashi@systemwarp.co.jp, nyta2326@gmail.com,
{s1891301cb,s1891302bd}@s.chibakoudai.jp
[2] Information Technology Center, The University of Tokyo, Tokyo, Japan
schuko@satolab.itc.u-tokyo.ac.jp
[3] Faculty of Science and Engineering, Hosei University, Tokyo, Japan
yoikana@hosei.ac.jp

Abstract. With the rapid progress of the Internet, security incidents are increasing and are becoming more sophisticated. Current trends in security incidents include not only cyber security threats such as viruses, malware, and unauthorized access, but also methods of stealing information, such as phishing, shoulder hacking, and electromagnetic wave eavesdropping. Thus, security incidents tend to diversify, and it is therefore important to consider physical countermeasures and psychological countermeasures other than the cyber security countermeasures typically taken. In this paper, in addition to the current cyber security countermeasures, we propose new multi-layer defense security countermeasures taking into consideration non-cyber security countermeasures that reflect physical viewpoints and psychological viewpoints. Specifically, we propose the concept of multi-layer defense security countermeasures based on dynamic reconfiguration multi-perimeter lines. Furthermore, a desktop simulation is performed with TPO (time, place, occasion) conditions in the office as specific perimeter lines, and the effectiveness of the concept is clarified. This contributes to the construction of a new paradigm of information security management in the digital transformation era.

1 Introduction

With the rapid progress of the Internet, security incidents are increasing and becoming more sophisticated. For example, cyber attacks represented by unauthorized access, malware, and recently targeted attacks are known. In addition, there are many cases of using phishing, "it's me" fraud, one-click fraud, etc. Among them, cheating fraud is increasing, which is a serious problem. This is a fraud that causes an unspecified number of people to transfer money to the bank account of the fraudulent party using

© Springer Nature Switzerland AG 2020
L. Barolli et al. (Eds.): NBiS-2019, AISC 1036, pp. 413–422, 2020.
https://doi.org/10.1007/978-3-030-29029-0_39

multiple means such as telephone and electronic mail. Companies are taking various countermeasures in response to such sophisticated and diversified security incidents. However, despite these countermeasures, security incidents have not decreased. Information leaks from security incidents in 2016 amounted to about 14 million, and estimated damages amounted to about ¥ 300 billion, and the damage was enormous [1].

As a trend of recent security incidents, "targeted attacks" targeting specific individuals, organizations, and information are increasing. In addition, "phishing scams" that steal security codes and credit card numbers by impersonating websites of financial institutions are also increasing [2].

There are many ways of social engineering. For example, a user's ID and password can be skillfully obtained by telephone. Other examples include "shoulder hacking" in trains and "trashing" to obtain information from the trash. As described above, security incidents have become more varied when social engineering is taken into consideration. This tendency cannot be prevented only by cyber security countermeasures. Furthermore, when internal fraud, etc., are considered, the current cyber security countermeasures against these various attacks are not enough.

As described above, in addition to the current cyber security countermeasures, it is important to consider non-cyber security aspects such as physical and psychological aspects, but current examination of those aspects is not sufficient.

In this paper, in addition to the current cyber security countermeasures, a concept model of multi-layered defense security countermeasures is proposed from the multi-faceted viewpoint that also includes non-cyber security aspects such as physical and psychological aspects. This contributes to the reduction of security incidents in the so-called "zero trust" era, which also includes internal fraud.

2 Current Status and Issues of Cyber Security

2.1 Changes in Cyber Attacks

Figure 1 shows an overview of the history of cyber attacks. As shown in Fig. 1, in the 1980s, cyber attacks were perpetrated by "fun-seeking criminals" that surprised other parties by such acts as displaying defects on the screen of a personal computer to a large number of other parties.

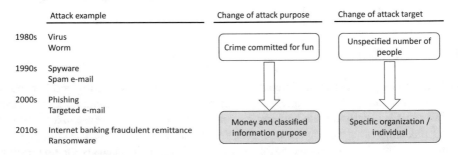

Fig. 1. Changes in cyber attacks [3]

Recently, there has been a tendency to exploit sensitive information and money in a more sophisticated manner against specific parties. In other words, attacks that used to be targeted at unspecified numbers have been changed to attacks made on specific individuals and organizations. For example, in a targeted attack, a specific department of a specific company is sent an e-mail loaded with malware. In this case, an aggressor carries out the preliminary survey of a target first. Victims are taken unaware when a particular e-mail subject or wording is used that is understood only by the persons concerned. As mentioned above, in some cases, the attacks are elaborate and take time. Thus, recent attackers are organized and commercialized [3].

2.2 Current Status of Cyber Security

2.2.1 Actual Status of Security Incidents

Figure 2 shows the number of new malware found. As shown in the figure, the number of new malware has continued to increase from 2015 and decreased once in 2016, but the number of newly discovered malware has increased in 2017 [4].

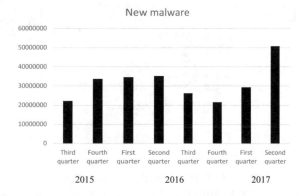

Fig. 2. Changes in cyber attacks [3]

2.2.2 Current Features of Cyber Security Incidents

Recent attacks, such as "phishing" and "it's me" fraud, include psychological tactics. Among such attacks, special scams targeting an unspecified number of people are increasing, which is a serious problem. These scams trick an unspecified number of people by multiple means such as telephone and e-mail to deposit funds into a criminal's bank account.

In this way, techniques related to human psychology are called social engineering. Social engineering is used to obtain important information such as passwords from administrators, users, etc., by "social" means such as eavesdropping and stealing [5]. As shown in Table 1, various social engineering attacks have been reported [6]. Furthermore, as shown in Fig. 3, when social engineering is classified in terms of attack period, it is broadly classified into three: those attacks performed in the short term, those performed over the medium and long term, and those not related to the period [7].

Table 1. Major social engineering attacks [6]

Main attack method	Contents
Shoulder hacking	A method for looking over the shoulder of a person at his or her PC screen to obtain important information such as passwords
Trashing	An attack that searches for configuration information from servers and routers discarded in the trash, network configuration diagrams, IP address lists, user names and passwords, etc.
Phishing	Phishing is "a fraudulent act of sending e-mail impersonating a real company or individual and trying to acquire confidential information such as credit card numbers, personal ID data, passwords, etc."

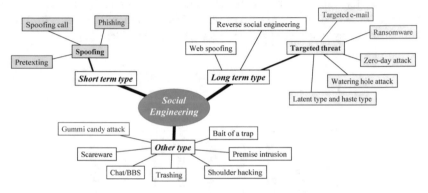

Fig. 3. Risk classification example of social engineering [7]

As mentioned above, various attack methods are expected to be created in the future, in view of the current number of types of cyber attacks and damage situations. Thus, it can be said that it is necessary to consider various countermeasures including cyber aspects and non-cyber aspects.

3 Multi-layer Defense Security

3.1 Summary of Multi-layer Defense Security

A schematic diagram of multi-layer defense security is shown in Fig. 4. Information is kept by arranging various defense systems in layers. In addition, the arranged security countermeasures are divided into three, "the countermeasure against an entrance", "the countermeasure against an exit", and again "the countermeasure against an exit" [8].

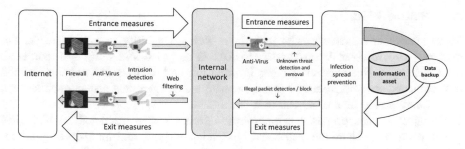

Fig. 4. Schematic diagram of multi-layer defense security [8]

3.2 Examples of Multi-layer Defense Security Model

3.2.1 Swiss Cheese Model

Figure 5 shows the outline of the Swiss cheese model, which is a model for controlling and reducing the risk of organizational accidents, proposed by James Reason, a psychology professor at the University of Manchester in the United Kingdom. The model is based on a "Swiss cheese" analogy of multiple defense countermeasures designed between risk factors and incidental damage. In the case of "holes", it is assumed that the occurrence of an accident is blocked by the overlapping of holes. Also, if an accident occurs, the risk will be revealed by tracing the holes in each layer; problems related to the attack will be apparent, and damage can be minimized [9].

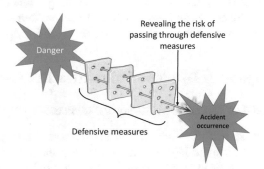

Fig. 5. Outline of Swiss cheese model [9]

3.2.2 Defense in Depth

Strict placement of defensive positions, as used in the design of European castles, is a strategy that mitigates an enemy's ability to reach the innermost parts of a castle without breaking through many walls and defenses. This strategy makes it possible for the attacking forces to be exhausted and the defenders to turn back to reach the innermost part of the castle. Furthermore, in order to defend against various attacks, the defense effect is raised by using different techniques and strategies. In addition, this technology is also used in Japanese castles [9].

3.3   Related Work

Many studies in connection with multi-layer defense security have been made. For example, a multi-layer defense system to APT is proposed in a study by Moon et al. This system is characterized by log information and various information through collection and analysis from each agent by installing an agent in a network appliance, a server, and the terminal of an end user [10].

Elhaj et al. has proposed a multi-layer defense system modeled on biological immune systems. There are distributed type, adapted type, and multi-layer type biological immune systems. In the network security system design by Elhaj et al., these characteristics are incorporated. A multi-layer defense system that specifically consists of two main layers and a layer with an inherency and adaptability is proposed [11].

Wang et al. has proposed a client puzzle as an effective countermeasure for DoS attacks. A puzzle auction and a congestion puzzle technique are proposed as an extension of the model in which puzzle technology is built into an IP layer [12].

Dasgupta refers to the information processing abilities of biological immune systems, such as feature extraction and the recognition, study, and memorization of patterns and their distributive characteristics. Specifically, he focuses on the construction of an autonomous defense system using the immunological metaphor for starting the correspondence related to information gathering, analysis, decision-making, and recognizing threats and attacks [13].

Each of these studies are characterized by the point of having multiple defense lines. However, despite the various studies on the multi-layer defenses of current cyber security countermeasures, by focusing on the "cyber-environment", only the surface has been scratched. Greater focus must be placed on the "non-cyber-environment", which is characterized by physical and psychological security aspects.

4   Concept Proposal of Dynamic Multi-layer Defense Security

4.1   Concept Model of Dynamic Multi-layer Defense Security

Here, the concept model of dynamic multi-layer defense security is proposed. The concept model of dynamic multi-layer defense security is shown in Fig. 6. As shown in Fig. 6 (1), defense is exercised in various hierarchies, not only within the cyber-layer, but also within non-cyber-layers, which include psychological, physical, and economic aspects. Furthermore, as shown in Fig. 6 (2), with the proposed multi-layer defense security countermeasures, the concept of a perimeter line (defense line) is introduced.

This enables the dynamic reconstruction of a perimeter line in a non-cyber-environment, for example, on the basis of TPO (time, place, occasion) conditions [14]. A concrete example is given. Generally, an office environment only consists of employees from the perspective of security. For this reason, countermeasures against physical attacks, such as shoulder hacking, is unnecessary. However, when a guest comes to an office, a countermeasure against shoulder hacking is needed.

Thus, according to conditions, by reconstructing a perimeter line dynamically, the cost of security countermeasures can be reduced and security policies can be somewhat relaxed, contributing to reducing the trade-off between the conveniences of security and safety.

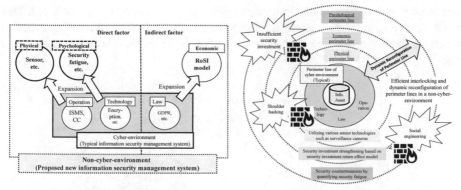

(1) Extension to non-cyber-environment   (2) Concept model of dynamic multi-layer defense security

Fig. 6. Concept model of dynamic multi-layer defense security

As an initial investigation, in the model of Fig. 6, the focus is placed on the psychological and physical aspects, which are direct factors, and the effect is clarified in this paper. The effect of the economic aspect, which is an indirect factor, is considered as a future subject.

4.2 Evaluation of Concept Model of Dynamic Multi-layer Defense Security

4.2.1 Evaluation of Concept Model

When examining the model that adds the hierarchies of the physical and psychological aspects of a non-cyber-environment, the threats and countermeasures for each hierarchy are simplified and defined in Table 2.

Table 2. Threat and countermeasure for each hierarchy

Hierarchy	Threat	Countermeasures
Layer 1: Cyber	Virus attack Unauthorized access	Firewall and anti-virus introduction Strict management of ID and PW
Layer 2: Physical	Shoulder hacking Trashing	Set filtering on screen Paper media is shredded
Layer 3: Psychology	Stress Lack of education	Regular stress check Re-education, implementation of training

Generally, it is difficult to perform all multi-layer defense countermeasures in regards to cost. In this paper, reduction of cost is proposed by fluctuating the quality and quantity of a countermeasure of each class in consideration of TPO conditions. As an example, as mentioned above, when a guest visits an office, risk increases. That is, countermeasures against a physical aspect, such as a countermeasure against shoulder hacking, become important. However, outside office hours, when guests usually are not permitted into the office, countermeasures against physical aspects can be relaxed. That is, the cost of such countermeasures for physical aspects can be reduced. The execution image of the perimeter line in the dynamic multi-layer defense security is shown in Fig. 7.

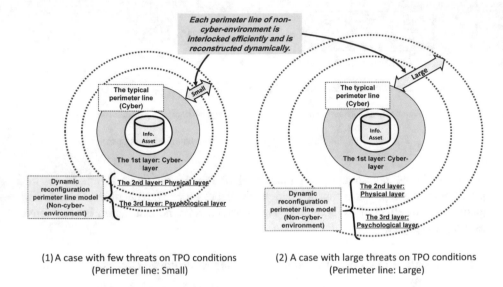

(1) A case with few threats on TPO conditions   (2) A case with large threats on TPO conditions
(Perimeter line: Small)   (Perimeter line: Large)

Fig. 7. Execution image of dynamic multi-layer defense security

4.2.2   Evaluation of Desktop Simulation

This desktop simulation targets the office space of a company. The countermeasures of each layer are dynamically changed according to the TPO pattern [15] in the office shown in Table 3.

Table 3. Example of TPO pattern [15]

TPO pattern	Safe	Danger
T (Time)	During business hours	After business hours
P (Place)	Within a company	Outside of a company
O (Occasion)	Only employees	Employee + guest

In the desktop simulation, based on the threat level of the TPO pattern shown in Table 3, the necessary countermeasure ratio for each layer was set. Here, as with the current countermeasures, the cyber security countermeasure is always 100% taken regardless of the threat level on the basis of the change factor of TPO. In non-cyber countermeasures, physical and psychological aspects are set as follows.

(1) **Threat side:** It is set up on a five-step level (5: maximum, 1: minimum) through subjective evaluation by the authors. The threat is assumed to be of a psychological and physical aspect on the basis of TPO conditions. In T (hours), work time is defined as "safe", and out of work time is defined as "danger". In P (place), the inside of the office is defined as safe, and the outside of the office is defined as dangerous. In O (Occasion), the environment for only employees is defined as safe, and the environment visited by guests is defined as dangerous.

(2) **Countermeasure side:** By subjective evaluation as (1) with the same countermeasure aspect, the countermeasure level was set to the display (100%: maximum, 0%: minimum) 100% here. First, the 1st layer (cyber-layer) was taken as a definition in which a regular countermeasure (100%) is performed as a typical countermeasure. In this situation, we decided to cope with it according to the threat (a guest's visit, external environment) of a physical aspect in the 2nd-layer (physical aspect) countermeasure. The 3rd-layer (psychological aspect) countermeasure was taken as the premise to which the countermeasure against a cyber-aspect and a physical aspect attack is carried out. For this reason, the third layer threat is set one level lower than the physical threat level.

The simulation results based on the above are shown in Table 4. As shown in the table, it can be seen that the cost of physical and psychological countermeasures is reduced on the basis of the TPO condition. In this way, it is clarified from the desktop simulation that low cost and efficient defense against multiple layers of security can be achieved by dynamically changing the countermeasure for each TPO pattern.

Table 4. Desktop simulation results

TPO Pattern	T (Time)	P (Place)	O (Occasion)	The 1st layer: Cyber-layer		The 2nd layer: Physical-layer		The 3rd layer: Psychological-layer		Sum of counter-measures
				Threat	Counter-measure	Threat	Counter-measure	Threat	Counter-measure	
①	During business hours	Within a company	Only employees	5	100	2	40	1	20	160
②	During business hours	Within a company	Employee + guest	5	100	3	60	2	40	200
③	During business hours	Outside	Only employees	5	100	3	60	2	40	200
④	During business hours	Outside	Employee + guest	5	100	4	80	3	60	240
⑤	After business hours	Within a company	Only employees	5	100	3	60	2	40	200
⑥	After business hours	Within a company	Employee + guest	5	100	4	80	3	60	240
⑦	After business hours	Outside	Only employees	5	100	4	80	3	60	240
⑧	After business hours	Outside	Employee + guest	5	100	5	100	4	80	280

Each pattern of the proposal system is reduced from the typical countermeasure (=300) against each-layer.

5   Conclusion and Future Work

In this paper, in addition to current cyber security countermeasures, a concept model of multi-layered defense security countermeasures is proposed from a multi-faceted viewpoint that includes non-cyber security aspects such as physical and psychological aspects. Furthermore, a desktop simulation revealed its effectiveness. As a result, the concept of a multi-layered defense security management model with dynamically reconfigured perimeter lines based on TPO conditions was established in consideration of non-cyber security aspects while considering physical and psychological aspects and cost reduction.

Our future work is to evaluate an implementation of the proposed concept model.

Acknowledgments. This work was supported by JSPS KAKENHI Grant Number JP 19H04098.

References

1. Japan Network Security Association: Survey Report on Information Security Incidents in 2018. http://www.jnsa.org/result/incident/. (in Japanese)
2. The Finance: Latest cyber attack and security countermeasures summary. https://thefinance.jp/fintech/150722. (in Japanese)
3. McAfee Blog: 20 Types of Cyber Attacks You Should Know to Protect Your Organization. https://blogs.mcafee.jp/cyber-attack-type-to-know. (in Japanese)
4. McAfee Blog: Why is it multilayer defense? Strongest security countermeasures that make a risk the minimum. https://blogs.mcafee.jp/defense-in-depth-multilayer-protection. (in Japanese)
5. Cyber security.com: What is social engineering? Consider countermeasures from specific methods. https://cybersecurity-jp.com/cyber-terrorism/14431. (in Japanese)
6. Tanimoto, S.: New paradigm of information security management in the digital transformation era, Keynote3. In: Proceedings of the 12th International Conference on Project Management (ProMAC2018), The Society of Project Management, Bangkok, pp. 45–76 (2018)
7. MIC: 2017 version information communication white paper. http://www.soumu.go.jp/johotsusintokei/whitepaper/ja/h29/index.html. (in Japanese)
8. Canon: Security solution site' Structure of the multilayer defense read and solved by illustration. https://www.canon-sas.co.jp/portal/security/securityinformation/deffences.html. (in Japanese)
9. Hoshi, T., et al.: Cyber attacker's behavior considerations for Defense in Depth implementation, IPSJ SIG Technical reports, 2015-CSEC-71, no. 2 (2015). (in Japanese)
10. Moon, D., et al.: MLDS: multi-layer defense system for preventing advanced persistent threats. Symmetry **6**, 997–1010 (2014). https://doi.org/10.3390/sym6040997
11. Elhaj, M.M.K., et al.: A multi-layer network defense system using artificial immune system. In: 2013 International Conference on Computing, Electrical and Electronic Engineering (ICCEEE), pp. 232–236 (2013)
12. Wang, X., et al.: A multi-layer framework for puzzle-based denial-of-service defense. Int. J. Inf. Secur. **7**(4), 243–263 (2008). https://doi.org/10.1007/s10207-007-0042-x
13. Dasgupta, D.: Immuno-inspired autonomic system for cyber defense. Inf. Secur. Tech. Rep. **12**(4), 235–241 (2007)
14. MONOist: How to protect a control system: "multilayer defense" and "status recognition". http://monoist.atmarkit.co.jp/mn/articles/1403/04/news005.html. (in Japanese)
15. Yoneda, S., et al.: A study of dynamic cooperation method between multi-clouds based on TPO conditions. IEICE Trans. Inf. Syst. (Jpn. Ed.) **J99-D**(10), 1045–1049 (2016)

Fault Detection of Process Replicas on Reliable Servers

Hazuki Ishii[1]($\boxtimes$), Ryuji Oma[1], Shigenari Nakamura[1], Tomoya Enokido[2], and Makoto Takizawa[1]

[1] Hosei University, Tokyo, Japan
{hazuki.ishii.5h,ryuji.oma.6r}@stu.hosei.ac.jp,
nakamura.shigenari@gmail.com, makoto.takizawa@computer.org
[2] Rissho University, Tokyo, Japan
eno@ris.ac.jp

Abstract. In order to make a system tolerant of faults, an application process is replicated on multiple servers. According to the advances of hardware and architecture technologies, each server can be considered to be free of fault, i.e. always proper. On the other hand, replicas of application processes easily suffer from faults, e.g. due to security attacks like virus and hacking. Even if a replica sends a proper reply, the replica may do faulty computation. It takes a longer or shorter time and a server to perform a faulty replica consumes more or smaller electric energy to perform a faulty replica. Such a faulty replica is referred to as implicitly faulty replica. In this paper, we discuss how to detect implicitly faulty replicas by using the power consumption and computation models of a server in addition to checking replies in a homogeneous cluster.

Keywords: Process faults · Implicitly faulty replica ·
Power consumption model · Computation model · Fault detection

1 Introduction

Information systems are composed of servers where application processes issued by clients are performed. Information systems have to be tolerant of faults. Systems can be more reliable and available by taking checkpoints [4,8] and replicating system components especially application processes [4,5,7].

Servers are recently getting more reliable and available according to the advances of hardware and architecture technologies like CPUs [1]. This means, each server can be assumed to be always proper, i.e. does not suffer from any fault. Even if a server gets faulty, the server is detected to be faulty and is recovered by itself. On the other hand, application processes performed on servers easily suffer from faults due to hacking and virus attacks [6]. A faulty replica may return an improper reply or no reply due to Byzantine fault [20]. Multiple replicas of an application process are performed on servers to make a system fault-tolerant and send replies. A majority one of replies from replicas can be

© Springer Nature Switzerland AG 2020
L. Barolli et al. (Eds.): NBiS-2019, AISC 1036, pp. 423–433, 2020.
https://doi.org/10.1007/978-3-030-29029-0_40

taken as a proper reply if fewer than the half of the replicas are faulty [20]. In addition to sending improper replies, a faulty replica may not properly behave, e.g. does different computation even if the replica sends a proper reply. Such a replica is referred to as *implicitly faulty*. In order to detect an implicitly faulty replica, further information is needed in addition to the correctness of a reply.

The macro-level power consumption models are proposed to show how much electric power [W] a server consumes to perform application processes without considering the power consumption of each hardware component like CPU [9–14,18]. In the SPC (Simple Power Consumption) model [11], a server consumes the maximum power [W] if at least one application process is performed. In the MLPC (Multi-level Power Consumption) model [10,18], the power consumed by a server depends on the numbers of active CPUs, cores, and threads. If no application process is performed, a server consumes the minimum power. The computation model of a server [10,13,18] gives the execution time of each application process performed on the server. If the minimum execution time of each application process on a server is *a prior* known like on-line transaction processing applications [18], the execution time of each application process and the energy consumption [J] of a server can be estimated by taking advantage of the power consumption and computation models.

In this paper, we assume each server including architecture and operating system is always proper, i.e. suffers from no fault. On the other hand, application processes may more easily suffer from faults due to security attacks like virus and hacking [6]. This means, each application process may stop by fault and may maliciously behave, i.e. does not follow the specification. If an application process is properly performed on a server, the application process returns a proper reply. In addition, the execution time of the application process can be estimated by using the computation model. Furthermore, we can estimate how much energy a server consumes to perform application processes based on the power consumption model and computation model. On the other hand, if an application process is faulty, the execution time of the application process may be shorter or longer and the server may consume more or less energy than expected even if every process may send a proper reply. In this paper, we discuss how to detect implicitly faulty processes on fault-tolerant servers by measuring the energy consumption of computers and execution time of each process in addition to checking replies.

In this paper, we consider a cluster composed of homogeneous servers. Each server is composed of the same hardware, architecture, and operating system. An application process issued by a client is replicated on each of the servers. Each replica may be faulty while each server is always proper. We newly propose a model to detect faulty replicas on each server in a cluster by using the power consumption and computation models. By monitoring not only the execution time of each replica but also the power computation of a server, implicitly faulty replicas are detected.

In Sect. 2, we present a system model. In Sect. 3, we discuss how to detect implicitly faulty processes.

2   System Model

2.1   Cluster of Servers

A cluster is composed of servers $s_1, ..., s_m$ $(m \geq 1)$ which are interconnected with a load balancer L (Fig. 1). A client issues a request to the load balancer L of the cluster. The load balancer L forwards the request to a server s_t in the cluster. An application process to handle the request is created and performed on the server s_t. In this paper, a term *process* means an application process to be performed on a server. The process on the server s_t sends a reply to the load balancer L. The load balancer L forwards a reply to the client. In this paper, we assume each server is always proper and the load balancer L is also proper. This means, the load balancer L can anytime get the electric power consumption [W] of each server and the execution time [sec] of each process performed on the each server.

If a process p_i gets faulty, the process p_i may send an improper reply or no reply to the load balancer L [4]. In order to be tolerant of the process fault, multiple replicas of a process are performed on servers, i.e. processes are actively replicated [4]. In this paper, we assume each process is deterministic [15]. A load balancer L collects replies from the replicas. If a replica only suffers from stop-fault, no faulty replica sends a reply and at least one replica is required to be proper. If a replica suffers from Byzantine fault [20], majority of replies from the replicas are taken as a reply. This means, more than half of the replicas are required to be proper in order to be tolerant of Byzantine faults of replicas.

Furthermore, a replica may do improper computation different from the specification even if the replica sends a proper reply. For example, a replica does different computation from the specification since the replica is infected with virus. The faulty replica wirestaps a reply of another replica which might be proper and just forwards the wiretapped reply to the load balancer L. Thus, a replica which sends proper replies but does different computation is referred to as *implicitly faulty*. Implicitly faulty replicas cannot be found in the traditional fault detection algorithms [4,16]. In this paper, we discuss how to detect faulty replicas on each server by monitoring the energy consumption of the server and the execution time of each process.

In this paper, we consider a homogeneous cluster which is composed of servers $s_1, ..., s_n$ $(n \geq 1)$ with the same architecture and operating system for simplicity. A client issues a process p_i to the load balancer L. The load balancer L creates a replica p_{it} of the process p_i on each server s_t and the replica p_{it} is performed on the server $s_t (t = 1, ..., n)$ (Fig. 1).

Then, the replica p_{it} sends a reply r_{it} to the load balancer L. By collecting replies from the replicas $p_{i1}, ..., p_{in}$, the load balancer decides on a reply r_t and sends the reply r_t to the client. Let P be a set of processes $p_1, ..., p_n$ $(n \geq 1)$ issued to a load balancer L. On each server s_t, a replica p_{it} of each process p_i is performed $(i = 1, ..., m)$. Let P_t be a set of replicas $p_{1t}, ..., p_{nt}$ of application processes $p_1, ..., p_n$, respectively, which are performed on a server s_t.

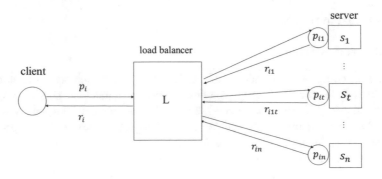

Fig. 1. Replicas of a process

2.2   Power Consumption Model

A server s_t is composed of np_t (≥ 1) homogeneous CPUs $cp_{t0}, \ldots, cp_{t,np_t-1}$. Each CPU cp_{tk} [1] is composed of nc_t (≥ 1) homogeneous cores $c_{tk0}, \ldots, c_{tk,nc_t-1}$. Each core c_{tkh} supports the same number ct_t of threads. Here, ct_t is one or two. A server s_t supports processes with totally nt_t ($= np_t \cdot nc_t \cdot ct_t$) threads usually. An *active* thread is a thread where at least one process is performed, otherwise *idle*. Let $CP_t(\tau)$ be a set of processes performed on a server s_t at time τ.

Processes are fairly allocated to threads on a server s_t in the round-robin (RR) algorithm [2,3]. Here, the electric power $NE_t(n)$ [W] consumed by a server s_t to concurrently perform n (≥ 0) processes is given as follows [17,19]:
[Power consumption for n processes] (Fig. 2)

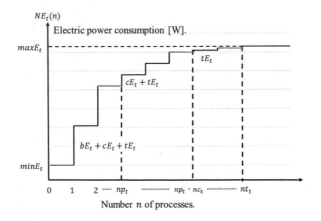

Fig. 2. MLPCM model.

$$NE_t(n) = \begin{cases} minE_t & \text{if } n = 0. \\ minE_t + n \cdot (bE_t + cE_t + tE_t) & \text{if } 1 \leq n \leq np_t. \\ minE_t + np_t \cdot bE_t + n \cdot (cE_t + tE_t) & \text{if } np_t < n \leq nc_t \cdot np_t. \\ minE_t + np_t \cdot (bE_t + nc_t \cdot cE_t) + nt_t \cdot tE_t & \text{if } nc_t \cdot np_t < n < nt_t. \\ maxE_t & \text{if } n \geq nt_t. \end{cases}$$

$$(1)$$

The electric power consumption $E_t(\tau)$ [W] of a server s_t at time τ is assumed to be $NE_t(|CP_t(\tau)|)$ in this paper.

2.3 Computation Model

Processes issued by clients are performed on servers in a cluster. Each process is at a time performed on a thread of a server. It takes T_{ti} time units [tu] to perform a process p_i on a thread of a server s_t. If only a process p_i is performed on a server s_t without any other process, the execution time T_{ti} of the process p_i is minimum, i.e. $T_{ti} = minT_{ti}$. Let $minT_i$ show a minimum one of $minT_{1i}, \ldots, minT_{mi}$, i.e. $minT_i = minT_{fi}$ on the fastest thread which is on a server s_f. A server s_f with the fastest thread is referred to as *fastest* in a cluster S. We assume one virtual computation step [vs] is performed on the thread of the fastest server s_f for one time unit [tu]. This means, the maximum computation rate $maxTCR_f$ of a thread in a fastest server s_f is one [vs/tu]. The maximum thread computation rate $maxTCR_t$ of a server s_t is defined to be $(minT_i/minT_{ti}) \cdot maxTCR_f = minT_i/minT_{ti}$ [vs/tu]. The maximum computation rate $maxSCR_t$ ($\leq nt_t$) of a server s_t is $nt_t \cdot maxCRT_t$ where nt_t is the number of threads of the server s_t. The total number $VC_i = minT_i$ [tu] $\cdot maxTCR_f$ [vs/tu] $= minT_i$ [vs] of virtual computation steps are performed by a process p_i. The maximum computation rate $maxPCR_{ti}$ of a process p_i on a server s_t is $VC_i/minT_{ti} = minT_i/minT_{ti}$ (≤ 1). Hence, for every pair of processes p_i and p_j on a server s_t, $maxPCR_{ti} = maxPCR_{tj} = maxTCR_t$.

The process computation rate $NPR_{ti}(n)$ ($\leq maxSCR_t$) [vs/tu] of a process p_i on a server s_t where n processes are concurrently performed τ is defined as follows:

[MLCM (Multi-level Computation with Multiple CPUs) model] [8,10,18]

$$NPR_{ti}(n) = \begin{cases} nt_t \cdot maxTCR_t/n & \text{if } n > nt_t. \\ maxTCR_t & \text{if } n \leq nt_t. \end{cases}$$

$$(2)$$

In this paper, the server computation rate $NSR_t(n)$ of a server s_t to perform n processes is given as $nt_t \cdot maxTCR_t (= maxSCR_t)$ for $n > nt_t$ and $n \cdot maxTCR_t$ for $n \leq nt_t$. The computation rate $PR_{ti}(\tau)$ of a process p_i at time τ is assumed to be $NPR_{ti}(|CP_t(\tau)|)$.

Figure 3 shows the computation rate $NPR_{ti}(n)$ of a process p_i on a server s_t where n processes are concurrently performed. If a fewer number n of processes than the number nt_t of threads are performed, the computation rate $NPR_{ti}(n)$

is $maxTCR_t$. If more number n of processes than nt_t are performed, more than one process is performed on a thread. Hence, the computation rate $NPR_{ti}(n)$ of each process p_i decreases. If $2 \cdot nt_t$ processes are performed on a server s_t, two processes are performed on each thread. Here, the computation rate $NCR_{ti}(n)$ of each process p_i is $maxTCR_t/2$.

Suppose a process p_i on a server s_t starts at time st and ends at time et. Here, $\sum_{\tau=st}^{et} PR_{ti}(\tau) = \sum_{\tau=st}^{et} NPR(|CP_t(\tau)|) = VC_i[\mathrm{vs}] = \min T_i$. Thus, $\min T_i$ shows the amount of computation of a process p_i.

[Computation model of a process p_i]

1. At time τ a process p_i starts, the computation residue R_{ti} of a process p_i is VC_i, i.e. $R_{ti} = VC_i$ $(=\min T_i)$;
2. At each time τ, the computation residue $R_{ti}(\tau)$ is decremented by the computation rate $PR_{ti}(\tau)$, i.e. $R_{ti} = R_{ti} - NPR_{ti}(|CP_t(\tau)|)$;
3. If R_{ti} (≤ 0), the process p_i terminates at time τ.

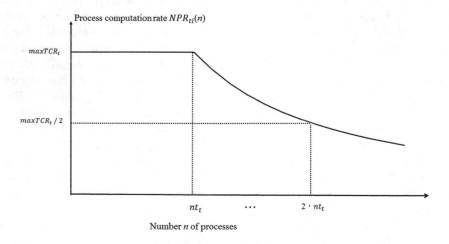

Fig. 3. Computation rate of a process.

3 Detection of Implicitly Faulty Processes

3.1 Estimation Model

A process p_i issued by a client is replicated on multiple servers $s_1, ..., s_n$ in a cluster. Let P be a set of processes $p_1, ..., p_n$ $(n \geq 1)$ issued to the cluster. Let p_{it} show a replica of a process p_i on a server s_t. One replica p_{it} of each process p_i is performed on each server s_t $(t = 1, ..., n)$. Each replica p_{it} starts on a server s_t at time $stime_{it}$. Thus, a set P_t of replicas $p_{1t}, ..., p_{nt}$ of the processes $p_1, ..., p_n$ in the process set P, respectively, are performed on each server s_t $(t = 1, ..., n)$.

At time τ, a variable R_{it} shows the computation residue of a replica p_{it}, i.e. how many virtual computation steps of a replica p_{it} are still to be performed. When a replica p_{it} starts on a server s_t, $R_{it} = minT_i$. At each time τ, the computation residue R_{it} is decremented by the process computation rate $NPR_{it}(|CP_t(\tau)|)$. A replica p_{it} terminates if the computation residue R_{it} gets zero. In this paper, we make the following assumptions;

1. The maximum energy consumption $maxE_t$ and minimum energy consumption $minE_t$ of each server s_t are known.
2. The minimum execution time $minT_{ti}$ of each replica p_{it} on a server s_t is assumed to be *a priori* known. This means, each process p_i is well defined like on-line transaction processing applications. Since the cluster is homogeneous, $minT_{ti} = minT_{ui} = minT_i$ for every pair of servers s_t and s_u and every process p_i.
3. Every replica p_{it} starts at the same time $stime_i$, i.e. $stime_{it} = stime_i$ on every server s_t. We can calculate the execution time ET_{it} of each replica p_{it} as $\tau - stime_{it}$ where τ shows current time.
4. We also can obtain how much electric power $E_t(\tau)$ each server s_t consumes and at how much process rate $NSR_{it}(n)$ of each replica p_{it} performs replicas $p_{it}, ..., p_{nt}$ in the replica set P_t at any time τ.

If a replica p_{it} is properly performed on each server s_t, the following conditions are satisfied;

1. The replica p_{it} returns a proper reply r_{it} to the load balancer L.
2. The execution time ET_{it} of the replica p_{it} is time estimated by the computation model. That is, a replica p_{it} terminates at time $etime_{it}$ obtained by the computation model.
3. The energy consumption EE_t of the server s_t is one estimated by the power consumption model and the computation model.

The execution time ET_{it} of each replica p_{it} and the energy consumption E_t of a server s_t to perform replicas in the replica set P_t are calculated by the estimation algorithm $EST(s_t.\tau)$ (Algorithm 1). The estimation procedure $EST(s, \tau)$ is shown in Algorithm 2. If a replica p_{it} starts at time τ, the replica p_{it} is added to the set P_t. A variable P_t shows a set of active replicas on the server s_t and $n = |P_t|$, i.e. number of active replicas on the server s_t at current time τ. If $n = 0$, E_t is incremented by the minimum power consumption $minE_t$. If $n > 0$, the energy consumption E_t is incremented by the power consumption $NE_t(n)$ [W]. For each replica p_{it}, the computation residue R_{it} is decremented by the process computation rate $NPR_{ti}(n)$. The amount AL_{it} of computation done in the replica p_{it} is incremented by $NPR_{ti}(n)$. If a proper replica p_{it} terminates, AL_{it} must be $minT_{ti}$ and $R_{it} = 0$. If $R_{it} \le 0$, the replica p_{it} terminates and is removed from the set replica P_t.

Algorithm 1. Estimation algorithm

---

1   $\tau = 0$; $EE_t = 0$;
2   **while do**
3   **for** each server s_t in a cluster **do**
4         $EST(s_t, \tau)$;   /* E_t = energy */
5         $EE_t = EE_t + E_t$;
6      $\tau = \tau + 1$;
7   **while end;**

Algorithm 2. Estimation; algorithm $EST(s_t, \tau)$

---

1   **input** : τ = current time;
2   s_t = server;
3   P_t = set of replicas performed on s_t;
4   R_{it} = computation residue of each replica p_{it} in P_t;
5   **output** : E_t = total energy consumption of s_t;
6   ET_{it} = execution time of each replica p_{it};
7   AL_{it} = amount of computation done in p_{it};
8        $E_t = 0$;   /* total energy consumption */
9   **for** each replica p_{it} which starts at time τ **do**
10      $R_{it} = minT_i$; $AL_{it} = 0$;
11      $stime_{it} = \tau$;
12      $P_t = P_t \cup \{p_{it}\}$;
13   **for end;**
14   $n = |P_t|$;   /* number of replicas performed at time τ */
15   **if** $n > 0$ **then**
16      $E_t = E_t + NE_t(n)$;   /* energy */
17   **while** $P_t \neq \phi$ **do**
18   **for** each replica p_{it} in P_t on a server s_t **do**
19            $R_{it} = R_{it} - NPR_{ti}(n)$;
20            $AL_{it} = AL_{it} + NPR_{ti}(N)$;
21   **if** $R_{it} \leq 0$; **then**
22   /* R_{it} terminations */
23               $P_t = P_t - \{p_{it}\}$;
24               $etime_{it} = \tau$;
25               $ET_{it} = etime_{it} - stime_{it} + 1$;   /* execution time */
26   **for end;**
27   **while end;**
28   **else**
29   /* no replica on s_t, i.e. s_t is idle. */
30      $E_t = E_t + minE_t$;

3.2   Detection of Faulty Replicas

By using the estimation procedure $EST(s_t, \tau)$, we can obtain the execution time ET_{it} of each replica p_{it} and the energy consumption E_t of a server s_t. There are totally n replicas $p_{i1}, ..., p_{in}$ of a process p_i in a cluster.

First, the load balancer L collects a reply of a replica p_{it} from every server s_t. The load balancer L decides on a proper reply by collecting a majority reply r_i from the replicas. If there is no majority reply, the load balancer L returns a reply $\perp$ to the cluster. The reply $\perp$ means more than half of the replicas are faulty and no proper reply can be obtained.

Then, suppose a proper reply r is obtained since more than half of the replicas send the proper reply r to the load balancer L. For each server which a replica p_{it} sends a proper reply r_i, it is checked each replica p_{it} is properly performed in terms of the execution time ET_{it} of the replica p_{it} and the energy consumption E_t of the server s_t. If the execution time ET_{it} and the energy consumption E_t are ones which are estimated by the algorithm EST, the replica p_{it} is decided to be proper. Otherwise, the replica p_{it} is implicitly faulty.

Suppose only one process p_i is issued to a cluster of servers $s_1, ..., s_n$. On each server s_t, only one replica p_{it} is performed. That is, $P = \{p_i\}$ and $P_t = \{p_{it}\}$ for each server s_t. If a replica p_{it} is issued, the replica p_{it} starts at time $stime_{it}$ ($=stime_i$) and returns a proper reply r_{it} at time $etime_{it}$.

If a pair of replicas p_{it} and p_{iu} are properly performed on servers s_t and s_u, respectively, a pair of the execution time $ET_{it} = etime_{it} - stime_i + 1$ [sec] and $ET_{iu} = etime_{iu} - stime_i + 1$ [sec] are the same, $ET_{it} = ET_t = ET$. A pair of the servers s_t and s_n consume the energy $E_t = \sum_{\tau=stime_i}^{etime_{it}} E_t(\tau) = ET_{it} \cdot NE_t(1)$ [J] and $E_u = \sum_{\tau=stime_i}^{etime_{iu}} E_u(\tau) = ET_{iu} \cdot NE_u(1)$ [J], respectively, and $E_t = E_u = E$. The amount of computation AL_{it} of a replica p_{it} is $\sum_{\tau=stime_i}^{etime_{it}} NPR_{it}(|CP_t(\tau)|)$, and $AL_u = \sum_{\tau=stime_i}^{etime_{iu}} NPR_{iu}(|CP_u(\tau)|)$. AL_{it} and AL_{iu} are $minT_i$.

If one replica p_{it} sends a proper reply p_{it} but is implicitly faulty, $ET_t \neq ET$ or $E_t \neq E$ and $AL_{it} \neq minT_i$.

The load balancer L behaves as follows;

1. On receipt of a request process p_i from a client, the load balancer L creates a replica p_{it} of the process p_i on each server s_t.
2. The load balancer L receives a reply r_{it} of a replica p_{it} of a process p_i from each server s_t.
3. Let R_i be a set $\{r_{it}, ..., r_{in}\}$ of replies from the replicas $P_{i1}, ..., P_{in}$. Here, if the load balancer L does not receive a reply from a replica p_{it}, r_{it} is $\perp$.
4. If the load balancer L gets a majority reply r from the reply set R_i, i.e. $|\{r_{it} \in R_i | r_{it} = r\}| > |R_i|/2$, the load balancer L checks the execution time ET_{it} and the computation amount AL_{it} of each reply p_{it} which sends the reply $r_{it} = r$ and the energy consumption E_t of the server s_t. A replica p_{in} which sends a reply r_{in} ($\neq r$) is improper.
5. If the amount AL_{it} of computation of a replica p_{it} is $minT_i$ and the server s_t consumes the expected energy, the replica p_{it} is proper. Otherwise, p_{it} is implicitly faulty.

6. Let PR_i be a subset of the replicas which are proper. If $|PR_i| > n/2$, a replica r in the set PR_i is proper and the load balancer L sends the reply r to the client. Otherwise, the load balancer L sends a reply $\bot$ to the client.

Thus, the load balancer L detects two types of faulty replicas. One is a faulty replica which sends an improper reply. This type of faulty replica can be detected by the traditional detection algorithm which take a majority reply. The second type is the implicitly faulty replica which sends a proper reply but does improper computation. In our detection algorithm, we can detect implicit faulty replica by taking usage of execution time of a replica and energy consumption of a server.

4   Concluding Remarks

In order to reliably perform application processes, multiple replicas of each process are performed on servers. Servers are considered to be fault-tolerant according to the advances of hardware are architecture technologies. On the other hand, replicas of processes still suffer from faults due to security attacks like virus and hacking. A faulty replica may send an improper reply or no reply. In addition, even if a replica sends a proper reply, the replica may do different computation from proper replicas. We newly introduced an implicitly faulty replica which sends a proper reply but does improper computation. In this paper, we discussed how to detect implicitly faulty replicas in a homogeneous cluster by taking advantage of the power consumption and computation models.

We are now evaluating the detection algorithm by measuring the execution time to check the energy consumption of each server and the execution time of each process. We are also modifying the detection algorithm to detect faulty replicas in heterogeneous clusters.

References

1. Intel Xeon processor 5600 series: The next generation of intelligent server processors, white paper (2010). http://www.intel.com/content/www/us/en/processors/xeon/xeon-5600-brief.html
2. Job scheduling algorithms in linux virtual server (2010). http://www.linuxvirtualserver.org/docs/scheduling.html
3. Linux operating systems. https://ja.wikipedia.org/wiki/Linux
4. Bernstein, P.A., Goodman, N.: The failure and recovery problem for replicated databases. In: Proceedings of the 2nd ACM Symposium on Principles of Distributed Computing, pp. 114–122 (1998)
5. Defago, X., Schiper, A., Sergent, N.: Semi-passive replication. In: Proceeding of IEEE the 17th Symposium on Reliable Distributed Systems, pp. 43–50 (1998)
6. Denning, D.E.R.: Cryptography and Data Security. Addison Wesley, Boston (1982)
7. Deplanche, A.M., Theaudiere, P.Y., Trinquet, Y.: Implementing a semi-active replication strategy in chorus/classix, a distributed real-time executive. In: Proceeding of IEEE the 18th Symposium on Reliable Distributed Systems, pp. 90–101 (1999)
8. Duolikun, D., Aikebaier, A., Enokido, T., Takizawa, M.: Energy-aware passive replication of processes. Int. J. Mob. Multimed. **9**(1,2), 53–65 (2013)

9. Duolikun, D., Enokido, T., Takizawa, M.: Dynamic migration of virtual machines to reduce energy consumption in a cluster. Int. J. Grid Util. Comput. **9**(4), 357–366 (2018)
10. Duolikun, D., Kataoka, H., Enokido, T., Takizawa, M.: Simple algorithms for selecting an energy-efficient server in a cluster of servers. Int. J. Commun. Netw. Distrib. Syst. **21**(1), 1–25 (2018)
11. Enokido, T., Aikebaier, A., Takizawa, M.: A model for reducing power consumption in peer-to-peer systems. IEEE Syst. J. **4**(2), 221–229 (2010)
12. Enokido, T., Aikebaier, A., Takizawa, M.: Process allocation algorithms for saving power consumption in peer-to-peer systems. IEEE Trans. Ind. Electron. **58**(6), 2097–2105 (2011)
13. Enokido, T., Ailixier, A., Takizawa, M.: An extended simple power consumption model for selecting a server to perform computation type processes in digital ecosystems. IEEE Trans. Ind. Inform. **10**(2), 1627–1636 (2014)
14. Enokido, T., Takizawa, M.: An integrated power consumption model for distributed systems. IEEE Trans. Ind. Electron. **60**(2), 824–836 (2013)
15. Fischer, M.J., Lynch, N.A., Paterson, M.S.: Impossibility of distributed consensus with one faulty process. In: Proceedings of the Second ACM SIGACT-SIGMOD Symposium on Principles of Database Systems, 21–23 March 1983, Colony Square Hotel, Atlanta, Georgia, USA, pp. 1–7 (1983)
16. Hayashibara, N., Takizawa, M.: Design of the notification system for failure detectors. Int. J. High Perform. Comput. Netw. **6**(1), 25–34 (2009)
17. Kataoka, H., Duolikun, D., Enokido, T., Takizawa, M.: Energy-efficient virtualisation of threads in a server cluster. In: Proceedings of the 10th International Conference on Broadband and Wireless Computing, Communication and Applications (BWCCA-2015), pp. 288–295 (2015)
18. Kataoka, H., Nakamura, S., Duolikun, D., Enokido, T., Takizawa, M.: Multi-level power consumption model and energy-aware server selection algorithm. Int. J. Grid Util. Comput. (IJGUC) **8**(3), 201–210 (2017)
19. Kataoka, H., Sawada, A., Duolikun, D., Enokido, T., Takizawa, M.: Energy-aware server selection algorithm in a scalable cluster. In: Proceedings of IEEE the 30th International Conference on Advanced Information Networking and Applications (AINA-2016), pp. 565–572 (2016)
20. Lamport, L., Shostak, R., Pease, M.: The byzantine generals problems. ACM Trans. Program. Lang. Syst. **4**(3), 382–401 (1992)

Deep Recurrent Neural Networks for Wi-Fi Based Indoor Trajectory Sensing

Hao Li[✉], Joseph K. Ng, and Junxing Ke

Department of Computer Science, Hong Kong Baptist University, Hong Kong, China
{cshaoli,jng}@comp.hkbu.edu.hk, csjxke@gmail.com

Abstract. Wi-Fi based sensing becomes more and more popular in ubiquitous computing with prevalence of Wi-Fi devices. In this paper, we propose a new task, indoor trajectory sensing, based on Received Signal Strength Indicator ($RSSI$) of Wi-Fi. Traditional distance measure based methods, like Dynamic Time Warping (DTW) based Nearest-Neighbor (1NN) method, have poor performance in indoor trajectory sensing due to that $RSSI$ of Wi-Fi in indoor environment is fluctuating, partially missing, time-varying and device-dependent. Recently, Recurrent Neural Networks (RNN) and its variants have strong abilities in learning the temporal dependency of sequence data since it can extract more meaningful features. Consequently, it is necessary to design an RNN model for the indoor trajectory sensing problem with relatively small size data. We adopt a passive way to collect Wi-Fi signals from the smart phone to ensure more data collected and generate multiple time series for each trajectory. For the recurrent neural network training, RNN and its variants are applied into our sequence data to find more meaningful patterns especially for different environment and devices. Series of real-world experiments have been conducted in our test bed and the results show that the deep based approach can achieve better performance than traditional methods with challenging environment and device factors.

Keywords: Ubiquitous computing · Wi-Fi sensing ·
Indoor trajectory · Recurrent neural networks

1 Introduction

With the increasing popularity of Wi-Fi technologies, not only wireless network services can be provided, but also many other Wi-Fi based applications have been proposed and applied, such as indoor localization, gesture detection and activity recognition. The basic principle behind these applications is to make full use of the wireless signal feature and recognize the pattern of different predefined labels, like the location, gesture and activity. Compared with other techniques, including cameras and motion sensors, Wi-Fi based techniques have advantages of privacy protection, low cost and acceptable accuracy. Thus more and more novel Wi-Fi based applications are promoted and in developing.

© Springer Nature Switzerland AG 2020
L. Barolli et al. (Eds.): NBiS-2019, AISC 1036, pp. 434–444, 2020.
https://doi.org/10.1007/978-3-030-29029-0_41

In this paper, we propose a new application based on Wi-Fi signal, indoor trajectory sensing. Most indoor areas are composed of rooms and corridors rather than an open space, people will leave different indoor trajectories following the floor plan when they have their indoor activities. At the same time, their smart phones will also leave or capture a corresponding $RSSI$ sequence data with the Wi-Fi function turned on. Figure 1 illustrates this with four Wi-Fi sensors and one moving smart phone. When the person in the figure goes through the blue trajectory, the mobile phone also interacts with surrounding Wi-Fi sensors and have an $RSSI$ sequence $R = \{R_1, R_2, ... R_n\}$. For the specific timestamp t, $R_t = [RSSI_{t_1}, RSSI_{t_2}, RSSI_{t_3}, RSSI_{t_4}]$, where $RSSI_{t_i}$ is the received signal strength indicator between i-th Wi-Fi sensor and the smart phone. Normally, the value of $RSSI$ is dependent on the indoor location of the smart phone. Hence each $RSSI$ sequence will have R_t in different values with different timestamp. Our motivation is to classify different $RSSI$ sequence into different indoor trajectories based on the pre-collected training data with known trajectory labels.

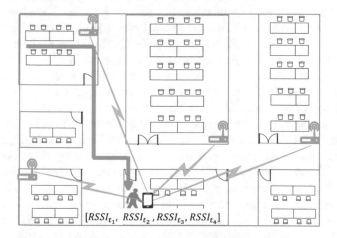

Fig. 1. Indoor trajectory sensing illustration (Blue line: trajectory; Cyan device: Wi-Fi sensor; Dark blue device: smart phone)

For this kind of time series classification problem, distance measure based classification algorithms are regarded as typical methods, including DTW-based 1NN [1], Euclidean-based 1NN [2] and Frechet-based 1NN [3]. However, for Wi-Fi based indoor trajectory task, the distance measure based methods cannot obtain good performance due to the following reasons:

– Different indoor trajectories may have overlapping, or they are very close to each other in some parts, which make the corresponding $RSSI$ sequences very similar.
– Under the effect of multipath fading in indoor environment, $RSSI$ reading fluctuates seriously [4] and could be in huge difference even in the same time stamp.

- Training data for trajectory sensing cannot cover all circumstances, mainly including the environment and the device. If the testing data is in different environment and devices, distance measure based methods could have a bad performance.
  - For different environment, the temperature and humidity are also different, which will cause the $RSSI$ reading to be time-varying [5].
  - Various smart phones have different types of Wi-Fi chips, which have their unique transmission and receiving characteristics. That is to say, $RSSI$ readings are device-dependent [6].
- There are always irregular missing data both in training data and testing data due to the Wi-Fi coverage and hardware problem of Wi-Fi sensors, which will the distance measure inaccurate.

Deep learning approach can automatically learn more meaningful features from data with multiple levels of representation [7], which has shown superior performance in many Wi-Fi sensing tasks. Xuyu Wang et al. propose to apply deep neural network into estimating the indoor location based on the Wi-Fi Channel State Information(CSI) [8]. Michal Nowicki et al. use the Wi-Fi scanning information (MAC + $RSSI$) to classify the building and floor with deep learning approach. Jeong-Sik Choi et al. propose an RNN model to identify ine-of-sight (LOS) and non-LOS channel conditions by using CSI series data in indoor environment. In [9], hand gestures can be recognized by $RSSI$ data with RNN and Long Short-Term Memory (LSTM).

For the indoor trajectory sensing task, Wei Hu et al. uses distance measure based method, DTW to classify short trajectories for localization purpose [10]. In [11], an efficient DTW based algorithm is proposed for short indoor trajectory sensing. In this paper, we identify that distance measure based methods may not be suitable for indoor trajectory sensing problem due to the previous mentioned practical challenges. We use deep learning approach, specifically recurrent neural network and its variants LSTM and Gated Recurrent Units (GRU) to extract more meaningful features from $RSSI$ time series data to overcome environment, device and missing data challenges.

To implement the deep approach, we adopt a passive way to collect Wi-Fi packets emitted from mobile phones. Compared with active scanning by the mobile phone, the passive way can obtain more $RSSI$ vectors per second. We further generate multiple $RSSI$ sequence for each trajectory by with randomly selecting one $RSSI$ vector from each second repeatedly, which ensures that our deep RNN approaches have enough data fed in for training. For the testing data, we design different sets of trajectories with short, medium and long size. Data is also collected in different environment and devices to verify whether RNN based approach can recognize the true labels with these challenges. Experimental results show that RNN based approach can achieve good performance in different size of trajectories, different environments and different devices when compared with distance measure based methods.

The rest of this paper is organized as follows: The methodology is given in Sect. 2. Section 3 compares the results from RNN based approaches with some other well-known methods in the series of experiments conducted in the test bed. And finally, in Sect. 4, we present our conclusions.

2   Methodology

In this section, we will describe the methodology in three parts, sequence data generation, recurrent neural networks and model architecture.

2.1   Sequence Data Generation

For each trajectory label. We want more data for training, while it is difficult to walk through indoor trajectories for multiple times. Hence we adopt a simple but reasonable way to generate many training data for each trajectory. A time series Wi-Fi $RSSI$ data can be divided into different segments according to times tamp by second. Each segment has many $RSSI$ data since normally Wi-Fi sensors can collect many packets from the mobile device in one second. We randomly select one $RSSI$ vector from each segment and combine all these selected vectors into an $RSSI$ time series. We can do it multiple times so that many time series can be generated for each trajectory. Figure 2 shows this process, where V is the randomly selected $RSSI$ vector in each second.

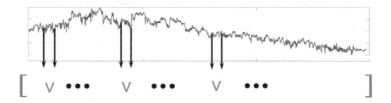

Fig. 2. Sequence data generation

2.2   Recurrent Neural Networks

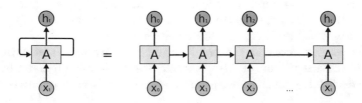

Fig. 3. Unrolled recurrent neural networks

Recurrent Neural Networks. Recurrent neural networks are capable to model sequential data for sequence recognition and prediction, due to its unique internal state (memory) to remember sequential inputs and find sequential patterns when compared artificial neural network. RNN uses a looping structure as shown in the left side of Fig. 3, where A is chunk of neural networks, X_t is the input value, and h_t is a output value forwarded to the next, which means RNN not only take their current input, but also utilize what they have perceived previously in time. We show the unrolled RNN in the right side of Fig. 3. The input is a sequence data $\{X_0, X_1, ..., X_t\}$, where $X_t \in \mathbb{R}^n$, n is the number of neurons in the input layer. $\{h_0, h_1, ..., h_t\}$ is the corresponding hidden layer, where where $h_t \in \mathbb{R}^m$, m is the number of neurons in the hidden layer. The hidden layer defines the state space of the whole system, and can be called memory:

$$h_t = f(W \cdot [h_{t-1}, x_t] + b) \tag{1}$$

where W and b is the weight and bias parameters in the hidden layer. f is the activation function.

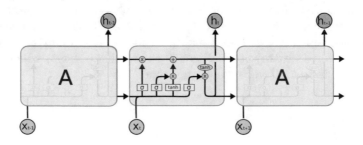

Fig. 4. Long short-term memory

Long Short-Term Memory. LSTM is a special kind of recurrent neural networks, which was firstly introduced by Hochreiter & Schmidhuber [12]. Many researchers improve and apply it into different applications. LSTM has superior performance than RNN on many tasks due to its strong power on learning the long-term dependency.

Figure 4 presents standard module of LSTM, which is composed of a cell state, a forget gate, an input gate and an output gate. The cell state memories the information from previous time line C_{t-1} and pass current information C_t to the future, which are implemented by three gates.

The forget gate looks at the previous hidden state h_{t-1} and current input x_t in time series. It will output f_t whose element is between 0 and 1 by sigmoid layer σ to decide how much information to keep for C_{t-1} with pointwise product.

$$f_t = \sigma(W_f \cdot [h_{t-1}, x_t] + b_f) \tag{2}$$

W_f and b_f is the weight and bias parameters in the forget layer. The input gate decides what new information to be stored into the cell state. The i_t will determine which values to be updated and $\overline{C}_t$ will give the candidates of new values. Then i_t and $\overline{C}_t$ can be combined by pointwise product to create an update to the state.

$$i_t = \sigma(W_i \cdot [h_{t-1}, x_t] + b_i)$$
$$\widetilde{C}_t = tanh(W_C \cdot [h_{t-1}, x_t] + b_C) \tag{3}$$

W_i, b_i, W_C, b_C are corresponding neuron parameters in the input layer. The cell state can be updated by adding the old values and the input values as shown below:

$$C_t = f_t * C_{t-1} + i_t * \widetilde{C}_t \tag{4}$$

The output gate will provide the current hidden state value for next module based on the current cell state value C_t. Firstly a sigmoid layer is used to give o_t, which determine what values will be the output.

$$o_t = \sigma(W_o \cdot [h_{t-1}, x_t] + b_o) \tag{5}$$

Then the cell state C_t will go through a *tanh* layer and multiply with o_t to provide the current hidden state.

$$h_t = o_t * tanh(C_t) \tag{6}$$

Compared to traditional RNN network, LSTM can learn the temporal dependency more effectively with the interior structure design, which put effort in the long periods information intentionally.

Gated Recurrent Units. Gated recurrent units (GRU) [13] are a gating mechanism in recurrent neural networks introduced by Kyunghyun Cho et al. GRU is very similar to LSTM but lacking the output gate. The efficiency of GRU is higher than LSTM computationally as that the interior structure is less complex. For the performance, GRUs perform better than LSTMs on certain smaller datasets [14], while LSTM can work well in more challenging tasks [15].

2.3   Model Architecture

Figure 5 presents the architecture of our network. The input layer is our generated *RSSI* sequence data, which are fed into multiple recurrent layers. For each recurrent layer, a dropout layer is added to prevent neural network from overfitting. The output hidden features are fed into a fully connected layer and then a softmax layer for trajectory prediction.

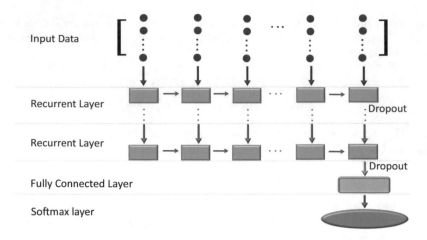

Fig. 5. Architecture of the deep neural network

3 Experiments Evaluation

In this section, we will show the experiments setup and experimental results of deep approach for indoor trajectory sensing.

3.1 Experiments Setup

The layout of an indoor public area is shown in Fig. 6(a), with the size of 20 m × 22 m. The place is a computer laboratory, which has many computers, monitors, chairs, and desks. In this testbed, multi-path interference is encountered nearly everywhere which makes signal fluctuated normally. Our sensors, 11 Wi-Fi sensors are deployed in different locations for collecting wireless packets from mobile devices shown in Fig. 6(a).

Three trajectory sets with different length levels are shown in Fig. 6(b), (c), (d). For each type of trajectories, there are 6 trajectories and some parts overlaps between each other. In the training data collection, we use Sony smart phone to emit packets and Wi-Fi sensors sniff these signals. In the testing data collection, three smart phones, including the same Sony, Samsung and Xiaomi, are adopted as device factors. To verify the effect of environment factors, the time gap between training and testing data collection is two weeks.

For the training data, 30 time series are generated for each trajectory. Based on the generated time series data and corresponding trajectory labels, we train RNN, LSTM and GRU models for different trajectory data sets. In the testing phase, 10 time series are generated for each measurement. Hence 60 testing instances are formed for specific trajectory size and smart phone. Then corresponding trained model can be utilized to predict the trajectory label of testing data.

3.2   Experiments Results

Table 1. The sensing accuracy for different sizes of trajectories

	Phones	1NN (Euclidean)	1NN (DTW)	1NN (Frechet)	RNN	LSTM	GRU
Long trajectories	Sony	**100%**	**100%**	73.33%	**100%**	**100%**	98.33%
	Samsung	50%	31.67%	21.67%	80%	80%	**83.33%**
	Xiaomi	33.33%	50%	60%	76.67%	**81.67%**	80%
Medium trajectories	Sony	31.67%	53.33%	23.3%	78.33%	**85%**	76.67%
	Samsung	16.67%	33.33%	38.33%	76.67%	**83.33%**	81.67%
	Xiaomi	16.67%	33.33%	50%	81.67%	**83.33%**	80%
Short trajectories	Sony	26.67%	60%	48.33%	73.33%	**88.33%**	73.33%
	Samsung	36.67%	16.67%	18.33%	56.67%	73.33%	**75%**
	Xiaomi	31.67%	33.33%	20%	76.67%	88.33%	**91.67%**

Table 1 presents the sensing accuracy of distance measure based methods and RNN based approaches. Generally speaking, RNN based approaches have stable performance (>70%) in different trajectory size, different environment and different devices. While distance measure based methods have low sensing accuracy, especially in medium and short trajectories. In the long trajectory data set, the unique features of different trajectories are obvious and hence all approaches can classify them correctly even in the different environments. When the devices are changed to other brands which have different Wi-Fi chips, the accuracy of distance measure based methods drops down a lot, but RNN based approaches can still obtain an acceptable performance. For the medium and short trajectories, since the ratio of overlapping part becomes bigger and they are also very close to each other, distance measure based methods is more easily effected by the complex environment and device factors. Figure 7 shows the confusion matrix for Xiaomi's trajectory sensing with distance measure based methods and RNN based approaches in short trajectory data set, we can see RNN based approaches can recognize the pattern of time series well despite of time-varying environment and transmission characteristics difference.

Apart from sensing accuracy, running time is another significant issue for real-time applications. RNN based approaches have a much higher speed than distance measure based methods since traditional methods need to compare the online time series with every sequence in training data set when using the nearest-neighbor algorithm. The training time for RNN based approaches is also acceptable. Figure 8 shows training accuracy of RNN approach with different sizes of trajectories. The RNN model can converge quickly and normally 100 epoch is enough.

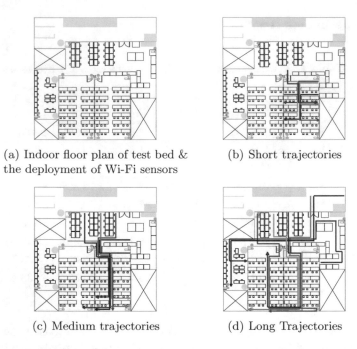

(a) Indoor floor plan of test bed & the deployment of Wi-Fi sensors

(b) Short trajectories

(c) Medium trajectories

(d) Long Trajectories

Fig. 6. Floor plan of test bed & different sizes of trajectories

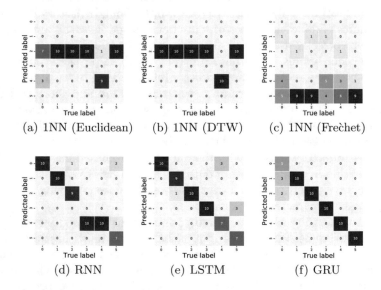

(a) 1NN (Euclidean)   (b) 1NN (DTW)   (c) 1NN (Frechet)

(d) RNN   (e) LSTM   (f) GRU

Fig. 7. Confusion matrix for Xiaomi's trajectory sensing with distance measure based methods and RNN based approaches in short trajectory data set.

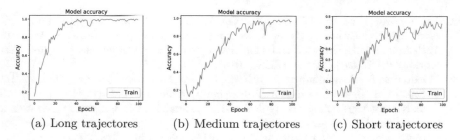

(a) Long trajectores   (b) Medium trajectores   (c) Short trajectores

Fig. 8. Train accuracy along training epochs of RNN approach with different sizes of trajectories for Xiaomi

4   Conclusions

In this paper, we focus on indoor trajectory sensing task. Practical challenges have been identified and traditional distance measure based methods have a bad performance in this task. We use deep recurrent neural networks to find more meaningful features to overcome environment, device and missing data problems. The experimental results shows that RNN based approaches can achieve stable performance in different data sets.

References

1. Jeong, Y.-S., Jeong, M.K., Omitaomu, O.A.: Weighted dynamic time warping for time series classification. Pattern Recognit. **44**(9), 2231–2240 (2011)
2. Wang, X., Mueen, A., Ding, H., Trajcevski, G., Scheuermann, P., Keogh, E.: Experimental comparison of representation methods and distance measures for time series data. Data Min. Knowl. Disc. **26**(2), 275–309 (2013)
3. Chouakria-Douzal, A., Nagabhushan, P.N.: Improved fréchet distance for time series. In: Data Science and Classification, pp. 13–20. Springer, Berlin (2006)
4. Chuan-Chin, P., Chung, W.-Y.: Mitigation of multipath fading effects to improve indoor RSSI performance. IEEE Sens. J. **8**(11), 1884–1886 (2008)
5. Zheng, V.W., Xiang, E.W., Yang, Q., Shen, D.: Transferring localization models over time. In: AAAI, pp. 1421–1426 (2008)
6. Li, H., Ng, J.K., Cheng, V.C.W., Cheung, W.K.: Fast indoor localization for exhibition venues with calibrating heterogeneous mobile devices. Internet of Things **3**, 175–186 (2018)
7. LeCun, Y., Bengio, Y., Hinton, G.: Deep learning. Nature **521**(7553), 436 (2015)
8. Wang, X., Gao, L., Mao, S., Pandey, S.: Deepfi: deep learning for indoor fingerprinting using channel state information. In: 2015 IEEE Wireless Communications and Networking Conference (WCNC), pp. 1666–1671. IEEE (2015)
9. Haseeb, M.A.A., Parasuraman, R.: Wisture: RNN-based learning of wireless signals for gesture recognition in unmodified smartphones. arXiv preprint arXiv:1707.08569 (2017)
10. Hu, W., Wang, Y., Song, L.: Sequence-type fingerprinting for indoor localization. In: International Conference on Indoor Positioning and Indoor Navigation (IPIN), Banff, Alberta, Canada (2015)

11. Ye, X., Wang, Y., Hu, W., Song, L., Gu, Z., Li, D.: Warpmap: accurate and efficient indoor location by dynamic warping in sequence-type radio-map. In: 2016 13th Annual IEEE International Conference on Sensing, Communication, and Networking (SECON), pp. 1–9. IEEE (2016)
12. Hochreiter, S., Schmidhuber, J.: LSTM can solve hard long time lag problems. In: Advances in Neural Information Processing Systems, pp. 473–479 (1997)
13. Cho, K., van Merriënboer, B., Gulcehre, C., Bahdanau, D., Bougares, F., Schwenk, H., Bengio, Y.: Learning phrase representations using rnn encoder-decoder for statistical machine translation. arXiv preprint arXiv:1406.1078 (2014)
14. Chung, J., Gulcehre, C., Cho, K., Bengio, Y.: Empirical evaluation of gated recurrent neural networks on sequence modeling. arXiv preprint arXiv:1412.3555 (2014)
15. Weiss, G., Goldberg, Y., Yahav, E.: On the practical computational power of finite precision rnns for language recognition. arXiv preprint arXiv:1805.04908 (2018)

The 10th International Workshop on Heterogeneous Networking Environments and Technologies (HETNET-2019)

A Model for Mobile Fog Computing
in the IoT

Kosuke Gima[1(✉)], Ryuji Oma[1], Shigenari Nakamura[1], Tomoya Enokido[2],
and Makoto Takizawa[1]

[1] Hosei University, Tokyo, Japan
{kosuke.gima.3r,ryuji.oma.6r}@stu.hosei.ac.jp,
nakamura.shigenari@gmail.com, makoto.takizawa@computer.org
[2] Rissho University, Tokyo, Japan
eno@ris.ac.jp

Abstract. In the IoT (Internet of Things), millions of devices with sensors and actuators are interconnected with clouds of servers in networks. In order to reduce the traffic of networks and servers in the IoT, types of the fog computing (FC) models are proposed, which are composed of fog nodes which process data from devices and forward the processed data to servers. In most FC models, fog nodes are located in fixed locations and do not move in networks. In this paper, we newly propose a mobile FC (MFC) model where fog nodes communicate with other nodes in wireless networks while moving like vehicles. Here, each fog node processes input data received from other nodes and devices and sends output data obtained by processing the input data to other fog nodes in the opportunistic way. In the opportunistic protocols, only data is exchanged among nodes. We propose an MFC protocol where not only data but also subprocesses are exchanged among nodes so that the electric energy to be consumed by the nodes can be reduced.

Keywords: IoT · Fog computing model ·
Mobile fog computing (MFC) model ·
Subprocess and Data transmission

1 Introduction

The IoT (Internet of Things) includes millions of sensor and actuator devices which are interconnected with clouds of servers in networks [6]. Due to the scalability of the IoT, networks are congested to forward sensor data to servers and servers are heavily loaded to process the sensor data. Fog computing (FC) models [7,13] are proposed to efficiently realize the IoT by reducing the traffic of networks and servers. The FC model is composed of fog nodes in addition to servers and devices. Sensor data is processed by a subprocess of a fog node and sent to other fog nodes to further process the data. In the FC models, each

© Springer Nature Switzerland AG 2020
L. Barolli et al. (Eds.): NBiS-2019, AISC 1036, pp. 447–458, 2020.
https://doi.org/10.1007/978-3-030-29029-0_42

fog node is fixed in a location, i.e. does not move. The TBFC (Tree-Based Fog Computing) model of the IoT is proposed, where fog nodes are structured to reduce the energy consumption of the fog nodes [8,12]. In order to make the TBFC model tolerant of faults, the FTBFC (Fault-tolerant TBFC) model is proposed [11].

In the IoT, mobile nodes like vehicles and smart devices are interconnected with wireless networks like V2V (vehicle-to-vehicle) networks in addition to fixed nodes like servers. Mobile nodes are communicating with other nodes in wireless ad-hoc networks. Since a node is moving, there may not be another node in the wireless communication range of the node ever if the node would like to send data. Thus, mobile nodes have to take advantage of opportunistic communication [2,14]. A node waits for opportunity that another node comes in the communication range. Once another node comes in the communication range of a source node, the source node forwards messages to the node to deliver to servers.

In this paper, we newly propose a mobile fog computing (MFC) model. Here, each fog node moves in networks and communicates with other fog nodes in wireless ad-hoc networks. Compared with routers, data carried by messages are processed and messages with processed data are sent to fog nodes by fog nodes. Thus, fog nodes are equipped with subprocesses to process input data. A fog node f_i has to send processed data to another fog node f_j which can process the data. Here, the node f_j is a target node of the fog node f_i. Suppose a fog node f_i finds another fog node f_j in the communication range. If the fog node f_j supports a subprocess to process data in the node f_i, i.e. f_j is a target node of f_i, the node f_i sends the processed data to the node f_j. Even if the node f_j is not a target node of the node f_i, the node f_i can ask the node f_j to deliver the data to a target node. In addition, if the target node f_j is overloaded or does not support enough storage to store data, the node f_j can send the subprocess to the node f_i and the data in the node f_i is processed by the subprocess. Thus, nodes exchange not only data but also subprocesses with other nodes while only data is exchanged in traditional opportunistic protocols. We propose message transmission (MT), subprocess transmission (ST), and message exchange (ME) ways to exchange data and subprocesses among nodes and to process data, so that the total energy consumption of the nodes can be reduced.

In Sect. 2, we present a system model. In Sect. 3, we present a model of a fog node. In Sect. 4, we discuss the behavior of mobile fog nodes.

2   System Model

A fog computing (FC) model is composed of clouds of servers, fog nodes, and devices with sensors and actuators [13]. In this paper, we consider mobile fog nodes in addition to fixed fog nodes which are located at a fixed location. Mobile fog nodes communicate with one another in wireless ad-hoc networks. Some fog node supports not only a subprocess but also sensor and actuator devices. A fog node receives data from sensors and issues actions to actuators. On receipt of input data from sensors or fog nodes, a fog node processes the input data and forwards the output data obtained by processing the input data to other fog nodes. Servers in clouds finally receive data processed by fog nodes. Then, the servers make a decision on actions to be done by actuators and deliver the actions to actuators via fog nodes. The subprocess F_i does the computation on input data from other nodes and generates output data to be sent to other nodes.

In the TBFC model [9,12], a fog node f_i is equipped with a subprocess F_i and memory storage M_i. Here, the node f_i is a host node of the subprocess F_i. Let $p(f_i)$ show a subprocess F_i supported by a node f_i. A message m is first created by a sensor to carry sensor data. The message m is received and processed by a fog node f_i with a subprocess F_i. The output data is obtained by the subprocess F_i through processing the input data. The output data has to be sent to a fog node which supports a subprocess F_j to process the output data. A subprocess F_i is referred to as *precede* a subprocess F_j (or F_i *follows* F_j) ($F_i \rightarrow F_j$) if the output message of the subprocess F_i is processed by the subprocess F_j. A message with processed data is sent to a fog node f_j with a subprocess F_j such that $F_i \rightarrow F_j$. Thus, messages are processed by a sequence of subprocesses. A node f_i is referred to as *precedes* a node f_j ($f_i \rightarrow f_j$) if $F_i \rightarrow F_j$ [Fig. 1]. A node f_j is a *target* node of a node f_i if $f_i \rightarrow f_j$. On receipt of a message m, a fog node f_i stores the message m in the message M_i. Let $maxM_i$ be the maximum size of the memory storage M_i of a fog node f_i, i.e. maximum number of messages which the node f_i can receive. Messages received and messages with output data are stoned in the memory storage M_i. If the memory storage M_i is fully engaged, the node f_i cannot receive any message from other nodes.

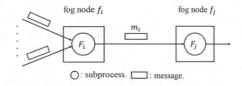

Fig. 1. f_i *precedes* f_j ($f_i \rightarrow f_j$ and $F_i \rightarrow F_j$).

A fog node f_i receives a message m_k from a fog node f_k such that $f_k \rightarrow f_i$. Let RM_i be a collection of messages stored in the memory M_i, which a node f_i receives from nodes preceding the node f_i. Then, the node f_i processes data in the messages RM_i and obtains the output data by a subprocess F_i. The fog node f_j creates a message m_i with the processed output data in the memory M_i. Then, the fog node f_i has to deliver the message m_i to a target fog node f_j with a subprocess F_j following the subprocess F_i of the fog node f_i ($F_i \rightarrow F_j$).

M_{ij} shows a set of messages in a fog node f_i, which are required to be sent to a fog node f_j with a subprocess F_j such that $F_i \rightarrow F_j$. A fog node f_i has to deliver the messages M_{ij} to a target node which supports a subprocess F_j.

3   Fog Nodes

Each fog node f_i moves and communicates with other fog nodes in a wireless network. A fog node f_i sends output messages M_{ij} in the memory M_i to a subprocess F_j following the subprocess F_i ($F_i \rightarrow F_j$). In the first way, if a target fog node f_j is in the communication range of a node f_i, the node f_i sends output messages M_{ij} to the node f_j. This is a *message transmission* (MT) way from a fog node f_i to a fog node f_j, which is used in conventional networks.

Another way is a *subprocess transmission* (ST) one. Here, a subprocess F_j following the subprocess F_i ($F_i \rightarrow F_j$) is sent to a fog node f_i. Then, the output messages M_{ij} of the subprocess F_i are furthermore processed by the subprocess F_j in the fog node f_i. The node f_i tries to send the output messages M_i of the subprocess F_j to a fog node f_k which supports a subprocess F_k following the subprocess F_j ($F_j \rightarrow F_k$).

In order to deliver messages to servers, a fog node f_i sends output messages M_{ij} to a fog node f_k even if $f_i \nrightarrow f_k$ in addition to holding the output messages M_{ij} in the memory M_i. Then, the fog node f_k tries to deliver the messages M_{ij} to a target fog node f_j supporting the subprocess F_j. This is a *message exchange* (ME) way.

A fog node f_i is assumed to follow the SPC (Simple Power Consumption) model [3–5]. If a subprocess F_i is performed on a node f_i, the node f_i consumes the maximum electric power $maxE_i$[W]. Otherwise, f_i consumes the minimum power $minE_i$[W]. The execution time $ET_i(x)$[sec] of a subprocess F_i depends on the number x of input messages M_{ij}. In this paper, we consider a pair of types of subprocesses with respect to the computational complexity $O(x)$ and $O(x^2)$ for the number x of input messages. $ET_i(x) = ct_i \cdot C_i(x)$ where ct_i is a constant and $C_i(x) = x$ or $C_i(x) = x^2$. The electric energy $EE_i(x)$[J] to be consumed by a fog node f_i to process the number x of input messages is given as $maxE_i \cdot ET_i(x) = ct_i \cdot C_i(x) \cdot maxE_i$.

A node f_i also consumes electric energy $se_i \cdot maxE_i$ and $re_j \cdot maxE_i$ [W] to send and receive messages, respectively. Here, $se_i \leq 1$ and $re_i \leq 1$ are constants of the node f_i [10]. Let $TT_i(x)$ and $RT_i(x)$ be time [sec] for a node f_i to send and receive the number x of messages, respectively. Here, the node f_i consumes the energy $se_i \cdot maxE_i \cdot TT_i(x)$ and $re_i \cdot maxE_i \cdot RT_i(x)$[J] to send and receive the number x of messages, respectively. The transmission and receiving time $TT_i(x)$ time $RT_i(x)$ are proportional to the number x of messages, i.e. $TT_i(x) = st_i \cdot x$ and $RT_i(x) = rt_i \cdot x$ where st_i and rt_i are constants [bit/sec].

A fog node f_i obtains output messages M_{ij} by processing input messages RM_i. The ratio $|M_{ij}|/|RM_i|$ is the output ratio ρ_i of the node f_i. If $\rho_i < 1$, a fewer number of output messages are generated than input messages.

As discussed, each fog node f_i is characterized by the energy consumption $maxE_i \cdot ET_i(x)$ and the output ratio ρ_i to process x input messages [Fig. 2]. The number of output messages is $\rho_i \cdot x$. The maximum power $maxE_i$ of a fog node f_i is 3.7[W] is realized in a Raspberry Pi3 node f_i, $maxE_i$ is 89.5[W] for a PC f_i with an Intel Core i7-6700K CPU. $f_i, re_i/se_i = 5$ according to our experiment [1].

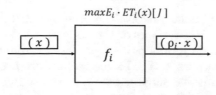

$$maxE_i \cdot ET_i(x)[J]$$

x = number of input messages.

Fig. 2. Model of a fog node.

4   Message and Subprocess Transmissions

4.1   Transmission Ways

Messages M_{ij} are not removed in the memory M_i of a node p_i even if the node f_i makes a success at delivering the messages M_{ij} to another node f_j since the node f_j might be unable to communicate with other nodes. In order to remove messages in the memory M_i, each message m carries the TTL (time to live) field $m.TTL$. Each time a message m is transmitted, $m.TTL$ is decremented by a constant c. If $m.TTL \leq 0$, the message m is removed in the memory M_i.

We propose algorithms to exchange messages and subprocesses among fog nodes. First, we present the message transmission (MT) way.

[Message transmission (MT)]. [Fig. 3] A source node f_i holding the output messages M_{ij} behaves as follows:

1. If a fog node f_i finds a target fog node f_j such that $f_i \rightarrow f_j$ in the communication range of the node f_i, the fog node f_i does the following procedure.
2. If $maxM_j$ - $(|M_{ij}| + |M_j|) > 0$, i.e. the node f_j supports enough memory M_j to receive the messages M_{ij}, the node f_i does the following procedure.
3. For each message m in the output message set M_{ij}, $m.TTL = m.TTL -$ a. The node f_i sends the output messages M_{ij} to the target node f_j. If $m.TTL \leq 0$, the message m is removed in the storage M_i of the node f_i.
4. In the fog node f_j, the messages in the set M_{ij} are processed by the subprocess F_j. Then, the output messages M_{jk} are stored in the memory M_j and the messages M_{ij} are removed in the node f_j.

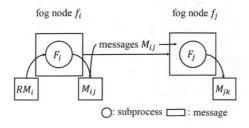

Fig. 3. Message transmission (MT).

If the target node f_j cannot receive the messages M_{ij} from the source node f_i because the memory M_j overflows or the node f_j is too overloaded to process the messages, the fog node f_i obtains a subprocess F_j from the host node f_j and processes the output messages M_{ij}. In the subprocess transmission (ST) way, the target node f_j sends the subprocess F_j to the source node f_i. Then, the messages M_{ij} are processed by the subprocess F_j in the source node f_i on behalf of the node f_j.

[Subprocess transmission (ST)]. [Fig. 4] The source node f_i behaves as follows;

1. If a fog node f_i finds a target fog node f_j following f_i, i.e. $f_i \rightarrow f_j$ in the communication range of the node f_i, the node f_i does the following procedure.
2. If $maxM_i - (|M_i| + |F_j|) > 0$, i.e. the source node f_i can receive the subprocess F_j from the node f_j and the node f_i is not too overloaded to perform the subprocess F_j, the node f_i does the following procedure.
3. A target fog node f_j sends the subprocess F_j to the source node f_i.
4. In the source fog node f_i, the messages M_{ij} are processed by the subprocess F_j. Then, the output messages M_{jk} are stored in the memory M_i. M_{jk} is a set of messages to be delivered to a fog node f_k supporting a subprocess F_k where $F_j \rightarrow F_k$. The subprocess F_j is removed in the node f_i.
5. If the target node f_j can receive the output messages M_{jk}, i.e. $maxM_j - (|M_{jk}| + |M_j|) > 0$, the source node f_i sends the messages M_{jk} to the node f_j and removes the messages M_{jk} in the memory M_i.
6. For each output message m in M_{ij}, $m.TTL = m.TTL - b$. If $m.TTL < 0$, the message m is removed in the memory M_i.

Here, the constant b is larger than the constant a in the message transmission ways, i.e. $b > a$.

Suppose a fog node f_i cannot send output messages M_{ij} to the node f_j due to the lack of memory of the node f_j. We also suppose the node f_i cannot receive the subprocess F_j from the node f_j due to the lack of memory and overload. Here, the nodes f_i and f_j exchange messages so that the usage ratios of the memories of f_i and f_j are the same. The memory usage ratio u_k of a node f_k is $|M_k| / maxM_k$.

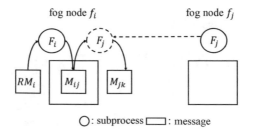

fog node f_i fog node f_j

F_i F_j F_j

RM_i M_{ij} M_{jk}

○ : subprocess ⬜ : message

Fig. 4. Subprocess transmission (ST).

[Message exchange (ME)]

1. The usage ratios u_i and u_j of memory M_i and M_j are $u_i = |M_i| / maxM_i$ and $u_j = |M_j| / maxM_j$, respectively.
2. If $u_i > u_j$, the node f_i sends $(maxM_i \cdot |M_j| - maxM_i \cdot |M_j|) / (maxM_i + maxM_j)$ messages to f_j [Fig. 5]. The messages are removed in the memory M_i.
3. If $u_i < u_j$, the node f_j sends $(maxM_j \cdot |M_i| - maxM_i \cdot |M_j|) / (maxM_i + maxM_j)$ messages to the node f_i. The messages are removed in the memory M_j.
4. Otherwise, the nodes f_i and f_j do nothing.

Fig. 5. Message exchange (ME).

4.2   Energy Consumption

We consider how much electric energy a fog node f_i consumes to exchange messages and a subprocess with a fog node f_j. Let $TT_i(x)$ and $RT_i(x)$ be a pair of time [sec] for a node f_i to transmit and receive x messages, respectively. $ET_i(x)$ shows time [sec] for a node f_i to process the x messages.

First, we consider the MT way. A source node f_i sends the output messages M_{ij} to a target node f_j. The target node f_j receives and processes the messages M_{ij} and generates output messages M_{jk}. A pair of the execution time TET_i and TET_j [sec] are given for a pair of a source node f_i and a target node f_j:

$$TET_i = TT_i(|M_{ij}|) = st_i \cdot |M_{ij}|. \tag{1}$$
$$TET_j = RT_j(|M_{ij}|) + ET_j(|M_{ij}|) = rt_j \cdot |M_{ij}| + ct_j \cdot C_j(|M_{ij}|). \tag{2}$$

A node f_i consumes energy to send and receive the number x of messages. In this paper, the power to send and receive messages is assumed to be $se_i \cdot maxE_i$ [W] and $re_i \cdot maxE_j$ for constants se_i and re_i, respectively. Here, $0 < se_i \le 1$ and $0 < re_i \le 1$. A pair of the energy consumption TEE_i and TEE_j [J] of the nodes f_i and f_j, respectively, are shown as follows:

$$TEE_i = se_i \cdot TT_i(|M_{ij}|) \cdot maxE_i = se_i \cdot st_i \cdot |M_{ij}| \cdot maxE_i. \tag{3}$$
$$TEE_j = re_j \cdot RT_j(|M_{ij}|) \cdot maxE_j + EE_j(|M_{ij}|) \tag{4}$$
$$= re_j \cdot rt_j \cdot |M_{ij}| \cdot maxE_j + ct_j \cdot C_j(|M_{ij}|) \cdot maxE_j.$$

The total energy MEE_{ij} consumed by the nodes f_i and f_j is $TEE_i + TEE_j$, where the node f_i sends messages M_{ij} to the node f_j.

In the ST way, a target fog node f_j sends the subprocess F_j to a source fog node f_i. $|F_j|$ shows the number of messages to carry the subprocess F_j. Let cr_{ij} be the ratio of the computation speed of a node f_i to a node f_j. The execution time TET_i and TET_j of the node f_i and f_j, respectively are given as follows:

$$TET_i = RT(|F_j|) + cr_{ij} \cdot ET_i(|M_{ij}|) = rt_i \cdot |F_j| + cr_{ij} \cdot ct_i \cdot C_i(|M_{ij}|). \quad (5)$$
$$TET_j = RT_j(|F_j|) = st_j \cdot |F_j|. \quad (6)$$

If the node f_i is slower than the node f_i, $cr_{ij} > 1$, i.e. it takes a longer time to perform the subprocess F_j on the node f_i than the node f_i. Otherwise, $cr_{ij} \leq 1$. The energy TEE_i and TEE_j to consumed by a source node f_i and a destination node f_j, respectively, are given as follow:

$$TEE_i = re_i \cdot RT_i(|F_j|) \cdot maxE_i + cr_{ij} \cdot ET_i(|M_{ij}|) \cdot maxE_i$$
$$= re_i \cdot rt_i \cdot |F_j| \cdot maxE_i + cr_{ij} \cdot ct_i \cdot C_i(|M_{ij}|) \cdot maxE_i. \quad (7)$$
$$TEE_j = se_j \cdot TT_j(|F_j|) \cdot maxE_j = se_j \cdot st_j \cdot |F_j| \cdot maxE_j. \quad (8)$$

The total energy consumption SEE_{ij} of the nodes f_i and f_j is $TEE_i + TEE_j$ where the subprocess F_j is sent to the fog node f_i from the node f_j.

Thirdly, we consider the ME way. For fog nodes f_i and f_j, the numbers x_{ij} and x_{ji} of messages are obtained as follow:

$$x_{ij} = (maxM_i \cdot |M_j| - maxM_j \cdot |M_i|)/(maxM_i + maxM_j) \quad (9)$$
$$x_{ji} = (maxM_j \cdot |M_i| - maxM_i \cdot |M_j|)/(maxM_i + maxM_j) \quad (10)$$

First, a fog node f_i sends x_{ij} messages to another fog node f_j. It takes time $TT_i(x_{ij})$ and $RT_j(x_{ij})$ for the nodes f_i and f_j to send and receive x_{ij} messages, respectively. The nodes f_i and f_j consume the energy TEE_i and TEE_j, respectively, as follows:

$$TEE_i = st_i \cdot TT_i(x_{ij}) \cdot se_i \cdot maxE_i = st_i \cdot se_i \cdot x_{ij} \cdot maxE_i. \quad (11)$$
$$TEE_j = rt_j \cdot RT_j(x_{ij}) \cdot re_j \cdot maxE_j = rt_j \cdot re_j \cdot x_{ij} \cdot maxE_j. \quad (12)$$

The total energy EEE_{ij} consumed by the nodes f_i and f_j is $TEE_i + TEE_j$ to send x_{ij} messages to the node f_j from the node f_i.

Next, a fog node f_j sends x_{ji} messages to another fog node f_j. The nodes f_i and f_j consume the energy TEE_i and TEE_j, respectively, as follows:

$$TEE_i = RT_i(x_{ji}) \cdot re_i \cdot maxE_i = rt_i \cdot re_i \cdot x_{ji} \cdot maxE_i. \quad (13)$$
$$TEE_j = TT_j(x_{ji}) \cdot se_j \cdot maxE_j = st_j \cdot se_j \cdot x_{ji} \cdot maxE_j. \quad (14)$$

The total energy EEE_{ji} to be consumed by the nodes f_i and f_j is $TEE_i + TEE_j$, where the number x_{ji} of messages are sent to the node f_i from the node f_j.

4.3   Selection Algorithms

We discuss how each fog node f_i behaves to deliver and process messages M_{ij}. Let FN_i be a set of fog nodes which are in the communication range of a fog node f_i. Let FF_i be a set of fog nodes in the set FN_i, which follow a node f_i, i.e. $FF_i = \{ f_j \in FN_i \mid f_i \to f_j \}$.

First, for each fog node f_j in the set FF_i, the total energy consumption MEE_{ij} and FEE_{ij} [J] to be consumed by the fog nodes f_i and f_j are obtained for the message transmission (MT) and subprocess transmission (ST) ways, respectively. If $MEE_{ij} < FEE_{ij}$, the MT way is taken for the target node f_j, i.e. $w_{ij} = MT$ and $TE_{ij} = MEE_{ij}$, else the ST way, i.e. $w_{ij} = ST$ are $TE_{ij} = FEE_{ij}$. A fog node f_j is taken where TE_{ij} is minimum in the set FF_i. The fog nodes f_i and f_j take the way w_{ij}.

Algorithm 1. Selection algorithms

---

1 **input** : f_i = source fog node,
2   f_j = target fog node;
3 **output** : E_i = total energy to be consumed by f_i and f_j,
4   w_i = way $\in \{MT, ST, ME\}$;
5 FF_i = set of fog nodes in the communication range;
6 $FN_i = \{f_i \in FF_i \mid f_i \to f_j\}$;
7 $E_i = \infty$; $w_i = \bot$;
8 **if** $FN_i \neq \phi$ **then**
9 $\quad$ **for** *each node f_j in FN_i* **do**
10 $\quad\quad$ **if** $maxM_j - (|M_{ij}| + |M_i|) > 0$, **then**
11 $\quad\quad\quad$ $MT_i = MEE_{ij}$;
12 $\quad\quad$ **if** $maxM_i - (|F_j| + |M_i|) > 0$, **then**
13 $\quad\quad\quad$ $ST_i = SEE_{ij}$;
14 $\quad\quad\quad$ **if** $MT_i < ST_i$, **then**
15 $\quad\quad\quad\quad$ $EE_i = MT_i$, $ww_i = MT$;
16 $\quad\quad\quad$ **else**
17 $\quad\quad\quad\quad$ $EE_i = ST_i$, $ww_i = ST$;
18 $\quad\quad$ **if** $EE_i < E_i$ **then**
19 $\quad\quad\quad$ $E_i = EE_i$;
20 $\quad\quad\quad$ $w_i = ww_i$;
21 $\quad\quad$ **for end**;
22 **else**
23 $\quad$ /* $FN_i = \phi$ */
24 $\quad$ $ww_i = ME$;
25 $\quad$ **if** $EEE_{ij} < EEE_{ji}$, **then**
26 $\quad\quad$ $E_i = EEE_{ij}$, $w_i = EM_{ij}$;
27 $\quad$ **else**
28 $\quad\quad$ $E_i = EEE_{ji}$, $w_i = EM_{ji}$;

If $FF_i = \phi$, the message exchange (ME) way is taken. Let NF_i be a set of fog nodes which are in the communication range of the node f_i. For each node f_j in the set NF_i, if $EEE_{ij} < EEE_{ji}$, $TE_i = EEE_{ij}$, else $TE_i = EEE_{ji}$. A fog node f_j where TE_i is minimum in the set NE_i is taken. If $EEE_{ij} < EEE_{ji}$, the node f_i sends the number x_{ij} of messages to the node f_j. Otherwise, the node f_j sends the number x_{ji} of messages to the node f_i.

5   Evaluation

We evaluate the selection algorithm for a pair of a source node f_i and target node f_j. The performance of the nodes f_i and f_j are the same. The source node f_i has messages M_{ij} to be sent to the node f_j. Here, the maximum electric power consumption $maxE_i$ is 3.7 [W]. We assume the electric power ratios re_i and se_i of a f_i are 0.729 and 0.676, respectively. We assume the execution time ratios rt_i, st_i and ct_i are 1, 0.222, and 1, respectively.

Figures 6 and 7 show the energy consumption and execution time of the fog nodes f_i and f_j for size x [B] of the messages M_{ij}. For $x \leq 1,500$, the MT way is better than the ST way. For $x > 1,500$, the ST way is better than the MT way.

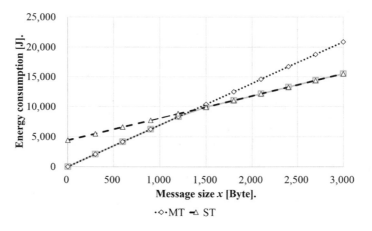

Fig. 6. Energy Consumption of fog nodes.

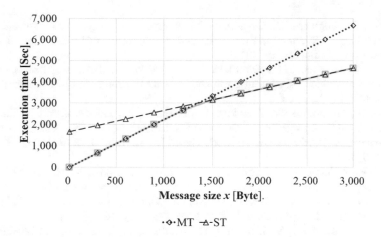

Fig. 7. Execution time of fog nodes.

6   Concluding Remarks

In this paper, we newly proposed the MFC (Mobile Fog Computing) model to efficiently realize the IoT. The MFC model is composed of mobile fog nodes which communicate with one another in wireless networks. A fog node receives data from other nodes and processes the data by a subprocess supported by the node. Then, the fog node delivers output data obtained by processing the data to a fog node. A fog node exchanges not only data but also a subprocess with another node. We newly proposed three ways, message transmission (MT), subprocess transmission (ST), and message exchange (ME) ways to realize the MFC model. We formalized how much energy every node consumes to transmit messages and to subprocess and process messages. We evaluated the algorithm.

References

1. Raspberry pi 3 model b. https://www.raspberrypi.org/products/raspberry-pi-3-model-b/
2. Dhurandher, S.K., Sharma, D.K., Woungang, I., Saini, A.: An energy-efficient history-based routing scheme for opportunistic networks. Int. J. Commun. Syst. **30**(7), e2989 (2015)
3. Enokido, T., Ailixier, A., Takizawa, M.: A model for reducing power consumption in peer-to-peer systems. IEEE Syst. J. **4**(2), 221–229 (2010)
4. Enokido, T., Ailixier, A., Takizawa, M.: Process allocation algorithms for saving power consumption in peer-to-peer systems. IEEE Trans. Ind. Electron. **58**(6), 2097–2105 (2011)
5. Enokido, T., Ailixier, A., Takizawa, M.: An extended simple power consumption model for selecting a server to perform computation type processes in digital ecosystems. IEEE Trans. Ind. Inform. **10**(2), 1627–1636 (2014)

6. Hanes, D., Salgueiro, G., Grossetete, P., Barton, R., Henry, J.: IoT Fundamentals: Networking Technologies, Protocols, and Use Cases for the Internet of Things. Cisco Press, Indianapolis (2018)
7. Oma, R., Nakamura, S., Duolikun, D., Enokido, T., Takizawa, M.: An energy-efficient model for fog computing in the internet of things (IoT). Internet of Things 1–2, 14–26 (2018)
8. Oma, R., Nakamura, S., Duolikun, D., Enokido, T., Takizawa, M.: Evaluation of an energy-efficient tree-based model of fog computing. In: Proceedings of the 21st International Conference on Network-Based Information Systems (NBiS-2018), pp. 99–109 (2018)
9. Oma, R., Nakamura, S., Duolikun, D., Enokido, T., Takizawa, M.: Energy-efficient recovery algorithm in the fault-tolerant tree-based fog computing (FTBFC) model. In: Proceedings of the 33rd International Conference on Advanced Information Networking and Applications (AINA-2019), pp. 132–143 (2019)
10. Oma, R., Nakamura, S., Duolikun, D., Enokido, T., Takizawa, M.: Evaluation of data and subprocess transmission strategies in the tree-based fog computing model (accepted). In: Proceedings of the 22nd International Conference on Network-Based Information Systems (NBiS-2019) (2019)
11. Oma, R., Nakamura, S., Duolikun, D., Enokido, T., Takizawa, M.: A fault-tolerant tree-based fog computing model (accepted). Int. J. Web Grid Serv. (IJWGS) 15(3), 219–239 (2019)
12. Oma, R., Nakamura, S., Enokido, T., Takizawa, M.: A tree-based model of energy-efficient fog computing systems in IoT. In: Proceedings of the 12th International Conference on Complex, Intelligent, and Software Intensive Systems (CISIS-2018), pp. 991–1001 (2018)
13. Rahmani, A.M., Liljeberg, P., Preden, J.S., Jantsch, A.: Fog Computing in the Internet of Things. Springer, Berlin (2018). https://doi.org/10.1007/978-3-319-57639-8
14. Spaho, E., Barolli, L., Kolici, V., Lala, A.: Evaluation of single-copy and multiple-copy routing protocols in a realistic vdtn scenario. In: Proceedings of the 10th International Conference on Complex, Intelligent, and Software Intensive Systems (CISIS-2016), pp. 285–289 (2016)

Converting Big Video Data into Short Video: Using 360-Degree Cameras for Searching Students Location and Judging Students Learning Style

Noriyasu Yamamoto[✉]

Department of Information and Communication Engineering,
Fukuoka Institute of Technology, 3-30-1 Wajiro-Higashi, Higashi-Ku,
Fukuoka 811-0295, Japan
nori@fit.ac.jp

Abstract. In general, at the universities the lecturers use the grades to evaluate the students' performance. However, the grades do not show the accumulated knowledge. Thus, the students can't understand whether they have studied enough to understand all the topics of the lecture. Recently, with the advancement with as IoT technology and applications, we can record students' study for every lecture using video images in all directions (360-degree). We can display high quality image for students. However, by recording many lectures, we will have big data. Therefore, we need to convert these data into short video and show to the students in order that students easily understand their study style. In this paper, we use 360-degrees cameras for searching the students' location and judge their learning style.

1 Introduction

In general, at the universities the lecturers use the grades to evaluate the students' performance. However, the grades do not show the accumulated knowledge. Thus, the students can't understand whether they have studied enough to understand all the topics of the lecture. Recently, with the advancement with as IoT technology and applications, we can record students' study for every lecture using sharp video image in all directions (360-degree). Also, we can display high quality image for students. However, by recording many lectures, we will have big data. Therefore, we need to convert these data into short video and show to the students in order that students easily understand their study style.

For reducing video time, we can use a fixed point camera. In Fig. 1 is shown an example of the reduced video by using fixed point camera. The 8 h car park video is reduced into 5 s. However, for different lectures, the students sit in different seats, so we can't use the conventional method for fixed point camera. We need some 360-degree cameras and we should consider the location point for cameras. Then, a target student can be automatically searched using these big (long) video data. In addition, we should convert the data into short video in order that a student can easily understand his learning style.

© Springer Nature Switzerland AG 2020
L. Barolli et al. (Eds.): NBiS-2019, AISC 1036, pp. 459–464, 2020.
https://doi.org/10.1007/978-3-030-29029-0_43

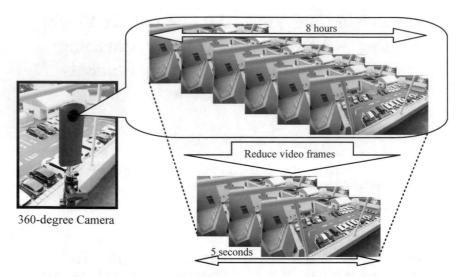

Fig. 1. An example of the reduced video by fixed point camera.

In this paper, we use 360-degrees cameras for searching the students' location and judge their learning style.

The paper structure is as follows. In Sect. 2, we introduce our implemented smartphone-based Active Learning System (ALS). In Sect. 3, we present the integration of ALS with 360-degree cameras. Finally, in Sect. 4, we give some conclusions and future work.

2   Smartphone-Based Active Learning System (ALS)

In this section, we present our implemented Smartphone-based Active Learning System (ALS), which was used to increase the students learning motivation [1–10]. Figure 2 shows the structure of ALS. By using ALS, the lecturer performs the interactive lecture by confirming the understanding degree of the student using their smartphone in real time. Prior to the lecture, the lecturer prepares "study points". "Small examination" refers to a mini quiz prepared for each study point. A mini quiz consists of simple multiple-choice questions. "Understanding level" is set by the result of the "Small examination". "Lecture speed" suggests whether students find the lecture progress too fast or too slow. By "Understanding level" and "Lecture speed" students' understanding degree can be judged. These two functions are used as feedbacks through the application on students' smartphone. During the lecture, these data are recorded in the database.

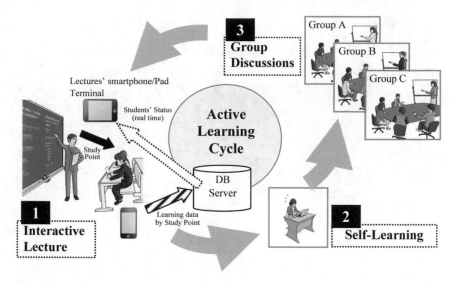

Fig. 2. Structure of ALS.

3 Integration of ALS and 360-Degree Cameras

In this section, we present the integration of smartphone-based Active Learning System (ALS) and 360-degrees cameras, which we use for searching the student location and judging the student learning style.

By using 360-degree camera, a target student can be automatically searched by using the recorded video data. But, the automatic search process takes times. To solve this problem, we use our implemented smartphone-based ALS []. In Fig. 2 is shown the implemented ALS during the lecture. The students use the smartphone which is connected to the Internet by Wi-Fi access points. The students' location is calculated by the distances from Wi-Fi access points (see Fig. 3) [7]. Figure 4 shows the video for the target student. We recorded q long video for the target student for every lecture. Then, we reduced the long video to the short video by using a conventional algorithm, which can reduce the number of frames.

By compressing the big video data recorded for 22 h, we get a five second digest video. Although, this digest video shows the students' study style, it's not easy to determine their study status such as studying, dazing and sleeping. For this reason, we integrate 360-degrees cameras with our smartphone-based ALS. In ALS, the students' learning status is classified in three types [7, 8].

[Type 1] Studying: The student concentrates on study.
[Type 2] Dazing: The student does not concentrate on study.
[Type 3] Sleeping: The student is sleeping without studying.

Using students' learning status information, the short video is changed to 3D image as shown in Fig. 5, which is easy to understand the students' learning situation.

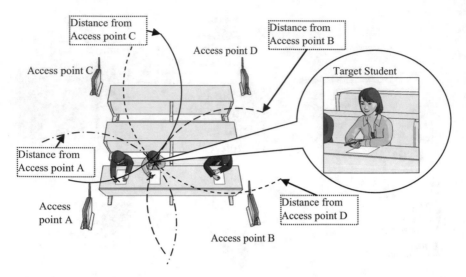

Fig. 3. Calculation of student location by ALS using the distance from Wi-Fi access points.

Fig. 4. Video recording for target student.

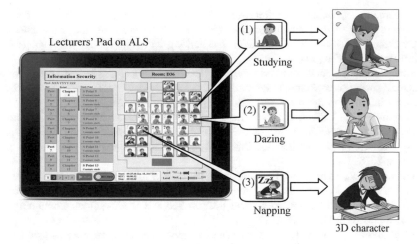

Fig. 5. The exchanged 3D character.

4   Conclusions

In this paper, we presented the integration of the smartphone-based ALS with 360-degree cameras. By recording students' learning situation for every lecture using video images in all directions (360-degree) will have big data. Therefore, we need to convert these data into short video and show to the students in order that students easily understand their study style. By using the proposed approach, the students' location can be calculated and learning style can be judged. In the future, we plan to perform extensive experiments with integrated system.

References

1. Yamamoto, N.: An interactive e-Learning system for improving students motivation and self-learning by using smartphones. J. Mob. Multimedia (JMM) **11**(1 & 2), 67–75 (2015)
2. Yamamoto, N.: New functions for an active learning system to improve students self-learning and concentration. In: Proceeding of the 18th International Conference on Network-Based Information Systems (NBiS-2015), pp. 573–576, September 2015
3. Yamamoto, N.: Performance evaluation of an active learning system to improve students self-learning and concentration. In: Proceeding of the 10th International Conference on Broadband and Wireless Computing, Communication and Applications (BWCCA-2015), pp. 497–500, November 2015
4. Yamamoto, N.: Improvement of group discussion system for active learning using smartphone. In: Proceeding of the 10th International Conference on Innovative Mobile and Internet Services in Ubiquitous Computing (IMIS-2016), pp. 143–148, July 2016
5. Yamamoto, N.: Improvement of study logging system for active learning using smartphone. In: Proceedings of the 11th International Conference on P2P, Parallel, Grid, Cloud and Internet Computing (3PGCIC–2016), pp. 845–851, November 2016

6. Yamamoto, N., Uchida, N.: Improvement of the interface of smartphone for an active learning with high learning concentration. In: Proceeding of the 31st International Conference on Advanced Information Networking and Applications Workshops (AINA-2017), pp. 531–534, March 2017

7. Yamamoto, N., Uchida, N.: Performance evaluation of a learning logger system for active learning using smartphone. In: Proceeding of the 20th International Conference on Network-Based Information Systems (NBiS-2017), pp. 443–452, August 2017

8. Yamamoto, N., Uchida, N.: Performance evaluation of an active learning system using smartphone: a case study for high level class. In: Proceeding of the 6th International Conference on Emerging Internet, Data & Web Technologies (EIDWT-2018), pp. 152–160, March 2018

9. Yamamoto, N., Uchida, N.: Dynamic group formation for an active learning system using smartphone to improve learning motivation. In: Proceedings of the 12th International Conference on Innovative Mobile and Internet Services in Ubiquitous Computing (IMIS-2018), pp. 183–189, March 2018

10. Yamamoto, N., Uchida, N.: Performance evaluation of a smartphone-based active learning system for improving learning motivation during study of a difficult subject. In: Proceeding of the 21th International Conference on Network-Based Information Systems (NBiS-2018), pp. 531–539, August 2018

Crowdsourcing Platform for Healthcare: Cleft Lip and Cleft Palate Case Studies

Krit Khwanngern[1,2]([✉]), Juggapong Natwichai[3], Suriya Sitthikham[1,2],
Watcharaporn Sitthikamtiub[2], Vivatchai Kaveeta[1,2], Arakin Rakchittapoke[4],
and Somboon Martkamjan[4]

[1] CMU Craniofacial Center, Faculty of Medicine, Chiang Mai University,
Chiang Mai, Thailand
{krit.khwanngern,suriya.sitthikham,vivatchai_k}@cmu.ac.th
[2] Center of Data Analytics and Knowledge Systhesis for Healthcare,
Chiang Mai University, Chiang Mai, Thailand
watcharaporn.sit@cmu.ac.th
[3] Department of Computer Engineering, Faculty of Engineering,
Chiang Mai University, Chiang Mai, Thailand
juggapong.n@cmu.ac.th
[4] MIMO Tech Company Limited, Bangkok, Thailand
{arakinr,somboonm}@ais.co.th

Abstract. In this paper, we present a design for the healthcare system which extends the existing medical cloud services in order to overcome the challenges of searching the new patients and following-up the existing patients. We explain that the existing platforms are working on different healthcare levels, with separate groups of users, and on different devices. We introduce additional design of the system which acts as a bridge between these platforms. Utilizing the aspect of crowdsourcing and online data sharing, the system can benefits contributors by significantly reduce the amount of time, cost and workload of all users in the system.

1 Introduction

1.1 Facial Cleft

Facial cleft is a type of facial birth defect when the anomalous splitting appears on the upper lip, gum, and palate. It may also cause other syndromes of facial asymmetry. In Thailand, the disorder incidence is approximately 1 to 1.5 in one thousand newborn babies. When considering the national birthrate, we can assume about 800 to 1200 infants born with cleft each year. With the collective in-treatment patients at over 60000 cases, a challenging aspect of the cleft treatment is coming from the long treatment period which extends from birth until at least 20 years of age. Throughout the entire treatment, multidisciplinary teams consisting of surgeons, dentists, medical technologists, nurses, speech therapists, etc. participation are required. With the limited number of specialists,

© Springer Nature Switzerland AG 2020
L. Barolli et al. (Eds.): NBiS-2019, AISC 1036, pp. 465–474, 2020.
https://doi.org/10.1007/978-3-030-29029-0_44

the individuals may be located in separate facilities under different authorities. Consequently, communicating and sharing critical patient information is an important challenge. On another hand, the patients had been forced to travel between locations which put off time-critical treatments due to the time and cost constraint.

1.2 Healthcare Levels in Thailand

There are four different levels in the healthcare structure of Thailand as shown in Fig. 1. From the left-hand side of the figure, the four levels are primary, secondary, tertiary and quaternary care. Each level represents different organizations with different expertise and tasks. The brief introduction for each level are as follows:

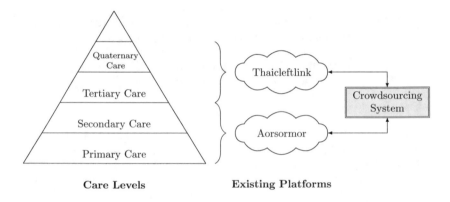

Fig. 1. Levels of healthcare in Thailand.

Primary Care. This level represents public health centers located in the center of the local population. They are usually crew by a small number of a medical professional with the support of a few dozen village health volunteers. Their services including general illness treatment, vaccination, health consulting, and health practice promoting. With the total number of almost 80,0000 in Thailand [2], they act as the medical first line of defence.

Secondary Care. This level represents local hospitals including both public and privates with the number of beds less than 100. They have more medical staff and equipment than the primary care unit. Able to provide the treatment for all general diseases and may including some complicate diseases.

Tertiary Care and Quaternary Care. Both of these levels represent large hospitals, usually provincial and regional, capable of treating highly specialized diseases. Requiring sub-specialty medical teams and sophisticated equipment. Patients with serious conditions can be transferred to one of these facilities in order to get the required treatment.

1.3   Information Platforms

The fact that these healthcare facilities are operating on different levels and are managed by different authorities, there is a separation in the information flow between them. For example, many small hospitals still relied on traditional hand-written documents or offline database system. On the other hand large hospitals tend to use privately developed platforms tuned to their special need. With the limited number of specialists in the field, some patient needs to be transferred between hospitals in order to receive the required treatments. Therefore the patient's medical data needed to be exchanged between the facility. But with the separation of the data platforms, the printed out documents were manually handed over by the hospital staff or even by the patients themselves. This process proves to be unreliable with many problems including incomplete data, time-consuming, and logistic cost. In recent years, few organizations stepped up to take on the task of developing the online medical information platforms. One of the challenges is to ensure that they are successfully following the regulations related to the privacy of patient information.

Following our mottoes of "Minimize the patient travel" and "Let the data travel instead of patient", Chiang Mai University's craniofacial center developed an online platform for collaboration between the multi-facilities members. "Thaicleftlink" initially deployed in the northern region of Thailand with currently over 1500 registered patients. Medical team members share patient data such as contact information, diagnosis, medical history, photo, and appointment. In a similar fashion, "Aorsormor Online" mobile application developed by Advanced Info Service, a Thailand mobile network operator, also intended as the platform to exchange the medical information and news between users.

2   Related Work

Electronic medical record system is one of the tools that modern medical facilities using to organize and provide patient care on a daily basis. However, most are offline systems which limit the data inside the facility. Many works were done on online platforms for medical information. Some are designed to be used by medical professionals, usually geared toward research data collecting, data analysis, or medical training [1,3,4]. Differently, other works were for data sharing between patients or between patients and professionals. [7] shows the benefit of a platform for sharing health information between epilepsy patient which helps to spread knowledge on the disorder. In [6] reviews the effect of the health care crowdsourcing system toward the physician-patient relationship.

For the crowdsourcing aspect, there are works on the crowdsourcing application for the medical field. [5] shows the benefit of distributed problem-solving to reduce the workload of the limited number of professionals. It is fair to say that there is no shortage of works related to online medical data systems. However, we believe that only a handful of them specifically looked at the system that interacts between two different platforms. Therefore in this work, we will look into a design of such system.

3 Crowdsourcing System Architecture

In this section, we will start by introducing two of the existing online medical information platforms. We discuss their differences in applications, functions and target users. Then, we discuss the need for a crowdsourcing system to enable the processes of searching and following up cleft patients. We go into detail about the internal process working, what kinds of data will be exchanged and how these processes can benefit all collaborators.

3.1 Existing Thaicleftlink Platform

Fig. 2. Thaicleftlink, a web-based medical data sharing platform for cleft patients. (Left) Patient list page with search and filter options (Profile images are replaced with placeholder icons). (Right) Cleft patient demographics, retrieved in May of 2019.

Thaicleftlink by Chiang Mai university's Craniofacial Center is a web-based platform used by medical professionals. As previously mentioned, its main focus is being a sharing platform between specialists in multiple hospitals for the treatment of cleft patients. The main users are specialists in the area of cleft treatment including surgeons, nurses, speech therapists, etc. In the initial phase, we deploy the platform to users in the northern region of Thailand. Current at more than 1600 in-treatment patients and about 50 active users originated from multiple hospitals. Figure 2 (Left) shows patient list page which shows the profile photo and basic information of the patients, (Right) shows the map of the cleft population on Thaicleftlink database.

As the system run as web-based service, the user can access it anywhere with an internet connection. Responsive design interface allows a wide range of different web browsers either on stationary and mobile devices to fit the pages to the screen. System modules gear toward managing medical information related to cleft treatment including patient basic information, diagnosis, medical history, gallery, appointment, and data visualization.

Fig. 3. Application logo and a screenshot of Aorsormor Online mobile application.

3.2   Existing Aorsormor Online Platform

Aorsormor Online is a mobile application developed by Advanced Info Service as community servicing information hub for healthcare personals. The application had been launched in 2015 and constantly being improved since. Main users are public health center staff and health volunteers who closely work collaboratively. The application provides the services for learning, receiving news, messaging, reporting volunteer works, reporting infectious disease occurrences and making appointments. Users can share any type of information such as image, audio, video, text, and location.

3.3   Extension of AIS System for Cleft Lip and Cleft Palate

Intentionally, both existing platforms are being utilized by the different groups of the user (Right side of Fig. 1). Thaicleftlink focused on tertiary care and quaternary care unit (Regional and provincial hospitals) and Aorsormor Online focused on primary and secondary care unit (Public health center and health volunteer). Currently, the ways to exchange data between these groups are manual phone calls, messaging applications or document handover which are tedious, time-consuming and costly. These are the reasons behind the development of a crowdsourcing system to act as the bridge between them.

There are two main functions which will greatly benefit the treatment of cleft patients. First is the process of searching for new patients. And second is the following up of the in-treatment patients. In the following sections, we will explain both of the processes in detail.

Searching Module. Searching for the new patient is the process of locating the new patient who is not already in a treatment program. This is one of the most important processes specifically for cleft patients, as the cleft treatment outcome are strongly depending on the correct time period of the treatment. For example, nasoalveolar molding (NAM) which can help reduce the gap of the cleft, should be used very early after birth increasing the success rate of the following surgeries. The delayed intervention may lead to undesirable treatment outcome and slow speech development.

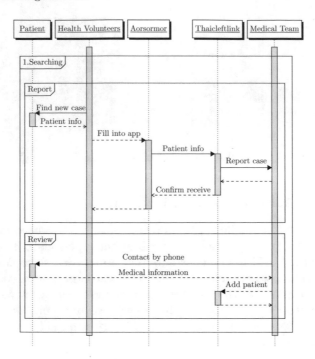

Fig. 4. Interactive diagram for searching module.

In the fortunate case that the cleft children were prenatally (before birth) diagnosed, the medical team can immediately provide them with the correct information, guiding them to the correct treatment plan as soon as possible. But in other cases, parents may not recognize that where who and when they should get the consultation. We hope that by employ the help of village health volunteers, there is a greater possibility that the patient will register into the treatment system. Because VHV is also living in the local community, they receive stronger trust and friendliness which encourage the parents to get the help required.

Figure 4 shows the entire interact in the searching process. At the top, there are five of the participants as 1. Patient, 2. Health volunteers, 3. Aorsormor application, 4. Thaicleftlink application and 5. Medical team. The interaction steps are as follows:

1. Health volunteers go on location to visit the potential cleft patient. And ask for the patient medical related information.
2. They fill the new patient form with basic information and type of the cleft (Fig. 5) into the searching module of Aorsormor application.
3. Aorsormor forward the new patient data to Thaicleftlink.
4. Thaicleftlink receives and store new patient data and report the case to the medical team.
5. Medical team's staff review and validate the data.

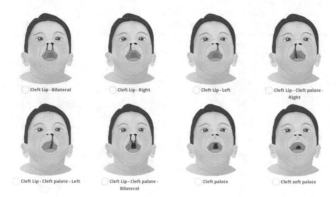

Fig. 5. Cleft type selection in the searching module.

6. Medical staff contact the patient by telephone to provide essential information and make an appointment.
7. Medical staffs register new patient into Thaicleftlink database.

Following Module. The second module is related to the following process. Following up is the process of contacting the patient for checking up their condition or getting an update of their information. There are four situations where the process is required as follows:

1. **Appointment confirm:** Confirm that the patient can travel to the booked appointment with the medical team at the hospital.
2. **Pre-surgery visit:** Confirm that the patient is in suitable health to undergo the surgery. Report if there is surgery preventing condition such as flu, fever, cough, infection, etc. Confirm the surgery dates and provide them with contact information.
3. **Post-surgery visit:** Update on the surgery outcome. Confirm that there is no split, fever higher than 38 degree Celsius, or abscess on the surgical area.
4. **Contact information update:** Make contact with the patient and their guardian to update new contact information.

Figure 6 shows the entire interact in the following up process. The interaction steps are as follows:

1. Medical team submits a request for the following up on Thaicleftlink system. The submitted information includes the patient identity, contact information and looking up reason and return data form. The user will need to select which public health center will receive the request.
2. Thaicleftlink forwards the information to Aorsormor application which displays the request to the appropriate target user.
3. Volunteer accepts the request and visits the patient on location.
4. Volunteer fills the report back into Aorsormor application.
5. Aorsormor forward the update information to Thaicleftlink.
6. Thaicleftlink shows the follow-up result to the medical staff.

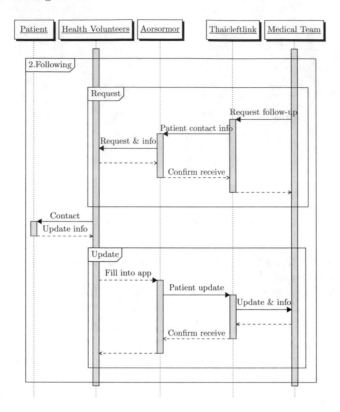

Fig. 6. Interactive diagram for following module.

4   Considerations

4.1   Data Exchange

While the patient information is being transferred in the system, the security of such data is a critical issue. The patient's medical data is protected under regulations and laws. We cannot simply forward all of them to the health centers and consecutively health volunteers. Many information is not required to perform search and follow-up functions such as diagnosis and medical history. They will be separated away from the data exchanging scheme. Another is the profile photos of the patient faces are other sensitive data, we cannot send or share them without identifying the person on receiving end. Additionally, Aorsormor server must not collect any log of patient data and images cache.

However, some key information is required for the process such as contact information, profile photo, and address. We need to balance the information transfer to include enough information for pinpoint the patient and limit them for patient privacy. All the data will need to be transferred via a secure channel. All API calls will be done as HTTPS request. As the communication between

Aorsormor and Thaicleftlink are made by static IP address server, we can apply strict IP filtering to prevent the malicious attempts.

4.2 Limited Response Time

Most of the following up process is required to be performed in a limited amount of time. For example, the patient's confirmation of doctor appointment or surgery date is required well in advance. In order to allocate some extra time for cancellation re-schedule. Therefore the system needs to keep the state of the jobs either it is active or inactive. The job state will be changed to inactive as soon as it was done or pass their predetermined period.

4.3 Benefits

Lastly, we like to point out the benefits of this system to every user groups.

Patient. They will experience a better chance of receiving advice about the cleft treatment process. They can reduce the time and cost associated with traveling to hospitals which can be located far away from their home.

Village Health Volunteers. They normally got payment reward by submitting the report of their works to the health center. Therefore cleft patient searching and following up actions can be included as one part of their work. They support the local population and build a strong bond to their community.

Medical Team Members. The system reduces their workload by shifting the works of manually contacting patients to the hands of local health units. Allow them to focus on improving the treatment process instead.

Platforms. We learn the procedure to connect two systems from a public and private organization. In the way that is secure and benefit every contributor. It also doubles as the promotional material to persuade new users to register into the platforms.

5 Conclusion

In this work, we have introduced the basic background of the cleft condition and two existing medical sharing platforms. We have discussed the limitation of the current platforms was based on the separate target users which make the communication between the time and cost consuming. We introduce the design of the crowdsourcing system to extend both platforms for cleft treatment. The system consists of two processes: searching and following. The detail interaction of each process is explained. Lastly, the benefits of such a system for all contributors have been explained.

Acknowledgement. This research was partially supported by Chiang Mai University and MIMO Tech Company Limited.

References

1. Chan, T.M., Thoma, B., Lin, M.: Creating, curating, and sharing online faculty development resources: the medical education in cases series experience. Acad. Med. **90**(6), 785–789 (2015)
2. Chuengsatiansup, K., Suksuth, P.: Health volunteers in the context of change: potential and developmental strategies. J. Health Syst. Res. 1(3-4), 268–275 (2007)
3. Cook, D.A., Steinert, Y.: Online learning for faculty development: a review of the literature. Med. Teacher **35**(11), 930–937 (2013)
4. Mohammed-Rajput, N.A., Smith, D.C., Mamlin, B., Biondich, P., Doebbeling, B.N., Investigators, O.M.C., et al.: Openmrs, a global medical records system collaborative: factors influencing successful implementation. In: AMIA Annual Symposium Proceedings. vol. 2011, p. 960. American Medical Informatics Association (2011)
5. Thawrani, V., Londhe, N.D., Singh, R.: Crowdsourcing of medical data. IETE Techn. Rev. **31**(3), 249–253 (2014)
6. Wald, H.S., Dube, C.E., Anthony, D.C.: Untangling the web–the impact of internet use on health care and the physician-patient relationship. Patient Educ. Couns. **68**(3), 218–224 (2007)
7. Wicks, P., Keininger, D.L., Massagli, M.P., de la Loge, C., Brownstein, C., Isojärvi, J., Heywood, J.: Perceived benefits of sharing health data between people with epilepsy on an online platform. Epilepsy & Behav. **23**(1), 16–23 (2012)

Jaw Surgery Simulation in Virtual Reality for Medical Training

Krit Khwanngern[1,2], Narathip Tiangtae[3(✉)], Juggapong Natwichai[3],
Aunnop Kattiyanet[3], Vivatchai Kaveeta[1,2], Suriya Sitthikham[1,2],
and Kamolchanok Kammabut[1]

[1] CMU Craniofacial Center, Faculty of Medicine, Chiang Mai University,
Chiang Mai, Thailand
{krit.khwanngern,vivatchai_k,suriya.sitthikham,kamolchanok.kum}@cmu.ac.th
[2] Center of Data Analytics and Knowledge Synthesis for Healthcare,
Chiang Mai University, Chiang Mai, Thailand
[3] Department of Computer Engineering, Faculty of Engineering,
Chiang Mai University, Chiang Mai, Thailand
{narathip.tiangtae,juggapong.n,aunnop_kattiyanet}@cmu.ac.th

Abstract. The cutting of the lower jaw bone is an important surgical operation in the treatment of craniofacial disorders. However, many of the craniofacial disorders are rare occurrences. As a result, many medical students may never have hands-on experience on the cutting process. They rely on recorded materials such as books and videos which are not reliable skills transferring methods. In this work, we developed a virtual reality application for simulating the mandibular bone surgery. The system allows visualization of an operation room in a highly realistic virtual reality environment. Multiple operation modes are available and the user can use motion controllers to grip, cut, drill, join and compare the 3D model of the skull. For the cutting and drilling process, the guidelines for optimal cutting path are shown on a virtual model. The progress of the operation is evaluated and displayed as percentage numbers on user interface. The system continuously tracks the cutting line and announces the failure when the error rate is more than 10%. Our proposed system was evaluated by specialists in the field. In general, they provided positive feedback with some improvement suggestions for future works.

1 Introduction

1.1 Craniofacial Disorders

Craniofacial disorders are birth defects in which the babies possess malformations on the facial and skull bones. There are many different classifiable types of diseases. Some types are relatively common occurrences such as cleft, but others can be very rare. Generally, the disease is not life-threatening. But as they are visually distinctive, the public attitude can have a huge impact on the patient mental health. They can negatively affect the patient's quality of life. For example, some craniofacial disorders can lead to speech problems which can eventually

© Springer Nature Switzerland AG 2020
L. Barolli et al. (Eds.): NBiS-2019, AISC 1036, pp. 475–483, 2020.
https://doi.org/10.1007/978-3-030-29029-0_45

lead to slow development. The main goal of the treatments is to minimize the effect of the disease and improve one's quality of life. The entire treatment process can be up to 20 years in total, during which multiple visits to the hospital are required.

One important procedure in the treatment of craniofacial disorder is jaw surgery. It is a required operation for patients with maxillary hypoplasia. The operation tries to correct the malformation of the jaw by moving the jaw bones into the proper position. The necessary movement can be both translation and rotation depending on the conditions. It can correct the teeth malocclusion (teeth not properly aligned). The procedure is not considered cosmetic surgery. However, the muscle tissue can move to the change of lower bone position, it will likely improve the overall facial appearance as well. The procedure requires highly experienced specialists and costly equipment. As a result, the number of operations that can be performed per year is very limited.

Maharaj Nakorn Chiang Mai Hospital is located in the northern region of Thailand. As a teaching hospital affiliated to Chiang Mai University, its important mission is to transfer medical knowledge to students. However, with the limited number of craniofacial cases and the number of specialists, some corrective operations are performed only a few times per year. Therefore, many students may never have hands-on experience. Instead, they rely on recorded materials such as books, images or videos. Eventually, they can be in a situation where they are unable to perform the required operation by themselves. This work tries to address this problem by simulating a surgical operation in a virtual reality environment.

1.2   Virtual Reality Devices

In recent years, there has been a new wave of affordable consumer virtual reality hardwares. With the help of the development of powerful graphic processing units and high quality screen technology for mobile devices, they have pushed visual fidelity, resolution and frame-rates up into a level that can sustain convincing and smooth virtual reality environments. Currently, 90 frames per second is generally recognized as the minimum number of framerate required for nausea-free VR experience. Lower framerate is associated with input delay and disorientation.

Virtual Reality Device Types. The VR devices can be differentiated into two groups. One is the tethered headsets which need to be connected to a computer either by wire or wireless. They usually require a few fixed position sensors in order to operate. At least 3 points of reference are required for room-scale virtual reality. All rendering and processing are done by computer and stream back to the head mounted display. As a result, the performance depends on the power of the connected computer. Another type is the mobile VR headsets. They are standalone devices which can perform the 3D location mapping and rendering by themselves. The inside-out sensors integrated with the headset allow them to operate without external sensors. Therefore, users can move around without the limited volume of sensor towers. However, the major downside is that they usually have lower processing power when compared with the tethered counterpart. In this work, we utilize the tethered VR head mount display because a large number of polygons is required for a realistic surgical simulation environment.

Degree of Freedom. Degree of Freedom (DoF) signifies the ability of the device to detect the user moving around the space. There are two sets of movements, one for translation (Forward-Backward, Up-Down and Left-Right) and rotation (Pitch, Yaw and Roll). 6 degrees of freedom (DoF) devices allow for the mobility in both translation and rotation. In contrast, 3 DoF devices are only restricted to rotation motion. Essentially the user can only look around but not move through space. This makes formal devices much more flexible to fit many types of applications. Consequently, we selected the 6 DoF VR headset for our proposed system.

1.3 Medical Applications of Virtual Reality

Considerable research on the applications of virtual reality in medicine has been carried out. A visual and haptic surgical simulator are evaluated in [5] for training purpose. [6] reviews the usage of virtual reality in neurosurgery simulation. They assess a critical trade off between the visual realism and real-time interactivity. Multiple techniques for tissue modeling, collision detection and haptic feedback were discussed. In [8] address the usefulness of VR simulations for patient specific treatment planning. They reviewed the advancement and validity of the simulators.

The authors in [3,6] proposed 3D modelling simulations for maxillofacial surgery. The models of the skull are made from the CT scan of the patient. In similar application, web based 3D simulation for operation planning of the craniofacial disorder is proposed in [7]. More specifically to craniofacial disorder, [4] shows the virtual craniofacial surgery with soft tissue mass spring simulation. [2] shows an interactable volumetric temporal bone simulation in which the user can perform virtual dissection. They show that experts can complete the tasks in virtual environment much faster than less experienced groups. [1] developed the augmented reality (AR) system to help with the maxillofacial surgery. They use AR head mounted display to display the pre-determined operation plan directly on the operating surgeon view.

2 Proposed System

In this work, we focus on an important action during jaw surgery, the process of cutting the mandibular bone (lower jaw). The cut requires high precision, and a small mistake can cause damage to the facial nerve which, in turn, can lead to paralyzation of facial muscle.

Fig. 1. Operation room model in virtual reality environment.

2.1   Operation Room Environment

To familiarize the user in the virtual reality, we simulate a operating room environment. Figure 1 shows the room which the user will see and interact with in our system.

2.2   Model Pre-processing

Fig. 2. Pre-cut models with the optimal path determined by the surgeon.

The skull model is prepared beforehand. First, the medical doctor marks the ideal cutting path on a physical 3D printed skull model as shown in Fig. 2 Right. We replicate the path on to the 3D model using modeling software, Fig. 2 Left. The paths are used as the guides to make the cuts in to the 3D model. As a result, the original 3D mesh is split into separate models.

2.3   Operation Modes

The system workflow consists of 5 different modes in our system as follows:

1. **Training mode**
   This is the first mode that the user interacts with before entering the main part of the application. The lesson will teach basic control and movements. The user can skip them anytime to the next modes. Figure 3 shows the dialog box for the training mode (Fig. 4).

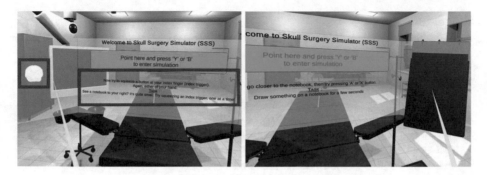

Fig. 3. In training mode, the user is introduced to application functions and how to follow the steps.

Fig. 4. (Left) The user selects the joining option. (Right) Joined bone object after the operation.

Fig. 5. In comparison mode the user can see the original (before) and operated (after) 3D models. They can pause and resume the turntable rotation.

2. **Regular mode**

   This is the main mode. The user can interact with tools and the skull. By using the controller, the skull can be grabbed for translation and rotation with a limited degree of freedom.

3. **Cutting and Drilling mode**

   The user can perform the cut in this mode. When the cutting path is successful, the skull model is split into meshes. However, if the cut was a failure, the skull model will be enlarged and the guideline will be shown. The user will need to repeat the cut until it is successful. The controlling process for this mode will be discussed in the next section.

4. **Joining mode**

   The user can join the separate bones into one mesh. This will keep the bone in the fixed relative position and can be moved around as a single object.

5. **Comparison mode**

   In this mode, the user can compare between the final cut model with the original model. As shown in Fig. 5, the models are displayed side by side with the turntable rotation. The rotation can be stopped in order to examine the specific part of the models.

2.4   Control Scheme

As previously mentioned, to accurately simulate the movement required by the operation, the user controller has to have 6 DoF so that the simulation can react to all types of the user's hand motion. There are 3 controls which the user can perform to the 3D skull models as follows:

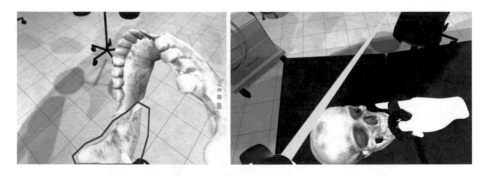

Fig. 6. The user can grip on the bone object by pointing the controller ray directly at the target, then press and hold the controller button.

Gripping. To move the 3D model in VR scene, the user will have to move the controllers to point at the objects. Two controllers have to grab both sides of the bone in order to grab and separate both objects from each other. Figure 6 shows a user performing the gripping action.

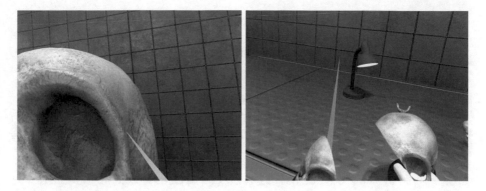

Fig. 7. (Left) The user determines the starting of the plane for cutting. (Right) Separated objects after plane cutting.

Plane Cutting. In this mode the skull model can be enlarged for easier access to the cutting area. The user can push and hold a button on the controller and drag it to create a cutting plane. The plane is used to cut the original mesh into two separate objects, one in front of the plane and another on the back side of it. Figure 7 shows the plane cutting in operation.

Fig. 8. (Left) The user performs the drilling operation. (Right) Drilling progress shown on the user interface.

Drilling. The user can drill into the skull by using either of the motion controllers. The cutting operation can be done simultaneously with cutting and gripping operation. When the user starts drilling, a dash guiding line will be shown to indicate the path which the cut should be performed. Figure 8 Left shows the drilling in action. During the drilling the user will hear the drill sound and some particles will be generated at the contact point simulating bone fragments. We indicate removed bone by changing the alpha value of the texture into the transparent level. The user has to drill until the path forms a complete loop conclude the drilling operation.

The progress of the operation is shown as two values on the interface panel. As shown in Fig. 8 Right, there are two percentage numbers. The first one is the achievement value which shows the drilling process percentage until completion. The second is the error percentage. This value goes up when the user cuts too far outside of the guideline. We fixed at the upper bound limit of 10% which immediately terminates the cut and deem the operation as a failure.

After the achievement value reaches 100%, the user hears the audio confirmation and the cutting guideline disappears. To simulate the bone separation, the original mesh is hidden and immediately replaced by two pre-processed separate bone objects in the exact locations. The process is very fast and visually unnoticeable by the user. The list of cut objects is updated on the interface. The cut bones will be separate objects and can be grabbed and moved by the user.

3   Results

The quality of our proposed system was evaluated by user testing. The testing candidates were a group of specialists in the field with various degrees of experience such as experienced surgeon, resident and medical student. They had never used our system beforehand. We asked them to give the feedback on both the positive and negative points focusing on the effectiveness of the system as a teaching platform for inexperienced individuals.

The comments were generally positive. All users saw the promising future of using the system as an introduction to the operation. They mentioned that the interactivity in virtual reality is more realistic than the current teaching materials. However, they also point out some limitations which should be improved in the subsequent works. The assessment can be summarized as follows:

1. The zoom in view when the user enters the cutting mode was positively received by the testers. It improved the visibility of the cutting region and help cutting point positioning.
2. However, the zoomed object should keep the height in relation to the floor.
3. The movement of the skull should be restricted to approximately 60 to 70° instead of freely movable.
4. The skull model should be made from the CT scans of the actual patients.

4   Conclusion

We developed a virtual reality system to simulate jaw bone operations for teaching purposes. The need to develop such system was based on the limited number of real cases of craniofacial disorders each year. We discussed the requirements for a comfortable VR experience. In addition, we described the details of the developed system, the operation mode and control scheme. Lastly, we tested the system by having a demonstration session with specialists with varying degree of skills. They mostly gave a positive feedback with some improvement suggestions.

Based on the results of the user testing, there are some limitations of the current system that we would like to address in future works. First, the cutting location on the 3D models is currently pre-determined. In subsequent works, we intend to implement a dynamic cut where the model can be separated by the actual path that the user created. Second, the 3D model should be based on the scan of actual patients. The medical scans of the craniofacial patients are readily available as the treatment relies on them. Volumetric information can be used to reconstruct accurate 3D meshes. The remaining challenge will be to optimize the 3D model within the level that can be handled in real-time in VR. Lastly, we intend to improve the visual fidelity of 3D models especially on the objects that the user needs to interact with.

References

1. Badiali, G., Ferrari, V., Cutolo, F., Freschi, C., Caramella, D., Bianchi, A., Marchetti, C.: Augmented reality as an aid in maxillofacial surgery: validation of a wearable system allowing maxillary repositioning. J. Cranio-Maxillofac. Surg. **42**(8), 1970–1976 (2014)
2. Khemani, S., Arora, A., Singh, A., Tolley, N., Darzi, A.: Objective skills assessment and construct validation of a virtual reality temporal bone simulator. Otol. & Neurotol. **33**(7), 1225–1231 (2012)
3. Marchetti, C., Bianchi, A., Bassi, M., Gori, R., Lamberti, C., Sarti, A.: Mathematical modeling and numerical simulation in maxillo-facial virtual surgery (visu). J. Craniofac. Surg. **17**(4), 661–667 (2006)
4. Meehan, M., Teschner, M., Girod, S.: Three-dimensional simulation and prediction of craniofacial surgery. Orthod. & Craniofac. Res. **6**, 102–107 (2003)
5. Morris, D., Sewell, C., Barbagli, F., Salisbury, K., Blevins, N.H., Girod, S.: Visuo-haptic simulation of bone surgery for training and evaluation. IEEE Comput. Graph. Appl. **26**(6), 48–57 (2006)
6. Robison, R.A., Liu, C.Y., Apuzzo, M.L.: Man, mind, and machine: the past and future of virtual reality simulation in neurologic surgery. World Neurosurg. **76**(5), 419–430 (2011)
7. Schendel, S.A., Montgomery, K.: A web-based, integrated simulation system for craniofacial surgical planning. Plast. Reconstr. Surg. **123**(3), 1099–1106 (2009)
8. Willaert, W.I., Aggarwal, R., Van Herzeele, I., Cheshire, N.J., Vermassen, F.E.: Recent advancements in medical simulation: patient-specific virtual reality simulation. World J. Surg. **36**(7), 1703–1712 (2012)

The 10th International Workshop on Intelligent Sensors and Smart Environments (ISSE-2019)

Fusion Method of Depth Images and Visual Images for Tire Inspection

Chien-Chou Lin[✉], Chun-Cheng Chang, Ching-Lung Chang,
and Chuan-Yu Chang

National Yunlin University of Science and Technology,
No. 123, University Rd., Section 3, Douliou 64002, Yunlin, Taiwan (R.O.C.)
linchien@yuntech.edu.tw

Abstract. In this paper, an alignment approach is proposed for a depth image and a visual image of a tire captured by a laser displacement sensor and a color camera respectively. While the global match usually misaligns the regular textures on a tire, the local match scheme proposed in this paper can exactly locate the logo textures on a tire. The result of the proposed method can be used for defect detection by 2D and 3D texture features. With 2D and 3D texture feature matching, the defect detection will be more accurate.

1 Introduction

Generally, tire defect inspection [1, 2] is a more difficult issue compared to other optical inspection issues since the tire and the defects are all black-colored, and the brightness variation induced by light reflection is not distinct, which makes it rather difficult for automatic tire defect detection to be conducted through image enhancement and identification technology. As a result, most tire defect inspection system adopts three-dimensional information and other information while processing. Common defects observed from the tire production process include serial number shedding, indentation, uneven thickness, bulge as shown in Fig. 1. Most types of defects cannot be easily detected by only using 2D image. Since fusion of 3D image and 2D image can enhance the quality of the 2D images by 3D information [3], further defect inspection can be conducted by integrating depth image and color cameras. Therefore, more multi-sensor systems are proposed for industrial inspection. In this paper, a fusion method which combines the range data and the visual image, is proposed for tire defect inspection system. The rest of this paper is organized as follows. The proposed algorithm is introduced in Sect. 2. The simulation result and the conclusion are in Sects. 3 and 4, respectively.

2 Template Matching Based Alignment

In Fig. 2, the proposed template matching based alignment has three main procedures. Firstly, the input depth data is transformed into depth image, enhanced by edge detection, and located by template matching [4]. In second part, the visual image is

© Springer Nature Switzerland AG 2020
L. Barolli et al. (Eds.): NBiS-2019, AISC 1036, pp. 487–492, 2020.
https://doi.org/10.1007/978-3-030-29029-0_46

captured by a single image from the video stream, enhanced by edge detection and located by template matching. In the third part, SIFT [5] is adopted to align both depth image and visual image.

In Fig. 3, the original depth data captured by a laser displacement sensor is shown in color which indicates the distance of a point from the laser source. It is difficult to transform a point cloud to gray-level image directly since the dynamic range of the depth data of a tire is large and the meaningful features are very small about 1–2 mm.

Fig. 1. Two types of tire defects.

Fig. 2. The proposed algorithm consists of three parts: In the first part, the deep data are transformed, enhanced to a depth image and the depth image is localized by template matching. In the second part, the visual image is enhanced and localized. In the third part, SIFT matching is adopted to find the transformation matrix between the depth image and the visual image.

Fig. 3. A depth data of a tire scanned by laser displacement sensor.

In order to enhance the features within the high dynamic range, histogram equalization is adopted in this paper. Thus, the proposed transformation from 3D point cloud to gray-level image can be expressed as

$$h_{(depth)} = \frac{(cdf_{(depth)} - cdf_{\min})}{cdf_{\max} - cdf_{\min}} .255 \tag{1}$$

where $h_{(depth)}$ is the obtained grayscale value and $cdf_{(depth)}$ is the cumulative sum of depth values, and cdf_{min} is the cumulative minimum.

For a gray-level image, the enhanced features are very useful for matching. In this paper, Scharr edge detection [6] is adopted to extracted the edge features. As shown in Fig. 4, the enhanced image after edge detection still contains many noises, which impact the subsequent alignment results. Therefore, it is hard to align the previous image with its visual image. Therefore, in this paper, some patterns are used to locate the same positions between two images. The visual image captured by a high-speed color camera is shown in Fig. 5. In order to make it into the same size as depth image, the lower part of the visual image is discarded. Like the depth image, Scharr edge detection is also applied to ensure that its features can be highlighted.

Fig. 4. The enhanced depth image by Scharr edge detection.

Fig. 5. The visual image captured by a high-speed color camera.

Fig. 6. The matching results of the depth image by the difference of square method.

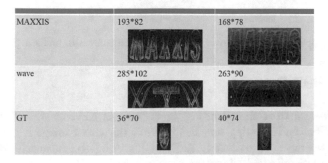

MAXXIS	193*82	168*78
wave	285*102	263*90
GT	36*70	40*74

Fig. 7. Templates of visual images (on the left-hand side) and depth images (on the right-hand side).

In this paper, alignment of two image is finding the corresponding positions of the same patterns firstly. The principle of finding a specific template in an image involves comparing each position in the image to the template to spot the similarities. A high similarity signifies the target position. Considering a template T, I is the input image. $T(x, y)$ and $I(x, y)$ are the pixels of the template and the input image, respectively. The larger the matching value, the weaker the match. In this paper, the correlation coefficient based matching has better results. Correlation coefficient is represented a matching measure that calculates the strength of the relationship between the images. The values range between -1.0 and 1.0. A correlation coefficient of -1.0 shows a perfect negative correlation, while a correlation coefficient of 1.0 shows a perfect positive correlation. $R(x, y)$ indicates the similarity at (x, y) of the input image. Matching results are illustrated in Fig. 6.

The tire templates used in this study are the most primary markings in the tire industry, namely MAXXIS, Wave and GT. The size of the visual image and depth image of MAXXIS markings are 193*82 and 168*78, Wave markings are 285*102 and 263*90, while GT markings are 36*70 and 40*74, as shown in Fig. 7.

After template places are allocated at two images, the transformation matrix between two images can be obtained by the corresponding pixel pairs. SIFT matching algorithm is adopted in this paper. Considering two corresponding pixels, $T(x,y)$ and $I(x',y')$, the transformation matrix can be expressed as

$$\begin{bmatrix} x' \\ y' \\ 1 \end{bmatrix} = \begin{bmatrix} a & b & t_x \\ c & d & t_y \\ 0 & 0 & 1 \end{bmatrix} \cdot \begin{bmatrix} x \\ y \\ 1 \end{bmatrix} \tag{2}$$

where t_x and t_y are the translation and a, b, c, and d are the coefficients of the rotation matrix and can be derived by least square error method.

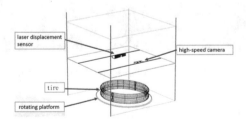

Fig. 8. The proposed system consists of laser displacement sensor and a high-speed camera.

3 Experiment Result

The proposed system is shown in Fig. 8. It consists Keyence Ljv7300, high-speed camera, light bar and a testing tire. In Fig. 9(a), the visual image is captured by the high-speed camera at approximately 225°. In Fig. 9(b), the enhanced depth image is shown. The result of SIFT alignment is shown in Fig. 9(C). Then, two images are

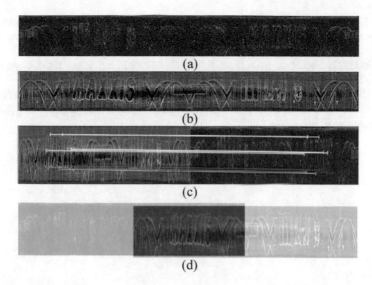

(a)

(b)

(c)

(d)

Fig. 9. The first tire is approximately 225°, the starting alignment results are (a) visual image (b) depth image (c) SIFT alignment results (d) overlay results (visual image on the left, depth image on the right)

merged together. The overlay result (visual image on the left, depth image on the right) is shown in Fig. 9(d).

4 Conclusion

In this paper, an alignment approach is proposed for a depth image and a visual image of a tire captured by a laser displacement sensor and a color camera respectively. While the global match usually misaligns the regular textures on a tire, the local match scheme proposed in this paper can exactly locate the logo textures on a tire. After alignment and fusion the two sensor data, the visual image can be enhanced by overlapping depth image and visual image for tire defect detection.

References

1. Zhao, G., Qin, S.: High-precision detection of defects of tire texture through X-ray imaging based on local inverse difference moment features. Sensors **18**(8), 191–201 (2018)
2. Zhang, Y., Lefebvre, D., Li, Q.: Automatic detection of defects in tire radiographic images. IEEE Trans. Autom. Sci. Eng. **14**(3), 1378–1386 (2017)
3. Klimentjew, D., Hendrich, N., Zhang, J.: Multi sensor fusion of camera and 3D laser range finder for object recognition. In: 2010 IEEE Conference on Multisensor Fusion and Integration (2010)

4. Tuzel, O., Porikli, F., Meer, P.: Region covariance: a fast descriptor for detection and classification. In: European Conference on Computer Vision (ECCV 2006), pp. 589–600 (2006)
5. Lowe, D.G.: Distinctive image features from scale-invariant keypoints. Int. J. Comput. Vision **60**(2), 91–110 (2004)
6. Jahne, B., Scharr, H., Korgel, S.: Principles of filter design. In: Computer Vision and Applications, vol. 2, Signal Processing and Pattern Recognition, Chapter 6, pp. 125–151. Academic Press, San Diego (1999)

A Research on Constructing the Recognition System for the Dynamic Pedestrian Traffic Signals Through Machine Vision

Chien-Chung Wu[(⊠)] and Yi-Chieh Hsug

Department of Computer Science and Information Engineering,
Southern Taiwan University of Science and Technology, No. 1, Nan-Tai Street,
Yongkang District, Tainan City 710, Taiwan (R.O.C.)
{wucc,ma4g0212}@stust.edu.tw

Abstract. In order to assist people with visually degraded or damaged to obtain the information from the dynamic pedestrian traffic signals at the intersection, this system captured the image in front of the pedestrian through the camera, and analyzed the content of the image to recognize the dynamic pedestrian traffic signal (traffic lights with the countdown seconds) at the intersection. Through the Haar-like with HSV color correction in the experiment, the average accuracy of the recognition result was 94.6%, and the average recall rate was 81.6%. In addition, as for the digit identification of the countdown seconds, the coordinates of the seconds were estimated first, the digit image was centered and then divided. Finally, the digit recognition was performed and then converted back to the number of seconds. The average precision of the recognition result was 95.2%, and the average recall rate was 77.2%.

1 Introduction

For the pedestrian traffic signals at the intersection, many countries have recently adopted the Dynamic Pedestrian Traffic Signals (pedestrian traffic lights and count-down display) to provide information for passersby, so that pedestrians can easily assess whether there is enough time to cross the intersection when crossing the road. However, people with visually degraded or damaged have not been significantly assisted by the design. These people can only be reminded through the rhythm of the sound made from the previous devices installed at the intersection. However, there is not a lot of intersections with sound-assisted devices. In order to solve this problem, this paper identified the information of the pedestrian signals by means of machine vision and provided information of the recognition results to pedestrians.

However, the mode to display the dynamic pedestrian traffic signal is performed by LED matrix scanning. This type of display technology is based on the characteristic of the persistence of vision transmitted to the human eye. The whole image appears in the human eye through the repeated scanning, but it is often unable to achieve a complete image of this type on a smartphone or a camera. Wu *et al.* [1] proposed to solve this problem by adjusting the shutter speed of the camera. It was mentioned in this paper that when the shutter speed was set to 20 ms or slower, the dynamic pedestrian traffic

L. Barolli et al. (Eds.): NBiS-2019, AISC 1036, pp. 493–501, 2020.
https://doi.org/10.1007/978-3-030-29029-0_47

signals could be captured completely. In the next experiment, the camera shutter speed was set at 1/50 s.

2 Design of System

In order to correctly recognize the dynamic pedestrian traffic signals at the intersection, the flow diagram is shown in Fig. 1. First, the image was captured (as shown in Fig. 1(a)). The system used the Haar-like feature to capture the dynamic pedestrian traffic signals (see Fig. 1(b)), but the correct rate of recognition was not ideal only through the Haar-like feature to find the signals. A large part of the framed target might be misjudged. In this paper, through the simple color detection, the misidentification was filtered out, which could greatly improve the correct rate of target image recognition (Fig. 1(c)).

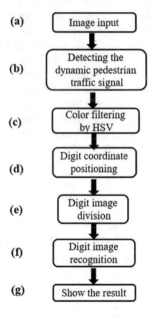

(a) Image input

(b) Detecting the dynamic pedestrian traffic signal

(c) Color filtering by HSV

(d) Digit coordinate positioning

(e) Digit image division

(f) Digit image recognition

(g) Show the result

Fig. 1. System flow diagram.

Next, the countdown seconds had to be identified, because the number of seconds was related to the relative position of the image of the dynamic pedestrian traffic light. After recognizing the image range and coordinates of the dynamic pedestrian traffic light, as long as the height of the traffic light was doubled and the upper half of the coordinate range was taken, the relative coordinate position of the countdown seconds in the image could be estimated. (Figure 1(d)) In order to simplify the digit recognition, the digit recognition system of this paper was only able to recognize the number 0–9, so it was necessary to divide the units digit and tens digit from the image. (Figure 1(e)) Finally, the numbers of different digits were separately identified, and then digits were restored by weight (as shown in Fig. 1(f)), and the result of the whole recognition was obtained (Fig. 1(g)).

The blocks in the system process flow will be explained next.

2.1   The Recognition System for the Dynamic Pedestrian Traffic Signals

In order to recognize the location of the dynamic pedestrian traffic signals in the input image quickly, this paper used the Haar-like algorithm [2], which was trained by the Haar-like classifier, using the OpenCV Train cascade method in the OpenCV Library.

After training, the collected positive/negative samples were designed, and the OpenCV Create samples method in the OpenCV Library was used to gray-scale the vec file, and then the training was started.

2.2   The Color Recognition System for the Dynamic Pedestrian Traffic Signal

After recognizing the traffic signal quickly through Haar-like, the recognition result would be filtered by HSV (Hue, Saturation, Value) and binarized by the adjusted HSV threshold. All the colors were filtered out except the dynamic pedestrian traffic light [3], and the pedestrian traffic light was left and finally framed.

2.3   Pedestrian Signal Digit Recognition System

After framing the coordinates and the length and width range of the pedestrian traffic light, by framing twice the coordinate position above the Y-axis through the calculation, the coordinates of the countdown seconds could be located. After capturing the image of the coordinates and the range of seconds, the method of binarization [4] was performed by color filtering, similar to the aforementioned HSV, leaving a white number with a black background (Fig. 2(a)).

Due to the discontinuous fracture problem in the image after the above treatment, the digit image would present a relatively continuous integrity after the expansion treatment and erosion treatment [5]. Next, in order to locate the position of each number correctly, the digit part had to be placed in the middle of the screen, by removing the blank part (as shown in Fig. 2(b)), and calculating the remaining of the white part.

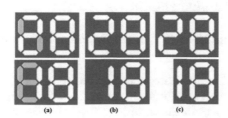

(a)   (b)   (c)

Fig. 2. Diagram of the digit processing of the countdown seconds, (a)The display of '28' and '18' in seven segments, and the white parts display the bright segments, (b) the diagram after expansion and erosion treatment, (c) the diagram of removing the blank part on both sides

When the digit image was centered, special processing would be required only when there was the units digit and tens digit was '1'. The problem will be explained next. The figures of '18' and '28' are specifically demonstrated in Fig. 2.

For processing centered digit images, it was the easiest to process as there was only the units digit left. However, the number was between 10 – 19 (as '18' shown in the lower part of Fig. 3), and there was '1' left in the tens digit of the second (as '18' shown in the lower part of Fig. 2(c)), although there were two digits, the width of the two digits was not the same, and the digit image couldn't be directly divided into half. In contrast, when the numbers were 20–99 (as '28' shown in the upper of Fig. 2) and the tens digit of the second was '2' (as '28' shown in the upper part of Fig. 2(c)), the digit image could be directly divided into half.

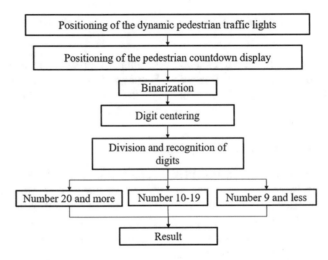

Fig. 3. Digit processing for the countdown seconds

Finally, the numerical values were put centered and divided into three different categories, and then they were recognized respectively. The entire digit processing flow is shown in Fig. 3.

3   Training and Implementation

3.1   The Collection and Training of the Dynamic Pedestrian Traffic Signals

As shown in Fig. 4, it is the procedure of collecting samples through a network or videos taken by ourselves. First, the video was converted into frames, and then to frame a large number of images. As shown in Fig. 4(a), after the video was converted into frames, the positive sample could be targeted and framed (as shown in Fig. 4(b)).

When the mouse was released, it would go directly to the next frame (Fig. 4(c)). To collect a large number of positive samples, and to normalize the samples after the collection was completed, then to collect negative samples, which were non-target images, and finally training.

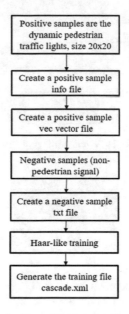

Fig. 4. Haar-like training process

The training file cascade.xml [6] could be obtained. The training procedure is shown in Fig. 4.

3.2 Color Capture of the Dynamic Pedestrian Traffic Lights

After the Haar-like screening, the image would be recognized by the HSV. The color recognition method was used to exclude the misjudgment, and all the colors except the dynamic pedestrian traffic lights were filtered into binarization and the white part was reserved [3]. HSV color filtering process is shown in Fig. 5.

3.3 Ways of Determinating Digits

The countdown seconds were mostly displayed in a font similar to a seven-segment display. Therefore, after capturing the digit image, the image processing was performed as described above, and the image was divided into seven regions for pattern recognition, and the regions were 'a', 'b',..., 'g', respectively. The relative coordinates of the seven regions would be calculated according to the size of the digit image, as shown in Fig. 6(a).

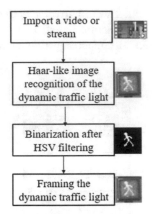

Fig. 5. HSV color filtering procedure

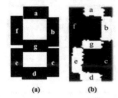

Fig. 6. HSV color filtering procedure, (a) the ideal digit image and the relative position of the regions, (b) the image captured based on the example of "2"

Figure 6(b) shows the number '2' as an example. The number was divided into seven regions, which were the regions from 'a' to 'g', and calculated the ratio of the sum of the white pixels to that of the all pixels in the region (Eq. 1), and set a threshold. When the ratio of the white pixels in the region exceeded the threshold, it was determined that the area was '1' instead of '0'. The threshold setting in the experiment was 50%.

$$White\_Ratio_{Region} = \frac{\sum\limits_{Region} Pixels_{white}}{\sum\limits_{Region} Pixels} \tag{1}$$

Taking Fig. 6(b) as an example, there were 'a', 'b', 'd', 'e', 'g', which was determined to be '1', and the rest of the region was '0'. The final result obtained with reference to Table 1 was the number '2'.

Table 1. Form of Digits

Digit	a	b	c	d	e	f	g
1	0	1	1	0	0	0	0
2	1	1	0	1	1	0	1
3	1	1	1	1	0	0	1
4	0	1	1	0	0	1	1
5	1	0	1	1	0	1	1
6	0	0	1	1	1	1	1
7	1	1	1	0	0	0	0
8	1	1	1	1	1	1	1
9	1	1	1	0	0	1	1
0	1	1	1	1	1	1	0

4 Experiments and Results

The videos in the experiment were recorded at several intersections in Tainan City. The camera specifications was SONY NEX-3N1, E PZ16–50 mm, F3.5–5.6 OSS. Training and testing video separately, positive and negative samples were taken through the video and then training. The detection and evaluation was based on the defined TP (Truth Positive), TN (Truth Negative), FP (False Positive), FN (False Negative), and the final result was evaluated by the precision and recall rate.

$$\text{precision} = \frac{TP}{TP + FP} * 100\%. \tag{2}$$

$$\text{Recall} = \frac{Tp}{TP + FN} * 100\%. \tag{3}$$

4.1 Recognition Result of the Dynamic Pedestrian Traffic Lights

The tested videos included Video1, Video2,..., Video10. The tested films were recorded in different weather, light and time. The test results of different videos varied greatly. After the data was substituted into Eqs. 2 and 3, the average accuracy of Haar-like was 41.2%, and the average recall rate was 83.4%. The experimental results are shown in Table 2.

4.2 Results Corrected by HSV

After the dynamic traffic signals was captured through Haar-like, the average accuracy after color filtering through HSV was 94.6%, the recall rate was 81.6%, the average accuracy was increased by 53.4%, and the recall rate was decreased by 1.8%. The experimental results are shown in Table 3.

Table 2. Haar-like recognition results

Video name	Precision (%)	Recall (%)
Video1	72.2	84.4
Video2	54.0	79.7
Video3	28.9	55.2
Video4	28.9	74.2
Video5	42.0	78.4
Video6	56.4	88.6
Video7	34.7	82.6
Video8	23.0	95.9
Video9	48.0	98.1
Video10	24.2	96.4
Average	41.2	83.4

Table 3. The result of the dynamic traffic signals corrected through HSV

Video name	Precision (%)	Recall (%)
Video1	98.6	84.4
Video2	97.7	75.7
Video3	88.9	55.2
Video4	92.6	73.8
Video5	94.1	77.8
Video6	99.4	88.6
Video7	81.4	81.7
Video8	100	94.5
Video9	95.7	97.9
Video10	97.3	86.7
Average	94.6	81.6

4.3   Digit Recognition Results of the Countdown Seconds

As to the recognition of numbers for the countdown seconds, different numbers had some differences in recognition. Especially, the number 9 and 7 were easily recognized as 7 or 1. The relevant data is shown in Table 4. The overall average precision was 95.2%, and the average recall was 77.2%.

Table 4. Results after filtering

Digit	Precision (%)	Recall (%)
0	91	78.5
1	100	64
2	98.3	65.6
3	90.4	68.7
4	100	81.1
5	93.3	63.6
6	100	91.5
7	88.7	90
8	94.3	83.1
9	95.6	86
Average	95.2	77.2

5   Conclusion

In this paper, the average accuracy of the dynamic pedestrian traffic signals in the first phase of Haar-like training was 41.2%, and the average recall rate was 83.4%. After using Haar-like plus HSV color correction in the second stage, the average accuracy result was 94.6%, and the average recall was 81.6%. In addition, the average accuracy of the digit recognition results in the countdown seconds was 95.2%, and the average recall rate was 77.2%.

Acknowledgments. The completion of this paper is very grateful to the Ministry of Science and Technology MOST 107-2221-E-218-023-MY2.

References

1. Wu, C.C., Hsu, Y.C.: The case study of capturing images for the dynamic pedestrian traffic sign. In: 2017 International Symposium on Novel and Sustainable Technology, pp. 84–85 (2017)
2. Viola, P., Jones, M.: Rapid object detection using a boosted cascade of simple features. In: IEEE Computer Society Conference on Computer Vision and Pattern Recognition, vol. 1, pp. 511–518 (2001)
3. Shanmugavadivu, P., Ashish, K.: Human skin detection in digital images using multi colour scheme system. In: 2017 Fourth International Conference on Image Information Processing (ICIIP), pp. 1–6 (2017)
4. Ranjit, G., Ayan, B.: An improved scene text and document image binarization scheme. In: 2018 4th International Conference on Recent Advances in Information Technology (RAIT), pp. 1–6 (2018)
5. Jin, H., Sun, H.: Rendering fake soft shadows based on the erosion and dilation. In: 2010 2nd International Conference on Computer Engineering and Technology, vol. 6, p. 236 (2010)
6. Meng, L.Y., Zaiqing, C., Feiyan, C.: Research on video face detection based on AdaBoost algorithm training classifier. In: 2017 First International Conference on Electronics Instrumentation & Information Systems (EIIS), pp. 1–6 (2017)

Learning Depth from Monocular Sequence with Convolutional LSTM Network

Chia-Hung Yeh[1,2(✉)], Yao-Pao Huang[2], Chih-Yang Lin[3],
and Min-Hui Lin[2]

[1] Department of Electrical Engineering, National Taiwan Normal University,
Taipei, Taiwan
`chyeh@ntnu.edu.tw`
[2] Department of Electrical Engineering, National Sun Yat-sen University,
Kaohsiung, Taiwan
[3] Department of Electrical Engineering, Yuan Ze University, Taoyuan, Taiwan

Abstract. Resolving depth from monocular RGB image has been a long-standing task in computer vision and robotics. Recently, deep learning based methods has become a popular algorithm on depth estimation. Most existing learning based methods take image-pair as input and utilize feature matching across frames to resolve depth. However, two-frame methods require sufficient and static camera motion to reach optimal performance, while camera motion is usually uncontrollable in most application scenarios. In this paper we propose a recurrent neural network based depth estimation network. With the ability of taking multiple images as input, recurrent neural network will decide by itself which image to reference during estimation. We train a u-net like network architecture which utilizes convolutional LSTM in the encoder. We demonstrate our proposed method with the TUM RGB-D dataset, where our proposed method shows the ability of estimating depth with various sequence lengths as input.

Keywords: Multi-view depth estimation · Deep learning ·
Convolutional LSTM · Recurrent neural network

1 Introduction

Depth estimation from image is a long-standing task in computer vision and robotics. Stereo cameras are widely used for gathering depth information. Depth can be easily estimated with triangulation when the relative distance of the two cameras is known. Although stereo systems achieve impressive results, monocular systems are much more widely used in daily application such as cell phones and web cams. Therefore, building a depth estimation system for monocular data is still a meaningful and important task.

Traditional methods such as REMODE [1] utilizes correspondences between two images to update probabilistic model, depth information is then estimated with the model. Recently, machine learning based method has been a popular solution for depth estimation, especially on monocular systems since machine learning is the best way to resolve scale ambiguous.

© Springer Nature Switzerland AG 2020
L. Barolli et al. (Eds.): NBiS-2019, AISC 1036, pp. 502–507, 2020.
https://doi.org/10.1007/978-3-030-29029-0_48

While learning based method is widely used for resolving depth, as training target, accurate dense depth map is hard to obtain since consumer depth camera such as Microsoft Kinect usually use structured in-far-red projector to obtain depth information. However, structured light system is sensitive to object texture and occlusions, the obtained depth map usually contains a lot of invalid values. Outdoor depth map is often obtained by lidar sensor, lidar has the advantage of high precision but it is usually highly priced, and the obtained depth map is very sparse. Therefore, training data of depth estimation is highly restricted. Some recent work such as [2] incorporates unsupervised learning techniques to reduce the required amount of training data. Lore et al. [3] use generative adversarial network for single image depth estimation. Singe image depth estimation has been a painful topic for supervised learning since the network needs to extract depth information base on the understating of the size of the learned object. Generative adversarial training helps generalizing through datasets, therefore not only reduces the needed amount of training data, but also reduces the need of network size.

Most existing state-of-the-art deep learning based methods proposed to use image pair with different view as input. Given more 3D structural information, deep learning network will utilize feature movement between inputs to give a more accurate estimation. DeMoN [4] is one of the state-of-the-art learning-based method which takes two images as input. DeMoN's network first estimates optical flow with the input pair, the optical flow is then concatenated with input image pair for the second part of its framework: iterative net to estimate depth and camera ego-motion. A refinement network further improves the output depth map. MVDepthNet [5] is another state-of-the-art that takes image pair as input. To further enhance performance and reduce computational cost, MVDepthNet also takes relative camera pose difference between the two images as input. The input poses are used to construct image matching cost for the network, which provides more 3D structural information and reduces the complexity of estimation. However, the application of MVDepthNet is also limited by the need of camera pose, where high accurate camera pose is hard to obtain in general, especially when only monocular RGB sequence is available.

Although multi-view system can achieve better results, it requires carefully pre-processed image paired to gather optimal performance. Since camera motion is usually uncontrollable at inference, input pairs may vary between insufficient camera movement and excessive camera movement, leading to undesirable estimation results. In this paper we propose a recurrent neural network based architecture, by incorporating convolutional LSTM [6], the network can decide by itself which image in the sequence to reference, therefore reduce the impact of inconsistence camera movement. During training, the proposed network is feed with a 10-image sequence, where it will estimate depth for the latest. Experimental results show that not only it is able to estimate depth with 10 image sequence, it is also capable of taking only one image as input, making our proposed method much more flexible for various situation.

2   Network Architecture

The overall network architecture is shown in Fig. 1. In this paper, we propose a simple architecture which combines U-net [7] and convolutional LSTM. The network takes 10 RGB images as input sequence and estimates the depth of the latest. The proposed network consists of 27 layers total. The encoder consists of 16 layers which alternate between normal convolution and convolutional LSTM and stride 1 and stride 2. The decoder consists of 11 layers that utilize transpose convolution for up-sampling. The first two convolution layers use kernel size of 7, followed by two layers using kernel size of 5, all the rest layers use kernel size of 3. Both input and output resolution are 256*192.

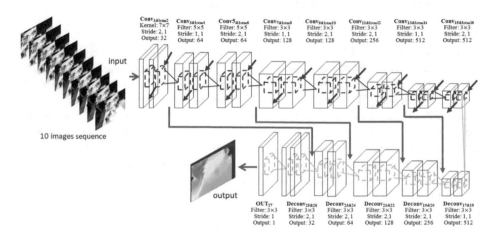

Fig. 1.   Network architecture

3   Experiments

The experiments are done with the TUM RGB-D dataset [8], which provides RGB images and depth maps with ground truth camera pose tracked by eight high speed tracking cameras. We use their validation split to separate train-test data. Our training set consist of 8036 RGB-D pairs where validation data consists of 967 pairs. All validations are done on the validation set Freiburg1_xyz_validation.

3.1   Error Metrics

Three measurements are used to evaluate the performance of the proposed method following DeMoN. The scale invariant error metric of [9] which indicates the overall structural accuracy of the estimated depth map is adopted as

$$sc - inv = \sqrt{\frac{1}{n}\sum_i z_i^2 - \frac{1}{n^2}\left(\sum_i Z_i\right)^2}, \tag{1}$$

where $z_i = \log d_i - \log \widehat{d}_i$, d_i is the estimated depth value and $\widehat{d}_i$ is the ground truth depth value. L1-rel computes the depth error relative to ground truth depth, which increases the importance of close objects. L1-inv represent the overall error of the depth map.

$$L1 - rel = \frac{1}{n}\sum_i \frac{\left|d_i - \widehat{d}_i\right|}{\widehat{d}_i}, \tag{2}$$

$$L1 - inv = \frac{1}{n}\sum_i \left|\frac{1}{d_i} - \frac{1}{\widehat{d}_i}\right|. \tag{3}$$

3.2 Results Comparison

The experimental results are shown in Table 1. The experimental results show that our proposed method can not only estimate depth with 10 image sequence, but also with only one image. Evaluation metrics prove that even with this simple and straight forward architecture, the propose method has the ability to give competitive accuracy comparing to state-of-the-art, especially on structure wise. Visualized comparison of depth map is shown in Fig. 2, the input RGB image is shown on the left, and the ground truth depth map is shown on the right. We provide both estimation result with one and ten images as input. It is shown that when estimated with only one image, our proposed method can still give acceptable estimation with minor structural errors, while using ten image sequence as input further enhances estimation accuracy.

Table 1. Evaluation of proposed method on the TUM RGB-D dataset

	Proposed RNN with one image input	Proposed RNN with ten image sequence input	DeMoN[]	MVDepthNet[]
L1-rel	0.140156	**0.130116**	**0.130**	**0.130**
L1-inv	0.120170	0.117848	**0.041**	0.069
Sc-inv	0.116459	**0.106782**	0.183	0.180

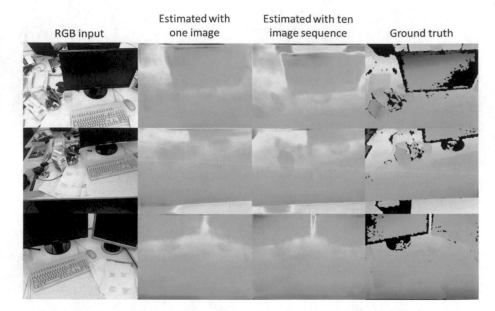

Fig. 2. Visualized depth estimation result on TUM RGB-D dataset

4 Conclusion and Future Work

In this paper, we propose a recurrent neural network based method to estimate depth information from monocular sequences. Comparing to existing methods, the proposed method does not require carefully pre-processed input therefore is less complex and more suitable for a variety of situations. We focus our work on making the proposed method to work in situations that only video sequence is available. The experimental results show that the proposed method can overcome the problem of uncontrollable camera movement, while also having the ability to estimate depth with only one image as input, therefore making the proposed method more applicable to different scenarios.

For the future work, more dataset can be added for training, such as outdoor data KITTI dataset [10] for further improvement on making the model more generalize. Also, more dataset can be added for evaluations. NYU dataset [11] is a public dataset which also provides RGB-D image pairs, which would be a good choice for augmenting indoor training data. We've proved that even with a simple architecture, recurrent neural networks can overcome the problem of uncontrollable camera movement. A more complex and fine-tuned architecture cane be used for better feature extraction and estimation accuracy.

Acknowledgments. The authors would like to thank the Ministry of Science and Technology, Taiwan, R.O.C. for financially supporting this research under grants MOST 107-2218-E-003-003-, MOST 107-2218-E-110-004-, MOST 105-2221-E-110-094-MY3 and MOST 106-2221-E-110-083-MY2.

References

1. Pizzoli, M., Forster, C., Scaramuzza, D.: REMODE: probabilistic, monocular dense reconstruction in real time. In: Proceedings of IEEE International Conference on Robotics and Automation, pp. 2609–2616, May 2014
2. Zhou, T., Brown, M., Snavely, N., Lowe, D.G.: Unsupervised learning of depth and ego-motion from video. In: Proceedings of the IEEE Conference on Computer Vision and Pattern Recognition, pp. 1851–1858, June 2017
3. Gwn Lore, K., Reddy, K., Giering, M., Bernal, E.A.: Generative adversarial networks for depth map estimation from RGB video. In: Proceedings of IEEE Conference on Computer Vision and Pattern Recognition Workshops, pp. 1177–1185 (2018)
4. Ummenhofer, B., Zhou, H., Uhrig, J., Mayer, N., Ilg, E., Dosovitskiy, A., Brox, T.: Demon: depth and motion network for learning monocular stereo. In: Proceedings of IEEE Conference on computer vision and pattern recognition, vol. 5, p. 6, Jul 2017
5. Wang, K., Shen, S.: MVDepthNet: real-time multiview depth estimation neural network. In: Proceedings of International Conference on 3D Vision (2018)
6. Shi, X., Chen, Z., Wang, H., Yeung, D.Y., Wong, W.K., Woo, W.C.: Convolutional LSTM network: a machine learning approach for precipitation nowcasting. In: Advances in Neural Information Processing Systems, pp. 802–810 (2015)
7. Ronneberger, O., Fischer, P., Brox, T.: U-net: convolutional networks for biomedical image segmentation. In: Proceedings of International Conference on Medical Image Computing and Computer-Assisted Intervention, pp. 234–241, October 2015
8. Sturm, J., Engelhard, N., Endres, F., Burgard, W., Cremers, D.: A benchmark for the evaluation of RGB-D SLAM systems. In: Proceedings of IEEE International Conference on Intelligent Robots and Systems, pp. 573–580, Oct 2012
9. Eigen, D., Puhrsch, C., Fergus, R.: Depth map prediction from a single image using a multi-scale deep network. In: Advances in Neural Information Processing Systems, pp. 2366–2374 (2014)
10. Geiger, A., Lenz, P., Urtasun, R.: Are we ready for autonomous driving? the kitti vision benchmark suite. In: Proceedings of IEEE Conference on Computer Vision and Pattern Recognition, pp. 3354–3361 (2012)
11. Silberman, N., Hoiem, D., Kohli, P., Fergus, R.: Indoor segmentation and support inference from RGBD images. In: Proceedings of European Conference on Computer Vision (2012)

Two-Dimensional Inductance Plane Sensor for Smart Home Door Lock

Wen-Shan Lin[1], Chao-Ting Chu[3], and Chian C. Ho[2]($\boxtimes$)

[1] Graduate School of Vocation and Technological Education, National Yunlin University of Science and Technology, Douliou 64002, Yunlin County, Taiwan
dl0343003@yuntech.edu.tw
[2] Department of Electrical Engineering, National Yunlin University of Science and Technology, Douliou, Yunlin County 64002, Taiwan
futureho@yuntech.edu.tw
[3] Internet of Things Laboratory, Chunghwa Telecom Laboratories, 99, Dianyan Rd., Yangmei District, Taoyuan City 32661, Taiwan
chaot@cht.com.tw

Abstract. This paper proposes two-dimensional inductance plane sensor for smart home door lock. Traditional inductance plane sensor adopts electronic circuit coil to detect one-dimensional distance position of the metal object composed of push-button switches, gear-tooth counting, and flow meters. Besides, the proposed inductance plane sensor has the advantages of lower power consumption and cheaper circuit cost. Technically, two-dimensional positioning coordinate estimation method is proposed to obtain the object's location and is integrated into the wearable device for unlocking smart home door.

1 Introduction

One of the most significant current discussions in smart home is the field of sensor control [1, 2]. Smart home system with some gateway portal to connect different home devices and manage device information transmission. Recently, literature has emerged that offers intelligent control findings about sensor detection [3, 4] to detect different sensor information for control system exhibition and reaction. Semnani et al. [3] proposes semi-flocking algorithm for motion control of mobile sensors in large-scale surveillance systems, and the surveillance systems use semi-flocking algorithm to track motion signal which have significant output response. Li et al. [4] designs and implements smart home control systems based on wireless sensor networks and power line communications so that smart home control systems make use of wireless sensor networks to monitor and feedback peripheral device situation.

Most studies in sensor control fields with capacitive components [5, 6] have many public literature. Capacitance sensors use the principle of capacitive effect to detect object movement without traditional button module cost. Zeng et al. [5] presents a capacitive sensor for the measurement of the departure from the vertical movement, and the experimental results show it works well that the capacitive sensor detects the departure from the vertical movement. In [6], a novel modified de-sauty autobalancing bridge-based analog interfaces for wide-range capacitive sensor applications have been

© Springer Nature Switzerland AG 2020
L. Barolli et al. (Eds.): NBiS-2019, AISC 1036, pp. 508–516, 2020.
https://doi.org/10.1007/978-3-030-29029-0_49

proposed by Mantenuto et al. Its analog interface can be integrated into different devices to show novel modified de-sauty autobalancing performance. In recent years, there is an increasing interest in the inductance sensors [7, 8]. The features of inductance sensor adopts non-contact methods and remote detection methods to obtain object signal. In [7], a design and analysis of a dual-slope inductance-to-digital converter for differential reluctance sensors have presented by Philip et al., and the dual-slope inductance differential reluctance sensors with novel algorithms can achieve remarkable output response. Mandal et al. [8] design a flow transmitter based on an improved inductance bridge network and a rotameter as sensors. Its experimental results show the inductance bridged network can detect the flow situation and feedback for real-time monitor.

To date, lots of methods have been developed and introduced to measure the object distance [9, 10]. However, the inductance sensor method to detect one-dimensional motion object has numerous limitations. This paper proposes two-dimensional inductance plane sensor for smart home door lock. The inductance plane sensor makes use of high-resolution inductance-to-digital converter of Texas Instruments LDC1614 and low-power microcontroller of MSP430 to calculate the object's location coordinate. In addition, the two-dimensional inductance plane sensor detects the object's location coordinate to obtain the object signal on smart home door lock. Therefore, the users can unlock the smart home door more easily than traditional sensor schemes. Experimental results show the microcontroller of MSP430 calculated object coordinate based on inductance sensor has remarkable output performance.

2 Hardware Architecture of Two-Dimensional Inductance Plane Sensor

Figure 1 is the function diagram of two-dimensional inductance plane sensor. This paper uses three inductance coils with the frequency converter core of LDC1614 to obtain electromagnetic induction strength between inductance coils and object. Then, This paper uses three inductance coils to calculate the object location coordinate based on the microcontroller of MSP430. Figure 2 shows the flowchart of two-dimensional inductance plane sensor for the microcontroller. The microcontroller obtains the strength signal of three inductance coils from the frequency converter core of LDC1614, and this paper analyzes the linear transformation relationship of signal strength to improve electromagnetic nonlinear issue. Therefore, this paper uses the signal strength of three inductance coils to establish the x-axis and y-axis location coordinate of the object. Finally, the two-dimensional inductance plane sensor structure detects different information, including the distance, electromagnetic signal strength, and movement, from the object.

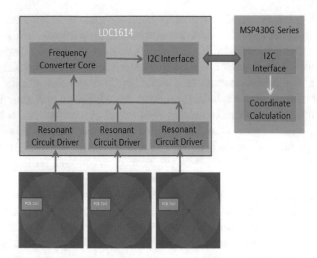

Fig. 1. Function diagram of two-dimensional inductance plane sensor.

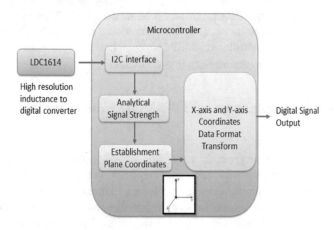

Fig. 2. Flowchart of two-dimensional inductance plane sensor for the microcontroller.

3   Algorithm Principle of Two-Dimensional Inductance Plane Sensor

3.1   System Deployment of Inductance Plane Sensor

The system deployment architecture of the inductance plane sensor is illustrated in Fig. 3. This paper applies the triangulation method to place three inductance coils, and the application scenario of the smart home door lock is shown in Fig. 4. The smart home door luck can utilize the NFC of wearable devices or other wireless technique to unlock door as normal. This paper further adds the security scheme into the two-dimension inductance plane sensor so that the user's hand must approach the security device and wave in the specific assigned way to unlock the door. Therefore, the door security lock is difficult to be cracked.

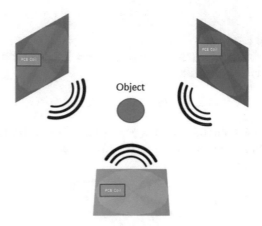

Fig. 3. System deployment architecture of the inductance plane sensor.

Fig. 4. Application scenario of the inductance plane sensor for the smart home door lock.

3.2 Least Squares Estimation

The electromagnetic characteristic of the inductance plane sensor has serious nonlinear issue so the inductance strength must be linearized before the object's location coordinate is calculated. In this section, this paper adopts least squares estimation method to improve the nonlinear issue of electromagnetic strength of the inductance plane sensor.

This paper must analyze the nonlinear equation to describe the object distance, so the assumption can be derived as (1):

$$y = a_2 x^2 + a_1 x + a_0 \tag{1}$$

where a_0, a_1, a_2 are constant.

Such a relationship between the nonlinear electromagnetic strength and the relative distance can be interpreted as (2):

$$(x_i, y_i) \quad for \quad i = 1, 2, 3, \ldots, m \tag{2}$$

where m is the total amount of estimated parameters. After substituting (2) to (1), the simultaneous equations can be derived as (3)–(6):

$$y_1 = a_2 x_1^2 + a_1 x_1 + a_0 \tag{3}$$

$$y_2 = a_2 x_2^2 + a_1 x_2 + a_0 \tag{4}$$

$$y_3 = a_2 x_3^2 + a_1 x_3 + a_0 \tag{5}$$

$$y_m = a_2 x_m^2 + a_1 x_m + a_0 \tag{6}$$

Then, (3)–(6) can be rewritten as (7):

$$W = \begin{bmatrix} x_1^2 & x_1 & 1 \\ x_2^2 & x_2 & 1 \\ x_3^2 & x_3 & 1 \\ & \vdots & \\ x_m^2 & x_m & 1 \end{bmatrix} \quad Y = \begin{bmatrix} y_1 \\ y_2 \\ y_3 \\ \vdots \\ y_m \end{bmatrix} \quad S_c = \begin{bmatrix} a_2 \\ a_1 \\ a_0 \end{bmatrix} \tag{7}$$

From (7), the parameters of nonlinear equation relationship between nonlinear electromagnetic induction strength and the relative distance can be interpreted as (8):

$$S_c = \left(W^T W \right)^{-1} W^T Y \tag{8}$$

3.3   Coordinate of the Object Positioning

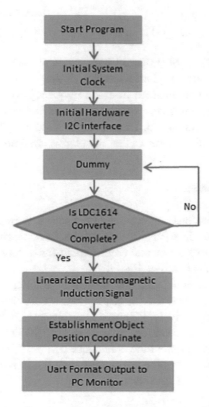

Fig. 5. Algorithm flowchart of proposed inductance plane sensor through the microcontroller.

The electromagnetic strength of the inductance plane sensor with least squares estimation method can be used to calculate the object's location coordinate. The relationship between the object and the inductance coil is interpreted as the following Eqs. (9)–(11) [11].

$$(x - x_a)^2 + (y - y_a)^2 = d_a \tag{9}$$

$$(x - x_b)^2 + (y - y_b)^2 = d_b \tag{10}$$

$$(x - x_c)^2 + (y - y_c)^2 = d_c \tag{11}$$

where x_a and y_a are the x-axis and y-axis coordinate of the first inductance coil. x_b and y_b are the x-axis and y-axis coordinate of the second inductance coil. x_c and y_c are the x-axis and y-axis coordinate of the third inductance coil. d_a, d_b, d_c are the relative distance of three inductance coil, respectively.

Therefore, this paper can deduce the relationship equation as (12):

$$\begin{bmatrix} x \\ y \end{bmatrix} = \begin{bmatrix} 2(x_a - x_c) & 2(y_a - y_c) \\ 2(x_b - x_c) & 2(y_b - y_c) \end{bmatrix}^{-1} \begin{bmatrix} x_a^2 - x_b^2 + y_a^2 - y_c^2 + d_c^2 - d_a^2 \\ x_b^2 - x_c^2 + y_b^2 - y_c^2 + d_c^2 - d_b^2 \end{bmatrix} \quad (12)$$

The algorithm flowchart of proposed inductance plane sensor through the micro-controller of MSP430G series is clarified as Fig. 5. The electromagnetic signal strength of the inductance coil can be obtained from the frequency converter core of LDC1614, and the microcontroller linearizes the electromagnetic signal strength before the positioning coordinate is calculated. Therefore, the situation of the object position and the electromagnetic signal strength of the inductance coil can be read in PC monitor.

4 Experimental Results

The proposed two-dimensional inductance plane sensor is implemented in the micro-controller of MSP430G2553 to achieve easy detection operating environment for the object's positioning. The experimental environment for testing is shown in Fig. 6, and three testing cases are illustrated as Fig. 7.

Figure 8 shows the received signal results of the inductance coil under the testing situation 3 (circles). Figure 8(a) and (b) display x-axis coordinate and y-axis coordinate of the received signal, respectively. Figure 8(c) displays two-dimensional coordinate of x-axis and y-axis of the received signal. Figure 8(d) displays the three-dimensional coordinate of the received signal of the inductance coil position. From the experimental result of Fig. 8, it is verified that the positioning coordinate estimation algorithm of the object indeed works well and immediately.

Fig. 6. Testing environment of proposed inductance plane sensor through the microcontroller.

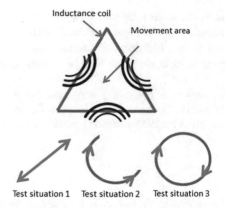

Test situation 1   Test situation 2   Test situation 3

Fig. 7. Three testing cases of proposed inductance plane sensor.

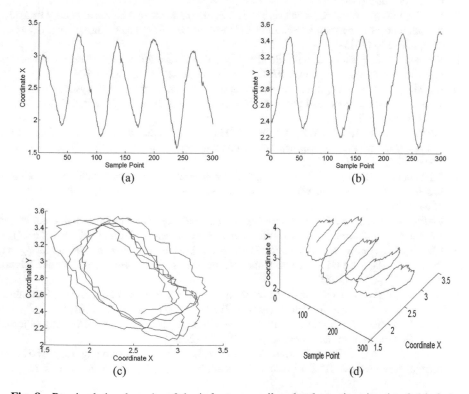

Fig. 8. Received signal results of the inductance coil under the testing situation 3 (circles).

5 Conclusions

This paper proposes and implements two-dimensional inductance plane sensor for smart home door lock. Firstly, this paper uses low-power microcontroller of MSP430G series to communicate the frequency converter core of LDC1614 and obtain the object signal strength. Secondly, the object signal strength adopts least squares estimation linearization and the equation of the signal strength transformation into the relative distance to calculate the object's location coordinate. Finally, the experimental results show the two-dimensional inductance plane sensor has remarkable performance.

Acknowledgments. This work was financially supported by the "Intelligent Recognition Industry Service Center" from The Featured Areas Research Center Program within the framework of Higher Education Sprout Project by Ministry of Education (MOE) in Taiwan.

References

1. Usman, M., Muthukkumarasamy, V., Wu, X.W.: Mobile agent-based cross-layer anomaly detection in smart home sensor networks using fuzzy logic. IEEE Trans. Consum. Electron. **61**(2), 197–205 (2015)
2. Liu, L., Liu, Y., Wang, L., Zomaya, A., Hu, S.: Economical and balanced energy usage in the smart home infrastructure: a tutorial and new results. IEEE Trans. Emerg. Top. Comput. **3**(4), 556–570 (2015)
3. Semnani, S.H., Basir, O.A.: Semi-flocking algorithm for motion control of mobile sensors in large-scale surveillance systems. IEEE Trans. Cybern. **45**(1), 129–137 (2015)
4. Li, M., Lin, H.J.: Design and implementation of smart home control systems based on wireless sensor networks and power line communications. IEEE Trans. Ind. Electron. **62**(7), 4430–4442 (2015)
5. Zeng, T., Lu, Y., Liu, Y., Yang, H., Bai, Y., Hu, P., Li, Z., Zhang, Z., Tan, J.: A Capacitive sensor for the measurement of departure from the vertical movement. IEEE Trans. Instrum. Meas. **65**(2), 458–466 (2016)
6. Mantenuto, P., De Marcellis, A., Ferri, G.: Novel modified de-sauty autobalancing bridge-based analog interfaces for wide-range capacitive sensor applications. IEEE Sens. J. **14**(5), 1664–1672 (2014)
7. Philip, N., George, B.: Design and analysis of a dual-slope inductance-to-digital converter for differential reluctance sensors. IEEE Trans. Instrum. Meas. **63**(5), 1364–1371 (2014)
8. Mandal, N., Kumar, B., Sarkar, R., Bera, S.C.: Design of a flow transmitter using an improved inductance bridge network and rotameter as sensor. IEEE Trans. Instrum. Meas. **63**(12), 3127–3136 (2014)
9. Moustafa, K.A.: What is the distance between objects in a data set: a brief review of distance and similarity measures for data analysis. IEEE Pulse **7**(2), 1364–1371 (2016)
10. Goel, S., Lohani, B.: A motion correction technique for laser scanning of moving objects. IEEE Geosci. Remote Sens. Lett. **11**(1), 225–228 (2014)
11. Wang, Y., Yang, X., Zhao, Y., Liu, Y., Cuthbert, L.: Bluetooth positioning using RSSI and triangulation methods. IEEE Consum. Commun. Netw. Conf. **11**(1), 837–842 (2013)

Wearable EMG Gesture Signal Acquisition Device Based on Single-Chip Microcontroller

Wen-Shan Lin[1], Chao-Ting Chu[3], and Chian C. Ho[2(✉)]

[1] Graduate School of Vocation and Technological Education, National Yunlin University of Science and Technology, Douliou 64002, Yunlin County, Taiwan
dl0343003@yuntech.edu.tw
[2] Department of Electrical Engineering, National Yunlin University of Science and Technology, Douliou 64002, Yunlin County, Taiwan
futureho@yuntech.edu.tw
[3] Internet of Things Laboratory, Chunghwa Telecom Laboratories, 99, Dianyan Rd, Yangmei District, Taoyuan City 32661, Taiwan
chaot@cht.com.tw

Abstract. This paper proposes a wearable gesture signal acquisition device based on single-chip microcontroller for Electromyography (EMG) monitoring. Specifically, this paper proposes an easy-to-detect ECG signal method and applies MCU of Texas Instruments MSP430I series featuring 24-bit sigma-delta ADC and low-power consumption. From experimental results, it still keeps remarkable performance on the wearable device. On the other hand, this paper designs a digital filter to reduce the environmental noise and skin interference through single-chip microcontroller.

1 Introduction

One of the most significant current discussions in electrical engineering is health care fields [1, 2]. Traditional health care device uses large-volume and cost-expensive medical equipment to detect body information. Recently, researchers have shown an increased interest in body signal information [3, 4] and have proposed many methods to detect body information. Dekens et al. [3] proposed a body-conducted speech enhancement method through equalization and signal fusion and the speech enhancement algorithm can detect the body pose effectively. The quantity of body signal data is very large so it must be collected into training database. Li et al. [4] presented a continuous biomedical signal acquisition system based on compressed sensing in body sensor networks. The body sensor networks collected different body information efficiency. A considerable amount of literature has been published in wearable device fields [5, 6]. These studies show the wearable device has qualified ability for real-time calculation. Su et al. [5] presented an integrated metal-frame antenna for smartwatch wearable device, in which the proposed mothed of smart watch antenna have remarkable performance in experimental results. In [6], a EMD-based electrocardiogram delineation for wearable low-power ECG monitoring device was proposed by Tan et al., and the body's ECG signal can be immediately detected to calculate the heart rate and other signal presentation information. More recently,

© Springer Nature Switzerland AG 2020
L. Barolli et al. (Eds.): NBiS-2019, AISC 1036, pp. 517–525, 2020.
https://doi.org/10.1007/978-3-030-29029-0_50

literature has emerged that offers health care findings about EMG [7, 8] signal analysis. The feature analysis of EMG signal can provide body muscle movement information such as gesture, fitness, sports fatigue and endurance training. In [7], a detection and compensation method for EMG disturbances through powered lower limb prosthesis control was proposed by Spanias et al., and its experimental results show the ECG detection algorithms to control lower limb prosthesis has remarkable performance. Benatti et al. [8] proposed a versatile embedded platform for EMG acquisition and gesture recognition, and the gesture can be effectively recognized by EMG.

Researches to date tend to focus on EMG algorithms [9, 10] rather than industry production's advantage. This paper proposes wearable gesture signal acquisition device based on single-chip microcontroller for EMG monitoring and further reduces the hardware cost and power consumption of the wearable device. Experimental results show MCU of Texas Instruments MSP430I series with digital filter to detect EMG signal has remarkable output performance to distinguish different gesture.

2   Hardware Architecture of Single-Lead ECG Monitor

Figure 1 shows the functional diagram of EMG gesture signal acquisition and thereby EMG signal is received by low-frequency electrode patch. The microcontroller applies high-resolution ADC to obtain digital EMG signal. Moreover, this paper designs a digital filter eliminating 60-Hz environmental noise and skin interference after EMG signal has been received. Finally, this paper adopts PC environment to analyze EMG signal data situation.

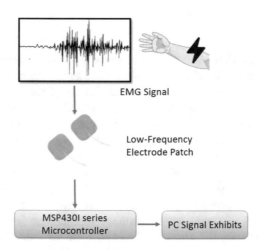

Fig. 1. Functional diagram of single-lead ECG monitoring.

Figure 2 shows the flowchart of EMG microcontroller. The MSP430I series accompany 60-Hz environmental noise, and the body EMG voltage is milli-level. Therefore, this paper applies high-resolution ADC to detect EMG signal. The MCU of

Texas Instruments MSP430I series has 24-bit sigma-delta ADC that combines programmable gain amplifier to zoom-in signal. The microcontroller utilizes a digital filter to remove the skin interference and other uncertainty after the programmable amplifier enlarges its gain. Finally, the EMG waveform data is transformed and transmitted to PC monitor.

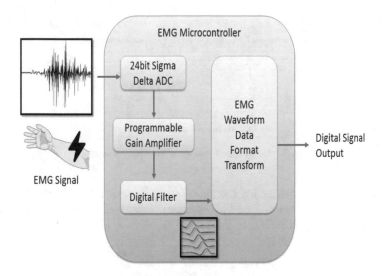

Fig. 2. Flowchart of EMG microcontroller.

3   ECG Signal Processing

3.1   Digital Filter Design

The MSP430I series have a hardware-based programmable gain amplifier which can adjust and enlarge the signal gain of sigma-delta ADC. However, MSP430I series detect signal which accompanies inevitably serious 60-Hz environmental noise, skin interference and hardware electrode interference so that this paper designs a digital filter to remove these disturbance. The digital filter is built into the MCU of MSP430I series whose computation cost and complexity are trivial. Moreover, the wearable device is usually battery-powered and the wearable device must take power consumption issue into account. Therefore, this paper chooses an infinite impulse response (IIR) digital filter to reduce its computation cost and complexity.

3.2   IIR Low Pass Filter

The electrode interference and electronic circuit board hardware issue that leads to high frequency interference. In this part, this paper designs an IIR low pass filter to remove high gain interference. Figure 3 is the diagram of IIR low pass filter that used iteration method to reduce high frequency vary. The output of IIR low pass filter is follow as (1):

$$F_{out,n} = \left(F_{in,n} - F_{in,n-1}\right) \cdot K + F_{in,n-1} \tag{1}$$

3.3 IIR Notch Pass Filter

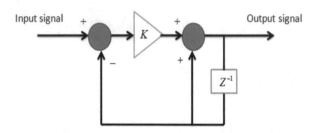

Fig. 3. Diagram of proposed low pass IIR filter.

The EMG monitors 60-Hz environmental noise issue which this paper designs an IIR notch pass filter to remove specific frequency band noise. The standard discrete equation of IIR notch pass filter is shown in (2):

$$G(z) = b_0 \frac{(z - e^{j\omega_0})(z - e^{-j\omega_0})}{(z - re^{j\omega_0})(z - re^{-j\omega_0})} \tag{2}$$

where b_0 means the gain, r represents the location of a pair of complex-conjugate pole. According to Euler's formula, (2) can be derived as (3):

$$
\begin{aligned}
G(z) &= b_0 \frac{z^2 - ze^{-j\omega_0} - ze^{j\omega_0} + 1}{z^2 - rze^{-j\omega_0} - rze^{j\omega_0} + r^2} \\
&= b_0 \frac{z^2 - z(\cos\omega_0 - j\sin\omega_0 + \cos\omega_0 + j\sin\omega_0) + 1}{z^2 - rz(\cos\omega_0 - j\sin\omega_0 + \cos\omega_0 + j\sin\omega_0) + r^2} \\
&= b_0 \frac{z^2 - z(2\cos\omega_0) + 1}{z^2 - rz(2\cos\omega_0) + r^2} \\
&= b_0 \frac{1 - (2\cos\omega_0)z^{-1} + z^{-2}}{1 - r(2\cos\omega_0)z^{-1} + r^2 z^{-2}} \\
&= b_0 \frac{1 - g_1 z^{-1} + z^{-2}}{1 - g_2 z^{-1} + g_3 z^{-2}}
\end{aligned} \tag{3}
$$

The flowchart of microcontroller calculation in MSP430I series is shown in Fig. 4. Firstly, this paper does not know EMG signal's happened time so this paper starts processing EMG signal after the microcontroller detects EMG voltage target to save energy in the wearable device. Secondly, this paper stops recording EMG data when the

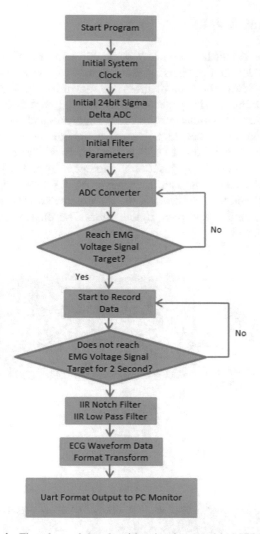

Fig. 4. Flowchart of the algorithm implemented in MSP430I.

microcontroller does not reach EMG voltage signal target. Thirdly, the EMG signal processed IIR notch to reduce 60-Hz environmental noise, and skin and hardware electrode interference are high frequency signal which applies IIR low pass filter to remove high frequency noise. Finally, the microcontroller transmits data format through Universal Asynchronous Receiver/Transmitter (UART) to PC monitor. Therefore, this paper adopts PC monitor to analyze and display data situation.

4 Experimental Results

This EMG monitor circuit has been implemented in the MCU of MSP430I2041 to achieve the smooth detection for EMG gesture signal acquisition. The experimental test environment and hardware function diagram are shown in Fig. 5, and the hand testing position is shown in Fig. 6. This paper sets 3 ms as the sampling time. Figure 7 shows hand testing position with various gesture cases. Figures 8 and 9 show the EMG signal data from various gesture cases, and this paper applies the discrete digital filter to improve the system distribution issue. Specifically, Figs. 8(a) and 9(a) show 60-Hz environmental noise issue is very serious. Figures 8(b) and 9(b) verifies that the environment and hardware noise can be removed by IIR notch digital filter, and furthermore, Figs. 8(c) and 9(c) verifies that the environment and hardware can be better removed by IIR notch and low pass digital filter.. The final output signal data have remarkable performance.

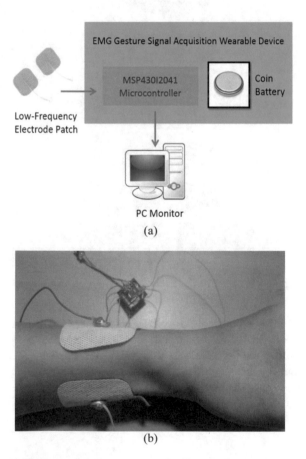

Fig. 5. (a) Hardware function diagram, (b) Experimental test environment.

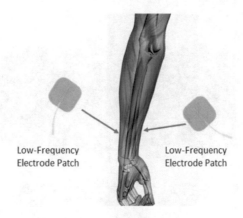

Fig. 6. Hand testing position.

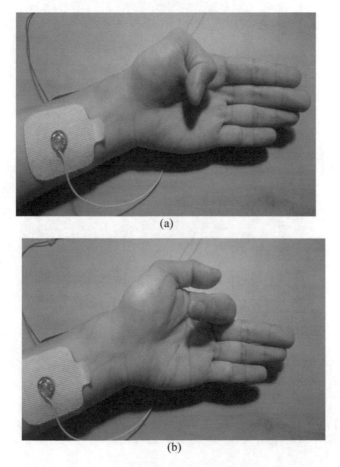

Fig. 7. Hand testing position with various gesture cases of (a) bent thumb finger and (b) bent index finger.

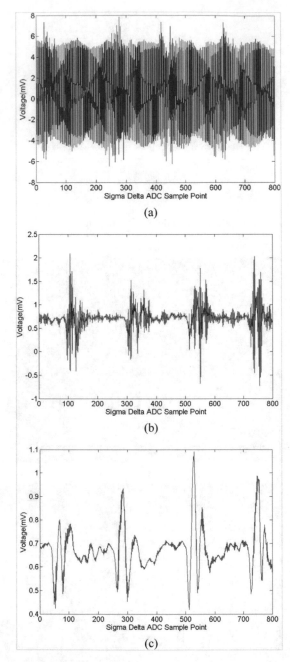

Fig. 8. EMG signal results of bent thumb finger gesture with improvement of (a) no digital filter (b) proposed IIR notch digital filter, (c) proposed IIR notch and low pass digital filter.

5 Conclusions

The results of this study indicate that this paper proposes and implements a wearable EMG gesture signal acquisition device based on single-chip microcontroller solution. This paper applies high-integration analog ADC microcontroller of MSP430I series featuring 24-bit sigma-delta ADC and programmable gain amplifier, and this paper designs discrete digital filter design to eliminate 60-Hz environmental noise and hardware electrode interference. Finally, the experimental results shows the hand EMG signals have remarkable performance through proposed discrete digital filter.

Acknowledgments. This work was financially supported by the "Intelligent Recognition Industry Service Center" from The Featured Areas Research Center Program within the framework of Higher Education Sprout Project by Ministry of Education (MOE) in Taiwan.

References

1. Islam, S.K., Fathy, A., Wang, Y., Kuhn, M., Mahfouz, M.: Hassle-free vitals: bio wireless for a patient-centric health-care paradigm. IEEE Microwave Mag. **15**(7), 25–33 (2014)
2. Kim, S., Jeon, Y.W., Kim, Y., Kong, D., Jung, H.K., Bae, M.K., Lee, J.H., Ahn, B.D., Park, S.Y., Park, J.H., Kwon, J.P., In, H., Kim, D.M., Kim, D.H.: Design and implementation of a web-service-based public-oriented personalized health care platform. IEEE Electron Device Lett. **43**(4), 62–64 (2013)
3. Dekens, T., Verhelst, W.: Body conducted speech enhancement by equalization and signal fusion. IEEE Trans. Audio Speech Lang. Process. **21**(12), 2481–2492 (2013)
4. Li, S., Xu, L.D., Wang, X.: A continuous biomedical signal acquisition system based on compressed sensing in body sensor networks. IEEE Trans. Industr. Inform. **9**(3), 1764–1771 (2013)
5. Su, S.W., Hsieh, Y.T.: Integrated metal-frame antenna for smartwatch wearable device. IEEE J. Biomed. Health Inform. **63**(7), 263–274 (2015)
6. Tan, X., Chen, X., Hu, X., Ren, R., Zhou, B., Fang, Z., Xia, S.: EMD-Based electrocardiogram delineation for a wearable low-power ECG monitoring device. Can. J. Electric. Comput. Eng. **37**(4), 212–221 (2014)
7. Spanias, J.A., Perreault, E.J., Hargrove, L.J.: Detection of and compensation for EMG disturbances for powered lower limb prosthesis control. IEEE Trans. Neural Syst. Rehabil. Eng. **24**(2), 326–333 (2016)
8. Benatti, S., Casamassima, F., Milosevic, B., Farella, E., Schönle, P., Fateh, S., Burger, T., Huang, Q., Benini, L.: A versatile embedded platform for EMG acquisition and gesture recognition. IEEE Trans. Biomed. Circ. Syst. **9**(5), 620–630 (2015)
9. Xu, Q., Quan, Y., Yang, L., He, J.: An adaptive algorithm for the determination of the onset and offset of muscle contraction by EMG signal processing. IEEE Trans. Neural Syst. Rehabil. Eng. **21**(1), 65–73 (2013)
10. Pasinetti, S., Lancini, M., Bodini, I., Docchio, F.: A novel algorithm for EMG signal processing and muscle timing measurement. IEEE Trans. Inf. Technol. Biomed. **64**(11), 2995–3004 (2015)

Ring-Based Routing for Industrial Wireless Sensor Networks

Ching-Lung Chang[✉], Hao-Ting Lee, and Chuan-Yu Chang

Department of CSIE, National Yunlin University of Science and Technology,
Douliu, Taiwan
{chang, chuanyu}@yuntech.edu

Abstract. Due to the dynamic routing and transmission collision in Ad Hoc networks, the data delivery time in IWSN is unpredictable. In this paper, a proactive routing which constructs the sensor nodes into a logical circular chain topology (i.e., ring topology) is adopted to avoid the data collision problem and to bound sensing data collecting time in industrial wireless sensor network (denoted IWSN). A load-balanced issue is considered to prolong the network lifetime. A Ping-Pong token mechanism is proposed to balance residual power of each sensor node and to prolong the network lifetime. Simulation results reveal that the linear programming scheme with Ping-Pong token has the best data delivery time, balanced the power consumption, and system life time.

1 Introduction

In recent years, wireless sensor networks (denoted WSN) are widely used in environment surveillance, factory surveillance [1, 2] and ocean exploration [3]. Due to the restricted communication range, the data delivery in WSN is usually performed Ad Hoc routing [9]. In the Destination-Sequenced Distance-Vector (DSDV) [6], each mobile host periodically advertises its neighbor interconnection topology with other mobile hosts. Thus, each mobile host has the whole network topology and uses shortest path routing for data transmission. The reactive routing protocol of Ad-hoc On-demand Distance Vector (AODV) [10] is proposed for hosts in an Ad Hoc network. For improving network performance, [11, 12] use multipath routing approach to efficiently utilize the available network resources. The routing of pair-wise directional geographical routing (PWDGR) [13] considers node geography information to make multipath to solve the problem that the node energy near sink becomes obviously higher than other nodes. Consider the load balancing and bandwidth aggregation, the directional geographical routing (DRG) [14] makes multiple disjointed paths for real-time video streaming over WSN.

The data transmission time becomes unpredictable in Ad-Hoc environment which may incur signal collision problem. How to bound the sensing data collection time in industrial wireless sensor network (denoted IWSN) is an important issue. For solving this problem, a Circular-Chain-Data-Forwarding mechanism (denoted CCDF) [4] is proposed, which constructs the sensor nodes in IWSN into logical circular chain (i.e., ring-based routing) and applies token-based medium access control to avoid data

© Springer Nature Switzerland AG 2020
L. Barolli et al. (Eds.): NBiS-2019, AISC 1036, pp. 526–536, 2020.
https://doi.org/10.1007/978-3-030-29029-0_51

collision problem. The concepts of CCDF is illustrates in Fig. 1. As the Fig. 1 depicts, the black node is sensor node which relies on the battery to provide power and only has wireless interface. The red node is sink node which has power line and not only have wireless interface but also have wire interface to connect with controller. Based on the wireless signal transmission power, the directly connect nodes of node i is defined as the nodes which can directly receive the node i signal. Thus, using the neighbor discovery technique the controller can obtain full network connectivity topology, as Fig. 1(a) depicts, according to collect the neighbor discovery results of each sensor node. Based on the network connectivity topology, the controller can construct the network topology into the logical circular chain, as Fig. 1(b) illustrates. In the IWSN, the logical chain may consist of many sub-chains. Each sub-chain is defined as a path from a sink node to its neighbor sink node. There are six sub-chains in the Fig. 1(b) illustration. Based on token-based medium access control, the controller periodically issues a token to sink 1 (denoted as S1). The S1 passes the token to its next node with clockwise. Once the token goes back to the S1, the S1 discards the token and waits for the new token issued by controller. The node has right to transmit packet only when it has a token.

The scheme of CCDF [4] did not take load-balanced into consideration. In this paper, we target on the load-balanced logical circular chain construction to bound the sensing data transmission delay and to prolong the network lifetime.

We propose two logical circular chain construction schemes in this paper. They are Traveling Salesman Problem (denoted as TSP-based) combined with genetic algorithm and linear programming (denoted as LP-based) combined with simulated anneal algorithm, respectively. In addition, a Ping-Pong token mechanism is proposed for the ring transmission to balance the residual power for each sensor node and to prolong the system life time.

The reset of this paper is organized as follows. In Sect. 2, we introduces the genetic algorithm with heuristic algorithm in TSP-based method to construct a load-balanced logical circular chain. The Sect. 3 introduces linear programming model for construction the load-balanced circular chain. The simulated annealing (SA) algorithm is used to find the optimal or near optimal solution of linear programming model. The performance comparison among different schemes is shown in Sect. 4. Finally, we make a bright conclusions.

2 TSP-Based Load-Balanced Logical Circular Chain Construction

The printing area is 122 mm × 193 mm. The text should be justified to occupy the full line width, so that the right margin is not ragged, with words hyphenated as appropriate. Please fill pages so that the length of the text is no less than 180 mm, if possible.

As Fig. 1(b) illustrates, the constructed logical chain is starting from a sink node and looping back to the same sink node. All of nodes including sink nodes and sensor nodes must be visited only one time. Thus, the IWSN's virtual circular chain construction can be transform to the traveling salesman problem (TSP) [5, 6, 8].

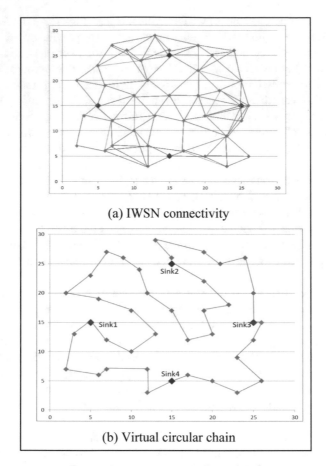

(a) IWSN connectivity

(b) Virtual circular chain

Fig. 1. Virtual circular chain for IWSN

For considering the load-balanced in the virtual circular chain construction, a balance index derived by [5], as Eq. (1), is use as the fitness value in GA.

$$LBI = \frac{\left(\sum B_i\right)^2}{\left(n \sum B_i^2\right)},$$ (1)

Where B_i is the length of sub-chain i and n is the number of sub-chain. Herein, the sub-chain length is defined as the number of nodes in the sub-chain. In the Eq. (1), if the LBI value is more near to one means the constructed logical circular chain is more balance.

The definition of balanced chain is that the variance of path length among each sub-chain is minimum. If we can build a more balanced chain, both of data collection time and power consumption can be minimized. For achieving this goal, we propose a heuristics algorithm for balanced chain construction as Fig. 2 shows.

Table 1. The moving matrix of sub-chain.

Sub-chain	S1	S2	S3
moving node	node1	node5	node4
	node2		
	node3		
	node4		

First, we calculate each sub-chain cost (or call sub-chain length). Among these sub-chain, the sub-chain with the highest cost is selected to build the associated moving matrix, as Table 1 shows. The moving matrix represents that which nodes in a sub-chain can move to which sub-chain. For example, the node 1, node 2, node 3, and node 4 in the sub-chain can move to the first sub-chain as Table 1 shows. Based on the moving matrix, we choose a node in the highest cost sub-chain moving to the lowest cost sub-chain, if the node can move. After adjusting the logical chain, we iterate the same process until no any node in the higher cost sub-chain can move to the lower cost sub-chain. The associated process flow of the heuristic algorithm is shown in Fig. 2.

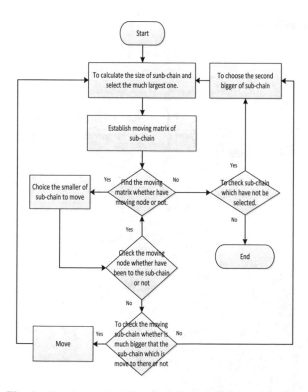

Fig. 2. Heuristics algorithm for balanced chain construction

3 LP-Based Chain Construction

In addition, the virtual logical circular chain construction is modeled to a min-max integer linear programming (denoted as LP-based) is employed in this paper. The formulation is described as follows:

Given Parameters:

$D = \{1, 2, ..., m\}$:   Index set of original destination (OD) pairs. The OD pair is composed by sink nodes. i.e., the path of an OD pair is a sub-chain. Herein, we assume the location and sequence of sink nodes are understood.

$N = \{1, 2, ..., n\}$:   Index set of nodes which include sensor nodes and sink nodes.

L: Index set of logic links. $(i, j) \in L$, means there is a direct connectivity between nodes j and i.

C_{ij}:   The cost of link (i, j).

Decision Variables:

x_{ij}^d: Binary variable. If this value is 1 means the link (i, j) is chosen in the routing path of OD pair d, $d \in D$, otherwise this value is 0.

u_k: Slack variable for preventing multiple logical circular chains, $k \in N$.

Objective Function:

$$\text{obj: min } z \tag{2}$$

Subject to:

$$\sum_{\{j:(ij)\in L\}} x_{ij}^d - \sum_{\{j:(ji)\in L\}} x_{ij}^d = \begin{cases} 1, & \forall i = \text{ source node} \\ -1, & \forall i = \text{ destination node} \\ 0, & \text{else} \end{cases} \tag{3}$$

$$\sum_{d\in D} \sum_{\{j:(ij)\in L\}} x_{ij}^d = 1, \forall i \in N \tag{4}$$

$$\sum_{d\in D} \sum_{\{j:(ji)\in L\}} x_{ji}^d = 1, \forall i \in N \tag{5}$$

$$\sum_{(ij)\in L} c_{ij}x_{ij}^d \leq z, \forall d \in D \tag{6}$$

$$u_i - u_j + nx_{ij} \leq n - 1, \; 2 \leq i \neq j \leq n \tag{7}$$

$$x_{ij}^d \in (0/1), \; d \in D, \; (ij) \in L, \; u_k \geq 0, \; k \in N \tag{8}$$

The flow conservation law is used to find a route for each OD pair in constraint (3). The constraints (4) and (5) limit all of nodes have to be visited and only visited once time. The maximum length among all of sub-chains is not great than variable z in the constraint (6). This is used to find the highest sub-chain length and to cooperate with object function to derive the load-balanced results. The objective function in (2) is to minimize the value of z for load-balanced chain construction. The constraint (6) is used to prevent multiple chains construction. Constraint (8) is an integer constraint.

The simulated annealing (SA) is applied to solve the solution of LP-based virtual circular logical chain construction, which has the ability to escape local minimum to find optimal or near optimal solution [7]. The pseudo code of SA algorithm is shown in Fig. 3. The parameter T stands for the temperature and I is the number of iteration. In each iteration, we find a neighbor solution. If the energy E of neighbor solution is lower than the current solution, we update the current solution to the neighbor solution. Otherwise, we take a probability to decide whether update the current solution or not. The temperature T is reduced 0.999 times in every L iterations. Herein, the energy E of a solution is calculated by,

$$E = \underset{d \in D}{Max} \sum_{(ij) \in L} c_{ij} x_{ij}^d \qquad (9)$$

```
procedure SA
1. begin
2. set_initia_order; best_order = current_order;
3. E^best =E(best_order) ; T=T^0;
4. while stopping_criterion not ture
5.   begin
6.      l = 0;
7.      while l < L do
8.         begin
9.         next_order = exchange_two_random_node (current_order);
10.        ΔE = E(next _ order) − E(current _ order);
11.        if ΔE ≤ 0 then
12.           begin
13.              current_order = next_order;
14.              if E(current_order) < E^best then
15.                 E^best = E(current_order);  best_order = current_order;
16.           end
17.        else if ranom(0,1) < e^{−ΔE / T} then
18.              current _ order = next _ order;  l = l +1;
19.        end
20.        T = T ×0.9999;
21.   end
22. end {procedure}
```

Fig. 3. Simulated annealing algorithm

4   Simulation Results

In this Section, we use computer simulation to evaluate the proposed algorithm for logical circular chain construction in IWSN. The sensor nodes are randomly distributed in the 30 m × 30 m area in the simulation.

First, we compare the performance of the SA algorithm with Lingo tool to evaluate whether the SA algorithm has the ability to find the optimal solution or not. Table 2 illustrates the comparison results. Herein, the symbol OPT stands for the Lingo optimal tool.

Table 2. Comparison between optimization and SA.

Number of sensor	Load Balancing Index (LBI)	
	SA	OPT
10	98%	98%
20	100%	100%
30	99%	99%
40	100%	100%

The simulated load-balanced against the number of sensor nodes among the methods of CCDF, TSP-based and LP-based is shown in Fig. 4. As the Fig. 4 depicts, the LP-based scheme has the best load-balanced performance compared to the CCDF and TSP schemes at all sensor node numbers.

For evaluation the network lifetime, the radio energy dissipation model of a node used in the simulation is shown in Fig. 5 [6]. The transmitter has energy dissipation in radio electronics and transmission power amplifier. The receiver dissipates energy to run the radio electronics.

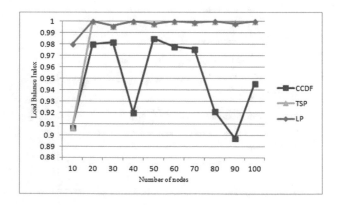

Fig. 4. The comparison of load balance index

Figure 6 plots the network lifetime against different sensor node numbers among CCDF, TSP-based and LP-based. At the Fig. 6 depicts the LP scheme enjoys the highest network lifetime than the others. It means that load-balanced consideration can prolong the network lifetime effectively.

Figure 7 illustrates the network lifetime against different sensor node numbers in the fixed number of node to number of sink ratio. Herein, the ratio is fixed in ten to one. As the Fig. 7 shows, the schemes of LP-based and TSP-based which consider the load-balanced chain construction have the better results than the CCDF.

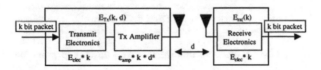

Fig. 5. Radio energy dissipation model

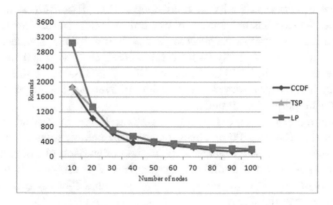

Fig. 6. Network lifetime in different number of nodes

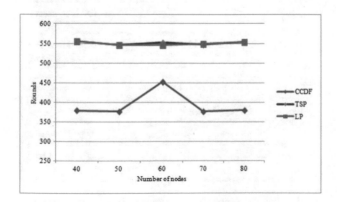

Fig. 7. Network lifetime in same ratio of node to sink

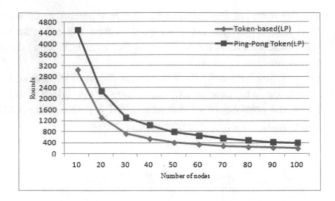

Fig. 8. The comparison of network lifetime in token based and ping-pong token based

In Token-based delivery mechanism, the amount of transmitted data depends on the location of node in the sub-chain. If a node locates at the end of the sub-chain, it needs to forward more other data for its previous nodes. Thus, it consumes more power. It stands that the scheme has unbalanced power consumption in each node. In order to solve this problem, we propose a Ping-Pong Token based delivery scheme. In the Ping-Pong Token, the controller firstly issues a token with clockwise and then it issues a token in the counterclockwise in next time and vise versa. Based on this scheme, each node transmits and receives same amount of data in average. Therefore, it can balance the power consumption in each node.

Figure 8 compares Token-based and Ping-Pong Token in network lifetime. The results shows that the Ping-Pong Token mechanism has significantly improved the network lifetime in all cases. Figure 9 depicts the comparison of network lifetime in different number of nodes with different mechanism. We also can see that when the number of sensor nodes is 40, Ping-Pong Token with LP-based can prolong 95% network lifetime than the CCDF.

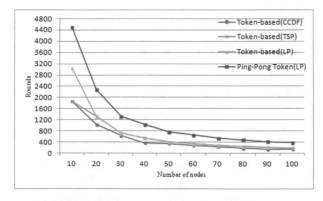

Fig. 9. The comparison of network lifetime in different

5 Conclusion

In the IWSN, this paper constructs a virtual load-balanced circular chain to meet the transmission time constrain and to reduce and to balance the power consumption to prolong the network lifetime. The virtual balanced chain construction is converted to the TSP problem and is modeled by the min-max linear programming problem. The simulated annealing algorithm is adopted to solve the min-max optimization problem. The simulation results reveal the LP-based scheme can achieve load-balanced chain construction and get better results in power consumption and network lifetime. In addition, a Ping-Pong Token scheme is proposed to achieve more balanced power consumption in each sensor node to prolong network lifetime. Cooperated with the LP-based, the Ping-Pong Token prolong 95% network lifetime than the CCDF in 40 sensor nodes condition.

References

1. Hou, L., Bergmann, N.W.: Novel industrial wireless sensor networks for machine condition monitoring and fault diagnosis. IEEE Trans. Instrum. Meas. **61**(10), 2787–2798 (2012)
2. Gungor, V.C., Hancke, G.P.: Industrial wireless sensor networks: challenges design principles and technical approaches. IEEE Trans. Ind. Electron. **56**(10), 4258–4265 (2009)
3. Roemmich, D., Johnson, G.C., Riser, S., Davis, R., Gilson, J., Owens, W.B., Garzoli, S.L., Schmid, C., Ignaszewski, M.: The Argo Program: observing the global ocean with profiling floats. Oceanography **22**(2), 34–43 (2009)
4. Toscano, E., Bello, L.L.: A novel approach for data forwarding in Industrial Wireless Sensor Networks. In: 2010 IEEE Conference on Emerging Technologies and Factory Automation (ETFA), September 2010
5. Or, I.: Traveling salesman-type combinatorial problems and their relation on the logistics of regional blood banking. Ph.D. thesis, Department of Industrial Engineering and Management Sciences, Northwestern University, Evanston IL (1976)
6. Laporte, G.: The traveling salesman problem: a overview of exact of approximate algorithms. Eur. J. Oper. Res. **59**, 231–247 (1992)
7. Chiu, D., Jain, R.: Analysis of the increase and decrease algorithms for congestion avoidance in computer networks. J. Comput. Netw. ISDN **17**, 1–14 (1989)
8. Heinzelman, W.B., Chandrakasan, A.P., Balakrishnan, H.: An application-specific protocol architecture for wireless microsensor networks. IEEE Trans. Wirel. Commun. **1**(4), 660–670 (2002)
9. Mantawy, A.H., Abdel-Magid, Y.L., Selim, S.Z.: A simulated annealing algorithm for unit commitment. IEEE Trans. Power Syst. **13**(I), 197–204 (1998)
10. Akkaya, K., Younis, M.: A survey on routing protocols for wireless sensor networks. Ad Hoc Netw. **3**, 325–349 (2005)
11. Perkins, C.E., Bhagwat, P.: Highly Dynamic Destination-Sequenced Distance-Vector Routing (DSDV) for mobile computers. ACM SIGCOMM Comput. Commun. Rev. **24**, 234–244 (1994)

12. Perkins, C.E., Royer, E.M.: Ad-hoc on-demand distance vector routing. In: Proceedings of the Second IEEE Workshop on Mobile Computer Systems and Applications, New Orleans, USA, pp. 90–100 (1999)
13. Radi, M., Dezfouli, B., Bakar, K.A., Lee, M.: Multipath routing in wireless sensor networks: survey and research challenges. Sensors 12(1), 650–685 (2012)
14. Li, B.-Y., Chuang, P.-J.: Geographic energy-aware non-interfering multipath routing for multimedia transmission in wireless sensor networks. Inf. Sci. 249, 24–37 (2013)
15. Wang, J., Zhang, Y., Wang, J., Ma, Y., Chen, M.: PWDGR: pair-wise directional geographical routing based on wireless sensor network. IEEE Internet Things J. 2(1), 14–22 (2015)

Proposal of Research Information Collection System

Takahiro Uchiya$^{(\boxtimes)}$, Ryoa Sugisaki, and Ichi Takumi

Nagoya Institute of Technology, Gokiso-chou, Showa-Ku,
Nagoya 460-8555, Japan
{t-uchiya,kajioka}@nitech.ac.jp,
sugisaki@uchiya.nitech.ac.jp

Abstract. In recent years, information about research progress has become more obtainable by virtue of internet technology and the promotion of open science. Nevertheless, many steps are necessary to collect such information adequately. Research summaries alone require much time and effort. For this study, we propose a system that collects information about individual laboratories, and which then analyzes and provides information as a single site automatically. As a prototype system, we developed important functions such as research keyword extraction, identifying laboratories that conduct similar research, and web site analysis of research abstract text. This report describes the experiment and evaluation of the proposed system.

1 Introduction

In recent years, the wider propagation of internet technology has made it possible for many people to collect information easily via a web browser. By accessing a website that provides search services and entering keywords, one can readily find websites that hold information such as research keywords. Furthermore, using blog sites, social networking services (SNS), etc., it has become easier for individuals to disseminate information. Academic information has become easier to collect and disseminate on the internet.

Today, many laboratories are building websites. Additionally, web services that provide information related to articles and books have been enhanced. People can obtain detailed academic information using these services. Moreover, the movement to Open Access (OA) of research results is progressing. In the United States and the United Kingdom, the obligatory OA provision of articles is being promoted gradually for research using public funds. Japan also promotes open science under the Fifth Science and Technology Basic Plan, which was formulated by the Cabinet on January 22, 2016. Based on that plan more scientific information is expected to be provided in the future.

Several web services provide information related to research information. CiNii Articles [1] provide information related to Japanese academic research reports. By searching using an author name or keyword, one can obtain information such as a report title, author information, collected publications, and the distribution destination URL. Cited references are also listed. Therefore, relevant research can be found.

© Springer Nature Switzerland AG 2020
L. Barolli et al. (Eds.): NBiS-2019, AISC 1036, pp. 537–544, 2020.
https://doi.org/10.1007/978-3-030-29029-0_52

In addition, Researchmap [2] includes researcher-related information. This service, which is linked with multiple research databases, lists the affiliation, dissertation information, author information, etc. for each researcher. In recent years, it has also become possible to obtain academic information from websites operated by the respective laboratories and from news on the web.

The means of obtaining academic information are increasing, but efforts to acquire, compare, and analyze large amounts of information are increasing concomitantly with those increased amounts of information. Particularly when the target information is not defined in detail, one must acquire diverse information, which makes it difficult to find information that meets users' specific intentions. Moreover, because information is provided separately on multiple web sites, time and effort must be taken to obtain such information comprehensively.

As described herein, we propose a system that automatically collects, analyzes, and summarizes internet-related laboratory information and provides it to users. By analyzing web pages and articles on the internet, we extract information that is important for understanding the research contents. Moreover, by providing information scattered on many web sites with a single web service, we support users' smooth accumulation of information.

2   Related Work

Nguyen proposed a system for the automatic extraction and visualization of researchers' research history by the clustering of research reports [3]. Meta-information of such reports acquired from the database of CiNii and KAKEN is vectorized. Then clustering is done using Cutting Plane Maximum Margin Clustering method. Subsequently, the research theme is set for each cluster. The research theme and the research period are visualized. By analyzing data from a report, a summary of the information can be produced, making it easy to ascertain a researcher's research activities without manually analyzing the information provided by the report. However, the adopted clustering method requires setting of the number of clusters in advance. Moreover, in the cluster, the number of reports tends to be the same. A great difference from manual classification is discernible. Therefore, solving these difficulties is important.

3   Proposed Method

3.1   Overview

The proposed mechanism analyzes and aggregates internet-related information and provides information related to a single web site, thereby allowing users to obtain information efficiently. This mechanism consists of the following four internal mechanisms (Fig. 1).

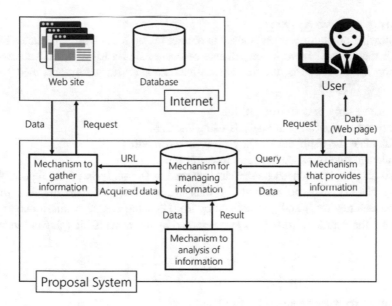

Fig. 1. System architecture.

3.2 System Architecture

A. Mechanism to gather information
This mechanism collects laboratories' research information from the internet. It obtains information on a regular basis from web sites and databases that have been registered in advance.

B. Mechanism to analyze information
This mechanism analyzes information. Information acquired from the internet includes much unnecessary information. Only useful information should be extracted.

C. Mechanism to manage information
This mechanism manages information collected by a mechanism to gather information and a mechanism to analyze information.

D. Mechanism to provide information
This mechanism provides information to users as a web site.

3.3 Function to Analyze Research Reports

As a prototype function of a mechanism to analyze information, we developed a system that analyzes research reports to extract the research contents of the respective laboratories. This function obtains keywords related to the research by analyzing the research report text. Additionally, the function searches for laboratories that conduct similar research by comparing the research report contents. An open source PDF viewer Xpdf [4] and morphological analysis engine MeCab [5] are used.

A. Extract research keywords

The following procedures must be used to extract research keywords. Distance between research reports is defined as the inverse of cosine similarity of tf-idf vectors. After clustering research reports, the function defines a noun with the highest tf-idf value in each cluster (Fig. 2).

(1) Retrieve only text from a pdf file.
(2) Perform morphological analysis using MeCab.
(3) Calculate the tf-idf vector for each research report.
(4) Calculate the distance between research reports.
     We adopted condensed hierarchical clustering for analysis of the report. We use the country average method for inter-cluster distance. The distance $d(j_1, j_2)$ between reports j_1 and j_2 is defined by Eq. (1) using cosine similarity $\cos(v_{j_1}, v_{j_2})$. Here, the feature of the report j is presented by the word tf-idf value as "vector v_j".

$$d(j_1, j_2) = \frac{1}{\cos(v_{j_1}, v_{j_2})} \tag{1}$$

(5) Cluster research reports for each laboratory.
(6) Extract research keywords.

B. Find laboratories that conduct similar research

By comparing clusters, this function finds laboratories conducting similar research. The procedures are described below.

(1) Calculate the average of tf-idf vectors of research reports.
(2) Calculate distances between clusters (inverse of cosine similarity of the average vector of tf-idf).
(3) Define the shortest distances between clusters as a distance between two laboratories.
(4) Define the nearest laboratories as those which conduct similar research at each laboratory.

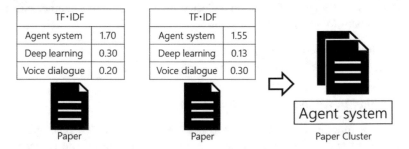

Fig. 2. Flow of analyzing research reports.

3.4   Web Site Analysis

As a function of the information analysis mechanism, we implemented a function to extract laboratory information by analysis of websites published by each laboratory. By analyzing the text information of the html file acquired in advance, we intend to extract the research outline and research activities of each laboratory.

By this function, the information of each laboratory's web site is expressed and analyzed as a multilayer graph, referring to the method used in research on the summary of discussion boards by Kitagawa et al. [6] (Fig. 3). Nodes have inclusion relations with nodes in other layers. Because analysis that assigns importance of the web page itself in addition to text information can be obtained, it is expected to extract sentences from more important web pages.

Each layer comprises directed graphs and undirected graphs. The edge of the page layer indicates a link by a tag. The edge of the word layer indicates co-occurrence in the same sentence. The sentence layer is edged between any two nodes and is weighted by the cosine similarity of the word frequency. The importance of these nodes is calculated using PageRank and LexRank for each layer. The final score of each sentence is calculated by being propagated to the sentence layer based on the inclusion relation.

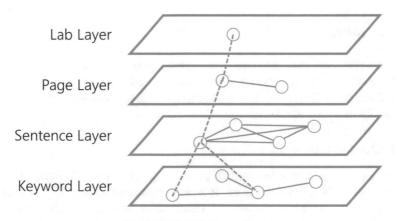

Fig. 3. Multiple layer graph.

3.5   Information Provision Mechanism

This mechanism provides users with information extracted in the information analysis mechanism (Fig. 4). This mechanism is implemented using a single web site. For pages that hold details of work being done at the respective laboratories, a user can browse the research keywords obtained using thesis analysis, the laboratory performing similar research, and abstract sentences obtained using website analysis. In addition to the titles of the articles that constitute the research keywords, a link to the article detail page in CiNii Articles is also provided.

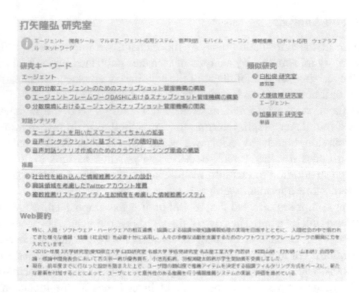

Fig. 4. Information provision web site.

4 Experiment

We tested a prototype system to analyze research reports. We also verified whether the proposed system extracts research keywords appropriately, or not, and whether it clusters them properly, or not. This evaluation experiment compares the dataset with the manually analyzed result.

4.1 Evaluation Data

For evaluation experiments, we used 50 proceedings of Nagoya Institute of Technology's Intelligent Program Graduation Research Presentation and 20 proceedings of Department of Information Engineering (Intellectual Science) Master thesis review board. We classified these research reports and set research keywords manually. Research reports were divided into 29 clusters. For the analyses described in the following section, this is taken as the correct answer.

4.2 Experiment 1: Clustering Accuracy

We compared clusters created by the system and those created manually. For the evaluation, we used the Rand Index [7], which reflects the similarity of two clustering results. This measure takes a value from 0 to 1. When clustering results match completely, the value is 1.

We calculated the Rand Index between clustering results done manually by a human and those done by the system. For comparison, the system clustered after defining distances between research reports as random numbers. We compared that result as well. This result, as presented in Table 1, demonstrates that the prototype function output is more effective than randomly clustering results. Nevertheless, this value is not extremely high. Therefore, this function can probably be improved.

Table 1. Rand index with manual clustering results

Number of clusters	Distance	
	Inverse of cosine similarity of tf-idf vectors	Random numbers
26	0.6223	0.4727
27	0.6168	0.4799
28	0.5968	0.4999
29	0.6039	0.5106
30	0.6061	0.5128
31	0.6218	0.5150
32	0.6328	0.5364

4.3 Experiment 2: Evaluation of Research Keyword Extraction Accuracy

To evaluate the research keyword extraction function, we compared keywords outputted by the system and set them manually in clusters. For 29 clusters outputted by the system, 8 matches were found. Some of the output matched contents, but some included no important meaning.

4.4 Experiment 3: Evaluation of Research Summary Website Analysis

To evaluate the summary using web site analysis, we used questionnaire responses. The 27 subjects were undergraduate students of Nagoya Institute of Technology. For each laboratory, the top three sentences with the highest final score were viewed together with the website of each laboratory. The questionnaire was answered. The evaluation items are the following three items.

Item 1: Users can ascertain the main point of research content

Item 2: No missing contents are needed to understand the research content

Item 3: Unnecessary content is not included in understanding of research content

Table 2 presents evaluation items and the evaluation results. Although ascertaining the contents of research was possible, it led to insufficient extraction of necessary information and elimination of unnecessary information. Furthermore, the contents of the extracted sentences, such as the introduction of research contents, commentary on the technology used for research, and information related to external dissertation, differ greatly among laboratories. Great variation in evaluation was found. The method of weighting specific information at the time of analysis must be considered.

Table 2. Evaluation of web site analysis

	Average	S.D.
Item 1	3.51	1.25
Item 2	3.10	1.08
Item 3	3.12	1.12

5   Conclusion

As described herein, we propose a mechanism to collect, analyze, summarize and provide laboratory information automatically from the internet to reduce the time necessary for collecting and assessing laboratory information. We also described an evaluation experiment for the proposed mechanism. Although the effectiveness of clustering in the analysis of reports has been demonstrated, some points related to the extraction of research keywords must be improved. The abstract extracted by the website analysis can present the gist of the research content, but necessary information might be missing. Moreover, unnecessary information might be mixed into it. For those reasons, it must be improved.

References

1. CiNii Articles. http://ci.nii.ac.jp/
2. Researchmap. http://researchmap.jp/
3. Nguyen, M.C., et al.: Automatic research history generation-visualization system based on metainformation. DEIM Forum 2011 C2-2, (2011)
4. Foo labs: Xpdf. http://www.foolabs.com/xpdf/
5. Kudo, T.: MeCeb. http://taku910.github.io/mecab/
6. Kitagawa, R., Fujita, K.: Extracting sentences for automated summarizing using multi-layer graph on online discussion forum. In: The 31st Annual Conference of the Japanese Society for Artificial Intelligence (2017)
7. Rand, W.M.: Objective criteria for the evaluation of clustering methods. J. Am. Stat. Assoc. **66**, 846–850 (1971)

Agricultural Pests Damage Detection
Using Deep Learning

Ching-Ju Chen[1]([⊠]), Jian-Shiun Wu[2], Chuan-Yu Chang[3],
and Yueh-Min Huang[2]

[1] Department of Bachelor Program in Interdisciplinary Studies,
National Yunlin University of Science and Technology, Yunlin, Taiwan
chen.chingju@gmail.com
[2] Department of Engineering Science, National Cheng Kung University,
Tainan, Taiwan
lover0258020@gmail.com, huang@mail.ncku.edu.tw
[3] Department of Computer Science and Information Engineering,
National Yunlin University of Science and Technology, Yunlin, Taiwan
chuanyu@yuntech.edu.tw

Abstract. In this study, a plurality of camera sensors distributed in the agricultural land was integrated into the Raspberry Pi, and photos were taken to observe whether the foliage of the crop was harmful or not. The image data were transmitted to the Alexnet, VGG-16 and VGG-19 convolutional nerves through deep learning methods. The network architecture extracts image features to detect the presence of pests and identifies the types of pests. Compared by the classification accuracy, training model and prediction time with a classifier based on a neural network, and a Support Vector Machine, the identified pest results will be immediately displayed on the farming management app as a timely epidemic prevention management of the farming.

1 Introduction

Most of the agricultural pests hide on the back of the leaves during the day and do not run onto the leaves until in the evening or at night. During the day, the pests cannot be observed from the surface of the leaves. When the farmers discover that the pests causing damage to the crops, they are often the largest number of pests reproduced, and the damage caused by them cannot be suppressed. Only a large amount of pesticides can be used to spray the crops to reduce the damage.

Crops can cause other bacterial infections due to pests, which can lead to large-scale crop diseases. To prevent such conditions, it is necessary to burn infected crops to prevent the spread of bacteria. This situation will lead to greater agriculture damage.

The deep learning Convolutional Neural Networks (CNN) is currently widely used for image recognition. This study first accumulated 2800 pest images to discover the features of the pest respectively through the transfer learning of the Alexnet, VGG-16, and VGG-19, classify the types of the pests. And verify the correct classification of the above methods by SVM.

© Springer Nature Switzerland AG 2020
L. Barolli et al. (Eds.): NBiS-2019, AISC 1036, pp. 545–554, 2020.
https://doi.org/10.1007/978-3-030-29029-0_53

Therefore, this study can warn farmers before pests and diseases begin to spread, and mark the location of pests so that farmers can spray pesticides accurately, which can save manpower and reduce the environmental impact of pesticides damage and increase the production of crops.

2 Related Works

2.1 The Traditional Method to Identify Pests Image

In the past, the pre-processed method used to identify pest images were tone correction, grayscale conversion and mark ROI (Region of Interest) with threshold intensity. Then different feature extraction algorithms are used to classify pest images. Common methods for feature extraction and classification are as follows:

(1) The characteristics of pests are different in biological behavior patterns [1].
(2) Based on the small size and clustering characteristics of the pests, the images of the leaves are analyzed to find out the areas affected by the pests [2].
(3) Through given the threshold to calculate the degree of association from the image features of the pest, it is used to determine whether the pest is present on the image [3].

2.2 Identifying and Classifying Pest Images by Machine Learning

In recent years, with the advancement of computer hardware, the large amount of image data used for reference have increased. Therefore, some scholars use machine learning methods to identify and classify pest images. The methods of machine learning for image recognition and classification can be divided into two categories: supervised learning and unsupervised learning:

(1) Using Support Vector Machine (SVM) [4] for supervised learning to identify images of pests.
(2) Classification of insect-infested leaves and pest-free leaves with gray-scale image features and regional features [5].
(3) Using the k-means algorithm for unsupervised learning to automatically classify the image features [6].

The above research methods can be successfully identified for pests such as whiteflies and moths. However, the color of the foliage and the light of the image can cause the misidentification of the pest. Therefore, how to extract enough features from a large amount of image data to identify different pest categories is the key to determining whether the judgment results are accurate.

2.3 Identifying and Classifying Pest Images by Artificial Intelligence Method

In recent years, due to a large number of image features can improve the accuracy of image recognition, research on image recognition has begun to use an artificial neural network (ANN) to calculate features in image data. Convolutional Neural Networks (CNN) is one of the deep learning methods widely used to identify images. Common CNN methods are:

(1) Using the famous VGG net for feature extraction and identification to detect moths from images [7].
(2) While detection and classification of 12 different pests by several CNN methods [8], and comparing the classification results with machine learning methods such as SVM and Fisher, the classification accuracy rate of the machine learning method is 80%, but the classification accuracy rate of the CNN method is 95%

3   Methods

In this study, three CNN nets (Alexnet, VGG-16, VGG-19) with better image classification were used to classify pests. Pest identification will use raspberry pi 3 and low-cost pi camera to capture images, which will be set up as multiple observation stations in agricultural land, and regularly obtain images of leaf back for pest classification in order to predict the types of pests. With that, it can provide the result of Identification of pests to farmers as management of agricultural land.

First, we collect images from the Raspberry Pi camera and then perform pre-processing image enhancement. Using through the learning algorithm for stochastic gradient descent, and the cross-entropy function as a loss function to classify and identify the processed images with the VGG-19 model. The flow chart of this study is as Fig. 1.

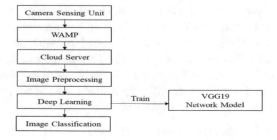

Fig. 1. The flow chart of article architecture.

3.1   Samples Collection

In this study, the four most common pests in the control of crop pests and diseases were selected: whitefly, thrips, moth larvae, and spider mites as experimental species. The characteristics of the sample will affect the accuracy of the training results. This system is used to identify the pests under the foliage, so we collected 140 sample images of each kind of pest appearing on the leaf back from Google image search engine. In addition, in order to determine that there were no pests at all, 140 leaf samples of pest-free were collected, and Fig. 2 is the example with 4 common pests in this study.

Whitefly	
Thrips	
Moth Larvae	
Spider Mites	
Pest-Free Leaves	

Fig. 2. The example with 4 common pests in this study.

3.2 Image Preprocessing

In the process of deep learning, the lack of training sample data will lead to over-fitting of the training module, and it will be difficult to obtain effective eigenvalues for accurate prediction. Therefore, in order to avoid over-fitting caused by insufficient training samples, this study will process each type of prediction image from 140 samples by rotating, resizing or flipping them, and then perform Data augmentation into 560 samples as training samples.

3.3 Deep Learning on Image Classification

The convolutional neural network is an efficient identification method that has been rapidly developed in recent years and has attracted widespread attention. In the 1960 s, when Hubel and Wiesel studied the local sensitive and directional selection of neurons in the cat's cerebral cortex, they found that their unique network structure can effectively reduce the complexity of the feedback neural network so they proposed the architecture of a convolutional neural network. This study will apply the two most convolutional neural network architectures currently used in image classification–Alexnet and VGGnet, for transfer learning [9] to retain the feature extraction layer and identify 5 categories to replace the original 1000.

(1) Alexnet [10] is the champion model of Imagenet in 2012, it was named from the first author Alex. Alexnet first proved that the Convolutional Neural Network (CNN) can effectively obtain classification results under complex models. The Alexnet deep learning layer architecture is shown in Fig. 2. Alexnet consisted of eight learning layers, five of which were convolutional layers and three fully connected layers, which used the GPU for data training during the 2012 Imagenet process. yielding 1000 classification results in an acceptable time.

(2) VGGNet [11] was the champion model of Imagenet in 2014, which was designed by Oxford's visual geometry group (VGG) team Karen Simonyan and Andrew Zisserman, VGGNet is mainly a network developed from AlexNet, mainly exploring the importance of depth for convolutional neural networks. VGGNet is characterized by a large number of convolutional layers and a considerable

amount of computation. If the depth is increased to 16–19 layers, the training effect of deep learning can be obviously improved. The deep learning architecture of VGG16 and VGG-19 is shown in Figs. 3 and 4.

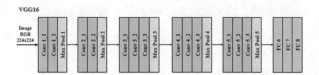

Fig. 3. The deep learning architecture of VGG16.

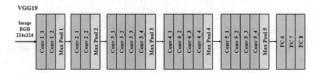

Fig. 4. The deep learning architecture of VGG19

The system performs transfer learning on the fc8 layer of VGG-16 and VGG-19, respectively retaining the first 38 layers and the first 44 layers of the VGG net feature, and changing the original 1000 output neurons to 5 types for what we want to identify.

3.4   Training Deep Learning Neural Networks

1. **Loss function**
   For multi-category classification, the cross-entropy function [12] is used to assign each input to one of the k mutually exclusive categories, so the loss function is defined as follows:

$$E(\theta) = \sum_{i=1}^{n} \sum_{j=1}^{k} t_{ij} \ln y_j(x_i, \theta) \tag{1}$$

Where θ is the vector parameter t_{ij} is used to indicate that the $i - th$ sample belongs to the $j - th$ category $y_j(x_i, \theta)$ is the output of the $i - th$ sample, which is the probability of i samples belonging to the $j - th$ class, which is $P(t_j = 1|x_i)$.

For more than two categories of classification, the output function will apply the softmax function, so that the measured probability is between 0 and 1. The softmax function is defined as follows:

$$y_r(x) = \frac{\exp(a_r(x))}{\sum_{j=1}^{k} \exp(a_j(x))} \tag{2}$$

Where $0 \le y_r \le 1$ and $\sum_{j=1}^{k} y_i = 1$

2. Learning Algorithm

This study uses Stochastic Gradient Descent [12] as the learning algorithm. The Stochastic Gradient Descent algorithm advances each update to a negative gradient through each weight update and parameter deviation so that the minimum error function value is finally obtained. Its formula is as follows:

$$\theta_{\ell+1} = \theta_\ell - \alpha \nabla E(\theta_\ell) \tag{3}$$

Where l is the number of iterations, α is the learning rate, which is set to 0.0001 in this experiment, θ is the parameter vector and $E(\theta)$ is the loss function.

The gradient $\nabla E(\theta)$ of the loss function is a subset of the data of the mini batch size (a subset of the training data), we use it to measure and update the gradient parameters. The process of updating and adjusting the gradient of the entire training data set is called an epoch.

In order to avoid GPU and memory overloaded, this experiment uses 32 samples as the size of the mini batch and the maximum epoch is set to 5. The training is stopped after the entire training data subset is repeated for 5 rounds.

Since the gradient algorithm oscillates in a steep gradient range, we add a momentum parameter γ to avoid such a situation. The formula is as follows:

$$\theta_{\ell+1} = \theta_\ell - \alpha \nabla E(\theta_\ell - \theta_{\ell-1}) \tag{4}$$

When α value is larger, it is expressing that the influence of the previous gradient on the current convergence effect is greater. In this experiment, γ was set to 0.9.

3 Samples Definition

In this study, before each training algorithm, 560 images of each category (2800 images in total) that had been amplified by data were randomly classified in each category in the experiment and 70% of the images were used as the training data set, the other 30% of the images as the test data set to calculate the accuracy of the training model.

3.5   Camera Sensing Unit and Experimental Design

The system integrates low-cost Pi-camera lens, evenly placed around the farmland, regularly collects images on the back of the leaf, and predicts the types of pests in the image by deep learning, so that the farmer can prepare for epidemic prevention.

The Pi-camera uses the CSI (Camera Serial Interface) interface and weighs only about 3 grams. The Pi-camera lens specifications are shown in Table 1, and its appearance is shown in Fig. 5. The resolution of the Pi-camera is 8 million pixels, which is enough to identify the types of pest and have good predictive results.

We use Simulink as a pre-processing of the signal input to integrate the Raspberry Pi camera with subsequent image predictions, as shown in Fig. 6.

Table 1. The Pi-camera lens specifications

Image sensor	Fixed-focus module with integral IR filter
Resolution	8 million pixels
Still picture resolution	3280 × 2464
Max image transfer rate	1080p30, 720p60, 640 × 480p90
Dimensions	25 × 23 × 9 mm
Weight (g)	3
Interface	CSI (Camera Serial Interface)

Fig. 5. Raspberry Pi Camera

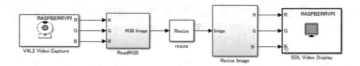

Fig. 6. Connection diagram of Raspberry Pi camera and Simulink

4 Results

Based on different parameter architectures, there is a big difference between Alexnet and VGG net. The first layer of Alex, VGG-16, and VGG-19 is shown in Fig. 7.

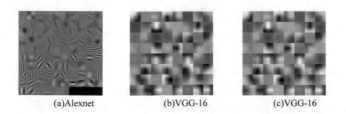

(a)Alexnet   (b)VGG-16   (c)VGG-16

Fig. 7. The first layer of Alex, VGG-16, and VGG-19

We will use the example image of pest in Fig. 8 to illustrate the three architectures of Alexnet, VGG-16, and VGG-19. The output of the first four layers (second to fifth) is shown in Figs. 9, 10 and 11.

Alexnet has fewer total output layers than VGG-16 and VGG-19. It can be seen from Fig. 9 that various features in the example image of Fig. 8 have been clearly expressed in the fifth layer output of Alexnet, but in the first few layers of VGG-16 (Fig. 10) and VGG-19 (Fig. 11), the features of sample images are not obviously apparent.

Fig. 8. The example image of pest

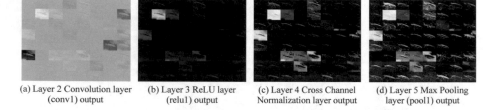

(a) Layer 2 Convolution layer (conv1) output (b) Layer 3 ReLU layer (relu1) output (c) Layer 4 Cross Channel Normalization layer output (d) Layer 5 Max Pooling layer (pool1) output

Fig. 9. The output of the first four layers (from second to fifty) of the AlexNet.

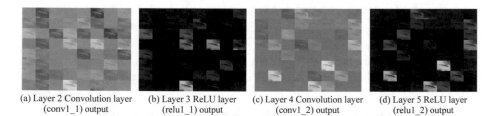

(a) Layer 2 Convolution layer (conv1_1) output (b) Layer 3 ReLU layer (relu1_1) output (c) Layer 4 Convolution layer (conv1_2) output (d) Layer 5 ReLU layer (relu1_2) output

Fig. 10. The output of the first four layers (from second to fifty) of the VGG-16.

The hardware used in this experiment is Intel Core i5-7500 quad-core CPU (3.40 GHz), GeForce GTX 1080 GPU (8 GB) and 8 GB RAM, and the operating system is Ubuntu16.04. The experimental results are shown in Table 2.

It can be seen from the results of this experiment that Alexnet, VGG-16, and VGG-19 have a good effect on feature extraction. The correct rate can be greater than 95% regardless of the SVM method of machine learning or the neural network method.

If you do not need high accuracy, you can use the SVM method to classify after feature extraction, which can greatly shorten the training time of the model.

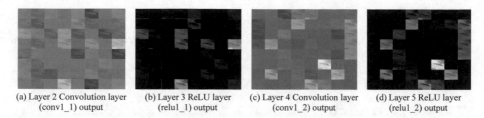

(a) Layer 2 Convolution layer (conv1_1) output	(b) Layer 3 ReLU layer (relu1_1) output	(c) Layer 4 Convolution layer (conv1_2) output	(d) Layer 5 ReLU layer (relu1_2) output

Fig. 11. The output of the first four layers (from second to fifty) of the VGG-19

Table 2. The results of experimental

Model	Training time (s)	Accuracy (%)	Classification time per image (s)
Alexnet	47.324	97.07	2.346
Alexnet+SVM	6.198	95.83	2.434
VGG-16	180.464	99.03	6.430
VGG-16+SVM	17.075	96.52	6.999
VGG-19	207.243	98.31	7.188
VGG-19+SVM	19.696	95.76	7.947

5   Conclusion

This study proposes a set of image recognition methods based on deep learning for pest and disease identification. In this paper, three kinds of convolutional neural network architectures-ALEXNET, VGG-16, and VGG-19 are compared in the feature extraction and classification of a small number of images and are also compared with the SVM classifier with the same feature extraction method. This study compared the various classification methods through the training time, model accuracy, and the time required for the model to predict an image.

This study does not consider the location of different pests on the leaf surface, image resolution or other factors, it only uses different pest images as testing and training samples then classifies the different types of pests.

We recommend that you use the DETECTOR to improve the recognition of pests. It will be able to accurately predict the number of pests on high-resolution images and to use the degree of pest damage as a decision-making consideration for agricultural management.

Acknowledgments. This research is supported by the Ministry of Science and Technology, Taiwan, R.O.C. under grant nos. MOST 107-2321-B-067F-001- and MOST 106-2119-M-309-002-MY2, which is also financially partially supported by the "Intelligent Recognition Industry Service Center" from The Featured Areas Research Center Program within the framework of the Higher Education Sprout Project by the Ministry of Education (MOE) in Taiwan.

References

1. Martin, V., Moisan, S.: Early pest detection in Greenhouses. In: International Conference on Pattern Recognition (2008)
2. Wang, K., Zhang, S., Wang, Z., Liu, Z., Yang, F.: Mobile smart device-based vegetable disease and insect pest recognition method. Intell. Autom. Soft Comput. **19**(3), 263–273 (2013)
3. Miranda, J.L., Gerardo, B.D., Tanguilig III, B.T.: Pest detection and extraction using image processing techniques. Int. J. Comput. Commun. Eng. **3**(3), 189–192 (2014)
4. Gondal, M.D., Khan, Y.N.: Early Pest Detection from Crop using Image Processing and Computational Intelligence
5. Cortes, C., Vapnik, V.: Support-vector networks. Mach. Learn. **20**(3), 273–297 (1995)
6. Faithpraise, F., Birch, P., Young, R., Obu, J., Faithpraise, B., Chatwin, C.: Automatic plant pest detection and recognition using k-means clustering algorithm and correspondence filters. Int. J. Adv. Biotechnol. Res. **4**(2), 189–199 (2013)
7. Ding, W., Taylor, G.: Automatic moth detection from trap images for pest management. Comput. Electron. Agric. **123**, 17–28 (2016)
8. Liu, Z., Gao, J., Yang, G., Zhang, H., He, Y.: Localization and classification of paddy field pests using a saliency map and deep convolutional neural network. Sci. Rep. **6**(1) (2016)
9. Pan, S.J., Yang, Q.: A survey on transfer learning. IEEE Trans. Knowl. Data Eng. **22**(10), 1345–1359 (2010)
10. Krizhevsky, I. Sutskever, Hinton, G.E.: ImageNet classification with deep convolutional neural networks. In: Advances in Neural Information Processing Systems, pp. 1097–1105 (2012)
11. Simonyan, K., Zisserman, A.: Very deep convolutional networks for large-scale image recognition. arXiv preprint arXiv:1409.1556 (2014)

The 10th International Workshop on Trustworthy Computing and Security (TwCSec-2019)

Analysis of Actual Propagation Behavior of WannaCry Within an Intranet (Extended Abstract)

Takanori Oikawa[✉], Masahiko Takenaka, and Yuki Unno

Fujitsu Laboratories Ltd., Kawasaki, Japan
oikawa.takanori@jp.fujitsu.com

Abstract. WannaCry is a ransomware that cannot only encrypt the victim's data but propagate laterally in the victim's local network. We present the results of analysis of some samples of WannaCry and actual propagation behavior of WannaCry in intranet environment.

1 Introduction

WannaCry (WannaCrypt0r 2.0) is a ransomware that targets Microsoft Windows operating system. WannaCry outbreak has being reported from May 12th, 2017. The damage amounted to 80 billion dollars, and in over 150 countries, over 300,000 terminals were infected by WannaCry [1]. Dialog box of WannaCry is available in 28 languages. It was reported that all messages except for Chinese and English have been made by using Google Translate [2] and messages in English have not been made by a native speaker [3].

Previous ransomwares are sent on e-mail as attached files or URLs, and it is executed by clicking. These ransomwares cause damage only to execution user's files. On the other hand, WannaCry can propagate laterally like worm malwares, so that WannaCry spread all over the world quickly. Unlike worm malwares, WannaCry causes direct damage by encrypting data, so that damage of WannaCry is more serious than that of worm malwares.

We divide WannaCry operation into 5 phases. Details of each phase are described as below.

Phase 1: DNS Lookup
First, WannaCry inquires a long and meaningless domain name such as "iuqerfsodp9ifjaposdf jhgosurijfaewrwergwea.com" to DNS server. In the case that WannaCry resolves the domain name, it suspends operation. On the other hand, in the case that WannaCry doesn't resolve the domain name, it moves to the next phase. This mechanism called "kill switch" is one of anti-analysis techniques. Marcus Hutchins detected kill switch of WannaCry and acquired the domain name. His work mitigated the threat of WannaCry [4].

© Springer Nature Switzerland AG 2020
L. Barolli et al. (Eds.): NBiS-2019, AISC 1036, pp. 557–559, 2020.
https://doi.org/10.1007/978-3-030-29029-0_54

Phase 2: Network Scan
Second, WannaCry searches and extracts terminals that open port 445 to use SMBv1 vulnerabilities. First, it scans IP addresses of the same domain. Then, it scans random IP addresses [5].

Phase 3: Propagation
Third, WannaCry tries to infect WannaCry to terminals extracted at Phase 2. If it's successful, the terminal runs from Phase 1. WannaCry sends broken SMB packets and inserts suspicious DLL to normal services. It uses two exploit kits, "EternalBlue" and "DoublePulsar". These exploit kits were leaked from NSA by the Shadow Brokers [6]. The vulnerability used by these exploit kits can be resolved by applying "MS17-010" patch that was published from March 2017 [7].

Phase 4: Encryption
After propagation, WannaCry encrypts files of the terminal. It connects to command and control (C&C) servers over TOR network and encrypts files of the terminal by using RSA 2048-bit and AES 128-bit [5]. More than 176 kinds of file extensions are encryption target of WannaCry [8].

Phase 5: Ransom Demand
After Encryption, WannaCry changes wallpaper and displays dialog box that shows the ransom amount and payee. WannaCry demands a ransom of $300 (three days later, $600) with bitcoin. The dialog box has a trial decrypting function that can decrypt some encrypted files randomly [5].

Thus, there are many reports about WannaCry. On the other hand, Kuzuno et al. has analyzed these reports and shown an overview of propagation of WannaCry on a worldwide scale [9].

However, little has been reported on propagation behavior of WannaCry within an intranet. We presents the results of analysis some samples of WannaCry and actual propagation behavior of WannaCry through the intranet by using "High-Speed Forensic Technology" [10].

References

1. Barlyn, S.: Insurance giant Lloyd's of London: global cyber attack could trigger $53 billion in losses—the same as Hurricane Sandy. Business Insider, 17 July 2017. http://www.businessinsider.com/r-global-cyber-attack-could-spur-53-billion-in-losses-lloyds-of-london-2017-7
2. Google: Google Translate. https://translate.google.com/
3. Condra, J., Costello, J., Chu, S.: Linguistic Analysis of WannaCry Ransomware Messages Suggests Chinese-Speaking Authors. Flashpoint, 25 May 2017. https://www.flashpoint-intel.com/blog/linguistic-analysis-wannacry-ransomware/
4. Solon, O.: Marcus Hutchins: cybersecurity experts rally around arrested WannaCry 'hero'. The Guardian, 11 August 2017. https://www.theguardian.com/technology/2017/aug/11/marcus-hutchins-arrested-wannacry-kronos-cybersecurity-experts-react
5. FireEye: Profile of WannaCry Malware (2017). https://www.fireeye.jp/company/press-releases/2017/wannacry-ransomware-campaign1.html

6. Wikipedia: The Shadow Brokers. https://en.wikipedia.org/wiki/The_Shadow_Brokers
7. Microsoft: Microsoft Security Bulletin MS17-010, 14 March 2017. https://docs.microsoft.com/en-us/security-updates/securitybulletins/2017/ms17-010
8. TrendMicro: Massive WannaCry/Wcry Ransomware Attack Hits Various Countries. 12 May 2017. https://blog.trendmicro.com/trendlabs-security-intelligence/massive-wannacrywcry-ransomware-attack-hits-various-countries/
9. Kuzuno, H., Inagaki, S., Magata, K.: Evaluation of Multiple WannaCry Reports from Various Organizations, CSS2017, 2B2-3 (2017)
10. Unno, Y., Oikawa, T.: High-speed forensic technology against targeted cyber attacks (extended abstract). In: NBiS 2017 (2017)

Proposal and Evaluation of Authentication Method Having Shoulder-Surfing Resistance for Smartwatches Using Shift Rule

Makoto Nagatomo[1], Kazuki Watanabe[1], Kentaro Aburada[2],
Naonobu Okazaki[2], and Mirang Park[1(✉)]

[1] Kanagawa Institute of Technology,
1030 Shimo-Ogino, Atsugi, Kanagawa 243-0292, Japan
mirang@nw.kanagawa-it.ac.jp
[2] University of Miyazaki,
1-1 Gakuen-Kibanadai-Nishi, Miyazaki, Miyazaki 889-2192, Japan

Abstract. Recently, mobile devices having small touchscreen such as smartwatches has been increasing due to miniaturization of electronic devices. Currently, PIN and pattern lock are used for personal authentication of these devices, but there is possibility of leakage of authentication information by shoulder-surfing attack. Many authentication methods having shoulder-surfing resistance are proposed until now. However, these methods are for smartphones or tablets having middle-size screen. Hence, when these authentication methods apply for smartwatches, the usability reduces because a user cannot touch the screen accurately. Therefore, in this paper, we propose personal authentication method having shoulder-surfing resistance for smartwatches. In this method, the user selects alternative icon to registered icon on 3×3 matrix using shift rule. In addition, we implemented the proposed method on smartwatch, and performed two experiments to confirm usability and shoulder-surfing resistance. As a result, average authentication time and authentication success rate was 13.8 s and 89.4%, and touch success rate was 96.2% when using shift rule. Also, the leakage rate of authentication information was 0.0%.

1 Introduction

Recently, devices having small touchscreen such as smartwatches has been increasing due to miniaturization of electronic devices. It is predicted that the shipments of smartwatches rise to 66 millon by 2022 [1]. Currently, many people use PIN code and pattern lock as personal authentication method for smartwatches. About PIN code, the smartwatch displays ten buttons on 4×3 matrix, but it is difficult for the user to see the buttons, and touch buttons correctly because the screen is small. This is called as "fat finger problem" [2], which causes the reduction of usability. In addition, it is possible to leak authentication information by peeking the screen. This is called "shoulder-surfing

© Springer Nature Switzerland AG 2020
L. Barolli et al. (Eds.): NBiS-2019, AISC 1036, pp. 560–569, 2020.
https://doi.org/10.1007/978-3-030-29029-0_55

attack" [3]. From the above, we set our goal of research to propose authentication method that meets two following conditions.

- Strong against shoulder-surfing attack.
- Easy to touch screen of small touchscreen device such as smartwatches.

Currently, many researchers proposed authentication method having shoulder-surfing resistance for mobile terminals [4–7]. These methods are for smartphone or tablets, so many small buttons are displayed on the screen. Hence, fat finger problem arises when these methods are applied to smartwatches due to the its small screen.

In order to solve the fat finger problem, there exists methods that has large region to touch. The method without GUI [8] is performed by tapping regions on four regions on the screen, and the method that the user taps two buttons from 2×2 buttons at the same time [9] are proposed. However, these methods are not taken into account shoulder-surfing resistance.

Therefore, in this paper, we propose an authentication method having shoulder-surfing resistance for small touchscreen terminal such as smartwatches. This method solves fat finger problem by displaying large authentication icons on 3×3 matrix, and has shoulder-surfing resistance by adopting shift rule, which the user touches alternative icon to registered icon. In addition, we implement the proposed method on smartwatch, and evaluate usability and shoulder-surfing resistance.

This paper is organized as follows. In Sect. 2, we introduce related work. In Sect. 3 and 4, we propose and implement our method, in addition, perform experiments and evaluate it in Sect. 5, finally, conclude this paper in Sect. 6.

2   Related Work

2.1   Authentication Method Having Shoulder-Surfing Resistance for Mobile Terminals with Touchscreen

Illusion Pin [4] is the method using hybrid images, which is combination of two images being able to see near and far from the screen. The user enters PIN image able to see near, but It is difficult for users to touch images correctly if this method applied to smartwatch due to displaying 10 images on the screen.

Tanaka et al. [6] proposed the authentication method using one position on a matrix and alphanumeric strings consisted of a–z and 0–9 as authentication information for smartphones. In authentication phase, two matrices are displayed. First matrix has colors on each cell at random to decide authentication color. Second matrix has alphanumeric character and background color on each cell. The user moves the authentication color of the registered position of first matrix into registered character in order to specify the character on second matrix. This method has shoulder-surfing resistance if the registered position does not leak. However, it is difficult to recognize colors and characters correctly when this method applies for small touchscreen devices because the second matrix displays many cells of a–z and 0–9.

Secret Tap with Double Shift (STDS) [7] is an authentication method using illustration icons and two shift numbers as authentication information. In authentication phase, this method displays a 4×4 matrix having icons on each cell and the user touches alternative icons to registered icons in accordance with shift rule. Shift rule separates a 4×4 matrix into four groups (four 2×2 matrix), then the user touches icon processed by the registered shift numbers between groups and in a group. This method has shoulder-surfing resistance because the user does not touch registered icons directly. However, if this method applies to smartwatches, fat finger problem arises due to difficulty of touching icons on 4×4 matrix.

2.2   Authentication Method for Small-Touchscreen Terminals

Knock Code [8] is an authentication method that does not use GUI. Authentication information is order of touch in four square regions on the screen on smartphone. This method does not need GUI because using relative touched region compared with first touch position as authentication position. In addition, this method can be applied to smartwatch due to each region is large. However, shoulder surfer can obtain the order of touch and touched region by peeking directly.

Personal Identification Chord (PIC) [9] uses single-tap or dual-tap of 2×2 buttons displayed on the smartwatch's screen. Authentication information is four taps, and the number of combinations of four taps is 10 as well as PIN code (single-tap: 4, dual-tap: 6). This method solves fat finger problem since the number of buttons is 4 (in the case of PIN code, the number of buttons is 10). However, shoulder surfer can obtain authentication information easily because the tapped buttons are specified as authentication information directly on the screen as well as Knock Code.

Higashikawa et al. [5] proposed the method improving pattern lock on 3×3 dots as challenge & response authentication. In authentication phase, 3×3 numbers are displayed at random, then the sequence of numbers on registered pattern becomes challenge. After 3×3 numbers are displayed at random again, the user draws the pattern in order of the challenge number sequence.

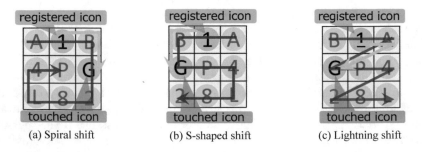

(a) Spiral shift   (b) S-shaped shift   (c) Lightning shift

Fig. 1. Shift rules and an example of shift when registered icon is 1 and shift number is 2

This method has shoulder-surfing resistance if the challenge does not leak. However, it is possible to leak challenge sequence by peeking and remembering the challenge is large memory burden for users.

3 Proposed Method

3.1 Overview

We present the authentication method that solves fat finger problem and has shoulder-surfing resistance. The user touches alternative icons to registered icons using shift rule. Authentication information is 0–9 and A–Z icons, and a shift rule. This method is improved version of STDS [7] for smartwatches. Change points are as follows.

- Simplification of registration icons
  STDS uses illustration icons as authentication information, but memory burden is high due to remembering illustration icons. We adopt 36 simple icons (A–Z and 0–9 icons) as registration icons for users to remember these icons easily.
- Size of matrix in authentication phase
  4×4 icons are displayed in authentication phase in STDS. In contrast, proposed method displays 3×3 icons in order to make it easy for the user to tap icons.
- Change of shift rule
  STDS uses shift rule, which the user touches alternative icons to registered icons using shift between group and shift in group. However, this rule is just for 4×4 matrix. Hence, we propose three shift rules for 3×3 matrix. The user touches icon proceeded from registered icon by shift number along each shift rule, as indicated in red arrow. When each shift rule reaches the end of the red arrow, shift rule returns to the start of the arrow. The three shift rules are as follows.
    - Spiral shift (see Fig. 1(a))
      This shift proceeds at a right angle when reaching the corners on matrix.
    - S-shaped shift (see Fig. 1(b))
      This shift proceeds in the opposite direction after going down one line when the shift reaches the corners on matrix.
    - Lightning shift (see Fig. 1(c))
      This shift proceeds from the left edge end after going down one line when reaching the corners on matrix.
  When registered icon is 1 and shift number is 2, the user touches the icon "G" proceeded by 2 from registered icon "1" in order to specify it. The user can decide shift number arbitrary using first registered icon in order to strengthen resistance against several shoulder-surfing attacks.

- Number of registration icons
  When the digit of registered icons is n, the proposed method has $(n-1)$-digit of authentication information because the proposed method uses first registered icon to decide shift number in authentication phase. Hence, probability of accidental authentication is $1/9^{(n-1)}$. We set the digit of the proposed method so that meets probability of accidental authentication is less than $1/10,000$, as well as 4-digit of PIN code, so set the digit of icons to register as 6 from the following equation.

$$1/9^{n-1} \leq 1/10,000 \rightarrow n \geq 6 \tag{1}$$

3.2   Registration Phase

Registration interface is 4×9 matrix having icons on each cell. Icons outside the screen appear when the user slides the screen down. The user registers the icons and shift rule as following procedure (see Fig. 2).

(1) The 4×9 matrix having 0–9 and A–Z icons is displayed on the screen.
(2) The user selects 6 icons or more.
(3) The user selects a shift rule out of four rules including non-shift rule.
(4) Finally, registration confirmation screen is displayed. Selected icons and shift rule are registered by touching registration button.

Note that the shift number is not decided in this registration phase, but is decided at first in authentication phase.

3.3   Authentication Phase

Input interface is a 3×3 matrix, which has icons on each cell. The user touches the icons different from the registered icons according to registered shift rule and shift number. Authentication procedure is as follows (See Fig. 3).

(1) 3×3 icons including first registered icon are displayed on the screen.
(2) The user decides shift number by tapping an icon proceeded from the first registered icon by arbitrary number along registered shift rule.
(3) 3×3 icons including second registered icon are displayed on the screen.
(4) The user taps icon proceeded from registered icon by the selected shift number.
(5) The user repeats step (3) and (4) four times.

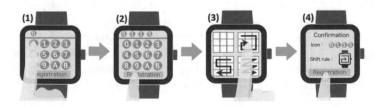

Fig. 2. Registration procedure

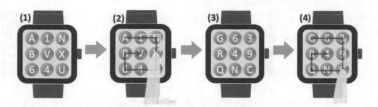

Fig. 3. Authentication procedure

4 Implementation

We implement the application of the proposed method using Java language on AndroidStudio. We show the state of user's wearing the SmartWatch 3 [10] deployed with our application in Fig. 4. The size of the smartwatch's screen is 1.6 in.

Registration screens are shown in Fig. 5. In icon select screen, touched icons are displayed on top of the screen. In addition, icons outside the screen are displayed by sliding the screen down. The user can delete selected icons by touching delete button (DEL.), and can register icons by touching registration button (REG.). We decided the number of registration icon was 6 or more in Sect. 3, but we set the number as 4 or more in developed application. The user can select 4 icons when putting more importance to usability than shoulder-surfing resistance. In shift rule select screen, the user can select shift rule by touching one shift rule out of non-shift, spiral shift, S-shaped shift, and lightning shift. In registration information confirmation screen, the user can reregister selected icons and selected shift rule by touching redo button (REDO), and complete registration by touching registration button (REG.).

Figure 6 shows screens of the authentication phase when the user registered icon $(0, 1, 2, 3)$. Each screen has registered icon in order (touch of first registered icon is used to decide shift number). After touching an icon, 3×3 icons are displayed again at random. The application finishes when authentication succeeds and redoes authentication again from the first step when the authentication fails.

Fig. 4. State of wearing smartwatch

Fig. 5. Implementation of registration

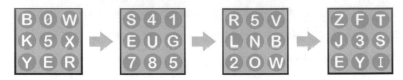

Fig. 6. Implementation of authentication

5   Experiment and Evaluation

We perform two experiments to confirm usability and shoulder-surfing resistance using SmartWatch 3 deployed with application in Sect. 4. Subjects wear the smartwatch on the arm opposite to dominant hand, and sit a chair during the experiments.

5.1   Usability Testing

In this section, we perform the experiment to confirm usability of proposed method using shift rule or not using it. Subjects are 17 university students in Kanagawa Institute of Technology. Measurement items during authentication in this experiment are authentication success rate, authentication time and touch success rate.

In addition, subjects answer a questionnaire after authentication. We create each item of questionnaire based on NASA-TLX [11]. The subject evaluates six items, which are mental demand (understanding), physical demand (easy to use), temporal demand (hurry), effort, performance (accomplishment) and frustration, as workload point from 0–100. This point is good when this value is low.

The subjects perform the experiment as following procedure:

(1) Receive explanation about the proposed method.
(2) Wear SmartWatch 3 on the arm opposite to dominant hand.
(3) Use the proposed method until getting used to it.
(4) Register 6 icons as authentication information.
(5) Perform authentication 5 times regardless of success or failure.
(6) Answer the questionnaire.

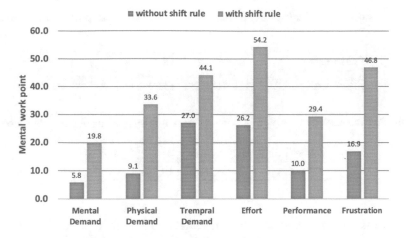

Fig. 7. Values of items of questionnaire

As a result, average authentication time was 5.3 s and 13.8 s when not using shift rule and using shift rule. This is because that subjects count shift number when using shift rule. In addition, authentication success rate was 98.8% and 89.4%, and touch success rate was 99.8% and 96.2% each. This result shows that the proposed method solves the fat finger problem since the touch success rate was high.

The average values of items on NASA-TLX are shown in Fig. 7. The all values with shift rule were larger than those without shift rule. The difference of the values between with and without shift rule were over 20 points except for Mental Demand and Temporal Demand. Especially, the difference is about 30 points on Effort and Frustration, but we expect that the two values decrease when assuming real life use.

5.2   Confirmation of Shoulder-Surfing Resistance

We perform confirmation experiment whether the proposed method has shoulder-surfing resistance or not by comparing the proposed method with pattern lock deployed with SmartWatch 3 by default. Subjects are 12 Kanagawa Institute of Technology students. Experiment is conducted as following procedure in pattern lock and the proposed method each:

(1) We organize a pair consisted of 2 subjects.
(2) We select a parent from the pair. The parent registers authentication information (pattern lock: pattern on 3 × 3 matrix, proposed method: 6 icons and shift rule).
(3) The parent performs 10 successful authentications. At the same time, the other subject (shoulder surfer) predicts authentication information by peeking the screen. Here, shoulder surfer can take note during the authentication.
(4) We switch the parent, and repeat step (2) and (3).

Table 1. Percentage of leakage of authentication information in pattern lock (after six, 0%)

Trial number	1	2	3	4	5	6
Percentage of leakage	58.3%	8.3%	16.7%	8.3%	0%	8.3%

Table 2. Percentage of leakage of authentication information in proposed method

Leaked digit	0	1	2	3, 4, 5, 6
Percentage	75.0%	8.3%	16.7%	0%

As a result, all attacks succeeded in pattern lock (the max number of attacks was 6). Table 1 shows percentage of leaked authentication information on pattern lock. We found that the information leaked at rate of over 90% until fourth attack the information leaked at rate of over 50% once attack. In addition, the result on the proposed method is shown in Table 2. The max leaked digit was 2, hence the all authentication information was never leaked once.

Therefore, we found that the proposed method has more shoulder-surfing resistance than patten lock deployed with SmartWatch 3 by default. Especially, all information of proposed method did not leak regardless of 10 attacks.

6   Conclusion

In this paper, we proposed personal authentication method for small touchscreen terminals such as smartwatches. In this method, the user performs authentication by touching alternative icon to registered icon by shift rule on 3×3 icons. In addition, we implemented the proposed method, and performed two experiments in order to evaluate usability of the proposed method and confirm whether the proposed method has shoulder-surfing resistance using SmartWatch 3.

As a result of usability testing, authentication time was 13.8 s, authentication success rate was 89.4% and touch success rate was 96.2%. This shows the user can touch icons correctly. As a result of questionnaire on NASA-TLX, we found that the proposed method add the user effort and frustration using shift rule, but it is possible to improve it assuming real life use. About the shoulder-surfing resistance experiment, we found that our method had more shoulder-surfing resistance than that in the pattern lock deployed in SmartWatch 3 by default. Especially, all authentication information did not leak in our method regardless of 10 peeking. Therefore, it is indicated that our method has high shoulder-surfing resistance.

Future work is to improve the proposed method so that the user can determine shift rule arbitrary as well as pattern lock, evaluate of recording attack, and perform continuous experiment assuming real life use.

Acknowledgements. This work was supported by JSPS KAKENHI Grant Numbers JP17H01736, JP17K00139.

References

1. Smartwatch Market Expected to Grow 41% in 2018. https://www.futuresource-consulting.com/press-release/consumer-electronics-press/smartwatch-market-expected-to-grow-41-in-2018/. Accessed 17 June 2019
2. Siek, K.A., Rogers, Y., Connelly, K.H.: Fat finger worries: how older and younger users physically interact with PDAs. In: Proceedings of the 2005 IFIP TC13 International Conference on Human-Computer Interaction, pp. 267–280 (2005)
3. Khan, H., Hengartner, U., Vogel, D.: Evaluating attack and defense strategies for smartphone PIN shouder surfing. In: Proceedings of the 2018 CHI Conference on Human Factors in Computing Systems, no. 164, 10 pages (2018)
4. Divyapriya, K., Prabhu, P.: Image based authentication using illusion pin for shoulder surfing attack. Int. J. Pure Appl. Math. **119**(7), 835–840 (2018)
5. Higashikawa, S., Kosugi, T., Kitajima, S., Mambo, M.: Shoulder-surfing resistant authentication using pass pattern of pattern lock. IEICE Trans. Inf. Syst. **E101**(1), 45–52 (2018)
6. Tanaka, M., Hiroyuki, I.: Proposal of improved background pattern slide authentication against shoulder surfing in consideration of convenience. J. Inf. Process. Soc. Jpn. **58**(9), 1513–1522 (2017). (in Japanese)
7. Kita, Y., Okazaki, N., Nishimura, H.: Implementation and evaluation of shoulder-surfing attack resistant users. IEICE Trans. Inf. Syst. **J97-D**(12), 1770–1784 (2014). (in Japanese)
8. The Galaxy S8 and Pixel Should Copy LG's Knock Code. https://www.forbes.com/sites/bensin/2017/03/02/the-galaxy-s8-and-pixel-should-copy-lgs-knock-code/. Accessed 17 June 2019
9. Oakley, I., Huh, J.H., Cho, J., Cho, G., Islam, R., Kim, H.: The personal identification chord: a four button authentication system for smartwatches. In: ASIACSS 2018 (2018)
10. SmartWatch 3. https://www.sony.jp/sp-acc/special/swr50style/. Accessed 13 May 2019
11. Hart, S.G., Staveland, L.E.: Development of NASA-TLX (task load index): result of empirical and theoretical research. Hum. Ment. Work. **1**(3), 139–183 (1988)

Evaluation of Manual Alphabets Based Gestures for a User Authentication Method Using s-EMG

Hisaaki Yamaba[1]([✉]), Shotaro Usuzaki[1], Kayoko Takatsuka[1],
Kentaro Aburada[1], Tetsuro Katayama[1], Mirang Park[2],
and Naonobu Okazaki[1]

[1] University of Miyazaki, Miyazaki, Japan
yamaba@cs.miyazaki-u.ac.jp
[2] Kanagawa Institute of Technology, Atsugi, Japan

Abstract. At the present time, since mobile devices such as tablet-type PCs and smart phones have widely penetrated into our daily lives, an authentication method that prevents shoulder surfing attacks comes to be important. We are investigating a new user authentication method for mobile devices that uses surface electromyogram (s-EMG) signals, not screen touching. The s-EMG signals, which are generated by the electrical activity of muscle fibers during contraction, can be detected over the skin surface, and muscle movement can be differentiated by analyzing the s-EMG signals. Taking advantage of the characteristics, we proposed a method that uses a list of gestures as a password in the previous study. In order to realize this method, we have to prepare a sufficient number of gestures that are used to compose passwords. In this paper, we adopted fingerspelling as candidates of such gestures. We measured s-EMG signals of manual kana of The Japanese Sign Language syllabary and evaluated their potential as the important element of the user authentication method.

1 Introduction

This paper presents an evaluation of manual alphabets as candidates of gestures used in the user authentication method for mobile devices by using surface electromyogram (s-EMG) signals, not screen touching.

An authentication method that prevents shoulder surfing, which is the direct observation of a users personal information such as passwords, comes to be important. At the present time, mobile devices such as tablet type PCs and smartphones have widely penetrated into our daily lives. So, authentication operations on mobile devices are performed in many public places and we have to ensure that no one can view our passwords. However, it is easy for people who stand near such a mobile device user to see login operations and obtain the users authentication information. And also, it is not easy to hide mobile devices from attackers during login operations because users have to see the touch screens of

© Springer Nature Switzerland AG 2020
L. Barolli et al. (Eds.): NBiS-2019, AISC 1036, pp. 570–580, 2020.
https://doi.org/10.1007/978-3-030-29029-0_56

their mobile devices, which do not have keyboards, to input authentication information. On using a touchscreen, users of a mobile device input their authentication information through simple or multi-touch gestures. These gestures include, for example, designating his/her passcode from displayed numbers, selecting registered pictures or icons from a set of pictures, or tracing a registered one-stroke sketch on the screen. The user has to see the touch screen during his/her login operation; strangers around them also can see the screen.

To prevent this kind of attack, biometrics authentication methods, which use metrics related to human characteristics, are expected. In this study, we investigated application of surface electromyogram (s-EMG) signals for user authentication. S-EMG signals, which are detected over the skin surface, are generated by the electrical activity of muscle fibers during contraction. These s-EMGs have been used to control various devices, including artificial limbs and electrical wheelchairs. Muscle movement can be differentiated by analyzing the s-EMG [1]. Feature extraction is carried out through the analysis of the s-EMGs. The extracted features are used to differentiate the muscle movement, including hand gestures.

In the previous researches [2–8], we investigate the prospect of realizing an authentication method using s-EMGs through a series of experiments. First, several gestures of the wrist were introduced, and the s-EMG signals generated for each of the motion patterns were measured [2]. We compared the s-EMG signal patterns generated by each subject with the patterns generated by other subjects. As a result, it was found that the patterns of each individual subject are similar but they differ from those of other subjects. Thus, s-EMGs can confirm ones identification for authenticating passwords on touchscreen devices. Next, a method that uses a list of gestures as a password was proposed [3,4]. And also, a series of experiments was carried out to investigate the performance of the method extracting feature values from s-EMG signals adopted in [5] and the methods to identify gestures using dynamic data warping (DTW) [6] and support vector machines (SVM) [7]. The results showed that the methods are promising to identify gestures using extracted feature values from s-EMG signals.

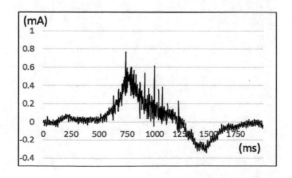

Fig. 1. A sample of an s-EMG signal

As for the selection of gestures used in the method, we introduced manual alphabets used in Japanese Sign Language [8]. In this paper, we evaluated their performances as the candidate of gestures used in the authentication method.

2   Characteristics of Authentication Method for Mobile Devices

It is considered that user authentication of mobile devices has two characteristics [2].

One is that an authentication operation often takes place around strangers. An authentication operation has to be performed when a user wants to start using their mobile devices. Therefore, strangers around the user can possibly see the user's unlock actions. Some of these strangers may scheme to steal information for authentication such as passwords.

The other characteristic is that user authentication of mobile devices is almost always performed on a touchscreen. Since many of current mobile devices do not have hardware keyboards, it is not easy to input long character based passwords into such mobile devices. When users want to unlock mobile touchscreen devices, they input passwords or personal identification numbers (PINs) by tapping numbers or characters displayed on the touchscreen. Naturally, users have to look at their touchscreens while unlocking their devices, strangers around them also can easily see the unlock actions. Besides, the user moves only one finger in many cases. So, it becomes very easy for thieves to steal passwords or PINs.

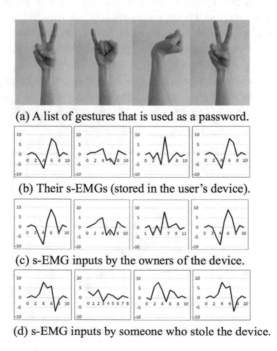

(a) A list of gestures that is used as a password.

(b) Their s-EMGs (stored in the user's device).

(c) s-EMG inputs by the owners of the device.

(d) s-EMG inputs by someone who stole the device.

Fig. 2. A list of gestures used as a password

To prevent shoulder-surfing attacks, many studies have been conducted. The secret tap method [9] introduces a shift value to avoid revealing pass-icons. The user may tap other icons in the shift position on the touchscreen, as indicated by a shift value, to unlock the device. By keeping the shift value secret, people around the user cannot know the true pass-icons, although they can still watch the tapping operation. The rhythm authentication method [10] relieves the user from looking at the touchscreen when unlocking the device. In this method, the user taps the rhythm of his or her favorite music on the touchscreen. The pattern of tapping is used as the password. In this situation, the users can unlock their devices while keeping them in their pockets or bags, and the people around them cannot see the tap operations that contain the authentication information.

3   User Authentication Using s-EMG

The s-EMG signals (Fig. 1) are generated by the electrical activity of muscle fibers during contraction and are detected over the skin surface [2]. Muscle movement can be differentiated by analyzing the s-EMG.

In the previous research, the method of user authentication by using s-EMGs that do not require looking at a touchscreen was proposed [3,4]. The s-EMG signals are measured, and the feature values of the measured raw signals are extracted. We estimate gestures made by a user from the extracted features. In this study, combinations of the gestures are converted into a code for authentication. These combinations are inputted into the mobile device and used as a password for user authentication.

1. At first, pass-gesture registration is carried out. A user selects a list of gestures that is used as a pass-gesture. (Fig. 2(a))
2. The user measures s-EMG of each gesture, extracts their feature values, and register the values into his mobile device. (Fig. 2(b))
3. When the user tries to unlock the mobile device, the user reproduces his pass-gesture and measures the s-EMG.
4. The measured signals are sent to his mobile device.
5. The device analyzes the signals and extracts the feature values.
6. The values are compared with the registered values.
7. If they match, the user authentication will succeed. (Fig. 2(c))
8. On the other hand, an illegal user authentication will fail because a list of signals given by someone who stole the device (Fig. 2(d)) will not be similar with the registered one.

Adopting s-EMG signals for authentication of mobile devices has three advantages. First, the user does not have to look at his/her device. Since the user can make a gesture that is used as a password on a device inside a pocket or in a bag, it is expected that the authentication information can be concealed. No one can see what gesture is made. Next, it is expected that if another person reproduces a sequence of gestures that a user has made, the authentication will not be successful, because the extracted features from the s-EMG signals are usually not the same between two people. And then, a user can change the list of gestures in our method. This is the advantages of our method against other biometrics based methods such as fingerprints, an iris, and so on. When authentication information, a fingerprint or an iris, come out, the user can't use them because he/she can't change his/her fingerprint or iris. But the user can arrange his/her gesture list again and use the new gesture list.

4 Collecting Gestures for the Authentication Method Using s-EMG

In order to realize the proposed authentication method using s-EMG, we have to prepare many gestures that represent characters of passwords. Such gestures have to be easy to tell each of them from others.

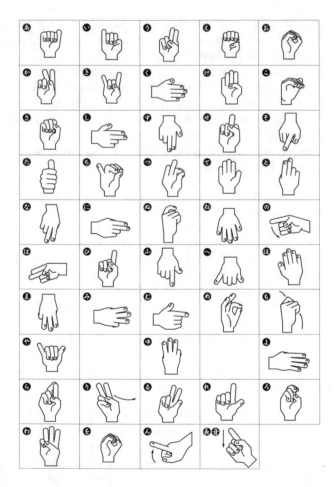

Fig. 3. The Japanese Sign syllabary. (https://upload.wikimedia.org/wikipedia/commons/d/dc/JSL-AIUEO.jpg)

We are planning to adopt such gestures referring to fingerspelling. Finger-spelling is the representation of the letters using hands. The set of manual signs is used to help the communication using sign languages. For example, proper nouns such as names of persons are represented by fingerspelling. Sign laguages such as American Sign Language, French Sign Laguage, British Sign Language, and so on, have there own manual alphabet.

We adopted the Japanese Sign syllabary (see Fig. 3) as the candidates of gestures used in our authentication method. They are called *yumimoji*, which means "finger letters." Comparing with manual alphabets representing Latin alphabet, there are larger numbers of manual kana in *yumimoji*, 46 letters.

By adopting gestures referring fingerspelling, a password that is made up of gestures corresponds to a string that is made up of letters. It is expected that this helps users to remember their passwords.

However, some *yubimoji* may not generate distinct s-EMG signals. To select *yumimoji* that are suitable for the use of this study, we examined their perfor-mances as pass-gesture by measuring their s-EMG signals.

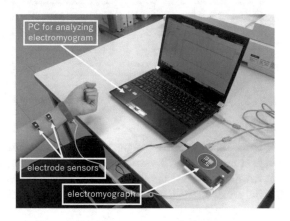

Fig. 4. Measuring an s-EMG signal

5   Expeeriments

5.1   Purpose

A series of experiments was carried out to investigate the prospect of the authen-tication method using *yubimoji*. Concretely, we investigated whether the signals of different persons were different from each other and whether the measured s-EMG signals of one experimental subject were similar. And also, we attempted to select prospect yubimoji and explore characteristics of such yubimoji. It is expected that characteristics obtained in these experiments can be used to devise efficient gestures. Finally, we introduced support vector machines to tell apart users using gestures based on *yubimoji*.

5.2 Conditions

In these experiments, the set of DL-3100 and DL-141 (S&M Inc.) that was an electromyograph used in the previous researches also used to measure the s-EMG of each movement pattern in this study (Fig. 4). Experimental subjects made gestures by their left hand. Subjects sit on an arm chair and put his forearm on the armrest. Two electrode sensors were put on the palm side of the forearm at the positions of 4 cm and 6 cm toward the elbow from the wrist (The former is called "m1" and the latter is called "m2" in the following sections.) The measured data were stored and analyzed on a PC.

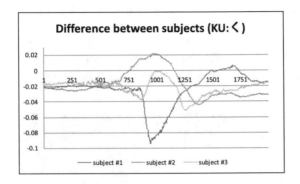

Fig. 5. S-EMG signals of "KU" of three subjects

Five students of University of Miyazaki participated as experimental subjects. And the 46 *yubimoji* shown in Fig. 3 were examined. However, since a subject put his forearm on an armrest, we arranged some *yubimoji*. When we make a *yubimoji* like *TE*, the palm faces forward. But in these experiment, the palm faces upward.

Before making each *yubimoji*, subjects clenched their left fists. S-EMG signals making *yubimoji* from a clenched fist were measured. The subjects repeated each *yubimoji* ten times and their s-EMG signals were recorded. This measurement was carried out 3 times and 30 signals were obtained for each subject and for each *yubimoji*.

5.3 Results

From the experiments, we obtained the three results shown below.

1. Signals are different between subjects.
2. Some gestures were classified into several groups.
3. Reproducibility was not noticeable.

5.3.1   Difference Between Subjects

Figure 5 shows the s-EMG signal of yubimoji "KU" of three subjects. Signal patterns of the three subjects are different from each other. From this results, it is expected that user authentication method using s-EMG can prevent unlock by illegal user who knows the pass-gesture.

5.3.2   Classification of Yubimoji Gestures

From the results of the experiments, we classified some gestures into five groups shown below.

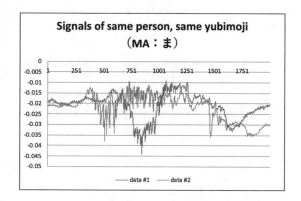

Fig. 6. S-EMG signals of "MA" measured in different days

Type 1: Yubimoji that bends one's wrist forth
   This kind of yubimoji bend one's wrist forth. Some of their fingers are extended pointing to the side. Amplitude of s-EMG signals are quite large.
Type 2: Yubimoji that points downward
   This kind of yubimoji bend one's wrist forth but their extended fingers point downward. Amplitude of s-EMG signals are large.
Type 3: Yubimoji that extends only one finger
   This kind of yubimoji extends only one finger. Also they don't bent one's wrist. Amplitude of s-EMG signals measured are very slight. It seems hard to use these letters as elements of pass-gestures.
Type 4: Yubimoji that extends more than one finger
   This kind of yubimoji extends more than one finger including index finger and middle finger. Also they don't bent one's wrist. Amplitude of s-EMG signals measured are slight.
Type 5: Yubimoji that turns the forearm
   This kind of yubimoji turns the forearm. They don't bent one's wrist. Amplitude of s-EMG signals are clear.

   Type 3 gestures that are simple one are not suitable because amplitude of vibration of signals generated by such gestures are quite small. Type 1 gestures that bends forth are promising because they generate larger amplitude. However, it has not been examined that we can distinguish gestures in this type. Type 1 gestures use muscles where electrode sensors are attached. So, it is expected that distinct signals can be obtained by attaching several sensors appropriate positions for each gesture.

5.3.3   Reproducibility

S-EMG signals of some yubimoji measured after a few days were not similar with the signals obtained in the first measurement. Figure 6 shows the s-EMG signals of yubimoji "MA" of the same subject. The blue line is signals obtained in the first measurement and the orange line is obtained a few days later. Especially, reproducibility of yubimoji like this was worse.

Table 1. Correct answer rates of some `yubimoji`

yubimoji	"fu"	"mu"	"ne"	"ta"
Correct answer rate (%)	0.50	0.62	0.54	0.50

Since participants paid attention to make yubimoji correctly, it seems that their arms were strained. The gestures used in the previous studies [2–8] were simple compared with gestures based on yubimoji. This means that gestures that are too complex are not suitable for candidates of user authentication method using s-EMG. On the other hand, simple gestures are not suitable because amplitude of vibration of signals generated by such gestures are quite small.

5.3.4   Introduction of Support Vector Machines

Support vector machines are one of the pattern recognition models of supervised learning. Linear SVM was proposed in 1963, and extended to non-linear classification in 1992. A support vector machine builds a classifier for sample data that belong to one of two classes. An SVM trains the separation plane that has the largest margin, and samples on the margin are called support vectors. An SVM is one of the recognition method that has the highest performance.

In this research, The programming language "R" was used. SVM function of the programming language R can classify data into several categories. One SVM is prepared for one *yumimoji* based gesture and trained by data of the gesture. This SVM selects one user for given s-EMG signals. We selected some gestures and trained their SVMs using some s-EMG signal data and other data were given to the corresponding SVM.

We used the pair of the maximum value and the minimum value of raw s-EMG signals as the feature value that was used in our previous study [4] for a brief investigation. The correct answer rates of some gestures are shown in Table 1. The performance of the simple features are not good. We will apply more complex sets of features used in [6, 7] in the future work.

6 Conclusion

We investigated a new user authentication method that can prevent shoulder-surfing attacks in mobile devices. To realize the authentication method using s-EMG, we examined the characteristics of the *yubimoji*, the Japanese Sign Language syllabary, as the candidate of the element of pass-gestures. A series of experiments was carried out to compare the measured s-EMG signals generated by making gestures based on *yubimoji*. Results of the experiments showed that s-EMG signals of generated by gestures based on *yubimoji* are different between subjects but reproducibility was not noticeable. Also, it is expected that which fingers are extended and whether a wrist is bent or not can be used as characteristics to identify *yubimoji* from s-EMG signals.

We are planning to examine characteristics of *yubimoji* using a lot of s-EMG data from many people and attempt to explore gestures that are good in the view point of reproducibility. Also, we would like to explore appropriate positions of measuring s-EMG.

Finally, We are planning to adopt other sets of features for training SVMs and to apply other machine learning methods such as Learning to identify gestures from s-EMG signals.

Acknowledgements. This work was supported by JSPS KAKENHI Grant Numbers JP17H01736, JP17K00139, JP17K00186, JP18K11268.

References

1. Tamura, H., Okumura, D., Tanno, K.: A study on motion recognition without FFT from surface-EMG. IEICE Part D **J90-D**(9), 2652–2655 (2007). (in Japanese)
2. Yamaba, H., Nagatomo, S., Aburada, K., et al.: An authentication method for mobile devices that is independent of tap-operation on a touchscreen. J. Robot. Netw. Artif. Life **1**, 60–63 (2015)
3. Yamaba, H., Kurogi, T., Kubota, S., et al.: An attempt to use a gesture control armband for a user authentication system using surface electromyograms. In: Proceedings of 19th International Symposium on Artificial Life and Robotics, pp. 342–345 (2016)
4. Yamaba, H., Kurogi, T., Kubota, S., et al.: Evaluation of feature values of surface electromyograms for user authentication on mobile devices. Artif. Life Robot. **22**, 108–112 (2017)
5. Yamaba, H., Kurogi, T., Aburada, A., et al.: On applying support vector machines to a user authentication method using surface electromyogram signals. Artif. Life Robot. (2017). https://doi.org/10.1007/s10015-017-0404-z
6. Kurogi, T., Yamaba, H., Aburada, A., et al.: A study on a user identification method using dynamic time warping to realize an authentication system by s-EMG. In: Advances in Internet, Data & Web Technologies (2018). https://doi.org/10.1007/978-3-319-75928-9_82
7. Yamaba, H., Aburada, A., Katayama, T., et al.: Evaluation of user identification methods for an authentication system using s-EMG. In: Advances in Network-Based Information Systems (2018). https://doi.org/10.1007/978-3-319-98530-5_64

8. Yamaba, H., Inotani, S., Usuzaki, S., et al.: Introduction of fingerspelling for realizing a user authentication method using s-EMG. In: Advances in Intelligent Systems and Computing (2019). https://doi.org/10.1007/978-3-030-15035-8_67
9. Kita, Y., Okazaki, N., Nishimura, H., et al.: Implementation and evaluation of shoulder-surfing attack resistant users. IEICE Part D **J97-D**(12), 1770–1784 (2014). (in Japanese)
10. Kita, Y., Kamizato, K., Park, M., et al.: a study of rhythm authentication and its accuracy using the self-organizing maps. In: Proceedings of DICOMO 2014, pp. 1011–1018 (2014). (in Japanese)

Analysis of the Reasons Affecting
the Simulation Results of SAR Imaging

Xing-Xiu Song[✉]

Department of Arms, The Training Base in Officers College of PAP,
Guangzhou 510440, GuangDong, China
wangxazjd@163.com

Abstract. This article expounds the original echo model of SAR and focus on the use of MATLAB software to the point target imaging simulation from the synthetic aperture radar (SAR) of the transmitted waveform, which directly reflects the principle of the R-D algorithm. Finally, through the simulation analysis, we summarize the factors that influence the simulation results in SAR imaging.

Keywords: Synthetic aperture radar · MATLAB simulation ·
Range Doppler algorithm

1 Introduction

Synthetic aperture radar (SAR) is a core technology in the field of imaging, and the simulation technology is developing rapidly along with the computer technology and the signal processing technology developing by leaps and bounds. The simulation result is close to the actual result, which makes the simulation became one of the foundations of modern radar design and researching, and MATLAB software is better than any other software in radar simulation [1].

SAR is one kind of high resolution imaging radar, which rely on relative motion between radar and target. SAR imaging has all-time and all-weather advantages compared with other imaging, which has widespread application in the national economy and national defense construction. We often analyze and verify the results of the study through the computer simulation, due to the expensive cost for researching stage of SAR. This article summarizes the factors that affect the simulation results of SAR imaging from the range Doppler algorithm of SAR imaging using MATLAB software imaging of point target imaging, and finally analysis the results.

2 Wave Simulation of SAR Emission

Synthetic aperture radar is an important tool for the remote sensing the earth. Image resolution depends on the size of the lighting footprint on the ground in traditional radar systems. However, the SAR technology can realize the all-time and all-weather imaging with high resolution. This is based on emitting wide bandwidth of linear

© Springer Nature Switzerland AG 2020
L. Barolli et al. (Eds.): NBiS-2019, AISC 1036, pp. 581–585, 2020.
https://doi.org/10.1007/978-3-030-29029-0_57

frequency modulation (LFM) signal in distance and using Doppler frequency shift information of orientation in the received signal.

The emission linear frequency modulation (LFM) signal of SAR is also called the Chirp signal, and it through the linear phase modulation to product bigger wide bandwidth, which is the earliest and most widely used pulse compression signal.

The LFM signal can be expressed as:

$$s_i(t) = Arect\left(\frac{t}{\tau}\right) \cos\left(\omega_c t + \frac{\mu t^2}{2}\right) \tag{1}$$

Among them,

$$rect\left(\frac{t}{\tau}\right) = \begin{cases} 1, & \left|\frac{t}{\tau}\right| \leq \frac{1}{2} \\ 0, & \left|\frac{t}{\tau}\right| > \frac{1}{2} \end{cases} \tag{2}$$

The envelope of linear frequency modulation signal is rectangular pulse width with τ width, and μ is adjustable frequency, and ω_i is instantaneous angular frequency, and the signal instantaneous frequency is linear change with time (Figs. 1, 2).

$$\omega_i = \frac{d\phi}{dt} = \omega_c + \mu t \tag{3}$$

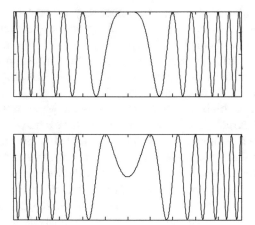

Fig. 1. LFM signal waveform

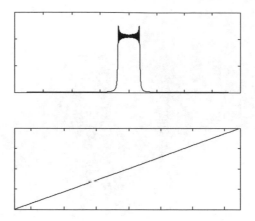

Fig. 2. Relationship between the amplitude and time frequency spectrum of the LFM signal

3   SAR Imaging Simulation

SAR transmits LFM signals in fighting, and it will meet the target and reflected echo signal, whose mathematical model is the basis of SAR imaging algorithm [2]. Echo signal is time delay of transmitted signal, and according to Eq. (1), we can be expressed the echo signal model as:

$$s_i(t) = Arect\left(\frac{t - t_0}{\tau}\right) \cos\left[\omega_c(t - t_0) + \frac{\mu(t - t_0)^2}{2}\right] \tag{4}$$

Among them, t_0 is time delay, and $\mu = \frac{\pi BW}{T}$ is linear frequency modulation rate, and BW is signal bandwidth, and t is fast time. The original echo signal waveform of 2 point target as shown in Fig. 3, and obtaining the baseband signal after mixing echo signal is as shown in Fig. 4.

$$m_{if}(t) = \frac{1}{4}\exp\left\{j\left[\omega_0 + \mu(t - t_0)^2 - \omega_c t_0\right]\right\} \tag{5}$$

Type in the $\omega_0 = \omega_c - \omega_d$, and ω_d is Intermediate frequency.

SAR point target imaging simulation is based on the echo model, and then make it become an image point and display the picture effect using the range Doppler (R-D) algorithm doing range and azimuth compression of simulated point target for echo data. The basic SAR imaging R-D algorithm has been discussed in the literature [4, 5], here no longer. The simulation results of MATLAB SAR point target imaging as shown in from Figs. 4, 5 and 6, it directly and representatively reflects the principle of R-D algorithm.

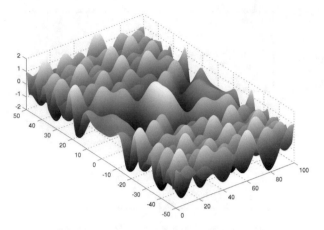

Fig. 3. Two point target echo signal of SAR

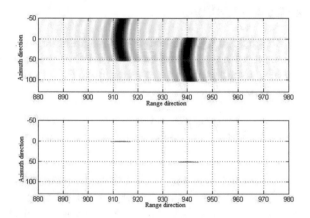

Fig. 4. Not correcting range migration of 2 point target imaging

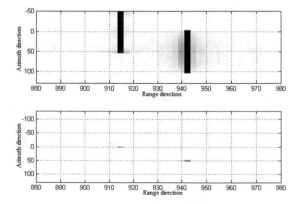

Fig. 5. Correcting range migration of 2 point target imaging

4   Simulation Analysis of Effect

In the SAR imaging simulation process, we analyze the factors which influence the simulation results.

(1)   The range resolution and azimuth resolution
      The SAR range resolution depends on bandwidth of transmitted signal, and the greater bandwidth, the better distance resolution. In the actual system, azimuth resolution is proportional to beam width of the antenna azimuth, and the resolution of synthetic aperture is inversely proportional to beam width of the azimuth.
(2)   Choosing of parameters of SAR system
      The most important parameters in SAR system is choosing the pulse repetition frequency, and PRF cannot be too low, otherwise it will cause the azimuth ambiguity, and it also cannot be too large, otherwise it will produce fuzzy distance.
(3)   Range migration
      RCM makes the range signal with frequency modulation characteristic, but if not corrects range migration, it will reduce the imaging resolution, especially deterioration of the range resolution.
(4)   Adding window of the distance and antenna beam of the azimuth
      A method of reducing the peak side lobe ratio is adding a window of the frequency matching filter in distance in frequency domain, which can smooth the spectrum. To Antenna beam can realize the spatial filtering, and it also can reflect the spatial filtering capability of radar.

5   Conclusion

The computer simulation has the extremely development prospects, but the factors impacting SAR imaging simulation results is various. In order to the better use of simulation software to analysis the SAR technology, this paper carries on MATLAB software to complete the point target imaging process. Through the simulation analysis, we summarize three important factors, which is seriously affecting the imaging quality.

References

1. Mahafza, B.R., Elsherbeni, A.Z.: Simulations for Radar Systems Design. Publishing House of Electronics Industry, Beijing, 10 (2009)
2. Meng, Q., Liao, J.: The computer simulation technology of synthetic aperture radar **4.2**(26), 203–207 (2003)
3. Duersch, M.I.: A very small, low-power LFM-CW synthetic aperture radar. Brigham Young University, 12 (2004)
4. Sun, H., Gu, H., Su, W., Liu, G.: Research on airborne synthetic aperture radar imaging algorithm **9**(23), 90–94 (2002)
5. Wei, Q.: Research on synthetic aperture radar imaging method and the method of jamming. Xi'an University of Electronic Science and Technology, pp. 18–25 (2006)
6. Cumming, L.G., Wong, F.H.: Algorithm and Implementation of Synthetic Aperture Radar Imaging, Publishing House of Electronics Industry, Beijing, 6 (2012)

Finite Element Simulation of Blasting Robote with ANSYS

Xing-Xiu Song[✉]

Department of Arms, The Training Base in Officers College of PAP,
Guangzhou 510440, GuangDong, China
419279223@qq.com

Abstract. In this paper, the explosive disposal robot arm is taken as the research object. Through the basic theory of robot kinematics, the motion characteristics of the explosive disposal robot are studied, the kinematic parameters of the explosive disposal robot are determined, and the problem of positive motion and inverse motion is solved. After the simulation of the motion of the explosive disposal robot, the finite element analysis is carried out by introducing the model into ANSYS, and the stress distribution of the explosive robot under static load is obtained. Based on the finite element results, the strength and weak points are can be find out. The finite element analysis has certain reference significance for material selection and structure optimization of explosive disposal robot.

Keywords: Explosive disposal robot · ANSYS · FEA

1   Introduction

Recent years has seen an increase in the number of vicious explosions in China. Ethnic separatists as well as a small number of criminals who are dissatisfied with the society hold explosives and create terrorist incidents more and more often, making extremely bad impact, killing quite a few policemen. With people's lives and property seriously threatened, the task of maintaining public security for the police is becoming increasingly heavy. In addition, with the continuous development of China's economy and the increase of large-scale international economic, political, cultural and sports activities, the prospect of demand for anti-terrorist and explosive disposal robots is incredibly broad. Therefore, China is in urgent need of upgrading anti-terrorism equipment, notably among which is the explosive disposal robot [1–7].

Explosive disposal robots are specialized equipment for explosive disposal personnel to dispose or destroy suspected explosives, thus avoiding unnecessary casualties. Being able to work in all kinds of complex terrains, the robots are mainly used to transport suspected explosives and other hazardous or dangerous goods and to destroy bombs using an explosive destructor, taking the place of explosive disposal personnel. It can also do field investigation and real-time transmission of video images, replacing the on-site security personnel. The robots can be equipped with shotguns to attack criminals when necessary, it can also be equipped with detection equipment to check

© Springer Nature Switzerland AG 2020
L. Barolli et al. (Eds.): NBiS-2019, AISC 1036, pp. 586–592, 2020.
https://doi.org/10.1007/978-3-030-29029-0_58

dangerous places as well as dangerous goods. Due to the characteristic of high technology content, explosive disposal robots are often extraordinarily expensive.

According to the method of operation, the explosive disposal robots are divided into two types. One is remote-controlled robots, which conduct artificial explosive disposal under visual conditions, with human playing the part of commander while robots the executive officer. The other is automatic robots, which can distinguish dangerous objects and get rid of the danger all by themselves. To do so, disks containing relevant programs written beforehand by programmers should be inserted into the robots, keeping the cost at a high level. Therefore, automatic robots are not often used unless under emergency.

Based on the explosive disposal robot arm and the basic kinematics theory, in this paper, the kinematics parameters of the explosive disposal robot are ascertained, and the problem of positive motion and inverse motion is solved. After the kinematics simulation of the explosive disposal robot is completed, the model is imported into ANSYS for finite element analysis, and the stress distribution of the explosive disposal robot under static load is obtained. Based on the FEA results, the intensity is checked and the weak point is found out. Finite element analysis has a good guiding significance for both material selection and structure optimization of explosive disposal robot.

2 Statics Theoretical Analysis of the Arm of Explosive Disposal Robot

The arm of the explosive disposal robot is composed of different rod-shaped parts, thus when we conduct the finite element analysis of the explosive disposal robot, we adopt the principle of statics of rod piece.

First, the whole member bar can be represented by 5 nodes and 4 units. In order to study the behavior characteristics of the rod piece, considering the deformation of the bar (cross-sectional area A, length l) under the action of external force F, the formula for the average stress of the bar is as follow:

$$\sigma = \frac{F}{A} \tag{1}$$

The average strain of the member σ is defined as the ratio of the original length F per unit to the change in length A after loading. According to Hooker theorem:

$$\sigma = E\xi \tag{2}$$

The elastic modulus of the material is denoted as E. Solve the simultaneous equations and we get:

$$F = (\frac{AE}{l})\Delta l \tag{3}$$

A bar of uniform cross-section under axial force can be considered as a spring, its equivalent stiffness is as follow:

$$K_{eq} = \frac{AE}{l} \tag{4}$$

The bar can be viewed as a model consisting of 4 springs in series. It satisfies the following equation:

$$f = K_{eq}(u_{i+1} - u_i) = \frac{A_{avg}E}{l}(u_{i+1} - u_i) = \frac{(A_{i+1} + A_i)E}{2l}(u_{i+1} - u_i) \tag{5}$$

The equivalent spring stiffness is:

$$K_{eq} = \frac{(A_{i+1} + A_i)E}{2l} \tag{6}$$

In the equation, A is the cross-sectional area of node i and node (i + 1). L is the length of the unit. E represents the length change at each node. Static equilibrium requires that the sum of forces at each node be 0, therefore there would be 5 equations (P is the applied force, R_1 is the reaction force):

Node1:

$$R_1 - k_1(u_2 - u_1) = 0$$

Node2:

$$k_1(u_2 - u_1) - k_2(u_3 - u_2) = 0$$

Node3:

$$k_3(u_4 - u_3) - k_4(u_5 - u_4) = 0 \tag{7}$$

Node4:

$$k_3(u_4 - u_3) - k_4(u_5 - u_4) = 0$$

Node5:

$$k_4(u_5 - u_4) - P = 0$$

A matrix can be obtained by separating the external force P and R_1 and recombining the equation:

$$\begin{pmatrix} k_1 & -k_1 & 0 & 0 & 0 \\ -k_1 & k_1+k_2 & -k_2 & 0 & 0 \\ 0 & -k_2 & k_3+k_2 & -k_3 & 0 \\ 0 & 0 & -k_3 & k_3+k_4 & -k_4 \\ 0 & 0 & 0 & -k_4 & k_4 \end{pmatrix} \begin{pmatrix} u_1 \\ u_2 \\ u_3 \\ u_4 \\ u_5 \end{pmatrix} = \begin{pmatrix} -R_1 \\ 0 \\ 0 \\ 0 \\ P \end{pmatrix} \qquad (8)$$

Separate the external force from the reaction force

$$\begin{pmatrix} k_1 & -k_1 & 0 & 0 & 0 \\ -k_1 & k_1+k_2 & -k_2 & 0 & 0 \\ 0 & -k_2 & k_3+k_2 & -k_3 & 0 \\ 0 & 0 & -k_3 & k_3+k_4 & -k_4 \\ 0 & 0 & 0 & -k_4 & k_4 \end{pmatrix} \begin{pmatrix} u_1 \\ u_2 \\ u_3 \\ u_4 \\ u_5 \end{pmatrix} - \begin{pmatrix} 0 \\ 0 \\ 0 \\ 0 \\ P \end{pmatrix} = \begin{pmatrix} -R_1 \\ 0 \\ 0 \\ 0 \\ 0 \end{pmatrix} \qquad (9)$$

And we can get:

[reaction force matrix] = [stiffness matrix][displacement matrix] − [load matrix]

When boundary condition $R_1 = 0$ is used, the unknown reaction force can be eliminated,

$$\begin{pmatrix} 1 & 0 & 0 & 0 & 0 \\ -k_1 & k_1+k_2 & -k_2 & 0 & 0 \\ 0 & -k_2 & k_3+k_2 & -k_3 & 0 \\ 0 & 0 & -k_3 & k_3+k_4 & -k_4 \\ 0 & 0 & 0 & -k_4 & k_4 \end{pmatrix} \begin{pmatrix} u_1 \\ u_2 \\ u_3 \\ u_4 \\ u_5 \end{pmatrix} = \begin{pmatrix} 0 \\ 0 \\ 0 \\ 0 \\ P \end{pmatrix} \qquad (10)$$

And we can get:

[stiffness matrix][displacement matrix] = [load matrix]

3 Establishment of Finite Element Model for Explosive Disposal Robot

Preprocessor of ANSYS software provides abundant material model, there are about 30 kinds of material models that can be directly input, which are divided into 7 categories: linear elastic model, nonlinear elastic model, elastoplastic model, foam model, composite material model, material model that acquire equation of state and other models. Under this specific situation, the linear elastic model is chosen [8–10].

The gridding process of entity model usually includes three steps: defining unit attributes, defining grid generation control and generating grid. First, set the unit properties of each part, which includes element type, real constant, material properties: mass density, modulus of elasticity, Poisson's ratio. After that, the MeshTool meshing command is used to mesh different parts. The partition method generally includes free partition and mapping partition. Free partitioning has no restrictions on the shape of the cell and has no specific rules while map partitioning is limited by cell shapes and specific rules, generating a more regular grid distribution. In this paper, the grid is divided using the build-in method of ANSYS, the result is as follow (Fig. 1):

Fig. 1. Finite element mesh of explosive disposal robot

In the process of analysis, we mainly analyzed the stress distribution of different bars and connectors, obtaining the finite element analysis results of the bar. According to the actual situation, we choose the material coefficient as follows: the material density to be kg/m^3, the elastic modulus to be MPa, and the Poisson's ratio to be 0.3.

4   Finite Element Analysis of the Model with ANSYS and Conclusion

After the pretreatment, the data obtained by ADAMS are analyzed. After input load, the stress distribution diagram of the bar under the simulation result can be obtained under ANSYS. As shown in Figs. 2 and 3.

It can be seen from the figure that the maximum stress value of the bar and the connector is far less than the dangerous stress of the material, thus failure of the material will not occur and affect the work of the explosive disposal robot. However, in actual situations, we should strengthen the structure where the stress is large, so as to prevent the fatigue stress caused by the long-time large stress sustained by the material. At the same time, the places with less stress can be appropriately optimized to reduce the material usage and the power consumption. It can also be seen from the figure that in the whole part of the bar and connector, the part that connects to the drive units subject to greater stress, therefore in the process of actual design, we should strengthen the structure of this part. We also noticed that the stress in the middle part is relatively average and small, so the structure can be optimized by reducing material, etc.

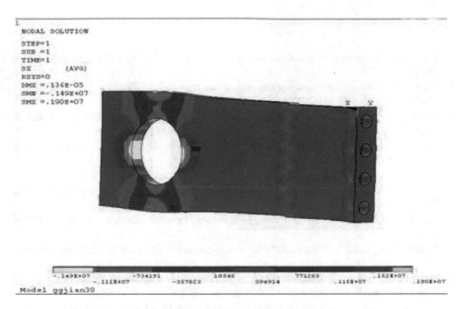

Fig. 2. Stress distribution of connector 2

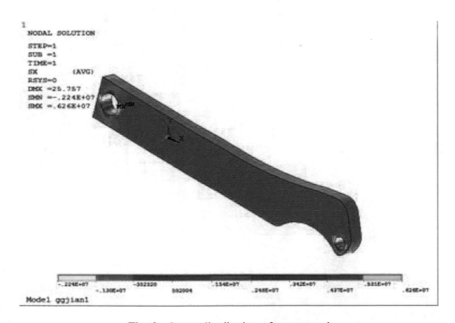

Fig. 3. Stress distribution of connector 1

References

1. Song, H.: The prospect of robot technology. J. Shanxi Coal Manag. Cadre Coll. **4**, 135–136 (2006)
2. Li, S.: The development and application of military robot. Electron. Eng. 33(5), 64–66 (2007)
3. Defense of robots for public safety-armed police equipment experts talk about anti-terrorism robots. Mod. Mil. **05**, 33–35 (2006)
4. Lin, Y.: Fire fighting, blasting, anti-terrorism, reconnaissance our robot appearance in Shanghai exhibition. Mod. Mil. **01**, 12–16 (2006)
5. Chan, D., et al: Police Robot, pp. 151–154. Science Press, Beijing (2008)
6. Liu, H., et al.: The Basis of Robot Technology, pp. 21–31. Metallurgical Industry Press, Beijing (2002)
7. Luo, J., et al.: Simulation of robot motion based on Matlab. J. Xiamen Univ. **5**(05), 640–644 (2005)
8. Studio, X.s.: The Latest Classic ANSYS and Workbench Tutorials, pp. 154–160. Electronic Industry Press, Beijing
9. Moaveni, S.: Finite Element Analysis-ANSYS Theory and Application, 2nd edn., pp. 6–11. Beijing Electronic Industry Press (2005). Edited by Song Wang and others
10. Da, S.h., et al.: ANSYS Finite Element Principle in Engineering Application. Tsinghua University Press, Beijing (2006)

The 9th International Workshop on Information Networking and Wireless Communications (INWC-2019)

Performance Evaluation of VegeCare Tool for Tomato Disease Classification

Natwadee Ruedeeniraman[1], Makoto Ikeda[2(✉)], and Leonard Barolli[2]

[1] Graduate School of Engineering, Fukuoka Institute of Technology,
3-30-1 Wajiro-higashi, Higashi-ku, Fukuoka 811-0295, Japan
mgm19108@bene.fit.ac.jp
[2] Department of Information and Communication Engineering, Fukuoka Institute
of Technology, 3-30-1 Wajiro-higashi, Higashi-ku, Fukuoka 811-0295, Japan
makoto.ikd@acm.org, barolli@fit.ac.jp

Abstract. The aged population in Japan is increased and some people start to work in agriculture after the retirement. Therefore, it is important to teach the skills to these agricultural beginners. Also, to improve the productivity of vegetables, advanced approaches based on Artificial Intelligence (AI) and sensing technology are needed. We focus on vegetable recognition using Deep Neural Network (DNN) as a method to visualize the knowledge in vegetable production. In this paper, we present the performance evaluation of VegeCare tool for tomato disease classification. We used 6 kinds of tomato diseases. The plant disease classification is one of the functions of the proposed VegeCare tool, which helps the growth of vegetables for farmers. The evaluation results show that the learning accuracy was more than 90%. We found that the tomato disease classification results was selected correctly for the tomato mosaic virus.

Keywords: Plant disease classification · VegeCare · Tomato disease · DNN

1 Introduction

When starting new farming, various agricultural machines are required and the initial cost is high. For this reason, the machines are provided by rental companies. There are also local areas where small groups collect money and jointly buy expensive agricultural machines. In these areas, the farmers are working together to address local issues. For example, people who can not take care of their fields rent the fields cheaply to a local group. Thus, they contribute to regional revitalization by growing vegetables and rice.

Smart IT agriculture can deliver high quality and safe food, which is very important for people's healthy and happy life. Recently, field environmental sensors have been commercialized and they can support the adjustment of water level in paddy fields in addition to temperature, solar radiation, humidity and

© Springer Nature Switzerland AG 2020
L. Barolli et al. (Eds.): NBiS-2019, AISC 1036, pp. 595–603, 2020.
https://doi.org/10.1007/978-3-030-29029-0_59

soil moisture. Therefore, the establishment of advanced data processing technology for handling the agriculture big data is required, and the realization of the next generation AI framework is expected.

Deep Neural Networks (DNNs) have been adopted in intelligent systems for multi-field environment and they can be applied to extremely difficult problems for humans [1,10,15]. Various cameras and sensors are scattered in our life. In some application areas, AI has made it possible to make decisions even better than humans [3,9,16,17].

In [12], we proposed a vegetable detection and classification system considering TensorFlow framework. Also, we proposed plant growth detection and insect/bug detection for managing the plant growth for a farmer. We presented the design of VegeCare tool, which considers these functions for the stable production of vegetables.

In this paper, we present the performance evaluation of VegeCare tool for tomato disease classification. We used 6 kinds of tomato diseases. The disease classification is one of the functions of the proposed VegeCare tool, which helps the growth of vegetables for farmers. For evaluation, we use as metrics the learning accuracy, loss and plant disease classification.

The structure of the paper is as follows. In Sect. 2, we give the related work. In Sect. 3, we describe the Neural Networks. In Sect. 4, we describe the proposed system. In Sect. 5, we provide the description of the simulation system and the evaluation results. Finally, conclusions and future work are given in Sect. 6.

2   Related Work

In recent years, new systems are used for rice cultivation, which automatically can measure the water level and water temperature of the paddy field. Also, a cultivation management system was proposed to collect the knowledge and skill of farmers. However, there are many problems in the field of biological information sensing such as the growing conditions and the physiological conditions.

Hokkaido Agricultural Research Center (HARC) is aiming for developing the unmanned system of agricultural machines (Tractor, rice planter, etc.) equipped with remote monitoring. The system is linked with the quasi-zenith satellite by applying the straight-line assist and object (human) detection [5].

In [2], the authors presented a framework for classifier fusion, which can support the automatic recognition of fruits and vegetables in a supermarket environment. The authors show that the proposed framework yields better results than several related works found in the literature.

In [11], the authors proposed a food recording system, which employs real-time eating action recognition and food categorization for the meal scene. The system can classify five categories of representative "Yakiniku" food items including meat, rice, pumpkin, bell pepper, and carrot. The system achieved 74.8% classification rate.

In [18], the authors presented the survey of different diseases classification methods used for plant leaf disease detection. They introduced an algorithm for image segmentation techniques that can be used for automatic detection as well as the classification of plant leaf diseases.

In [14], the authors presented a tomato leaf disease detection and classification method based on Convolutional Neural Network (CNN) with the learning vector quantization algorithm. The proposed classification method misdiagnosed the diseased tomato leaves as healthy.

3 Neural Networks

Deep Learning (DL) has a deep hierarchy that connects multiple internal layers for feature detection and representation learning. Representation learning is used to express the extracting essential information from observation data in the real world. Feature extraction uses trial and error by artificial operations. However, DL uses the pixel level of the image as an input value and acquires the characteristic that is most suitable to identify it [4,7]. The simplest kind of Neural Network (NN) is a single-layer perceptron network, which consists of a single layer of output, the inputs are fed directly to the outputs. In this way, it can be considered the simplest kind of feed-forward network.

The learning method in a CNN uses the backpropagation model like a conventional multi-layer perceptron. To update the weighting filter and coupling coefficient, CNN uses stochastic gradient descent. In this way, CNN recognizes the optimized feature by using convolutional and pooling operations [6,8]. For the task of vegetable classification, Rectified Linear Unit (ReLU) is used in CNN to speed up training. CNN has been applied to object recognition.

4 Proposed System

The structure of our proposed system is shown in Fig. 1. Also, we designed VegeCare, which is a mobile application for managing the plant growth for the farmer.

The processing system runs on Ubuntu Linux 14.04. The CPU is equipped with Intel Core i3 3.3 GHz, NVIDIA GeForce GTX 1060 for GPU, and RAM for 16 GB. Based on plant disease classification characteristics and challenges, we consider the accuracy and loss, which is computed by the TensorFlow framework (version 1.13.1).

4.1 VegeCare Functions

The mobile application called VegeCare has three functions:

1. Plant disease classification,
2. Vegetable classification,
3. Insect classification.

The mobile application runs on the Android terminal.

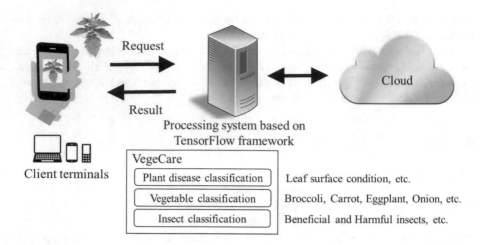

Fig. 1. System model.

4.2   Plant Disease Classification

The growth management of plants is important for increasing the productivity of vegetables. We proposed a monitoring system of plant growth using AI [12]. The monitoring targets are the surface condition of leaves and color/shape of plants. Since the color and shape of plants change during the growth process, we will improve detection accuracy by managing and utilizing growth information.

4.3   Vegetable Classification

The vegetable classification system using AI supports the farmers for selling vegetables on the Internet. We proposed an EC site and VegeShop tool in our previous work [13]. The performance results of our vegetable classification tool were good, but for leafy vegetables it had some issues [12].

4.4   Insect Classification

We coexist with animals and plants on earth, and plants coexist with insects. Some predatory species of insects are the natural enemies of common pests. Beneficial insects increase the ecological diversity of fields and defend vegetables from harmful insects. Beneficial insects have positive effects on a field, they aid in pollination and some cases serve as natural pesticides. The detection system considers insect classification such as beneficial insects and harmful insects. The system can warn about the problem of harmful insects. We assume to use a camera in a mobile device rather than fixed cameras. The user can take a picture of insects using the mobile application and can determine if the insects are harmful or harmless to vegetables and plants.

5   Performance Evaluation Results

5.1   Evaluation Setting

We evaluate the performance of our proposed tool for 6 kinds of tomato diseases. For training, we use totally 3, 080 images. We prepared 330 or 550 images of each disease (Bacterial spot, Healthy, Late blight, Septoria leaf spot, Tomato mosaic virus, and Yellow leaf curl virus). We show a list of tomato diseases and the number of images for training, validation, and test in Table 1. The example images from each disease are shown in Fig. 2.

Table 1. List of diseases and the number of images for training, validation and test.

#	Label	Training	Validation	Test
1	Bacterial spot	550	40	15
2	Healthy	550	40	15
3	Late blight	550	40	15
4	Septoria leaf spot	550	40	15
5	Tomato mosaic virus	330	40	15
6	Yellow leaf curl virus	550	40	15

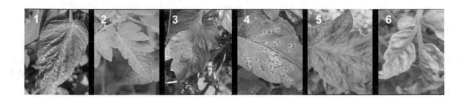

Fig. 2. Example images from each disease of tomato leaf.

1. The bacterial spot is a serious disease, which is especially severe in warm, wet and windy conditions, causing plants to lose their leaves.
2. Healthy indicates that the condition of the leaf is good.
3. The symptom of late blight on tomato leaf is irregularly shaped water-soaked lesions that can be observed on a young leaf of the plant. Under humid conditions, lesions become brown, then the leaves shrivel and become necrotic and die.
4. Septoria leaf spot spreads upwards from oldest to youngest growth. In general, there are many spots on the leaf. If there are many leaf lesions, the leaves turn slightly yellow, then brown and wither.
5. The tomato mosaic virus shows the mottled coloring and pattern, with alternating yellowish and darker green areas, the latter often appearing thicker and raised giving a blister-like appearance. The younger leaves may be twisted.

6. At the early stage of the yellow leaf curl virus, younger leaves are discolored from the leaf edges, and then between the veins become yellowed and shrunk.

Table 2. List of hyperparameters.

Function	Values
Epoch	200
Batch size	32
Filter sizes for convolution layer	3×3
Activation function	ReLU
Loss function	Categorical cross-entropy
Optimizer	RMSprop
Dropout	0.5

We show a list of hyperparameters of CNN in Table 2. In order to create a learning model, we used 200 epochs and batch size is 32. The image size has been resized to 256×256. The network is based on the sequential model, which consists of three convolutional layers, three pooling layers, and four fully connected layers.

5.2   Learning Results

For training and validation, we used 3,080 and 240 images, respectively. Data sets should be diverse to prevent over-learning and improve classification results. We use various leaf images of tomato diseases, which are converted from original images by using random zooming and shear conversion. For testing, we use 15 images for each disease (totally 90 images).

The results of learning accuracy and loss are shown in Fig. 3. The results of accuracy are increased with the increase of epochs. We observed that the results of accuracy for training are more than 90% after 35 epochs. The training loss for more than 35 epochs is less than 0.5. Also, we confirmed that validation accuracy and validation loss improve with the increase of epochs, but they have some oscillations.

5.3   Classification Results

We evaluated the disease classification of tomato leaf based on the learning result. In Table 3 are shown the results of disease classification. The classification result for all diseases is higher than 60%. For the tomato mosaic virus, our proposed classification tool accuracy was 100%. On the other hand, for classifying the healthy leaves, the accuracy of our tool is 60%. This is because some images of healthy condition leaf have green and darker green areas.

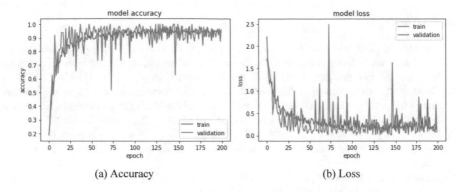

(a) Accuracy   (b) Loss

Fig. 3. Training and validation results.

The performance of our proposed tool is decreased for the yellow leaf curl virus. In this case, our tool mistakenly selected the septoria leaf spot at about 33%. This is because the leaf color and leaf shape are similar. If leaf lesions proceed, both leaves turn yellow, then wither or shrunk. In addition, our proposed classification tool classified the diseased tomato leaves correctly.

Table 3. Classification results.

#	Label	1	2	3	4	5	6	Accuracy
1	Bacterial spot	11	0	0	3	0	1	73.33
2	Healthy	0	9	0	1	5	0	60.00
3	Late blight	1	0	14	0	0	0	93.33
4	Septoria leaf spot	0	0	0	13	2	0	86.67
5	Tomato mosaic virus	0	0	0	0	15	0	100.00
6	Yellow leaf curl virus	0	0	0	5	1	9	60.00

6   Conclusions

In this paper, we proposed a plant disease classification tool using the TensorFlow framework. We evaluated the performance of learning accuracy and loss of our proposed tool for tomato diseases. From the evaluation results, we found that our proposed classification tool selected correctly the tomato mosaic virus.

In future work, we would like to develop an insect classification system to help the growth of vegetables for farmers. Moreover, we will consider different structures of CNN.

References

1. Arridha, R., Sukaridhoto, S., Pramadihanto, D., Funabiki, N.: Classification extension based on IoT-big data analytic for smart environment monitoring and analytic in real-time system. Int. J. Space-Based Situated Comput. **7**(2), 82–93 (2017)
2. Faria, F.A., dos Santos, J.A., Rocha, A., da S. Torres, R.: Automatic classifier fusion for produce recognition. In: Proceedings of the 25th International Conference on Graphics, Patterns and Images (SIBGRAPI-2012), pp. 252–259 (2012)
3. Gentile, A., Santangelo, A., Sorce, S., Vitabile, S.: Human-to-human interfaces: emerging trends and challenges. Int. J. Space-Based Situated Comput. **1**(1), 3–17 (2011)
4. Hinton, G.E., Osindero, S., Teh, Y.W.: A fast learning algorithm for deep belief nets. Neural Comput. **18**(7), 1527–1554 (2006)
5. Hokkaido Agricultural Research Center, NARO: HARC brochure (2017). http://www.naro.affrc.go.jp/publicity_report/publication/files/2017NARO_english_1.pdf
6. Kang, L., Kumar, J., Ye, P., Li, Y., Doermann, D.: Convolutional neural networks for document image classification. In: Proceedings of 22nd International Conference on Pattern Recognition 2014 (ICPR-2014), pp. 3168–3172, August 2014
7. Le, Q.V.: Building high-level features using large scale unsupervised learning. In: Proceedings of IEEE International Conference on Acoustics, Speech and Signal Processing 2013 (ICASSP-2013), pp. 8595–8598, May 2013
8. Lee, H., Grosse, R., Ranganath, R., Ng, A.Y.: Convolutional deep belief networks for scalable unsupervised learning of hierarchical representations. In: Proceedings of the 26th Annual International Conference on Machine Learning, pp. 609–616, June 2009
9. Mahesha, P., Vinod, D.: Support vector machine-based stuttering dysfluency classification using GMM supervectors. Int. J. Grid Util. Comput. **6**(3/4), 143–149 (2015)
10. Mnih, V., Kavukcuoglu, K., Silver, D., Rusu, A.A., Veness, J., Bellemare, M.G., Graves, A., Riedmiller, M., Fidjeland, A.K., Ostrovski, G., Petersen, S., Beattie, C., Sadik, A., Antonoglou, I., King, H., Kumaran, D., Wierstra, D., Legg, S., Hassabis, D.: Human-level control through deep reinforcement learning. Nature **518**, 529–533 (2015)
11. Okamoto, K., Yanai, K.: Real-time eating action recognition system on a smartphone. In: Proceedings of the IEEE International Conference on Multimedia and Expo Workshops (ICMEW-2014), pp. 1–6 (2014)
12. Ruedeeniraman, N., Ikeda, M., Barolli, L.: TensorFlow: a vegetable classification system and its performance evaluation. In: The Proceedings of the 13th International Conference on Innovative Mobile and Internet Services in Ubiquitous Computing (IMIS-2019), July 2019
13. Sakai, Y., Oda, T., Ikeda, M., Barolli, L.: VegeShop tool: a tool for vegetable recognition using DNN. In: Proceedings of the 11th International Conference on Broad-Band Wireless Computing, Communication and Applications (BWCCA-2016), pp. 683–691, November 2016
14. Sardogan, M., Tuncer, A., Ozen, Y.: Plant leaf disease detection and classification based on CNN with LVQ algorithm. In: Proceedings of the 3rd International Conference on Computer Science and Engineering (UBMK-2018), pp. 382–385, September 2018
15. Shoji, S., Koyama, A.: A fast search and classification method of isomorphic polygons in LSI design data using geometric invariant feature value. Int. J. Space-Based Situated Comput. **6**(4), 199–208 (2016)

16. Silver, D., Huang, A., Maddison, C.J., Guez, A., Sifre, L., van den Driessche, G., Schrittwieser, J., Antonoglou, I., Panneershelvam, V., Lanctot, M., Dieleman, S., Grewe, D., Nham, J., Kalchbrenner, N., Sutskever, I., Lillicrap, T., Leach, M., Kavukcuoglu, K., Graepel, T., Hassabis, D.: Mastering the game of Go with deep neural networks and tree search. Nature **529**, 484–489 (2016)
17. Silver, D., Schrittwieser, J., Simonyan, K., Antonoglou, I., Huang, A., Guez, A., Hubert, T., Baker, L., Lai, M., Bolton, A., Chen, Y., Lillicrap, T., Hui, F., Sifre, L., van den Driessche, G., Graepel, T., Hassabis, D.: Mastering the game of Go without human knowledge. Nature **550**, 354–359 (2017)
18. Singh, V., Misra, A.: Detection of plant leaf diseases using image segmentation and soft computing techniques. Inf. Process. Agric. **4**(1), 41–49 (2017)

Performance Analysis of WMNs by WMN-PSODGA Simulation System Considering Load Balancing: A Comparison Study for Exponential and Weibull Distribution of Mesh Clients

Seiji Ohara[1(✉)], Admir Barolli[2], Shinji Sakamoto[3], and Leonard Barolli[4]

[1] Graduate School of Engineering, Fukuoka Institute of Technology,
3-30-1 Wajiro-Higashi, Higashi-Ku, Fukuoka 811-0295, Japan
seiji.ohara.19@gmail.com

[2] Department of Information Technology, Aleksander Moisiu University of Durres,
L.1, Rruga e Currilave, Durres, Albania
admir.barolli@gmail.com

[3] Department of Computer and Information Science, Seikei University,
3-3-1 Kichijoji-Kitamachi, Musashino-shi, Tokyo 180-8633, Japan
shinji.sakamoto@ieee.org

[4] Department of Information and Communication Engineering, Fukuoka Institute
of Technology, 3-30-1 Wajiro-Higashi, Higashi-Ku, Fukuoka 811-0295, Japan
barolli@fit.ac.jp

Abstract. Wireless Mesh Networks (WMNs) are becoming an important networking infrastructure because they have many advantages such as low cost and increased high-speed wireless Internet connectivity. In our previous work, we implemented a Particle Swarm Optimization (PSO) based simulation system, called WMN-PSO, and a simulation system based on Genetic Algorithm (GA), called WMN-GA, for solving node placement problem in WMNs. Then, we implemented a hybrid simulation system based on PSO and distributed GA (DGA), called WMN-PSODGA. Moreover, we added in the fitness function a new parameter for the load balancing of the mesh routers called NCMCpR (Number of Covered Mesh Clients per Router). In this paper, we consider Exponential and Weibull distributions of mesh clients and carry out a comparison study. The simulation results show that the performance of the Weibull distribution is better compared with the Exponential distribution.

1 Introduction

The wireless networks and devices are becoming increasingly popular and they provide users access to information and communication anytime and anywhere [3, 8–11, 14, 20, 26, 27, 29, 33]. Wireless Mesh Networks (WMNs) are gaining a lot of attention because of its low-cost nature that makes it attractive for providing wireless Internet connectivity. A WMN is dynamically self-organized and

© Springer Nature Switzerland AG 2020
L. Barolli et al. (Eds.): NBiS-2019, AISC 1036, pp. 604–619, 2020.
https://doi.org/10.1007/978-3-030-29029-0_60

self-configured, with the nodes in the network automatically establishing and maintaining mesh connectivity among itself (creating, in effect, an ad hoc network). This feature brings many advantages to WMN such as low up-front cost, easy network maintenance, robustness and reliable service coverage [1]. Moreover, such infrastructure can be used to deploy community networks, metropolitan area networks, municipal and corporative networks, and to support applications for urban areas, medical, transport and surveillance systems.

Mesh node placement in WMNs can be seen as a family of problems, which is shown (through graph theoretic approaches or placement problems, e.g. [6,15]) to be computationally hard to solve for most of the formulations [37]. In fact, the node placement problem considered here is even more challenging due to two additional characteristics:

(a) locations of mesh router nodes are not pre-determined, in other wards, any available position in the considered area can be used for deploying the mesh routers.
(b) routers are assumed to have their own radio coverage area.

We consider the version of the mesh router nodes placement problem in which we are given a grid area where to deploy a number of mesh router nodes and a number of mesh client nodes of fixed positions (of an arbitrary distribution) in the grid area. The objective is to find a location assignment for the mesh routers to the cells of the grid area that maximizes the network connectivity and client coverage.

Node placement problems are known to be computationally hard to solve [12,13,38]. In some previous works, intelligent algorithms have been recently investigated [4,7,16,18,21–23,31,32].

In [24], we implemented a Particle Swarm Optimization (PSO) based simulation system, called WMN-PSO. Also, we implemented another simulation system based on Genetic Algorithm (GA), called WMN-GA [19], for solving node placement problem in WMNs. Then, we designed and implemented a hybrid simulation system based on PSO and distributed GA (DGA). We call this system WMN-PSODGA. We considered the load balancing problem. Different from our previous work, we add in the fitness function a new parameter called NCMCpR (Number of Covered Mesh Clients per Router).

In this paper, we present the performance analysis of WMNs by WMN-PSODGA system considering Exponential and Weibull distribution of mesh clients.

The rest of the paper is organized as follows. The mesh router nodes placement problem is defined in Sect. 2. We present our designed and implemented hybrid simulation system in Sect. 3. The simulation results are given in Sect. 4. Finally, we give conclusions and future work in Sect. 5.

2 Node Placement Problem in WMNs

For this problem, we have a grid area arranged in cells we want to find where to distribute a number of mesh router nodes and a number of mesh client nodes of fixed positions (of an arbitrary distribution) in the considered area. The objective is to find a location assignment for the mesh routers to the area that maximizes the network connectivity and client coverage. Network connectivity is measured by Size of Giant Component (SGC) of the resulting WMN graph, while the user coverage is simply the number of mesh client nodes that fall within the radio coverage of at least one mesh router node and is measured by Number of Covered Mesh Clients (NCMC).

An instance of the problem consists as follows.

- N mesh router nodes, each having its own radio coverage, defining thus a vector of routers.
- An area $W \times H$ where to distribute N mesh routers. Positions of mesh routers are not pre-determined and are to be computed.
- M client mesh nodes located in arbitrary points of the considered area, defining a matrix of clients.

It should be noted that network connectivity and user coverage are among most important metrics in WMNs and directly affect the network performance.

In this work, we have considered a bi-objective optimization in which we first maximize the network connectivity of WMNs (through the maximization of the SGC) and then, the maximization of the NCMC.

In fact, we can formalize an instance of the problem by constructing an adjacency matrix of the WMN graph, whose nodes are router nodes and client nodes and whose edges are links between nodes in the mesh network. Each mesh node in the graph is a triple $v = <x, y, r>$ representing the 2D location point and r is the radius of the transmission range. There is an arc between two nodes u and v, if v is within the transmission circular area of u.

3   Proposed and Implemented Simulation System

3.1   Particle Swarm Optimization

In PSO a number of simple entities (the particles) are placed in the search space of some problem or function and each evaluates the objective function at its current location. The objective function is often minimized and the exploration of the search space is not through evolution [17].

Each particle then determines its movement through the search space by combining some aspect of the history of its own current and best (best-fitness) locations with those of one or more members of the swarm, with some random perturbations. The next iteration takes place after all particles have been moved. Eventually the swarm as a whole, like a flock of birds collectively foraging for food, is likely to move close to an optimum of the fitness function.

Each individual in the particle swarm is composed of three $\mathcal{D}$-dimensional vectors, where $\mathcal{D}$ is the dimensionality of the search space. These are the current position $\vec{x}_i$, the previous best position $\vec{p}_i$ and the velocity $\vec{v}_i$.

The particle swarm is more than just a collection of particles. A particle by itself has almost no power to solve any problem; progress occurs only when the particles interact. Problem solving is a population-wide phenomenon, emerging from the individual behaviors of the particles through their interactions. In any case, populations are organized according to some sort of communication structure or topology, often thought of as a social network. The topology typically consists of bidirectional edges connecting pairs of particles, so that if j is in i's neighborhood, i is also in j's. Each particle communicates with some other particles and is affected by the best point found by any member of its topological neighborhood. This is just the vector $\vec{p}_i$ for that best neighbor, which we will denote with $\vec{p}_g$. The potential kinds of population "social networks" are hugely varied, but in practice certain types have been used more frequently. We show the pseudo code of PSO in Algorithm 1.

In the PSO process, the velocity of each particle is iteratively adjusted so that the particle stochastically oscillates around $\vec{p}_i$ and $\vec{p}_g$ locations.

3.2   Distributed Genetic Algorithm

Distributed Genetic Algorithm (DGA) has been used in various fields of science. DGA has shown their usefulness for the resolution of many computationally hard combinatorial optimization problems. We show the pseudo code of DGA in Algorithm 2.

Population of individuals: Unlike local search techniques that construct a path in the solution space jumping from one solution to another one through local perturbations, DGA use a population of individuals giving thus the search a larger scope and chances to find better solutions. This feature is also known as "exploration" process in difference to "exploitation" process of local search methods.

Fitness: The determination of an appropriate fitness function, together with the chromosome encoding are crucial to the performance of DGA. Ideally we would construct objective functions with "certain regularities", i.e. objective functions that verify that for any two individuals which are close in the search space, their respective values in the objective functions are similar.

Selection: The selection of individuals to be crossed is another important aspect in DGA as it impacts on the convergence of the algorithm. Several selection schemes have been proposed in the literature for selection operators trying to cope with premature convergence of DGA. There are many selection methods in GA. In our system, we implement 2 selection methods: Random method and Roulette wheel method.

Crossover operators: Use of crossover operators is one of the most important characteristics. Crossover operator is the means of DGA to transmit best genetic features of parents to offsprings during generations of the evolution process. Many methods for crossover operators have been proposed such as Blend Crossover (BLX-α), Unimodal Normal Distribution Crossover (UNDX), Simplex Crossover (SPX).

Mutation operators: These operators intend to improve the individuals of a population by small local perturbations. They aim to provide a component of randomness in the neighborhood of the individuals of the population. In our

Algorithm 1. Pseudo code of PSO.

/* Initialize all parameters for PSO */
Computation maxtime:= Tp_{max}, $t := 0$;
Number of particle-patterns:= m, $2 \leq m \in N^1$;
Particle-patterns initial solution:= P_i^0;
Particle-patterns initial position:= x_{ij}^0;
Particles initial velocity:= v_{ij}^0;
PSO parameter:= ω, $0 < \omega \in R^1$;
PSO parameter:= C_1, $0 < C_1 \in R^1$;
PSO parameter:= C_2, $0 < C_2 \in R^1$;
/* Start PSO */
Evaluate(G^0, P^0);
while $t < Tp_{max}$ **do**
    /* Update velocities and positions */
    $v_{ij}^{t+1} = \omega \cdot v_{ij}^t$
            $+ C_1 \cdot \text{rand}() \cdot (best(P_{ij}^t) - x_{ij}^t)$
            $+ C_2 \cdot \text{rand}() \cdot (best(G^t) - x_{ij}^t)$;
    $x_{ij}^{t+1} = x_{ij}^t + v_{ij}^{t+1}$;
    /* if fitness value is increased, a new solution will be accepted. */
    Update_Solutions(G^t, P^t);
    $t = t + 1$;
end while
Update_Solutions(G^t, P^t);
return Best found pattern of particles as solution;

Algorithm 2. Pseudo code of DGA.

/* Initialize all parameters for DGA */
Computation maxtime:= Tg_{max}, $t := 0$;
Number of islands:= n, $1 \leq n \in N^1$;
initial solution:= P_i^0;
/* Start DGA */
Evaluate(G^0, P^0);
while $t < Tg_{max}$ **do**
    for all islands **do**
        Selection();
        Crossover();
        Mutation();
    end for
    $t = t + 1$;
end while
Update_Solutions(G^t, P^t);
return Best found pattern of particles as solution;

---

system, we implemented two mutation methods: uniformly random mutation and boundary mutation.

Escaping from local optima: GA itself has the ability to avoid falling prematurely into local optima and can eventually escape from them during the search process. DGA has one more mechanism to escape from local optima by considering some islands. Each island computes GA for optimizing and they migrate its gene to provide the ability to avoid from local optima (See Fig. 1).

Convergence: The convergence of the algorithm is the mechanism of DGA to reach to good solutions. A premature convergence of the algorithm would cause that all individuals of the population be similar in their genetic features and thus the search would result ineffective and the algorithm getting stuck into local optima. Maintaining the diversity of the population is therefore very important to this family of evolutionary algorithms.

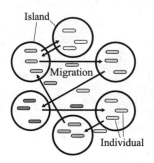

Fig. 1. Model of migration in DGA.

3.3   WMN-PSODGA Hybrid Simulation System

In this subsection, we present the initialization, particle-pattern, fitness function and client distributions. Also, our implemented simulation system uses Migration function as shown in Fig. 2. The Migration function swaps solutions among lands including PSO part.

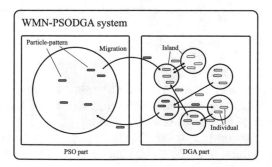

Fig. 2. Model of WMN-PSODGA migration.

Initialization

We decide the velocity of particles by a random process considering the area size. For instance, when the area size is $W \times H$, the velocity is decided randomly from $-\sqrt{W^2 + H^2}$ to $\sqrt{W^2 + H^2}$. In our system, many kinds of client distributions are generated. In this paper, we consider Exponential and Weibull distribution of mesh clients.

Particle-Pattern

A particle is a mesh router. A fitness value of a particle-pattern is computed by combination of mesh routers and mesh clients positions. In other words, each particle-pattern is a solution as shown is Fig. 3.

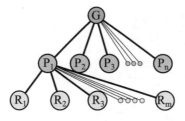

G: Global Solution
P: Particle-pattern
R: Mesh Router
n: Number of Particle-patterns
m: Number of Mesh Routers

Fig. 3. Relationship among global solution, particle-patterns and mesh routers in PSO part.

Gene Coding

A gene describes a WMN. Each individual has its own combination of mesh nodes. In other words, each individual has a fitness value. Therefore, the combination of mesh nodes is a solution.

Fitness Function

WMN-PSODGA has the fitness function to evaluate the temporary solution of the router's placements. The fitness function is defined as:

$$Fitness = \alpha \times SGC(\boldsymbol{x}_{ij}, \boldsymbol{y}_{ij}) + \beta \times NCMC(\boldsymbol{x}_{ij}, \boldsymbol{y}_{ij}) + \gamma \times NCMCpR(\boldsymbol{x}_{ij}, \boldsymbol{y}_{ij}).$$

This function uses the following indicators.

- SGC (Size of Giant Component)
  The SGC is the maximum number of the routers constructing in the same network. The SGC indicator means the connectivity of the routers.
- NCMC (Number of Covered Mesh Clients)
  The NCMC is the number of the clients belong to the network constructed by the SGC's routers. The NCMC indicator means the covering rate of the clients.
- NCMCpR (Number of Covered Mesh Clients per Router)
  The NCMCpR is the number of clients covered by each router. The NCMCpR indicator means load balancing.

WMN-PSODGA aims to maximize the value of the fitness function in order to optimize the placements of the routers using the above three indicators. The fitness function has weight-coefficients α, β, and γ for SGC, NCMC, and NCM-CpR. Moreover, the weight-coefficients are implemented as $\alpha + \beta + \gamma = 1$.

Router Replacement Methods
A mesh router has x, y positions and velocity. Mesh routers are moved based on velocities. There are many router replacement methods, such as:

Constriction Method (CM)
 CM is a method which PSO parameters are set to a week stable region ($\omega = 0.729$, $C_1 = C2 = 1.4955$) based on analysis of PSO by Clerc et al. [2,5,35].
Random Inertia Weight Method (RIWM)
 In RIWM, the ω parameter is changing ramdomly from 0.5 to 1.0. The C_1 and C_2 are kept 2.0. The ω can be estimated by the week stable region. The average of ω is 0.75 [28,35].
Linearly Decreasing Inertia Weight Method (LDIWM)
 In LDIWM, C_1 and C_2 are set to 2.0, constantly. On the other hand, the ω parameter is changed linearly from unstable region ($\omega = 0.9$) to stable region ($\omega = 0.4$) with increasing of iterations of computations [35,36].
Linearly Decreasing Vmax Method (LDVM)
 In LDVM, PSO parameters are set to unstable region ($\omega = 0.9$, $C_1 = C_2 = 2.0$). A value of V_{max} which is maximum velocity of particles is considered. With increasing of iteration of computations, the V_{max} is kept decreasing linearly [30,34].
Rational Decrement of Vmax Method (RDVM)
 In RDVM, PSO parameters are set to unstable region ($\omega = 0.9$, $C_1 = C_2 = 2.0$). The V_{max} is kept decreasing with the increasing of iterations as

$$V_{max}(x) = \sqrt{W^2 + H^2} \times \frac{T - x}{x}.$$

Where, W and H are the width and the height of the considered area, respectively. Also, T and x are the total number of iterations and a current number of iteration, respectively [25].

3.4   WMN-PSODGA Web GUI Tool and Pseudo Code

The Web application follows a standard Client-Server architecture and is implemented using LAMP (Linux + Apache + MySQL + PHP) technology (see Fig. 4). Remote users (clients) submit their requests by completing first the parameter setting. The parameter values to be provided by the user are classified into three groups, as follows.

- Parameters related to the problem instance: These include parameter values that determine a problem instance to be solved and consist of number of router nodes, number of mesh client nodes, client mesh distribution, radio coverage interval and size of the deployment area.
- Parameters of the resolution method: Each method has its own parameters.

- Execution parameters: These parameters are used for stopping condition of the resolution methods and include number of iterations and number of independent runs. The former is provided as a total number of iterations and depending on the method is also divided per phase (e.g., number of iterations in a exploration). The later is used to run the same configuration for the same problem instance and parameter configuration a certain number of times.

We show WMN-PSODGA Web GUI tool in Fig. 5. The pseudo code of our implemented system is shown in Algorithm 3.

4   Simulation Results

In this section, we present the simulation results for Exponential and Weibull distribution of mesh clients. Table 1 shows the common parameters for simulations.

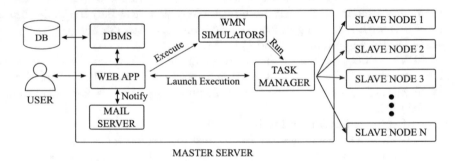

Fig. 4. System structure for web interface.

Fig. 5. WMN-PSODGA Web GUI Tool.

Moreover, there are two cases for each distribution to evaluate the effect of load balancing. In the first case are shown the simulation results of WMN-PSODGA system when the weight-coefficients are $\alpha = 0.1$, $\beta = 0.9$, $\gamma = 0$. In the second case are shown the simulation results of WMN-PSODGA system when the weight-coefficients are $\alpha = 0.1$, $\beta = 0.8$, $\gamma = 0.1$.

Algorithm 3. Pseudo code of WMN-PSODGA system.

Computation maxtime:= T_{max}, $t := 0$;
Initial solutions: P.
Initial global solutions: G.
/* Start PSODGA */
while $t < T_{max}$ **do**
　　Subprocess(PSO);
　　Subprocess(DGA);
　　WaitSubprocesses();
　　Evaluate(G^t, P^t)
　　/* Migration() swaps solutions (see Fig. 2). */
　　Migration();
　　$t = t + 1$;
end while
Update_Solutions(G^t, P^t);
return Best found pattern of particles as solution;

Table 1. The common parameters of the simulations.

Parameters	Values
Number of Mesh Clients	48
Number of Mesh Routers	16
Radius of a Mesh Router	2.0–3.5
Number of GA Islands	16
Number of Migrations	200
Evolution Steps	9
Selection Method	Roulette Wheel Method
Crossover Method	SPX
Mutation Method	Uniform Mutation
Crossover Rate	0.8
Mutation Rate	0.2
Replacement Method	LDVM
Area Size	32.0×32.0

4.1 Exponential Distribution

Figure 6 shows the visualization results after the optimization for Exponential distribution. Figure 7 shows the number of covered clients by each router. Figure 8 shows the transition of the standard deviations. The value of r means the correlation coefficient. In Fig. 8(a), when the load balancing is not considered, the standard deviation is increased with the increase of the number of updates. On the other hand, in Fig. 8(b), when the load balancing is considered, the standard deviation is keeping almost the same value, which shows that better optimization is achieved by considering load balancing.

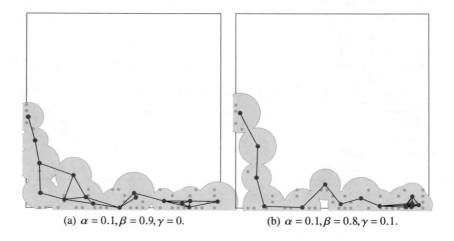

(a) $\alpha = 0.1, \beta = 0.9, \gamma = 0.$ (b) $\alpha = 0.1, \beta = 0.8, \gamma = 0.1.$

Fig. 6. Visualization results after the optimization for Exponential distribution.

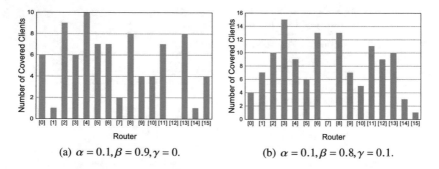

(a) $\alpha = 0.1, \beta = 0.9, \gamma = 0.$ (b) $\alpha = 0.1, \beta = 0.8, \gamma = 0.1.$

Fig. 7. Number of covered clients by each router after the optimization for Exponential distribution.

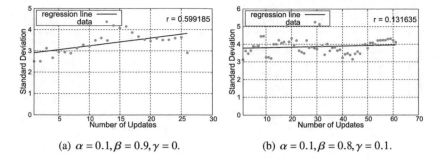

(a) $\alpha = 0.1, \beta = 0.9, \gamma = 0$.

(b) $\alpha = 0.1, \beta = 0.8, \gamma = 0.1$.

Fig. 8. Transition of the standard deviations for Exponential distribution.

4.2 Weibull Distribution

Figure 9 shows the visualization results after the optimization for Weibull distribution. Figure 10 shows the number of covered clients by each router. Figure 11 shows the transition of the standard deviations. In Fig. 11(a), when the load

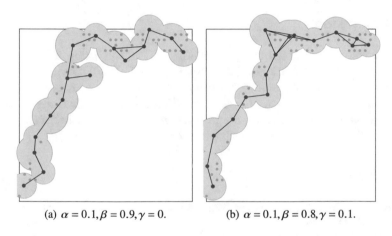

(a) $\alpha = 0.1, \beta = 0.9, \gamma = 0$.

(b) $\alpha = 0.1, \beta = 0.8, \gamma = 0.1$.

Fig. 9. Visualization results after the optimization for Weibull distribution.

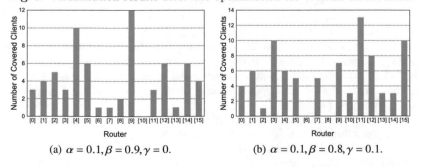

(a) $\alpha = 0.1, \beta = 0.9, \gamma = 0$.

(b) $\alpha = 0.1, \beta = 0.8, \gamma = 0.1$.

Fig. 10. Number of covered clients by each router after the optimization for Weibull distribution.

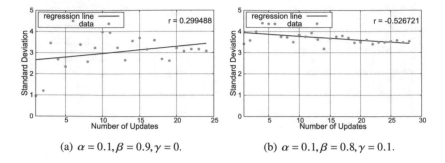

(a) $\alpha = 0.1, \beta = 0.9, \gamma = 0$.   (b) $\alpha = 0.1, \beta = 0.8, \gamma = 0.1$.

Fig. 11. Transition of the standard deviations for Weibull distribution.

balancing is not considered, the standard deviation is increased with the increase of the number of updates. On the other hand, in Fig. 11(b), when the load balancing is considered, the standard deviation is decreased with the increase of the number of updates, which shows that better optimization is achieved by considering load balancing.

5   Conclusions

In this work, we evaluated the performance of WMNs using a hybrid simulation system based on PSO and DGA (called WMN-PSODGA). We considered the Exponential and Weibull distribution of mesh clients. From the simulation results, we found that the performance of Weibull distribution was better compared with the Exponential distribution when using WMN-PSODGA. In future work, we will consider other distributions of mesh clients.

References

1. Akyildiz, I.F., Wang, X., Wang, W.: Wireless mesh networks: a survey. Comput. Netw. **47**(4), 445–487 (2005)
2. Barolli, A., Sakamoto, S., Ozera, K., Ikeda, M., Barolli, L., Takizawa, M.: Performance evaluation of WMNs by WMN-PSOSA simulation system considering constriction and linearly decreasing Vmax methods. In: International Conference on P2P, Parallel, Grid, Cloud and Internet Computing, pp. 111–121. Springer (2017)
3. Barolli, A., Sakamoto, S., Barolli, L., Takizawa, M.: Performance analysis of simulation system based on particle swarm optimization and distributed genetic algorithm for WMNs considering different distributions of mesh clients. In: International Conference on Innovative Mobile and Internet Services in Ubiquitous Computing, pp. 32–45. Springer (2018)
4. Barolli, A., Sakamoto, S., Ozera, K., Barolli, L., Kulla, E., Takizawa, M.: Design and implementation of a hybrid intelligent system based on particle swarm optimization and distributed genetic algorithm. In: International Conference on Emerging Internetworking, Data & Web Technologies, pp. 79–93. Springer (2018)
5. Clerc, M., Kennedy, J.: The particle swarm-explosion, stability, and convergence in a multidimensional complex space. IEEE Trans. Evol. Comput. **6**(1), 58–73 (2002)

6. Franklin, A.A., Murthy, C.S.R.: Node placement algorithm for deployment of two-tier wireless mesh networks. In: Proceedings of Global Telecommunications Conference, pp. 4823–4827 (2007)
7. Girgis, M.R., Mahmoud, T.M., Abdullatif, B.A., Rabie, A.M.: Solving the wireless mesh network design problem using genetic algorithm and simulated annealing optimization methods. Int. J. Comput. Appl. **96**(11), 1–10 (2014)
8. Goto, K., Sasaki, Y., Hara, T., Nishio, S.: Data gathering using mobile agents for reducing traffic in dense mobile wireless sensor networks. Mob. Inf. Syst. **9**(4), 295–314 (2013)
9. Inaba, T., Elmazi, D., Sakamoto, S., Oda, T., Ikeda, M., Barolli, L.: A secure-aware call admission control scheme for wireless cellular networks using fuzzy logic and its performance evaluation. J. Mob. Multimed. **11**(3&4), 213–222 (2015)
10. Inaba, T., Obukata, R., Sakamoto, S., Oda, T., Ikeda, M., Barolli, L.: Performance evaluation of a QoS-aware fuzzy-based CAC for LAN access. Int. J. Space-Based Situated Comput. **6**(4), 228–238 (2016)
11. Inaba, T., Sakamoto, S., Oda, T., Ikeda, M., Barolli, L.: A testbed for admission control in WLAN: a fuzzy approach and its performance evaluation. In: International Conference on Broadband and Wireless Computing, Communication and Applications, pp. 559–571. Springer (2016)
12. Lim, A., Rodrigues, B., Wang, F., Xu, Z.: k-Center problems with minimum coverage. Theor. Comput. Sci. **332**(1–3), 1–17 (2005)
13. Maolin, T., et al.: Gateways placement in backbone wireless mesh networks. Int. J. Commun. Netw. Syst. Sci. **2**(1), 44–50 (2009)
14. Matsuo, K., Sakamoto, S., Oda, T., Barolli, A., Ikeda, M., Barolli, L.: Performance analysis of WMNs by WMN-GA simulation system for two WMN architectures and different TCP congestion-avoidance algorithms and client distributions. Int. J. Commun. Netw. Distrib. Syst. **20**(3), 335–351 (2018)
15. Muthaiah, S.N., Rosenberg, C.P.: Single gateway placement in wireless mesh networks. In: Proceedings of 8th International IEEE Symposium on Computer Networks, pp. 4754–4759 (2008)
16. Naka, S., Genji, T., Yura, T., Fukuyama, Y.: A hybrid particle swarm optimization for distribution state estimation. IEEE Trans. Power Syst. **18**(1), 60–68 (2003)
17. Poli, R., Kennedy, J., Blackwell, T.: Particle swarm optimization. Swarm Intell. **1**(1), 33–57 (2007)
18. Sakamoto, S., Kulla, E., Oda, T., Ikeda, M., Barolli, L., Xhafa, F.: A comparison study of simulated annealing and genetic algorithm for node placement problem in wireless mesh networks. J. Mob. Multimed. **9**(1–2), 101–110 (2013)
19. Sakamoto, S., Kulla, E., Oda, T., Ikeda, M., Barolli, L., Xhafa, F.: A comparison study of hill climbing, simulated annealing and genetic algorithm for node placement problem in WMNs. J. High Speed Netw. **20**(1), 55–66 (2014)
20. Sakamoto, S., Kulla, E., Oda, T., Ikeda, M., Barolli, L., Xhafa, F.: A simulation system for WMN based on SA: performance evaluation for different instances and starting temperature values. Int. J. Space-Based Situated Comput. **4**(3–4), 209–216 (2014)
21. Sakamoto, S., Kulla, E., Oda, T., Ikeda, M., Barolli, L., Xhafa, F.: Performance evaluation considering iterations per phase and SA temperature in WMN-SA system. Mob. Inf. Syst. **10**(3), 321–330 (2014)
22. Sakamoto, S., Lala, A., Oda, T., Kolici, V., Barolli, L., Xhafa, F.: Application of WMN-SA simulation system for node placement in wireless mesh networks: a case study for a realistic scenario. Int. J. Mob. Comput. Multimed. Commun. (IJMCMC) **6**(2), 13–21 (2014)

23. Sakamoto, S., Oda, T., Ikeda, M., Barolli, L., Xhafa, F.: An integrated simulation system considering WMN-PSO simulation system and network simulator 3. In: International Conference on Broadband and Wireless Computing, Communication and Applications, pp. 187–198. Springer (2016)

24. Sakamoto, S., Oda, T., Ikeda, M., Barolli, L., Xhafa, F.: Implementation and evaluation of a simulation system based on particle swarm optimisation for node placement problem in wireless mesh networks. Int. J. Commun. Netw. Distrib. Syst. **17**(1), 1–13 (2016)

25. Sakamoto, S., Oda, T., Ikeda, M., Barolli, L., Xhafa, F.: Implementation of a new replacement method in WMN-PSO simulation system and its performance evaluation. In: The 30th IEEE International Conference on Advanced Information Networking and Applications (AINA-2016), pp. 206–211 (2016)

26. Sakamoto, S., Obukata, R., Oda, T., Barolli, L., Ikeda, M., Barolli, A.: Performance analysis of two wireless mesh network architectures by WMN-SA and WMN-TS simulation systems. J. High Speed Netw. **23**(4), 311–322 (2017)

27. Sakamoto, S., Ozera, K., Barolli, A., Ikeda, M., Barolli, L., Takizawa, M.: Implementation of an intelligent hybrid simulation systems for WMNs based on particle swarm optimization and simulated annealing: performance evaluation for different replacement methods. Soft Comput. **23**(9), 3029–3035 (2017)

28. Sakamoto, S., Ozera, K., Barolli, A., Ikeda, M., Barolli, L., Takizawa, M.: Performance evaluation of WMNs by WMN-PSOSA simulation system considering random inertia weight method and linearly decreasing Vmax method. In: International Conference on Broadband and Wireless Computing, Communication and Applications, pp. 114–124. Springer (2017)

29. Sakamoto, S., Ozera, K., Ikeda, M., Barolli, L.: Implementation of intelligent hybrid systems for node placement problem in WMNs considering particle swarm optimization, hill climbing and simulated annealing. Mob. Netw. Appl. **23**(1), 27–33 (2017)

30. Sakamoto, S., Ozera, K., Ikeda, M., Barolli, L.: Performance evaluation of WMNs by WMN-PSOSA simulation system considering constriction and linearly decreasing inertia weight methods. In: International Conference on Network-Based Information Systems, pp. 3–13. Springer (2017)

31. Sakamoto, S., Ozera, K., Oda, T., Ikeda, M., Barolli, L.: Performance evaluation of intelligent hybrid systems for node placement in wireless mesh networks: a comparison study of WMN-PSOHC and WMN-PSOSA. In: International Conference on Innovative Mobile and Internet Services in Ubiquitous Computing, pp. 16–26. Springer (2017)

32. Sakamoto, S., Ozera, K., Oda, T., Ikeda, M., Barolli, L.: Performance evaluation of WMN-PSOHC and WMN-PSO simulation systems for node placement in wireless mesh networks: a comparison study. In: International Conference on Emerging Internetworking, Data & Web Technologies, pp. 64–74. Springer (2017)

33. Sakamoto, S., Ozera, K., Barolli, A., Barolli, L., Kolici, V., Takizawa, M.: Performance evaluation of WMN-PSOSA considering four different replacement methods. In: International Conference on Emerging Internetworking, Data & Web Technologies, pp. 51–64. Springer (2018)

34. Schutte, J.F., Groenwold, A.A.: A study of global optimization using particle swarms. J. Glob. Optim. **31**(1), 93–108 (2005)

35. Shi, Y.: Particle swarm optimization. IEEE Connect. **2**(1), 8–13 (2004)

36. Shi, Y., Eberhart, R.C.: Parameter selection in particle swarm optimization. In: Evolutionary Programming VII, pp. 591–600 (1998)

37. Vanhatupa, T., Hannikainen, M., Hamalainen, T.: Genetic algorithm to optimize node placement and configuration for WLAN planning. In: Proceedings of the 4th IEEE International Symposium on Wireless Communication Systems, pp. 612–616 (2007)
38. Wang, J., Xie, B., Cai, K., Agrawal, D.P.: Efficient mesh router placement in wireless mesh networks. In: Proceedings of IEEE Internatonal Conference on Mobile Adhoc and Sensor Systems (MASS-2007), pp. 1–9 (2007)

Evaluation of 13.56 MHz RFID System Considering Tag Magnetic Field Intensity

Yuki Yoshigai and Kiyotaka Fujisaki[✉]

Fukuoka Institute of Technology, 3-30-1 Wajiro-higashi, Higashi-ku,
Fukuoka 811-0295, Japan
mgm19107@bene.fit.ac.jp, fujisaki@fit.ac.jp

Abstract. RFID is an automatic recognition tool for correctly inputting information into the management system. In recent years, RFID has become widespread in various situations because it is very convenient. However, RFID is affected by the environment in the vicinity and communication performance changes. In the previous work, we have shown the possibility to expand the communication distance between the reader and the target tag when some interference tags come close to each other. In this paper, after increasing tags as the interference source, we evaluate the communication distance. Furthermore, we observe the magnetic field intensity on the position of the target tag by using the spectrum analyzer when some interference tags come close to each other.

1 Introduction

Radio Frequency Identification (RFID) realize the contactless data exchange by using the induction field or the radio wave between a communication terminal and tags. In recent years, the rapid spread of the system based on RFID technique is remarkable. The RFID is applied to train tickets, car driver's license, passport, electronic keys, electronic money and so on.

RFID technique is the useful tool for management of a large amount of goods. For example, by using RFID system, we expects the efficiency of the following services to library users: (1) rental of book and the return, (2) collection inventory, (3) search of the book, (4) access control of users [4, 13]. Furthermore, in the future, the library will track books and users in a library and users may get the useful information from the library.

A lot of applications using RFID technique have been proposed [8, 12, 14, 16]. Furthermore, the development of RFID device and the performance evaluation using RFID system were performed to implement reliable RFID system [1–3, 5–7, 9–11, 15]. In [5], using a 13.56 MHz RFID system based on the international standard of ISO15693, the influence of paper and other RFID tag on the resonant frequency of the RFID tag was evaluated. In [6], the basic performance of a table type RFID reader was evaluated and the effect of metallic plate to the reading rate of RFID system was investigated when the RFID reader was placed on the metallic plate.

© Springer Nature Switzerland AG 2020
L. Barolli et al. (Eds.): NBiS-2019, AISC 1036, pp. 620–629, 2020.
https://doi.org/10.1007/978-3-030-29029-0_61

In our previous work [7], using 13.56 MHz RFID system, we evaluated the resonant frequency of RFID tag and the communication distance between the reader and the target tag when some tags becoming as the interference sources came close to each other. Furthermore, using the handmade type reader, we evaluated the communication distance between the reader and the target tag and showed the possibility to have a long communication distance compared with the single tag case when the tags becoming as the interference sources sandwiched between the reader and the target tag.

In this paper, in order to show the effect of the interference tags to the communication performance, we increase the number of interference tags and evaluate the communication distance between the reader and the target tag. Furthermore, by observing the magnetic field at the position of the target tag, we evaluate the characteristics of magnetic field made by interference tags which stand in a single line with the target tag.

The paper structure is as follows. In next section, we introduce the RFID system. Then, the communication distance and magnetic intensity of the RFID tag system are evaluated. Finally, we conclude the paper.

2   RFID System

An RFID system is one of the most important tool for the automatic identification. The system uses the wireless communication technique to get the ID from the tag without touching it. An RFID system consist of two components as shown in Fig. 1. One is called the RFID tag and sticks it on the object which we want to manage. Another is the reader/writer (interrogator) and is controlled from a management system.

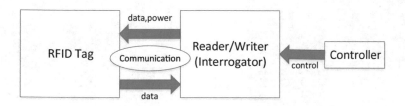

Fig. 1. Construction of RFID system

The shape of the tag is changed according to the purpose of use. For example, the card type RFID can be used for member's card and the stick type RFID tag is used for the management of animals. On the other hand, RFID reader/writer can be stationary type or portable type.

According to whether the tag has a power supply, RFID tag is classified into three kinds of tags: active tag, passive tag, and semi-active tag. Because the active tag needs a battery, the communication distance between RFID tag and RFID reader/writer become long. For this system, periodical battery exchange is needed. For this reason, the cost of maintenance management is increased. On the other hand, in order that a passive tag receives an electric power supply from a reader, the communication range is short, but the cost of maintenance management is cheap. Since use of RFID system in a library is assumed in our research, we use the passive tag and consider the method for the performance improvement.

3   Evaluation of 13.56 MHz RFID System

The 13.56 MHz RFID system communicates by electromagnetic induction. The tag is a passive tag that does not have a power supply. Thus, the tag does not move if the induced voltage generated by the radiation field from the reader is lower than the minimum voltage required to drive the tag. Therefore, the magnetic field radiated from the reader to the tag is considered to affect the communication performance.

In this paper, we evaluate the communication distance between the RFID tag of the target and the RFID reader and the intensity of magnetic field on the RFID tag of the target when there were several interference tags aligned with the target tag. In this experiment, the 13.56 MHz RFID system based on the ISO 15693 International Standard was used. Figure 2 shows the handmade type RFID reader which is made using ISO 15693 IC tag kit sold by SOFEL. The communication antenna of this reader is a loop coil with a diameter of about 45 mm. Figure 3 shows a 13.56 MHz RFID tag used in the experiment. This tag is composed of a coil as an antenna and an IC chip.

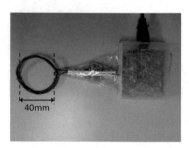

Fig. 2. Photo of 13.56 MHz RFID handmade type reader

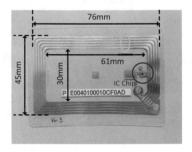

Fig. 3. Photo of example of 13.56 MHz RFID tag in practical use

3.1   Evaluation of Communication Distance Between Reader and Target Tag

In [7], the communication distance between the reader and the target tag was evaluated when some tags becoming as the interference sources are placed between the reader and the target tag. In this experiment, the number of interference tags changed from 1 to 6. Here, the interference sources is further increased and the communication distance between the reader and the target tag is observed when several interference tags line up with the target tag.

Figure 4 shows a measurement image of the communication distance between the reader and the target tag. The number of interference tags is changed from 1 to 12 and the distance between adjacent tags is set to d [mm]. The communication distance between the reader and the target tag is measured five times, and the average value is calculated.

Figure 5 shows the experimental results. In this figure, the horizontal axis indicates the number of interference tags N, and the vertical axis indicates the communication distance L [mm] between the reader and the target tag. The broken line indicates the communication distance between the reader and the target tag when there is no interference tag, and $L = 104$ mm. From Fig. 5, when the gap between tags is a certain distance, the communication distance between the reader and the target tag is extended as the interference tag is increased.

In this experiment, when $d = 20$ mm and $N = 11$, we could expand the communication distance to 2.4 times. From this, it can be considered that the coil which is the antenna of the interference tag relays the power radiated from the reader to the target tag. On the other hand, the smaller the gap between tags, the shorter the communication distance between the reader and the target tag. This is because the resonance frequency of the target tag becomes lower as the distance between tags becomes narrow. As the result, when the induced voltage in the transmission frequency of the reader is lower than the minimum of the voltage necessary to drive a tag, the tag does not work. In order to expand the communication length, the setting of an appropriate gap is necessary between tags.

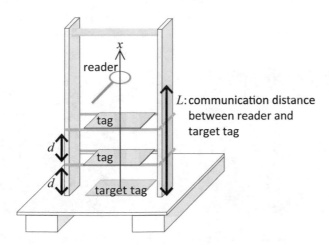

Fig. 4. Measurement image of communication distance between reader and target tag

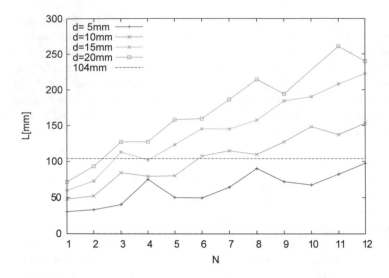

Fig. 5. Communication distance L between reader and target tag vs the number of interference tags

3.2   Evaluation of Magnetic Field Intensity Emitted from Reader

In the previous section, we showed the interference tag's effects to the communication performance of the RFID system. When between tags there is an appropriate gap, the communication distance becomes long. In this section, the influence that an interference tag gives to the magnetic field emitted by a reader is evaluated. In this experiment, we observe the magnetic field intensity on the position of the target tag by using the spectrum analyzer when some interference tags come close to each other.

Fig. 6. Photo of spectrum analyzer MS2601B of Anritsu Corp

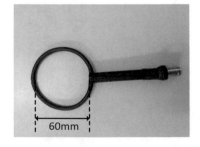

Fig. 7. Photo of probe EM-6993 of Electro-Metrics Corp

Figure 6 shows spectrum analyzer MS2601B of Anritsu Corp. Figure 7 shows the probe EM-6993 of Electro-Metrics Corp having 6 cm in a diameter of loop antenna. Because this probe has a similar cross section of target tag, the observed power have the same electricity to be supplied to the target tag, if the tag resonates at 13.56 MHz. Figure 8 shows a measurement image of received power. The target tag in Fig. 4 is replaced to the probe of Fig. 7 and the received power is measured by the spectrum analyzer while moving the reader in the x-axis direction.

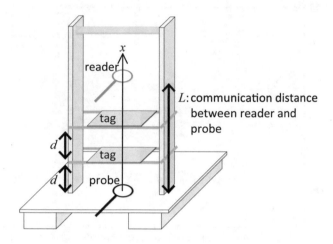

Fig. 8. Measurement image of received power

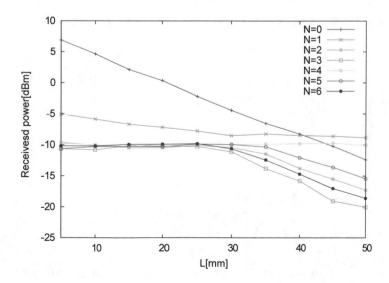

Fig. 9. Received power vs reader-probe distance: L when $d = 10$ mm

In this experiment, the number of tags is changed from 1 to 6 and the gap between tags is set to $d = 10, 20, 30\,\text{mm}$ and we evaluate the received power of each distance L. The results are shown in Figs. 9, 10 and 11. In these figures, the horizontal axis represents the distance L [mm] between the reader and the probe, and the vertical axis represents the received power [dBm]. The result when $N = 0$ shows no interference tag's case. In this case, the received power is decreased with increase of distance L. On the other hand, when some interference

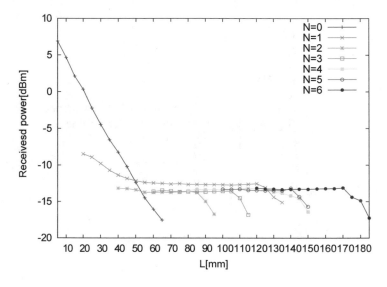

Fig. 10. Received power vs reader-probe distance: L when $d = 20\,\text{mm}$

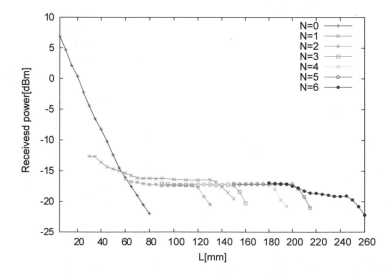

Fig. 11. Received power vs reader-probe distance: L when $d = 30\,\text{mm}$

tags were in a single line with the probe, the received power decreased compared with the result in case of $N = 0$. But there was a zone with stable received power achieved by the coupling between tags. The zone becomes long when the gap between tags and/or the number of tags are increased.

From Figs. 9, 10 and 11, we see that the coil antenna of the tag relays the magnetic power which radiated from the reader. For example, when there is no interference source between the reader and the probe (see Fig. 12), the magnetic flux emitted from the reader spreads widely. As the result, the received power of the probe decreases as the distance between the reader and the probe increases because the magnetic induction decreases. On the other hand, when some interference tags exist between the reader and the probe (see Fig. 13), an electric current is caused on the coil of the interference tag by the magnetic flux via the inside of the antenna coil, and the electric current produces a new magnetic field again. As a result, these magnetic fluxes are connected, and an expanse of the magnetic flux is suppressed and comes to strongly emit it forward. Therefore, the communication distance can be expanded by the interference tags.

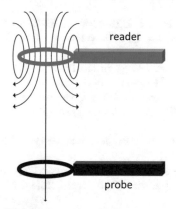
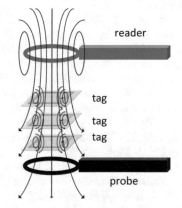

Fig. 12. Image of magnetic flux without interference tag between reader and probe

Fig. 13. Image of magnetic flux when three interference tags are aligned between reader and probe

4   Conclusion

In this paper, we increased the number of interference tags and evaluated the communication distance between the reader and the target tag. Furthermore, by observing the magnetic field at the position of the target tag, we evaluated the characteristics of magnetic field made by interference tags which stand in a single line with the target tag.

As a result, when the tag's gap becomes a certain distance, the communication distance between the reader and the target tag could be extended as the interference tag was increased. Moreover, when some interference tags existed in a single line with the probe, it was shown that an expanse of the magnetic flux is suppressed and comes to strongly emit it forward. In other words, it was clearly shown that the coil antenna of the tag can relay the magnetic power which is radiated from the reader.

In the future work, we will change the position of interference tags in various situations and evaluate the communication performance of RFID system and consider a method to improve the performance.

References

1. Basat, S.S., Kyutae, L., Laskar, J., Tentzeris, M.M.: Design and modeling of embedded 13.56 MHz RFID antennas. In: Proceedings of 2005 IEEE Antennas and Propagation Society International Symposium, pp. 64–67 (2005). https://doi.org/10.1109/APS.2005.1552740
2. Bolomey, J.C., Capdevila, S., Jofre, L., Romeu, J.: Electromagnetic modeling of RFID-modulated scattering mechanism. Application to tag performance evaluation. Proc. IEEE **98**(9), 1555–1569 (2010). https://doi.org/10.1109/JPROC.2010.2053332
3. Cantatore, E., Geuns, T.C.T., Gelinck, G.H., et al.: A 13.56-MHz RFID system based on organic transponders. IEEE J. Solid-State Circuits **42**(1), 84–92 (2007). https://doi.org/10.1109/JSSC.2006.886556
4. Fujisaki, K.: The implementation of the RFID technology in the library, and electromagnetic compatibility. Mon. EMC **183**, 86–94 (2003). (in Japanese)
5. Fujisaki, K.: Implementation of a RFID-based system for library management. Int. J. Distrib. Syst. Technol. **6**(3), 1–10 (2015). https://doi.org/10.4018/IJDST.2015070101
6. Fujisaki, K.: Evaluation and measurements of main features of a table type RFID reader. J. Mob. Multimed. **11**(1–2), 21–33 (2015)
7. Fujisaki, K.: Evaluation of 13.56 MHz RFID system performance considering communication distance between reader and tag. J High Speed Netw. **25**(1), 61–71 (2019). https://doi.org/10.3233/JHS-190603
8. Ha, O.K., Song, Y.S., Chung, K.Y., et al.: Relation model describing the effects of introducing RFID in the supply chain: evidence from the food and beverage industry in South Korea. Pers. Ubiquitous Comput. Arch. **18**(3), 553–561 (2014). https://doi.org/10.1007/s00779-013-0675-x
9. Kuoa, S.K., Hsub, J.Y., Hungb, Y.H.: A performance evaluation method for EMI sheet of metal mountable HF RFID tag. Measurement **44**(5), 946–953 (2011). https://doi.org/10.1016/j.measurement.2011.02.018
10. Li, N., Gerber, B.B.: Performance-based evaluation of RFID-based indoor location sensing solutions for the built environment. Adv. Eng. Inform. **25**(3), 535–546 (2011). https://doi.org/10.1016/j.aei.2011.02.004
11. Potyrailo, R.A., Morris, W.G., Sivavec, T., et al.: RFID sensors based on ubiquitous passive 13.56-MHz RFID tags and complex impedance detection. Wirel. Commun. Mob. Comput. **9**(10), 1318–1330 (2009). https://doi.org/10.1002/wcm.711
12. Prasad, N.R.K., Rajesh, A.: RFID-based hospital real time patient management system. Int. J. Comput. Trends Technol. **3**(3), 509–517 (2012)
13. Sing, J., Brar, N., Fong, C.: The State of RFID applications in libraries. Inf Technol Libr **25**(1), 24–32 (2006). https://doi.org/10.6017/ital.v25i1.3326

14. Symonds, J., Seet, B.C., Xiong, J.: Activity inference for RFID-based assisted living applications. J. Mob. Multimed. **6**(1), 15–25 (2010)
15. Uysal, D.D., Gainesville, F., Emond, J., Engles, D.W.: Evaluation of RFID performance for a pharmaceutical distribution chain: HF vs. UHF. In: Proceedings of 2008 IEEE International Conference on RFID, pp. 27–34 (2008). https://doi.org/10.1109/RFID.2008.4519382
16. Zhonga, R.Y., Dai, Q.Y., Qu, T., et al.: RFID-enabled real-time manufacturing execution system for mass-customization production. Robot. Comput.-Integr. Manuf. **29**(2), 283–292 (2013). https://doi.org/10.1016/j.rcim.2012.08.001

Numerical Analysis of Fano Resonator in 2D Periodic Structure for Integrated Microwave Circuit

Hiroshi Maeda[1]([✉]), Naoki Higashinaka[2], and Akihito Ochi[1]

[1] Department of Information and Communication Engineering,
Fukuoka Instute of Technology, 3-30-1 Wajiro-Higashi, Fukuoka 811-0295, Japan
hiroshi@fit.ac.jp
[2] Graduate School of Engineering, Fukuoka Institute of Technology,
3-30-1 Wajiro-Higashi, Fukuoka 811-0295, Japan

Abstract. Resonance in Fano resonator in two-dimensional pillar-type periodic structure was numerically investigated by Constrained Interpolated Profile (CIP) method. The model is designed in microwave frequency range around $4\,\mathrm{GHz}$. Obvious resonance in a point-defect cavity, situated in the periodic structure, was confirmed.

1 Introduction

Photonic crystal structures or electromagnetic band gap structures have periodic distribution of material constants in it and are applied into practical use in optical components for signal generation, transmission and reception, because of its unique and sensitive characteristics with respect to the signal frequency. Those characteristics are based on photonic band gap (PBG) phenomena [1–4]. In signal procession and transmission utilizing PBG devices in optical integrated circuits, high density multiplexing in frequency domain is expected due to its sensitivity with respect to optical wavelength. This is important to improve capacity of information transmission in photonic network with dense multiplexing technique of signal in wavelength domain.

The behavior of electromagnetic wave in periodic structure can be controlled by selecting material constants, designing periodic profile of the structure and the frequency spectrum range of the signal. For various kinds of materials and for various frequency ranges of purposes, PBG might be found by designing the structure with fundamental unit lattice. This means that, by setting the parameters appropriately, confinement and transmission of electromagnetic wave along line-defect in the structure is possible for desired range of frequency from microwave to optical domain. In this meaning, we examined the propagation and filtering characteristics of two dimensional photonic crystal waveguide and cavities with triangular lattice of dielectric pillar in microwave frequency around $4\,\mathrm{GHz}$. In the experiment [5–8], authors have used ceramic rods as dielectric pillar. For its quite low-loss property and high dielectric constant of $\varepsilon_r = 36.0$,

L. Barolli et al. (Eds.): NBiS-2019, AISC 1036, pp. 630–637, 2020.
https://doi.org/10.1007/978-3-030-29029-0_62

ceramic is suitable to confine electromagnetic field tightly when periodic structure is composed with less numbers of layers.

As a useful numerical analysis technique, finite different time domain (FDTD) method [9] is powerful and widely applicable, for enabling to design various boundary shape of structure with multi-dimensional problems. However, it is known that FDTD shows physically incorrect behavior for problems including large gap of material constants at the boundary. It is possible to avoid such behavior by setting smaller cells, however, it increases cell numbers for the entire analysis region with increase of memory and time computation. This means that we should pay attention to guarantee reliable results to choose the discrete cells within reasonable computation time.

On the contrary, constrained interpolated profile (CIP) method has been proposed by Yabe [10], with advantage of preventing such spurious behavior in FDTD method. Because the CIP method hires cubic polynomials to express the profile in a cell, it is possible to renew not only the profile at each discretized point but also the first order spatial derivatives of the profile. Authors have been applied the CIP method for analysis of wave propagation in periodic structures composed by ceramic pillars in air background (Fig. 1).

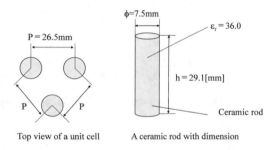

Fig. 1. Top view of fundamental triangular lattice of photonic crystal and side view of ceramic rod with parameters.

In this paper, filtering characteristics of Fano cavity [11], situated as point defect along line-defect waveguide, is numerically investigated by CIP method [12–16]. In the simulation, band-limited wave with time evolving envelope of sampling function is given as input. The resonant frequency peaks were obtained by fast Fourier transform (FFT) of output electric field in time domain. The results showed that obvious resonant peaks are observed in Fourier transformed domain.

2   CIP Method to Solve Two-Dimensional Maxwell's Equations

Maxwell's curl equations for electric field vector $\mathbf{E}$ and magnetic field vector $\mathbf{H}$ in lossless, isotropic, non-dispersive and non-conductive material are given as follows;

$$\nabla \times \mathbf{H} = \varepsilon \frac{\partial \mathbf{E}}{\partial t}, \tag{1}$$

$$\nabla \times \mathbf{E} = -\mu \frac{\partial \mathbf{H}}{\partial t}, \tag{2}$$

where ε is permittivity and μ is permeability of the space, respectively.

Assuming two dimensional uniform space along with z axis (i.e. $\partial/\partial z = 0$), Maxwell equations are decomposed into two sets of polarization. We analyze TE mode or E-wave which includes (H_x, H_y, E_z) as the component and propagates to x axis. For TE mode, Maxwell's equations are reduced to following sets;

$$\frac{\partial E_z}{\partial y} = -\mu \frac{\partial H_y}{\partial t}, \tag{3}$$

$$\frac{\partial E_z}{\partial x} = \mu \frac{\partial H_y}{\partial t}, \tag{4}$$

$$\frac{\partial H_y}{\partial x} - \frac{\partial H_x}{\partial y} = \varepsilon \frac{\partial E_z}{\partial t}. \tag{5}$$

From Eqs. (3) to (5), we obtain an vector partial differential equation as;

$$\frac{\partial}{\partial x} \mathbf{A} \mathbf{W} + \frac{\partial}{\partial y} \mathbf{B} \mathbf{W} + \frac{\partial}{\partial t} \mathbf{C} \mathbf{W} = 0, \tag{6}$$

where

$$\mathbf{W} = \begin{pmatrix} H_x \\ H_y \\ E_z \end{pmatrix}, \tag{7}$$

$$\mathbf{A} = \begin{pmatrix} 0 & 1 & 0 \\ 0 & 0 & 1 \\ 0 & 0 & 0 \end{pmatrix}, \tag{8}$$

$$\mathbf{B} = \begin{pmatrix} -1 & 0 & 0 \\ 0 & 0 & 0 \\ 0 & 0 & 1 \end{pmatrix}, \tag{9}$$

$$\mathbf{C} = \begin{pmatrix} 0 & 0 & -\varepsilon \\ 0 & -\mu & 0 \\ \mu & 0 & 0 \end{pmatrix}. \tag{10}$$

Split step procedure [10] is used for solving Eq. (6). Transformation into advection equations, procedures of split step solution for advection equations are same as our previous work [16] and is omitted here, however they are described in Refs. [10] and [12] in detail.

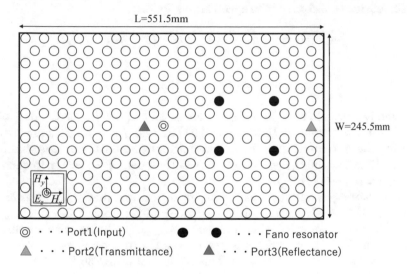

Fig. 2. Top view of 2D periodic triangular lattice with a line-defect waveguide and 2 pairs of Fano resonators.

3 Two-Dimensional, Pillar-Type Photonic Crystal Structure, Line-Defect Waveguide and Fano Cavity

In Fig. 2, top view of periodic triangular lattice with a line-defect is shown, together with electromagnetic field components. The longitudinal axis of the cylinder corresponds to polarization direction of electric field E_z of TE mode. Material of the cylinder is ceramic with relative dielectric constant $\varepsilon_r = 36.0$ at frequency $f = 4.0$ GHz. In the simulation, input frequency band is from 3.9 to 4.2 GHz, the dielectric loss is as negligibly small as 10^{-6} and the real part of dielectric constant can be assumed to be uniform in the frequency range. The ceramic rods are fabricated and supplied by Kyocera company in Japan for general use as microwave circuit elements. The lattice period $P = 26.5$ mm was designed so that the line defect structure shows PBG for frequency range from 3.6 to 4.2 GHz in the experiment. Following the design, the incident wave is guided along with defect without penetrating into periodic structures.

In Fig. 2, TE mode with components (H_x, H_y, E_z) is excited at port #1. The electric field E_z has Gaussian profile along y-axis with full beam waist $w_0 = 24.8$ mm. Also shown in Fig. 2, two pairs of Fano resonator with defect cavity are situated along waveguide to achieve filtering circuits.

4   FFT Analysis of Outputs Electric Field in Cavity

In numerical analysis, the discretization for space and time are set to be $\Delta x = \Delta y = 0.75\,\text{mm}$ and $\Delta t = 2.5 \times 10^{-13}\,\text{s}$, respectively. Supposing band-limited spectrum with square profile, the time evolving input wave $f(t)$ is given by inverse Fourier transform of the spectrum as follows;

$$Re\{f(t)\} = -f_L \times Sa(2\pi f_L t) + f_U \times Sa(2\pi f_U t), \tag{11}$$

where

$$Sa(x) = \frac{\sin(x)}{x} \tag{12}$$

is a sampling function, f_L and f_U are lower and upper frequency [Hz] of the limited band, respectively. Here, $f_L = 3.9\,\text{GHz}$ and $f_H = 4.2\,\text{GHz}$ are used for obtaining the flat spectrum. The maximum input amplitude in the simulation comes at time $t = 100/f_C$ [s], where $1/f_C$ is time period for center frequency of the range and $f_C = (f_L + f_U)/2$.

For obtaining frequency resolution to be comparable with experimental results, the time evolving data from CIP method is sampled every $\Delta t_{sample} = 100\Delta t$. Therefore, sampled time interval $\Delta t_{sample} = 25.0\,\text{s}$ with numbers of sample data $N_{sample} = 4096$ was set to obtain frequency resolution $\Delta f = (\Delta t_{sample} \times N_{sample})^{-1} \simeq 9.77\,\text{MHz}$. This resolution brings $300/9.77 \simeq 30$ points in the input frequency range.

In Fig. 3(a)–(c), power spectrum of output port 2, reflection port 3, and center of Fano resonator is shown. Figure 3(a) is output power spectrum to show that output power decreases as the resonator space moves narrower. On the other hand, output becomes larger when the resonator is original size as is in Fig. 2 or the resonator no longer exists (+1.0P). This means that the output decreases when Fano resonator exists along waveguide. Figure 3(b) shows reflected power spectrum at port 3 in Fig. 2 to show all the curves overlap each other. Also, amount of the spectrum is quite smaller than the output spectrum. This means reflected spectrum exists but the power is constant for all the cases. Figure 3(c) shows power spectrum trapped into Fano resonator. For original resonator and non-resonator (+1.0P) cases, the electromagnetic wave does not penetrate in to the resonator for all frequency range. As resonator length changes from +0.2P toward +0.4P, the trapped power increases as the resonant peak moves to higher frequency because the resonator length goes shorter. Also, there exists a trade-off between output power and trapped power in Fano resonator, as for +0.4P, output power is minimum while the trapped power shows maximum in the resonator.

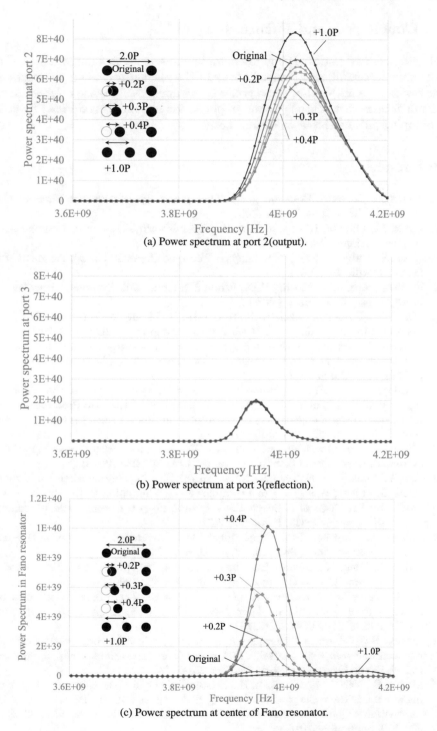

(a) Power spectrum at port 2(output).

(b) Power spectrum at port 3(reflection).

(c) Power spectrum at center of Fano resonator.

Fig. 3. Power spectrum at (a) port2, (b) port 3, and (c) center of Fano resonator. The parameters are shift of left-hand side rod of Fano resonator toward the right-hand side rod, as shown in the insets.

5   Conclusion and Future Subject

Filtering characteristics of Fano resonator is numerically demonstrated. The output power spectrum and trapped power in the resonator changes as function of the resonator length. Also, those power spectrum shows a trade-off relationship. As our future work, output power amplification should be confirmed by giving a pumping pulse as input to assume active (gain) device.

References

1. Yasumoto, K. (ed.): Electromagnetic Theory and Applications for Photonic Crystals. CRC Press, Boca Raton (2006)
2. Inoue, K., Ohtaka, K. (eds.): Photonic Crystals - Physics, Fabrication and Applications. Springer, New York (2004)
3. Noda, S., Baba, T. (eds.): Roadmap on Photonic Crystals. Kluwer Academic Publishers, Dordrecht (2003)
4. Joannopoulos, J.D., Meade, R.D., Winn, J.N.: Photonic Crystals. Princeton University Press, New Jersey (1995)
5. Maeda, H., Inoue, S., Nakahara, S., Hatanaka, O., Zhang, Y., Terashima, H.: Experimental study on X-shaped photonic crystal waveguide in 2D triangular lattice for wavelength division multiplexing system. In: Proceedings of 26th International Conference on Advanced Information Networking and Applications (AINA-2012), pp. 629–632, March 2012
6. Maeda, H., Zhang, Y., Terashima, H.: An experimental study on X-shaped branching waveguide in two-dimensional photonic crystal structure. In: Proceedings of 6th International Conference on Complex, Intelligent, and Software Intensive Systems (CISIS-2012), pp. 660–664, July 2012
7. Maeda, H.: Four-branching waveguide in 2D photonic crystal structure for WDM system. J. Space-Based Situated Comput. $3(4)$, 227–233 (2013)
8. Bao, Y., Maeda, H., Nakashima, N.: Studies on filtering characteristics of X-shaped photonic crystal waveguide in two-dimensional triangular lattice by microwave model. In: Proceedings of International Symposium on Antenna and Propagation (ISAP2015), pp. 842–845, November 2015
9. Taflove, A.: Advances in Computational Electrodynamics – The Finite-Difference Time-Domain Method, 3rd edn. Artech House (2005)
10. Yabe, T., Feng, X., Utsumi, T.: The constrained interpolation profile method for multiphase analysis. J. Comput. Phys. **169**, 556–593 (2001)
11. Yu, Y., Xue, W., Semenova, E., Yvind, K., Mork, J.: Demonstration of a self-pulsing photonic crystal Fano laser. Nat. Photonics **11**, 81–84 (2016). https://doi.org/10.1038/NPHOTON.2016.248
12. Maeda, H.: Numerical technique for electromagnetic field computation including high contrast composite material. In: Optical Communications, Chap. 3, pp. 41–54. InTech Open Access Publisher, October 2012
13. Maeda, H., Yasumoto, K., Chen, H., Tomiura, K., Ogata, D.: Numerical and experimental study on Y-shaped branch waveguide by post wall. In: Proceedings of 16th International Conference on Network-Based Information Systems (NBiS 2013), pp. 508–512, September 2013

14. Jin, J., Bao, Y., Chen, H., Maeda, H.: Numerical analysis of Y-shaped branch waveguide in photonic crystal structures and its application. In: Proceedings of 7th International Conference on Broadband, Wireless Computing, Communication and Applications (BWCCA-2014), pp. 362–365, November 2014
15. Maeda, H., Ogata, D., Sakuma, N., Toyomasu, N., Nishi, R.: Numerical analysis of 1×4 branch waveguide in two dimensional photonic crystal structure. In: Proceedings of International Conference on Advanced Information Networking and Applications (AINA2015), pp. 366–369, March 2015
16. Maeda, H., Cada, M., Bao, Y., Jin, J., Tomiura, K.: Numerical analysis of transmission spectrum of X-shaped photonic crystal waveguide for WDM system. In: Proceedings of International Conference on The Tenth International Conference on Complex, Intelligent, and Software Intensive Systems (CISIS-2016), pp. 186–189, July 2016

The 8th International Workshop on Advances in Data Engineering and Mobile Computing (DEMoC-2019)

Evaluation of I/O Performance Regulating Function with a Virtual Machine

Takashi Nagao[1,3]([✉]), Nasanori Tanabe[1,4], Kazutoshi Yokoyama[2],
and Hideo Taniguchi[1]

[1] Graduate School of Natural Science and Technology, Okayama University,
3-1-1 Tsushima-naka, Kita-ku, Okayama-shi 700-8530, Japan
ghede6845@gmail.com
[2] School of Information, Kochi University of Technology,
185 Miyanoguchi, Tosayamada-cho, Kami-shi 782-8502, Japan
[3] Hitachi, Ltd., 1-6-6 Marunouchi, Chiyoda-ku, Tokyo-to 100-8280, Japan
[4] NTT Data Corporation, 3-3-3 Toyosu, Koto-ku, Tokyo-to 135-6033, Japan

Abstract. A function that keeps the execution speed of a process constant despite the operation of other processes improves computer operational convenience. In this paper, we have described an I/O performance regulating function, which regulates execution time of a process's I/O request (I/O time) according to the performance level that has been specified. Specifically, the proposed function limits the number of I/O requests for an I/O device, in order to guarantee that the I/O response time of the target process will achieve a specified time. Furthermore, the proposed function delays waking up the process until the time corresponding to the specified performance, in order to regulate the I/O time of the process. Regarding the usage form on the computer, borrowing virtual machines is widespread, and in this paper, we show that the proposed function can regulate the process I/O time on a virtual machine with high accuracy. In practice, the introduced function cannot regulate the timing of individual I/O actions particularly well, because the I/O time on the virtual machine is too short; however, the function can regulate with high accuracy over a longer term by carrying forward the deviations from each individual regulating events.

1 Introduction

Generally, a user uses multiple software on a computer. For example, the user makes anti-virus software scan files, uses a media player software to play a video, and composes a document using an editor software. In such situations, the user will feel uncomfortable if a software that the user is directly using runs slower because of activity by the other software. The computer Operating System (OS) controls the execution of software by a process—specifically, the OS controls allocation time of the processor, and the processing order of I/O requests for each process. If the OS either allocates a long time for a process, or processes an I/O request from another process, the original process will run slower.

© Springer Nature Switzerland AG 2020
L. Barolli et al. (Eds.): NBiS-2019, AISC 1036, pp. 641–649, 2020.
https://doi.org/10.1007/978-3-030-29029-0_63

A function that keeps the execution speed of the process constant, despite the influence of other processes, improves computer convenience. We have therefore proposed performance regulating functions for a processor [1], and for the I/O function [2]. The I/O performance regulating function creates an I/O device with virtual user-specified performance, and allows the process to use the I/O device exclusively. Specifically, the function regulates process I/O request execution time (I/O time) according to the specified performance. The function has the following sub-functions:

(1) It limits the number of I/O requests for an I/O device, in order to guarantee the I/O device executes the process I/O request according to the specified performance.
(2) It delays waking up the process until the time corresponding to the specified performance, in order to regulate I/O time of the process.

With performance improvement, a computer can run multiple virtual machines. In addition, a cloud provider—such as Amazon® and Google®—provides rental service of virtual machines, in a trend which is prompting transition from computer ownership to use of virtual machines.

In this paper, we show that the I/O performance regulating function can regulate I/O process time in a virtual machine with high accuracy. In fact, while this function cannot regulate each I/O action time well—because I/O time on a virtual machine is too short—it can regulate I/O process time with high accuracy over the longer run, by carrying deviations in each regulating step forward.

2 I/O Flow

A process calls the I/O system (an I/O system call) in order to access data in the I/O device—with the flow of an I/O system call process shown in Fig. 1. The OS manages I/O requests that are in an I/O device queue by using an I/O scheduler (for example, Linux has noop, deadline, anticipatory, cfq [3]). The process waits for completion of the I/O requests, and the OS monitors the state of the I/O device. If the I/O device is ready to receive a new I/O command, the OS makes the device driver create an I/O command from the next request in the waiting queue, and sends the I/O command to the I/O device. The I/O device reads or writes the storage cell according to the I/O command, and then notifies the OS that it has completed the I/O command. The time for processing an I/O command is called the real I/O time, and the OS wakes up the process and sends the next I/O command if the I/O device indicates that it is ready to receive. Based on this sequence, I/O time can be prolonged by time spent in the waiting queue. Finally, the process confirms the completion message, and terminates the I/O system call.

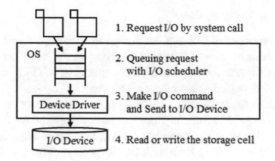

Fig. 1. I/O overview.

3 I/O Performance Regulating Function

3.1 Basic Flow

The user designates a process identifier, the required I/O performance, and an I/O device identifier, for the I/O performance regulation function (RF). Here the required I/O performance is a percentage, based on I/O device processing capacity. The RF presented here (this RF) calculates target I/O regulation time, based on the required I/O performance. Figure 2 shows the basic control flow for this regulation function.

(1) A process makes a system call.

(2) This RF decides whether to process or defer the I/O request of this process, based on a threshold called Tolerance for Waiting Time (TWT). If this RF decides to process, then it executes step (4), and if not, it executes step (3).

(3) This RF manages I/O requests in a waiting queue, which is sorted into descending order, based on the required I/O performance and request arrival order.

(4) This RF passes the I/O request to the device driver, and the OS sends the I/O command created by the device driver to the I/O device.

(5) This RF recognizes the number of I/O commands in the I/O device, selects the lead I/O request in the waiting queue, and executes (2).

(6) This RF delays waking up the process until the target I/O time, in order to regulate I/O process time.

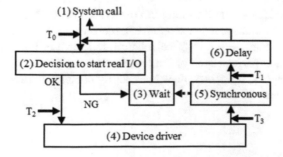

Fig. 2. Basic control flow of the I/O performance regulating function.

3.2   Tolerance for Waiting Time

This RF limits the number of I/O requests for an I/O device, in order to guarantee that the I/O device executes the I/O request in compliance with the target I/O time. Specifically, this RF limits the number of I/O requests to < TWT, which is calculated from the required I/O performance of each process.

First, we consider the calculation method applied when there is one target process for regulation. The target I/O time includes the processing time for its own I/O request. Thus, the difference between the target I/O time and the processing time for its own I/O request is the maximum waiting time limit for the process. This RF defines the number of I/O requests that can be processed by the I/O device within this maximum waiting time limit as TWT. For example, if the user specified 20% as the required performance, the target I/O time would be five times the real I/O time. The difference between the target I/O time and the processing time for its own I/O request is four times the real I/O time, and so the TWT is 4, and this RF keeps TWT to at least 1, so as to avoid stopping the I/O. In this case, the TWT calculation equation is given as shown in Eq. (1), where P means the required I/O performance of the target process.

$$Max\,(1, 100/P - 1) \tag{1}$$

Next, we consider the calculation method applied when there are multiple target processes for regulation. All I/O requests of the target processes need to be completed within each target I/O time. For this purpose, this RF calculates TWT from the sum of the required I/O performances. In this system, it may be that the TWT calculated from some high I/O performance requirement may be enough to achieve all target I/O times— and this can be explained by reference to two processes, which have required I/O performances of 10% and 20%, respectively, as an example. This RF calculates TWT from just the required I/O performance of 20%, so the TWT is 4 (= max (1, 100/20 − 1)). Furthermore, this RF takes out the I/O request of the process having the required I/O performance of 20% before the I/O request of the other process, so as a result, waiting time of the process having the required I/O performance of 20% is less than four times the real I/O time. Thus, this RF guarantees this process's I/O time is less than the target I/O time (which is five times the real I/O time). Similarly, waiting time of the process having a required I/O performance of 10% is less than five times the real I/O time, as this process can wait for an I/O request of the process having a required I/O performance of 20%, and four I/O requests allowed by TWT. This RF therefore can guarantee that this process's I/O time is less than the target I/O time (which is ten times the real I/O time).

In this way, we introduced a tolerance parameter, k, to control the number of required I/O performances, for TWT calculation, as shown in Eq. (2). Here, P_i means that the required I/O performance of the target process, with the i th highest required performance among the processes, does not wait for completion of the I/O request; that is, it can make a system call.

$$Max\left(1, 100/\sum{}^{k} P_i - 1\right) \qquad (2)$$

The tolerance parameter k is set by the user.

Table 1. Measurement environment.

Bare machine	Processor	2.5 GHz 4-core
	Memory	8 GB
	I/O Device	5400 rpm SATA/600 HDD
Virtual machine (bhyve)	Processor	1 core
	Memory	1 GB

3.3 Regulation of I/O Time

This RF delays waking up the process until the target I/O time, in order to regulate the I/O time of the process. Delay time, T_s, is calculated from the target I/O time and the elapsed time from when the process makes a system call ($T_1 - T_0$ in Fig. 2), as shown in Eq. (3).

$$T_s = 100/P \times \text{Real I/O time} - (T_1 - T_0) \qquad (3)$$

Each I/O device has a different performance and real I/O time, so this RF measures the real I/O time of each I/O device. In the case of one I/O request in the I/O device, the real I/O time is the time from sending an I/O command to the I/O device until notification of command completion, that is, $T_3 - T_2$, in Fig. 2. Otherwise, the I/O device processes I/O commands continuously, so the real I/O time is the completion notification interval. Therefore, the real I/O time is calculated as shown in Eq. (4).

$$\text{Real I/O time} = min(T_3 - T_2, T_3 \text{ interval}) \qquad (4)$$

This RF uses *sleep()*, which monitors elapsed time through timer interruptions, so that the actual delay time may deviate from T_s. Therefore, this RF improves long-term regulating accuracy, by carrying over this deviation time to the next I/O time regulation.

4 Evaluation

4.1 Configuration

We evaluated the accuracy of this RF using a regulation ratio, as shown Eq. (5).

$$\text{Regulation ratio} = \text{Actual I/O time}/\text{Target I/O time} \qquad (5)$$

When the regulation ratio = 1, the accuracy of regulation is the highest: that is, the regulated I/O time is too long if the regulation ratio > 1. Conversely, the regulated I/O time is too short if the regulation ratio < 1.

We used FreeBSD ver 11.2 + bhyve ver 1.10.5 for measurement, as shown in Table 1.

The measurement program repeated processor processing and issuance of I/O requests, and the number of I/O requests was set at 1,000. In addition, the measurement program could change processor processing time, so we measured two patterns: one in which the processor processing time was 0, and the other in which the time was the same as the I/O time.

We adapted the I/O time from 100–900, for evaluation, as, either immediately after the I/O start, or just before the end, the real HDD I/O time may become unstable, based on the system spinning down for power saving.

4.2   Experimental Results

Figure 3 shows the regulation ratio averages, and shows that the regulating method can regulate I/O performance on the virtual machine with as high a regulating accuracy as the bare machine. It was also shown that, when the required performance was over 50%, the regulating accuracy of the virtual machine was better than the bare machine.

Figure 4 shows the regulation ratio frequency, and it can be seen that the regulating accuracy for each I/O on the virtual machine was lower than that of the bare machine. Specifically, the number of regulation ratios with high accuracy, between 0.9 and 1.1, was 591 (74%) on the bare machine. The highest regulation ratio was 3.9.

On the other hand, the 'almost' regulation ratio was between 0.1 and 0.3 on the virtual machine, which represented cases where the regulating function didn't regulate I/O time, and the number of 'other' regulation ratios—ratios between 17.4 and 37.1— was 21. This regulation ratio polarization resulted from carrying over the deviation time, as described in Sect. 3.3, and overall, it was clear that I/O time on the virtual machine was shorter than that for the bare machine, and that the delay time, T_s, on virtual machine was also shorter.

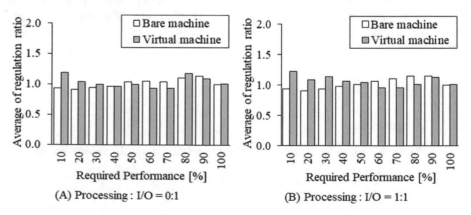

Fig. 3.   Regulation ratio average.

In contrast, the minimum time to sleep on the virtual machine was longer than it was for the bare machine, and so the actual delay time exceeded the requested delay time, T_s, on the virtual machine. This deviation time was due to the increased regulation ratio and carry over time. Furthermore, the RF doesn't regulate I/O time in order to decrease carryover time, so for this reason, the 'almost' regulation ratio was low.

Overall, the accuracy of regulating I/O performance per I/O process, was low, on the virtual machine; however, the RF regulated I/O performance with high accuracy over the long term, in the virtual machine.

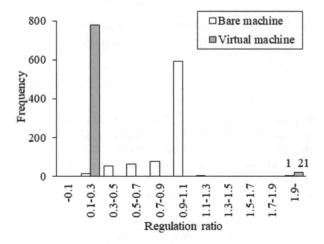

Fig. 4. Regulation ratio frequency. When the ratio of processing time / I/O time = 0/1, the required performance is 20%.

5 Related Work

For the purpose of improving I/O throughput, collective I/O has been a widely used approach [4, 5]. Yu et al. [6] adopted a polling method for notification of completion from the I/O device, in order to reduce the processing interruption overhead. As for Solid State Devices (SSD), the I/O device must lock in accessing memory cells, and thus I/O requests are sorted to distribute access areas, in order to avoid access conflicts [7–10]. Additionally, Zhang et al. [11] proposed a processor scheduler that controlled allocation time, to avoid conflicts in the timing of making I/O system calls. However, these methods cannot control the I/O time increase of important processes, as they do not distinguish between important and non-important processes.

On the other hand, priority control [12, 13], deadline control [14–19], and proportional allocation control [20, 21] have been proposed for the purpose of I/O performance guarantee. In addition, Merchant et al. [22] proposed a combined process of priority control and proportional allocation control. Additionally, Wu et al. [23] proposed an I/O scheduler that limited the amount of resources available per unit of time for important processes. However, the I/O time of important processes can be changed by the waiting time incurred by the I/O requests coming from non-important processes.

6 Conclusion

We have described the workings of an I/O performance regulating function and evaluated the function's accuracy using a virtual machine. Our evaluation showed that the I/O performance regulating function was able to regulate I/O process time on a virtual machine with a high degree of accuracy. In fact, the introduced function cannot regulate the timing of individual I/O actions particularly well, because I/O time on the virtual machine was too short; however, the function can regulate with high accuracy over the longer term, by carrying forward the deviations from each individual regulating event.

Further evaluations with multiple processes are needed, in order to clarify the I/O performance regulation effect in a more realistic environment.

References

1. Taniguchi, H.: A process schedule mechanism for regulating service processing time. IEICE Trans. Inf. Syst. **J81-D-I**(4), 386–392 (1998). (Japanese Edition)
2. Nagao, T., Taniguchi, H.: Implementation and evaluation of mechanism for regulating the service time based on controlling the number of I/O requests. IEICE Trans. Inf. Syst. **J94-D** (7), 1047–1057 (2011). (Japanese Edition)
3. Seelam, S., Romero, R., Teller, P., Buros, B.: Enhancements to Linux I/O scheduling. In: Linux Symposium (2005)
4. Kim, J., Seo, S., Jung, D., Kim, J., Huh, J.: Parameter-Aware I/O management for solid state disks (SSDs). IEEE Trans. Comput. **61**(5), 636–649 (2012)
5. Son, Y., Yeom, H.Y., Han, H.: Optimizing I/O operations in file systems for fast storage devices. IEEE Trans. Comput. **66**(6), 1071–1084 (2017)
6. Yu, Y.J., Shin, D.I., Shin, W., Song, N.Y., Choi, J.W., Kim, H.S., Eom, H., Yeom, H.Y.: Optimizing the block I/O subsystem for fast storage devices. ACM Trans. Comput. Syst. (TOCS) **32**(6), 1–48 (2014)
7. Park, S., Shen, K.: FIOS: a fair, efficient flash I/O scheduler. In: 10th USENIX Conference on File and Storage Technologies (2012)
8. He, S., Wang, Y., Sun, X., Huang, C., Xu, C.: Heterogeneity-aware collective I/O for parallel I/O systems with hybrid HDD/SSD servers. IEEE Trans. Comput. **66**(6), 1091–1098 (2017)
9. Jo, M.H., Ro, W.W.: Dynamic load balancing of dispatch scheduling for solid state disks. IEEE Trans. Comput. **66**(6), 1034–1047 (2017)
10. Won, Y., Jung, J., Choi, G., Oh, J., Son, S., Hwang, J., Cho, S.: Barrier-enabled IO stack for flash storage. In: 16th USENIX Conference on File and Storage Technologies, FAST 2018, pp. 211–226 (2018)
11. Zhang, X., Davis, K., Jiang, S.: Opportunistic data-driven execution of parallel programs for efficient I/O services. In: 2012 IEEE 26th International Parallel and Distributed Processing Symposium (IPDPS), pp. 330–341 (2012)
12. Betti, E., Bak, S., Pellizzoni, R., Caccamo, M., Sha, L.: Real-Time I/O management system with COTS peripherals. IEEE Trans. Comput. **62**(1), 45–58 (2013)
13. Kim, S., Kim, H., Lee, J., Jeong, J.: Enlightening the I/O path: a holistic approach for application performance. In: 15th USENIX Conference on File and Storage Technologies, FAST 2017, pp. 345–358 (2017)

14. Liu, C.L., Layland, J.W.: Scheduling algorithms for multiprogramming in a hard-real-time environment. J. ACM **20**(1), 46–67 (1973)
15. Povzner, A., Kaldewey, T., Brandt, S., Golding, R., Wong, T.M., Maltzahn, C.: Efficient guaranteed disk request scheduling with Fahrrad. In: EuroSys 2008: Third ACM European Conference on Computer Systems, pp. 13–25 (2008)
16. Han, S., Chen, D., Xiong, M., Lam, K., Mok, A.K., Ramamritham, K.: Schedulability analysis of deferrable scheduling algorithms for maintaining real-time data freshness. IEEE Trans. Comput. **63**(4), 979–994 (2014)
17. Zhang, Q., Feng, D., Wang, F., Xie, Y.: An interposed I/O scheduling framework for latency and throughput guarantees. J. Appl. Sci. Eng. **17**(2), 193–202 (2014)
18. Kang, D., Jung, S., Tsuruta, R., Takahashi, H.: Range-BW: I/O scheduler for predicable disk I/O bandwidth. In: 2010 2nd International Conference on Computer Engineering and Applications (ICCEA), pp. 175–180 (2010)
19. Povzner, A., Sawyer, D., Brandt, S.: Horizon: efficient deadline-driven disk I/O management for distributed storage systems. In: Proceedings of 19th ACM International Symposium on High Performance Distributed Computing (2010)
20. Tsai, C., Huang, T., Chu, E., Wei, C., Tsai, Y.: An efficient real-time disk-scheduling framework with adaptive quality guarantee. IEEE Trans. Comput. **57**(5), 634–657 (2008)
21. Valente, P., Checconi, F.: High throughput disk scheduling with fair bandwidth distribution. IEEE Trans. Comput. **59**(9), 1172–1186 (2010)
22. Merchant, A., Uysal, M., Padala, P., Zhu, X., Singhal, S., Shin, K.: Maestro: quality-of-service in large disk arrays. In: Proceedings of the 8th ACM International Conference on Autonomic computing (ICAC), pp. 245–254 (2011)
23. Wu, Y., Jia, B., Qi, Z.: IO QoS: a new disk I/O scheduler module with QoS guarantee for cloud platform. In: 2012 4th International Symposium on Information Science and Engineering (ISISE), pp. 441–444 (2012)

A Continuous Media Data Broadcasting Model for Base Stations Moving Straight

Tomoki Yoshihisa[1(✉)], Yusuke Gotoh[2], and Akimitsu Kanzaki[3]

[1] Cybermedia Center, Mihogaoka 5-1, Ibaraki, Osaka 567-0047, Japan
yoshihisa@cmc.osaka-u.ac.jp
[2] Graduate School of Natural Science and Technology, Okayama University,
Tsushima-naka, Kita-ku, Okayama 700-8530, Japan
[3] Institute of Science and Technology, Academic Assembly, Shimane University,
Nishikawatsu-cho 1060, Matsue, Shimane 690-8504, Japan

Abstract. Due to the recent development of robotics technology, movable robots work in some shopping malls or plants. They can generally communicate with other robots or computers and have a large possibility to serve as communications base stations. However, it is difficult to keep communications with them for long time since they move for their main services. Especially for the communications for receiving continuous media data such as video or audio, this is a large problem because the communication times tend to be long and the clients cannot receive the whole data in the cases that the robots move away in the middle of the communications. Slowing down their moving speeds shrinks a merit of using robots. Hence, in this paper, we create a communication model for movable base stations and investigate the possibility for robots to serve as base stations.

1 Introduction

Due to the recent development of robotics technology, some movable robots work in various places. For example, UAV (Unmanned aerial vehicles) such as drones fly on a city and take aerial videos. For another example, humanoid robots move in a shopping mall and guide routes for visitors. To enable their autonomous moving function, they equip with powerful resources for their communications and computations. They can generally communicate with other robots or computers while moving.

Conventional base stations for client machines (smart phones or laptops) such as the 5G cellular base stations or the Wi-Fi access points are located at fixed positions and the communication ranges covered by them are fixed. The utilization of movable robots as base stations enables moving communication ranges and enlarges the covered communication area. For example, an UAV equipped with a Wi-Fi access point can be regarded as a base station for communications. The communication area covered by the UAV is almost the same as its flying area. Therefore, movable robots have a large possibility to serve as communications base stations.

© Springer Nature Switzerland AG 2020
L. Barolli et al. (Eds.): NBiS-2019, AISC 1036, pp. 650–657, 2020.
https://doi.org/10.1007/978-3-030-29029-0_64

However, it is difficult to keep communications with them for long time since they move for their main services. Especially for the communications for receiving continuous media data such as video or audio, this is a large problem because the communication times tend to be long and the clients cannot receive the whole data in the cases that the robots move away in the middle of the communications. One of the simple solutions for this short communication time problem is slowing down the moving speeds of the robots. But, this shrinks one of the merits of using movable robots and it is better for them to be able to move faster. For example, it takes a longer time to distribute a video over a city by an UAV with a slower moving speed. In addition, it takes a longer time for a humanoid robot to go around a shopping mall with a slower moving speed.

Hence, in this paper, we propose a communication model for movable base stations (MBSs) with keeping a faster moving speed. In our proposed model, a continuous media data is divided into some segments and distributed by some MBSs. Each MBS broadcasts one segment cyclically and follows the MBS that broadcasts the previous segments. They can move with a faster speed so as not to interrupt their communications. In this paper, we also investigate the possibility for robots to serve as base stations by evaluating its performance by our developed computer simulation.

The remainder of the paper is organized as follows. In Sect. 2, we introduce related work. We explain our assumed MBSs in Sect. 3 and our proposed communication model in Sect. 4. In Sect. 5, we show some evaluation results for our proposed model and finally conclude the paper in Sect. 6.

2 Related Work

In [3], the authors have proposed a method that periodically monitors a certain set of locations in a target region using multiple UAVs. This method assigns a part of locations to each UAV and sets the moving path of each UAV in which the UAV visits all of its assigned locations. By setting the moving path of each UAV in such a way that every locations is monitored by a UAV within a predetermined time interval, this method achieves the periodical monitoring.

In [4], the authors have proposed a method that monitors the entire of a target region using multiple UAVs owned by different organizations and individuals. This method introduces direct wireless communication between UAVs for sharing information on frequency of monitoring at each location in the target region. Using shared information, each UAV adjusts its own moving path so as to move to locations where no UAVs have visited for a long time. By doing so, this method achieves frequent monitoring at every locations in the entire target region.

On the other hand, the methods proposed in [1,2] control the moving path of each UAV in order to achieve the efficient monitoring of a target region with smaller energy consumption. In [5], the authors have proposed a locations exploration algorithm for a disaster scenario.

These methods aim to control the moving path of each UAV so that a certain set of locations in a target region can be monitored by UAVs as frequently as possible. However, these do not assume the distribution of a large continuous media data and do not consider the communication time. For the distribution of continuous media data, UAV can use our previously proposed system in [6]. But, the system does not consider movable base stations which we target in this paper.

3 Movable Base Stations

In this section, we explain our assumed system and our target problem.

3.1 Assumed System

Some MBSs travel along a fixed route cyclically. They store one same continuous media data such as video or audio in their storages before they start traveling. They can communicate with the clients such as smart phones or laptops in their communication ranges and can send their stored data to the clients.

The clients start receiving the continuous media data that the MBSs have immediately when they can communicate with one of the MBSs. Once they start receiving the continuous media data, they play it until the end without moving. In the cases that a MBS goes away in the middle of the reception, the clients try to receive the subsequent data from other MBSs.

If they do not finish receiving a part of the continuous media data at the time to play it, an interruption occurs. The interruptions of playing continuous media data annoy the users of the clients and the number of the interruptions degrades the quality of the service. Therefore, we assume that the system should enable that the clients can play the continuous media data without interruptions.

3.2 Target Problem

To avoid interruptions, the clients should receive each part of continuous media data before they start playing it. In the cases that the communication times with each MBS are short, it takes long time for the clients to receive the data since they sometimes cannot communicate with the MBSs and the probability to occur interruptions increases. Therefore, slowing down the moving speeds of MBSs is effective to avoid interruptions. However, this shrinks one of the merits of using movable robots. Hence, the problem we tackle in this paper is avoiding interruptions for continuous media data distributions with a faster speed of MBSs.

4 Proposed Model

In this section, we explain our proposed model for solving the target problem explained in the previous section.

4.1   Data for MBSs

In our proposed model, a continuous media data is divided into some segments with equal durations. They are distributed by the same number of the MBSs. Each MBS broadcasts one segment cyclically and follows the MBS that broadcasts the previous segments. The clients can receive each segment from other MBSs since the data is divided into segments. Therefore, each MBS can move with a faster speed compared with the case that one MBS distributes the continuous media data. We can calculate the maximum speed for each MBS under the condition that the interruptions do not occur.

4.2   Mathematical Analysis

In this subsection, we assume that each MBS moves straight to calculate the maximum speed. We define the maximum difference of the clients' positions from the straight line of the route for MBSs by Y as shown in Fig. 1. That is, the clients exist in the gray area in the figure can play the continuous media data without interruptions. Let D denote the duration of the continuous media data and the data is divided into N segments. The number of the MBSs is also N. The consumption rate of the continuous media data is C. In the case that the continuous media data is video data, C means the bitrate of the video. B_i $(i = 1, \cdots, N)$ is the communication bandwidth between the clients and the ith MBS. V_i is the moving speed of the ith MBS and R_i is the radius of the communication range. The communication time between the ith MBS and the clients that exist on y far from the line of the straight route and is:

$$\frac{2\sqrt{R_i^2 - y^2}}{V_i}. \tag{1}$$

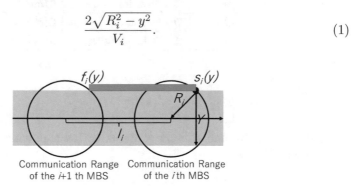

Communication Range   Communication Range
of the i+1 th MBS   of the ith MBS

Fig. 1. The variables for analysis

The data amount of one segment is CD/N. To finish receiving the ith segment from the ith MBS within its communication time, the following inequality should be established:

$$\frac{CD}{B_i N} < \frac{2\sqrt{R_i^2 - y^2}}{V_i}. \tag{2}$$

Hence,

$$V_i < \frac{2B_i N \sqrt{R_i^2 - y^2}}{CD}. \tag{3}$$

Thus, the maximum speed of our proposed model is $2B_i N \sqrt{R_i^2 - y^2}/(CD)$.

Moreover, the clients should be able to start the communication with the ith MBS before finishing playing $i-1$th segment. This condition leads the maximum distance among MBSs. Let I_i denote the distance between the ith MBS and the $i+1$th MBS. The elapsed time to start the communication with the ith MBS from the time to start the communication with the first MBS $s_i(y)$ is given by the following equation.

$$s_i(y) = \frac{\sum_{j=1}^{i-1} I_j + \sqrt{R_1^2 - y^2} - \sqrt{R_i^2 - y^2}}{V_i} \tag{4}$$

The time to start playing ith segment is:

$$\frac{D(i-1)}{N}. \tag{5}$$

To satisfy the above condition, $(4) \leq (5)$ should be established. In the case that the radiuses and the communication bandwidths for all MBSs are equivalent, the maximum speed is also equivalent and the value V is:

Table 1. Evaluation parameters

Symbol	Description	Value
D	The duration	1 [min.]
C	The consumption rate	1 [Mbps]
B	MBS's bandwidth	10 [Mbps]

$$V = \frac{2BN\sqrt{R^2 - y^2}}{CD}. \tag{6}$$

Here, R is the radius and B is the communication bandwidth. The value of $s_i(y)$ in this case $s_i^*(y)$ is:

$$s_i^*(y) = \frac{\sum_{j=1}^{i-1} I_j}{V}. \tag{7}$$

The maximum distances among MBSs are given when $s_i^*(y) = (5)$. In this case,

$$\frac{\sum_{j=1}^{i-1} I_j}{V} = \frac{D(i-1)}{N} \tag{8}$$

$$\sum_{j=1}^{i-1} I_j = \frac{D(i-1)}{N} V = \frac{2B(i-1)\sqrt{R^2 - y^2}}{C} \tag{9}$$

Thus, I_i is constant and is $\frac{2B\sqrt{R^2-y^2}}{C}$.

5   Evaluation

In this section, we evaluate the performance of the MBSs under our proposed model and investigate the possibility for robots to serve as base stations.

For the evaluation, we assume that the MBSs are drones and fly on a city. They move straight as shown in Fig. 1 and communicate with the clients based on our proposed model. Their communication type is Wi-Fi and they distribute a commercial movie to the clients on the grounds. The evaluation parameters used in this evaluation is shown in Table 1. We use our developed computer simulation to get evaluation results.

5.1   Maximum Velocity

To investigate the maximum velocity of MBSs, we measured them changing the value of y. y is the maximum distance of the clients from the line of the straight route, that can play the continuous media data continuously. We set the communication radius of MBSs to 50 [m]. The result is shown in Fig. 2.

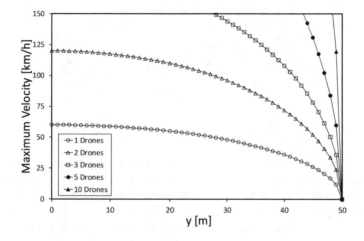

Fig. 2. The maximum velocity and the value of y

In the figure, the horizontal axis is the value of y and each line shows the maximum velocity under a different number of the drones. The maximum velocity decreases as y increases since the communication times for the clients existing a farer position from the route of the drones decreases. To enable continuous play of the continuous media data for them, the drones should move slowly. The maximum velocity increases as the number of the drones increases since a larger number of drones gives a wider total communication range. For example, when y is 48 [m], the maximum velocity under the case of just one drone is 16.8 [km/h] and is proportional to the number of the drones. In the case of 10 drones, they can move with 168 [km/h]. The maximum speed of general drones is about 80 [km/h]. The drones that moves with this maximum speed can distribute the continuous media data without any interruptions when the number of the drones is more than 7.

5.2 Distances Among MBSs

To investigate the distance among MBSs (I_i), we measured them changing the value of y. The distance does not depend on the number of the drones as shown in Eq. (9). The result is shown in Fig. 3.

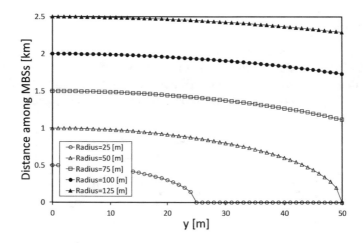

Fig. 3. The distance among MBSs and the value of y

In the figure, the horizontal axis is the value of y and each line shows the distance under a different communication radius. The distance decreases as y increases since the time spans that the clients cannot communicate with the drones lengthen as their positions are farther from the route of the drones. Therefore, the next drone should cover the clients sooner after the communications finish and the distance among the MBSs should be shorter. The distance increases as the communication radius increases since the communication time lengthens as the radius enlarges. This is because the clients can receive a larger amount of the continuous media data and the drones can lengthen the distance among MBSss. For example, when y is 48 [m], the distance can be 280 [m] under two drones and be 1.15 [km] under three drones. A few number of drones can provide distributions of continuous media data without interruptions and it is realistic for the drones to serve as MBSs under our proposed model.

6   Conclusion

UAV or humanoid robots can be movable base stations. However, it is difficult to keep communications with them for long time since they move for their main services. In this paper, we proposed a communication model for MBSs and investigated the possibility for robots to serve as base stations. In our proposed model, a continuous media data is divided into some segments and each MBS broadcasts one segment cyclically following the MBS that broadcasts the previous segments. Our simulation evaluation revealed that our proposed model gives a realistic situation that the robots serve as MBSs. In the future, we will create more complicated model for movable base stations and investigate the performance.

Acknowledgements. This research was supported by a Grants-in-Aid for Scientific Research(C) numbered JP17K12673 and JP18K11316, and by I-O DATA Foundation.

References

1. Cesare, K., Skeelke, R., Yoo, S.-H., Zhang, Y., Hollinger, G.: Multi-UAV exploration with limited communication and battery, prod. In: International Conference on Robotics and Automation, ICRA 2015, pp. 2230–2235 (2015)
2. Franco, C.D., Buttazzo, G.: Energy-aware coverage path planning of UAVs. In: Proceedings of International Conference on Autonomous Robot Systems and Competitions, ICARSC 2015, pp. 111–117 (2015)
3. Ghaffarkhah, A., Mostofi, Y.: Dynamic networked coverage of time-varying environments in the presence of fading communication channels. ACM Trans. Sensor Netw. **10**(3), Article no 45 (2014)
4. Kanzaki, A., Akagi, H.: A UAV-collaborative sensing method for efficient monitoring of disaster sites. In: Proceedings of International Conference on Advanced Information Networking and Applications, AINA 2019, pp. 775–786 (2019)
5. Sanchez-Garcia, J., Reina, D.G., Toral, S.L.: A distributed PSO-based exploration algorithm for a UAV network assisting a disaster scenario. Future Gener. Comput. Syst. **90**, 129–148 (2019)
6. Yoshihisa, T., Gotoh, Y., Kanzaki, A.: A system to select reception channel by machine learning in hybrid broadcasting environments. In: Proceedings of International Workshop on Advances in Data Engineering and Mobile Computing, DEMoC 2017, pp. 833–840 (2018)

A Support System for Second Tourism

Yusuke Gotoh[✉]

Graduate School of Natural Science and Technology, Okayama University,
Okayama, Japan
gotoh@cs.okayama-u.ac.jp

Abstract. Due to the recent popularization of the regional creation based on the growth strategy of Japan, the challenge to the sightseeing industry is attracted great attention. In particular, researchers have proposed a support system to analyze users' tourism and provide tourist information. In second tourism, users need to set the tourist route by specifying the visiting place based on their memories and photos in their first tourism. In addition, when users uses the location-based application that has saved the tourist route in their previous sightseeing, it is difficult to judge whether the tourist spot on this route wants to go for the second time or not. In this paper, we propose a system for supporting users' second tourism based on their travel lifelogs stored in multiple applications. In our proposed system, we develop an algorithm to collect information about the spots in the second tourism based on the history of searching words, the calendar information, and the playback history on YouTube. Next, we design and implement a web application to control our proposed system. In the evaluation by the questionnaire after using the proposed system, users can easily remember the detail of their previous tourism by checking the search history.

1 Introduction

Due to the recent popularization of the regional creation based on the growth strategy of Japan, the challenge to the sightseeing industry is attracted great attention [1]. In particular, researchers have proposed a support system to analyze users' tourism and provide tourist information.

As a problem in providing tourist information, when a user revisits a tourist spot visited in the past, it is difficult to provide a tourist route where the user can obtain high satisfaction. In second tourism, users need to set the tourism route by specifying the visiting place based on their memories and photos in their first tourism. In addition, when users uses the location-based application that has saved the tourist route in their previous sightseeing, it is difficult to judge whether the tourist spot on this route wants to go for the second time or not.

In this paper, we propose a system for supporting users' second tourism based on their travel lifelogs stored in multiple applications. In our proposed system, we develop an algorithm to collect information about the spots in the

© Springer Nature Switzerland AG 2020
L. Barolli et al. (Eds.): NBiS-2019, AISC 1036, pp. 658–668, 2020.
https://doi.org/10.1007/978-3-030-29029-0_65

second tourism based on the history of searching words, the calendar information, and the playback history on YouTube. Next, we design and implement a web application to control our proposed system.

The remainder of the paper is organized as follows. We explain the tourism in Sect. 2. Related works are introduced in Sect. 3. We explain our proposed broadcasting system in Sect. 4 and evaluate it in Sect. 5. Finally, we conclude in Sect. 6.

2 Tourism

2.1 Outline

A tourism is classified into two types. One is a tourism behavior as a trip for pleasure, and the other is a tourism phenomenon as a social activity. In addition, a tourism behavior is classified into three steps. The first step is a tourism behavior before traveling. In planning a sightseeing, the user searches for the sightseeing spot, and sets the tourist route, the restaurant, the transportation, and the hotel. The second step is a tourist behavior while traveling. The third step is a tourism behavior after traveling. Users look back on their trip by confirming the reaction to their articles posted on the Social Networking Service (SNS).

The tourist information is mainly considered before and during travel. Users had acquired tourist information via television, newspapers, and magazines. Recently, the number of users who acquire tourist information by searching on the Internet using a computer or a mobile device increases.

2.2 Problem in Conventional Tourism

We consider the case of revisiting a tourist spot that we visited once. Usually, users create a sightseeing plan using the information of guidebook, sightseeing information site, and the word of mouth. On the other hand, when users revisit a tourist spot once visited, they use a location-based application that stores photographs taken at the tourist spot and the information of a tourist route. However, since the user needs to search from a large number of photographs taken at that time, the time for creating a sightseeing plan is lengthened. In addition, when the user uses a location-based application, it is difficult to remember the tourism at that time with information only on tourist routes.

3 Related Works

In the lifelog system using the photograph [2], the user can recall an event by displaying the search query used when the photo was taken. The search query allows the user to recall the date and the location of the photo. This system uses search queries to recall the past tourist routes. In the proposed system, the user recalls the past tourism based on the search query.

When users share My Map that integrates geographic information originally created by them, a My Map searching system [3] using Point of Interest (POI) has been proposed. In this system, users can easily understand the types and locations of POIs registered in My Map. In addition, the user can search the past My Maps made by other users.

A recommendation method [4] based on the popularity of POIs ranks the POI using five classes classified according to geographical location and location conditions, the number of visitors, and the length of stay, Next, this method recommends a tourist route to get higher rank.

A curation system [5] for sightseeing movie using Consumer Generated Media (CGM) makes a tourist route by collecting the information of tourist spots. Next, this system makes a movie which moves the tourist route based on the CGM.

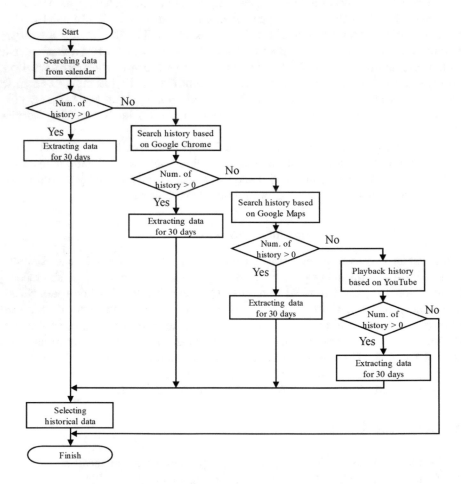

Fig. 1. Flowchart of word search

4   Proposed System

4.1   Outline

In this study, we propose a system to support the tourism when revisiting a tourist spot. The proposed system presents the candidate of the sightseeing spots in the second tourism by remembering the previous tourism with the sightseeing history.

4.2   Assumed Environment

In this system, we assume that users use this application at the stage of planning a trip. Therefore, users can use the proposed system not on mobile devices but on computer devices that can operate on Windows OS.

Table 1. List of keywords

Hotel	Inn	Spa	Lodge
Meal	Breakfast	Lunch	Dinner
Travel	Sightseeing	One-day trip	Family trip
Viewing	Shrines	Temples	Castles
Traveling alone	Sacred place	Photogenic	Walk and eat

There are four types of history information used in the proposed system: the schedule based on the calendar, the search history based on Google Chrome, the search history based on Google Maps, and the playback history on YouTube. The proposed system sets the priority in this order.

4.3   Word Search

Searching methods of history are classified into three types. In the word search, the user searches the sightseeing history based on its input words. We show the process flow of searching word in Fig. 1. Firstly, the system checks whether the calendar schedule has a history including the input word. If the calendar schedule has this history, the system extracts the history data for 30 days.

Next, after searching for history data using both input words and keywords related to the tourism, the system deletes the data that does not contain its history. Table 1 shows the list of keywords.

Secondly, if the calendar schedule has no history data, the system checks the search history of Google Chrome with the input word. If the system has the history of Google Chrome, the system extracts the search history of Google Chrome for 30 days, and deletes the candidate history that has a low degree of association with the input word.

Thirdly, if the search history on Google Chrome has no history data, the system checks the search history of Google Maps with the input word. If the system has the history of Google Maps, the system extracts them for 30 days, and deletes the candidate history that has a low degree of association with the input word.

Fourthly, if the search history on Google Maps has no history data, the system checks the playback history on YouTube with the input word. If the system has the playback history on YouTube, the system extracts them for 30 days, and deletes the candidate history that has a low degree of association with the input word.

Finally, if the playback history on YouTube has no history data, since the search word is not included in all types of history data, the search process is finished.

4.4   Data Search

In the date search, the system extracts all historical data between the start date and the end date entered by the user. If the user enters only the start the or the end date, the system extracts 10 days of historical data.

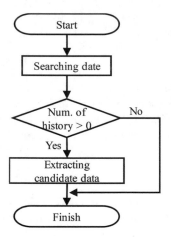

Fig. 2. Flowchart of word and data searches

4.5   Word and Data Searches

Figure 2 shows the processing flow of word and data searches. Firstly, the system extracts data for a period specified by the user. Next, the system deletes the candidate history that has a low degree of association with the input word. On the other hand, if there is no history in the date search, the search process is finished.

4.6   System Configuration

Figure 3 shows the configuration of the proposed system. Arrows in the figure indicate the flow of data between processing units.

We explain the processing units that compose the system. In the analyzing unit, the system transmits the searching method and the search condition set by the user to each processing unit.

The database unit manages the history data. The system stores four types of data in XML format: the calendar schedule, the search history on Google Chrome, that on Google Maps, and the playback history on YouTube.

The data conversion unit converts the data in XML format received from the database unit into the data that can be processed by a program.

The history search unit searches the history data under conditions set by the user and analyzes them. The sightseeing unit searches for the spots of tourism proposed by the system. The recommendation unit integrates the spots that the user wants to visit again with the spots proposed by the system.

The interface unit is composed of three types of functions which cooperate between users and computers. The input unit receives the search condition such as the spot information entered by the user, and sends it to the analyzing unit. The historical output unit outputs the historical information based on to the search result. The output unit shows the spots to visit on the second tourism proposed by the system.

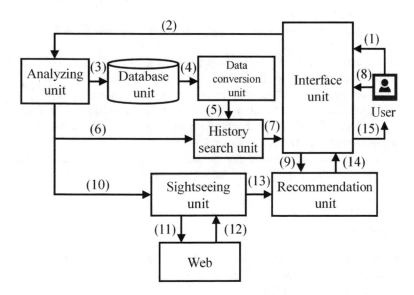

Fig. 3. System configuration

4.7   Searching Process

We show the process from receiving input information of users to presenting recommended spots.

(1) The user enters a searching method and searching conditions in the input unit.
(2) The input unit sends the input information of the user to the analyzing unit.
(3) The analyzing unit requests the database unit to acquire the history information.
(4) After receiving a request from the analyzing unit, the database unit sends a data file in XML format to the data conversion unit.
(5) The data conversion unit sends the deserialized object of the XML format to the history search unit.
(6) The analyzing unit sends the analyzed information of searching conditions to the history search unit.
(7) The history search unit searches the history information and extracts it based on the data received in steps (5) and (6) and the search conditions. Next, the history search unit sends data to the interface unit.
(8) The user selects the history of interest from its information sent to the interface unit.
(9) The interface unit sends the history information selected by the user to the recommendation unit.
(10) The analyzing unit sends the spot information received from the input unit to the sightseeing unit.
(11) The sightseeing unit requests the HTML data to the tourism site based on the spot information received from the analyzing unit.
(12) The server sends the request data to the sightseeing unit.
(13) The sightseeing unit extracts the tourism information using Web scraping based on HTML data. Next, The sightseeing unit send the extracted information to the recommendation unit.
(14) The recommendation unit sends the recommended spot to the output unit.
(15) The output unit proposes the recommended spot in second tourism to the user.

4.8   Implementation

Based on the search method of history information and the system procedure, we implemented the proposed system on a Windows form by the application on .NET Framework in Visual C#.

A screen shot of the proposed system is shown in Fig. 4. Figure 4 shows a case where the user inputs "Kyoto" in a word search. If the user selects "Fushimi Inari", "Kinkakuji", and "Kiyomizu-dera" based on the history information, the system displays them in blue letters on the recommendation area. Also, the system displays the recommendation spots in red letters.

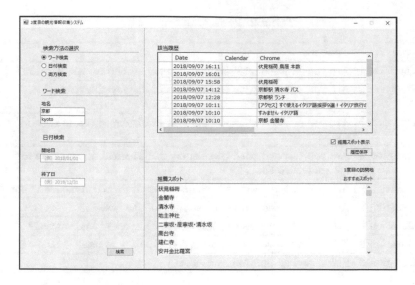

Fig. 4. Screenshot of proposed system

5   Evaluation

5.1   Evaluation Environment

We evaluate the availability of the proposed system based on the questionnaire by the examinees.

We collected a total of 570 historical data for three months from September 5, 2018 to December 5, 2018. The examinees were 8 males and 8 females in their twenties.

5.2   Evaluation Items

The questionnaire items for the examinees are shown in Table 2. Examinees answered six questions on a scale of 1 to 5. In addition, examinees answered questions 7 to 9 in an essay style.

5.3   Evaluation Results

The results of Questions 1 to 6 are shown in Table 3. Table 3 shows the number of questions and the weighted average for each question.

Figure 5 shows the answer to Question 7. In necessary historical information, six examinees selected the search history in Google Chrome and two examinees selected it in the Google Maps. On the other hand, in unnecessary historical information, all examinees selected the playback history on YouTube.

Table 2. Items of questionnaire

Items	Contents
Question 1	Are you satisfied with support for remembering past tourism?
Question 2	Are you satisfied with the amount of displayed historical information?
Question 3	Are you satisfied with the amount of displayed recommended spots?
Question 4	Are you satisfied with the display of historical information and recommended spots?
Question 5	Are you satisfied with the smooth search for recommended spots?
Question 6	Are you satisfied with the usability of the system?
Question 7	What are the necessary/unnecessary historical information?
Question 8	What other historical information should you use?
Question 9	Please add system improvements

Table 3. Results of Questions 1 to 6

Items	Num. (Percentage of total)					Ave.
	1	2	3	4	5	
Question 1	0 (0.0%)	0 (0.0%)	0 (0.0%)	2 (25.0%)	6 (75.0%)	4.8
Question 2	0 (0.0%)	0 (0.0%)	0 (0.0%)	5 (62.5%)	3 (37.5%)	4.4
Question 3	0 (0.0%)	3 (37.5%)	3 (37.5%)	0 (0.0%)	2 (25.0%)	3.1
Question 4	0 (0.0%)	0 (0.0%)	3 (37.5%)	5 (62.5%)	0 (0.0%)	3.6
Question 5	0 (0.0%)	0 (0.0%)	4 (50.0%)	3 (37.5%)	1 (12.5%)	3.6
Question 6	0 (0.0%)	1 (12.5%)	4 (50.0%)	3 (37.5%)	0 (0.0%)	3.3

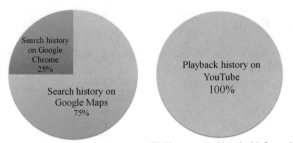

(A) Necessary historical information   (B) Unnecessary historical information

Fig. 5. Result of Question 7

Table 4 shows the answer to Question 8. As the additional candidates of the history item, examinees mentioned the photos, the movies, the historical information by the application, and the communication tool such as LINE and Twitter.

Table 4. Result of Question 8

Historical candidates	Reasons
Photos and Movies	Taking photos and videos often while traveling
Historical information on application	Containing travel plan on application
Communication tools such as LINE and Twitter	Acquiring spot information about where the user went or could not go

Table 5 shows the answer to Question 9. The examinees mentioned many improvements on the user interface of the system.

Table 5. Result of Question 9

- It would be useful if users can enlarge the area of historical information. - Users want a function to access the Web page by clicking the item of the recommendation spot. - Users want to display a map with recommendation spots. - It is difficult to understand how to use the system. - Users want a function to check whether the historical information has been saved. - Users want a function to display recommended spots according to the degree of recommendation.

6 Conclusion

In this paper, we designed and implemented a system for supporting the second tourism. In the proposed system, users can look back on their past tourism by extracting necessary information from historical information stored by multiple applications. In addition, since the system recommends the recommendation spots based on the past visiting spots, users can make plans for second tourism.

In the evaluation, the examinees answered the questionnaire after using the proposed system. In the evaluation result, the examinees were able to look back the past tourism. In addition, the examinees were able to recall their tourism in the past by confirming the search history.

In the future, we will add the types of historical information and improve the accuracy of information extraction.

Acknowledgement. This work was supported by JSPS KAKENHI Grant Number 18K11265.

References

1. WHITE PAPER Information and Communications in Japan (2018). http://www. soumu.go.jp/johotsusintokei/whitepaper/eng/WP2018/2018-index.html
2. Kubota, A., Tominaga, T., Hijikata, Y., Sakata, N.: Adding search queries to picture lifelogs for memory retrieval. In: Proceedings of 2016 IEEE/WIC/ACM International Conference on Web Intelligence, WI 2016, pp. 689–694 (2016)
3. Kanehira, T., Arakawa, Y., Yasumoto, K., Wada, T.: CURAP: CURating georelated information on a mAP. In: Proceedings of 2016 IEEE International Conference on Consumer Electronics, ICCE 2016, pp. 325–326 (2016)
4. Sun, J., Zhuang, C., Ma, Q.: Travel route recommendation by considering user transition patterns. e-Rev. Tour. Res. (eRTR) **16**(2/3), 73–83 (2019)
5. Kanaya, Y., Kawanaka, S., Hidaka, M., Suwa, H., Arakawa, Y., Yasumoto, K.: Preference-aware video summarization for virtual tour experience. In: International Workshop on Smart Sensing Systems, IWSSS 2018, vol. 23, pp. 244–253 (2018)

Omotenashi Robots: Generating Funny Dialog Using Visitor Geographical Information

Kazuki Haraguchi$^{(\boxtimes)}$, Satoshi Aoki, Tomohiro Umetani, Tatsuya Kitamura, and Akiyo Nadamoto

Konan University, Kobe, Japan
m1924009@s.konan-u.ac.jp

Abstract. In this paper, we propose an automatic scenario generation method which is based on funny dialogue. The scenario is for visitors who come to the tourist destination and the two robots present a funny dialogue based on our generated scenario. We call these robots "Omotenashi-robots". Specifically, we propose a method of automatically generate funny dialogue based scenario for the Omotenashi Robot using the place name given by the visitor. In this scenario, we generate three types of funny dialogue components which are area-related information component, place name misunderstanding component and land classification component. Area-related information component and place name misunderstanding component are based on the place names and their area-related information. Land classification component is based on general geographical information. Our generated scenario is kinds of a personalized scenario for a user based on the geographical information of the user's input place name.

1 Introduction

Recently, the rapid progress of the various communication robots, people use many communication robots in the tourism industry. For example, there is a hotel that performs reception work using robots[1]. In this way, robots that communicate with people such as receptionists are expected to increase in the near future. With the increase of inbound tourists, tourist guidance by robots can be expected. In this paper, we propose a guide robot for sightseeing spot like reception service.

Nevertheless, it is difficult for robots to communicate smoothly with humans. Siri[2] and Alexa[3] have good communication with people. However, they do not always communicate smoothly with all dialogues. Then we describe ways in which people feel friendly to robots merely by watching the dialogue between robots. We specifically examine the dialogue between robots. Then we create

---

[1] Henna hotel https://www.hennahotel.com/.
[2] Siri https://apple.com/siri.
[3] Alexa https://alexa.amazon.com.

© Springer Nature Switzerland AG 2020
L. Barolli et al. (Eds.): NBiS-2019, AISC 1036, pp. 669–679, 2020.
https://doi.org/10.1007/978-3-030-29029-0_66

communication robots to communicate with people naturally, smoothly, and with familiarity according to their dialogue. However, it is difficult to generate a conversation between robots automatically. Robots need some dialogue scenario to support a conversation. It is nevertheless difficult to create a scenario for the first meeting at a travel destination. Therefore, we propose the automatic generation of a dialogue-based scenario for the robots.

On the other hand, in the case of communication between people, they can have a friendly dialogue by asking "Where did you come from?" as the first meeting dialogue at a travel destination[4]. People naturally have favor with a new person who is interested in them. This is a method of dialogue that asks the new person in the dialogue "Where did you come from?" We focus on this method of dialogue and start with a conversation based on such visitor's geographic information. In this time, we consider that the visitor can not familiar to the robot only using their geographic information as a first dialogue. In this paper, we call visitor as user. We have already proposed a means of generating Manzai scenarios automatically [5,6]. Manzai scenario is consists of funny dialogue. Then we consider that this funny dialogue generation method could be applied to the first meeting between the traveler and the robot. In this paper, we propose an automatic scenario generation method which is based on funny dialogue for the robots. The scenario is for visitors who come to the tourist destination and the two robots present a funny dialogue based on our generated scenario. We call these robots "Omotenashi-robots".

Our proposed method of automatically generate funny dialogue scenario based on using the place name given by user. We call scenario "Omotenashi-robots-scenario". In the Omotenashi-robots-scenario, we generate three types of funny dialogue components which are area-related information component, place name misunderstanding component and land classification component. Area-related information component and place name misunderstanding component use the geographical information based on user's input place name. Our generated scenario is kinds of a personalized scenario for a user based on the geographical information of the user's input place name. The omotenashi robot scenario is an introduction part of communication between a robot and a traveler. After talking about the omotenashi robot scenario, the robot performs route guidance and restaurant guidance. In this research, we propose the omotenashi robot scenario which is the introduction part of this communication.

The remainder of this paper is organized as explained below. First, Sect. 2 presents a summary of related works. Section 3 presents the method of generating three types of components. Section 4 presents discussion related to experimental evaluation in addition to evaluation results. Finally, in Sect. 5, we end this report with conclusions and expectations for future work.

---

[4] Asahi https://dot.asahi.com/webdocu/2013112800001.html.

2   Related Work

Numerous studies have presented Manzai and joke analysis. Klaus et al. [8] have proposed a real-time adaptation of a robotic joke teller based on human social signals, namely facial smiles and vocal laughs. They implemented an entertainment robot and shown to learn jokes that were in accordance with the user's preferences without explicit feedback. Tsutsumi [7] has conducted a conversation analysis of "boke-tsukkomi" in Manzai. In addition, he conducted experiments to show English speakers the Manzai translated to English and showed that laughter was transmitted to the English speakers with the translated Manzai. Hayashi et al. [4] have proposed a Robot Manzai system using a pair of robots as a passive social medium. They conducted a comparison between "passive medium" and "passive social medium". They conducted a comparison experiment between "Robot Manzai" and "Manzai" by a human in videos, and they demonstrate the usefulness of "Robot Manzai" as entertainment. They specifically examine the comparison between "Robot Manzai" and "Manzai", but we specifically examine the funny dialogue for visitors who come to the tourist destination. Mashimo, Aoki, et al. [3,5,6] have purposed automatically generating Manzai scenario, in contrast our purposed method is automatic generating funny dialogue for visitors who come to the tourist destination.

Numerous studies also have presented and assessed tourism site recommendations. Almeida [2] has created a tourist site recommendation system for the purpose of finding a tour plan suited to the user. Anacleto et al. [1] have made a mobile application a tourism decision support system that recommends a tourist destination of interest from the user's profile. This makes it possible to change the tourism plan according to the situation at the tourist site. They specifically examine recommend tourist sites for the user, in contrast we have purposed automatically generating funny dialogue based on the place name.

3   How to Create Omotenashi-Robots-Scenario

In this paper, we generate Omotenashi-robots-scenario automatically using the user's input hometown name. There are three types of a component of Omotenashi-robots-scenario. The components consist of funny dialogue. The Omotenashi-robots-scenario is one component, the system determines which components used depends on the condition of the hometown name.

The flow of the omotenashi-robots-scenario is as follows.

1. User inputs land name where hare/his hometown.
2. System identify the hometown name because there are multiple station/town/city/prefecture may have the same name. In this time, we use land name database which we create to identify the hometown name.
3. If the condition of the area-related information component matches to the user input's hometown, the system creates the component.

4. If the condition of the place name misunderstanding component matches the user input's hometown, the system creates the component.
5. If the hometown name does not match the condition of (2) and (3), the system creates a land classification component.

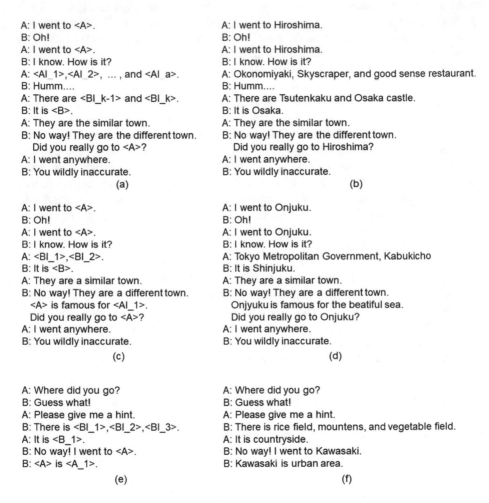

A: I went to <A>.
B: Oh!
A: I went to <A>.
B: I know. How is it?
A: <AI_1>,<AI_2>, ... , and <AI_a>.
B: Humm....
A: There are <BI_k-1> and <BI_k>.
B: It is <B>.
A: They are the similar town.
B: No way! They are the different town.
   Did you really go to <A>?
A: I went anywhere.
B: You wildly inaccurate.
(a)

A: I went to Hiroshima.
B: Oh!
A: I went to Hiroshima.
B: I know. How is it?
A: Okonomiyaki, Skyscraper, and good sense restaurant.
B: Humm....
A: There are Tsutenkaku and Osaka castle.
B: It is Osaka.
A: They are the similar town.
B: No way! They are the different town.
   Did you really go to Hiroshima?
A: I went anywhere.
B: You wildly inaccurate.
(b)

A: I went to <A>.
B: Oh!
A: I went to <A>.
B: I know. How is it?
A: <BI_1>,<BI_2>.
B: It is <B>.
A: They are a similar town.
B: No way! They are a different town.
   <A> is famous for <AI_1>.
   Did you really go to <A>?
A: I went anywhere.
B: You wildly inaccurate.
(c)

A: I went to Onjuku.
B: Oh!
A: I went to Onjuku.
B: I know. How is it?
A: Tokyo Metropolitan Government, Kabukicho
B: It is Shinjuku.
A: They are a similar town.
B: No way! They are a different town.
   Onjyuku is famous for the beatiful sea.
   Did you really go to Onjuku?
A: I went anywhere.
B: You wildly inaccurate.
(d)

A: Where did you go?
B: Guess what!
A: Please give me a hint.
B: There is <BI_1>,<BI_2>,<BI_3>.
A: It is <B_1>.
B: No way! I went to <A>.
B: <A> is <A_1>.
(e)

A: Where did you go?
B: Guess what!
A: Please give me a hint.
B: There is rice field, mountens, and vegetable field.
A: It is countryside.
B: No way! I went to Kawasaki.
B: Kawasaki is urban area.
(f)

Fig. 1. Scenario framework and example of scenario

3.1   The Area-Related Information Component

People sometimes dialogue about their hometown as innocent boastful talk. In this dialogue, they say their local information such as specialty products and landmarks. In this paper, we call such local information "area-related information". We determine area-related information is specialty products, sightseeing spots, people with connection, and landmarks. In this component, the system makes a willful mistake of the hometown name to another location name which has some similar area-related information to the hometown. The system generates funny dialogue based on location mistake. We call hometown name which user input is "A" and another location name which the system makes a willful mistake is "B".

The framework and the example of the area-related information component in Figure 1(a), (b). In this example. the user input location name A is "Hiroshima", and B is "Osaka". Osaka and Hiroshima have similar area-related information of Okonomiyaki and Skyscraper, but Tsutenkaku and Osaka castle is only Osaka's area-related information. The user initially thinks that the robot is talking about his hometown Hiroshima, after that he realizes that something is wrong. In this way, a little misunderstanding makes the user laugh.

How to Determine Area-Related Information of A

A is the user's hometown name which user inputs. We determine area-related information of A using Web search. The following is the flow of extracting area-related information of A.

1. The system searches the web using the query that is "A is" or "specialty products of A" or "sightseeing spots of A" or "people with a connection of A" or "landmarks of A".
2. It extracts snippet of the web pages whose title includes A.
3. It extracts terms whose term frequency is more than a threshold α from the snippet. The terms are the candidate for area-related information of A. We call it as $AIpre_x(x = 1, \ldots, n)$.
4. It searches the web using $AIpre_x(x = 1, \ldots, n)$ as query and extracts the snippet of the results pages.
5. It extracts the nouns and proper nouns whose term frequency is more than a threshold α from the snippet again. It becomes area-related information of A. It means we search bidirectional search. We determine the area-related information of A as $AI_j(j = 1, \ldots, m)$.

How to Determine B and Area-Related Information of B

B is a system willful mistake place-name and it has the same area-related information of A. The following is the flow of determining B.

1. The system searches Web sites using AI_j.
2. It extracts place-names except for A from search results of snippets. The land names become the candidate of B.
3. The candidate of B which has the most number of AI_j becomes B.

After the system determines B, it extracts the area related information of B which we call $BI_k(k = 1, \ldots, p)$. The extraction procedure of BI_k is the same as (1)–(4) of B extraction procedure. In this time the system uses B and candidate of BI_k as a query. After the procedure of (4), the system extracts the nouns and proper nouns whose term frequency is more than a threshold α from the snippet. A word containing B but not A in spinet and becomes BI_k.

3.2   The Place-Name Misunderstanding Component

There is not always user input place-name whose area-related information is similar to the other local name area-related information. In this time, we consider that we create a funny dialogue with a slight mistake. Then we focus on place-name mistake to generate a slight mistake dialogue. We call this dialogue component "the place-name misunderstanding component". Figure 1(c), (d) shows the framework and an example of place-name misunderstanding component. In this component, we use two place-names which are user input place-name and pronunciation similar place-name, and their area-related information. In Fig. 1(d), Onjuku is user input place-name and Shinjuku is pronunciation similar place-name. When we determine pronunciation similar place-name, we change the character of user input place name one by one and create a word. Next, we match the words we created with the place name database and use the matched words as the place names. The method of area-related information is as same in Sect. 3.1.

3.3   The Land Classification Component

In the area-related information component or the place-name misunderstanding component, we generated the dialogue using two names which are input by the user and determined by the system by using two methods. However, a user input place-name may not match either. In this case, we consider the method which uses attribute information of the place-name. We regard the attribute of the National Land Numerical Informations[5] as the attribute of the place-name. Figure 1(e), (f) shows the framework and example of the tree structure of the attribute of the place-name. First, the system determines the attribute information in the user input place-name. Next, the system extracts the node that is the grandchild node of the parent node on two steps in the tree structure as the attribute information of the related information (show Fig. 2). If there are multiple candidates, it is randomly acquired.

4   Experiments

4.1   Experiment 1: Demonstrate Geographical Information Extraction Method

We conducted an experiment to ascertain whether our geographical information extraction method.

Condition
Table 1 shows the place name lists using the experiment. We determined these place names randomly. We regard the correct answer is the information which is written in the each tourist association of municipality. In addition, when the

---

[5] The National Land Numerical Informations http://nlftp.mlit.go.jp/ksj-e/index. html.

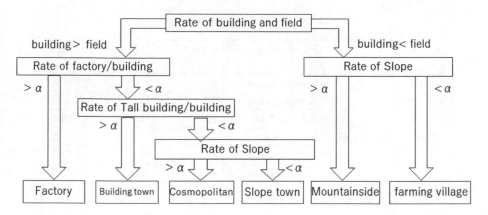

Fig. 2. Example of attribute

Table 1. Place names

Place name (Formal name)	Place name (Abbreviation)	Prefectures
Kanonji city	Kanonji	Kagawa prefecture
Iwakuni city	Iwakuni	Yamaguchi prefecture
Kotohira town	Kotohira	Kagawa prefecture
Kure city	Kure	Hiroshima prefecture
Hiroshima prefecture	Hiroshima	-
Hiroshima city	-	Hiroshima prefecture
Kagawa prefecture	Kagawa	-
Takamatsu city	Takamatsu	Kagawa prefecture
Kurasiki city	Kurasiki	Okayama prefecture
Osaka prefecture	Osaka	-
Osaka city	-	Osaka prefecture
Onomichi city	Onomichi	Hiroshima prefecture

geographical information extracted by the system is a passage of the correct answer, we regard the information is false data.

Result and Discussion

Table 2 shows the accuracy of the experimental result. The results of "A is" as a query is little bad, but the results of "sightseeing spots of A" is good.

In a case of "A is", an example of the correct area-related information is "Shukaen" and "Shimanami Kaido" for "Onomichi" and "Senkoji" for "Onomichi City". These spots are famous sightseeing spots and restaurants located in Onomichi City, and each has a strong relationship with Onomichi City. The example of an incorrect answer in the case of "A is", "rare sugar" and "insect" for "Kagawa". We consider the reason that "rare sugar" is included in

Table 2. Result of experiment 1

Snippet	Place name	Num of extraction	Num of collect	Fit rate
A is	Formal name	25	17	68.00%
	Abbreviation	27	15	55.56%
	Sum and average	52	32	61.54%
Sightseeing spot of A	Formal name	21	18	85.71%
	Abbreviation	21	18	85.71%
	Sum and average	42	36	85.71%
Sum and Average		94	68	72.34%

Wasanbon sugar produced in Kagawa. Furthermore, there are many terms of "insect" in the event and parks in Kagawa, then the system extracted them. In a case of "sightseeing spot of A", an example of the correct answer is that "Kaiyukan", "Tsutenkaku" and "Namba Grand Kagetsu" for "Osaka". These are all tourist destinations representing Osaka. As an example of an incorrect answer of "Tourist spot of A", there are "Heiwaki" for "Hiroshima". This is a part of "Heiwa kinen shiryou kan". It is a mistake of morphological analyzer. We have to strengthen the dictionary of morphological analyzers.

4.2   Experiment 2: User Experiments of Proposed Method

We conducted an experiment to ascertain whether our method of automatically generate funny dialogue based scenarios for the Omotenashi Robot.

Condition
The subjects are eight men and women of the 20th. We compare our proposed method with baseline. The baseline is the Manzai scenario which is already we proposed. It generates automatically Manzai scenario based on a news article. We use same query both our proposed method and baseline. The query is "Kobe" which is one of the famous city name in Japan. The Manzai scenario used in the existing method is the news of Kobe city, "Kobe invites World Para Athletics 2021 tournament. Kobe and Paris are fighting by invitation. It will be decided in April." The scenario of using our proposed method is shown in Fig. 3. In addition, we used "PaPeRo i" of NEC Platforms, Inc. shown in Fig. 4 for the experiment. First, we presented our proposed method and baseline to the subjects. Next, we asked subjects three questions which are "Are the robots familiar for you?", "Can you understand the content?", and "Is it funny?". The evaluation was made in 5 steps which are "suitable", "almost suitable", "about the same", "little not suitable", and "not suitable". Figure 5(a) shows the result of "Are the robots familiar for you?", Fig. 5(b) the result of "Can you understand the content?", and Fig. 5(c) the result of "Is it funny?".

Result and Discussion

Figure 5 shows the results of our experiment. First, we discuss the result of the question about familiar with the generated content. Half of the subjects answered positive responses in both the baseline and our proposed method. From the result, our proposed method gave the user some familiarity with the robots. On the other hand, there is no negative answer in the baseline but 17% subjects answered negative answer in our proposed method. The reason is that subjects do not know well about Muroto which the system extracted. From this result, we have to consider the user's recognition of place-name.

We discuss the result of the question about understanding and funny of the generated content. In this question, the result of our proposed method is better than the result of baseline. We consider the reason that generated the content of our proposed method is shorter than the baseline. Furthermore, baseline uses news article but our proposed method uses place-names and their area-related information. The baseline contains stiff style sentences but our proposed method is easy to understand dialogues.

```
Mary I went to Kobe.
Bob   Oh!
Mary I went to Kobe.
Bob   I know. How is it?
Mary Residential area and fashionable shops.
Bob   Humm....
Mary There are Dolphin center and Cape Muroto.
Bob   It is Muroto city.
Mary They are similar town.
Bob   No way! They are the different town.
      Did you really go to Kobe?
Mary I went anywhere.
Bob   You wildly inaccurate.
```

Fig. 3. Scenario generated proposed method

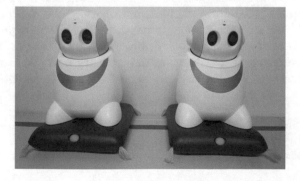

Fig. 4. PaPeRo i

From the results, our proposed method is understandable and funny. However, we did not get enough results with regard to the friendliness we value most. The reason is that the content of the baseline is longer than our proposed method. We consider that people feel familiarity for robots, they need contact time to robots. Then we have to consider the generated content length.

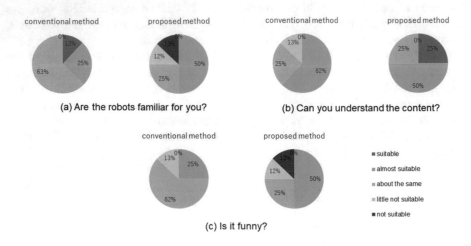

(a) Are the robots familiar for you?   (b) Can you understand the content?

(c) Is it funny?

Fig. 5. Result of experiment 2

5   Conclusion

We proposed automatic funny dialogue based scenario generation for the Omotenashi Robot using the place name given by the visitor. The scenario consists of three components which are area-related information component, place name misunderstanding component and land classification component. Area-related information component and place name misunderstanding component are based on the place names and their area-related information. Land classification component is based on general geographical information. We conducted an experiment to assess the benefits of our proposed method. The experiment results indicate some benefits of our proposed method. In the near future, we expect to conduct experiments which compare the ordinary dialogue at reception with our proposed method. Furthermore, we expect to create a smoothly interactive dialogue between visitors and robots.

Acknowledgment. This work was partially supported by Research Institute of Konan University, and by JSPS KAKENHI Grant Number 17K00430. NEC platforms support this research for rental the PaPeRo-i robots.

References

1. Anacleto, R., Figueiredo, L., Luz, N., Almeida, A., Novais, P.: Recommendation and planning through mobile devices in tourism context. In: Novais, P., Preuveneers, D., Corchado, J.M. (eds.) Ambient Intelligence - Software and Applications, pp. 133–140. Springer, Heidelberg (2011)
2. Anacleto, R., Luz, N., Figueiredo, L.: Personalized sightseeing tours support using mobile devices. In: Forbrig, P., Paternó, F., Mark Pejtersen, A. (eds.) Human-Computer Interaction, pp. 301–304. Springer, Heidelberg (2010)
3. Aoki, S., Umetani, T., Kitamura, T., Nadamoto, A.: Generating Manzai-scenario using entity mistake. In: Barolli, L., Enokido, T., Takizawa, M. (eds.) Advances in Network-Based Information Systems, pp. 1007–1017. Springer, Cham (2018)
4. Hayashi, K., Kanda, T., Miyashita, T., Ishiguro, H., Hagita, N.: Robot Manzai - robots' conversation as a passive social medium. In: 5th IEEE-RAS International Conference on Humanoid Robots, pp. 456–462, December 2005. https://doi.org/10.1109/ICHR.2005.1573609
5. Mashimo, R., Umetani, T., Kitamura, T., Nadamoto, A.: Automatic generation of Japanese traditional funny scenario from web content based on web intelligence. In: Proceedings of the 17th International Conference on Information Integration Web-Based Applications & Services, pp. 165–173 (2015)
6. Mashimo, R., Umetani, T., Kitamura, T., Nadamoto, A.: Human-robots implicit communication based on dialogue between robots using automatic generation of funny scenarios from web. In: Proceedings of the eleventh ACM/IEEE International Conference on Human Robot Interaction, pp. 327–334 (2015)
7. Tsutsumi, H.: Conversation analysis of Boke-Tsukkomi exchange in Japanese comedy. New Voices 5, 147–173 (2011). https://doi.org/10.21159/nv.05.07
8. Weber, K., Ritschel, H., Lingenfelser, F., André, E.: Real-time adaptation of a robotic joke teller based on human social signals. In: Proceedings of the 17th International Conference on Autonomous Agents and MultiAgent Systems, AAMAS 2018, International Foundation for Autonomous Agents and Multiagent Systems, Richland, SC, pp. 2259–2261 (2018). http://dl.acm.org/citation.cfm?id=3237383.3238141

The 8th International Workshop on Web Services and Social Media (WSSM-2019)

Web Service for Searching Halal Compliant Restaurants

Imama[1], Masaki Kohana[2], and Masaru Kamada[1(✉)]

[1] Ibaraki University, Hitachi, Ibaraki 316-8511, Japan
19nm704f@vc.ibaraki.ac.jp, m.kamada@mx.ibaraki.ac.jp
[2] Chuo University, Shinjuku, Tokyo 162-8473, Japan
kohana@tamacc.chuo-u.ac.jp

Abstract. We present a web service for searching Halal compliant restaurants in order to help the increasing visitors and students from the Muslim culture to Japan find places where Halal food is served. This service geographically presents the Halal compliant restaurants on Google Maps that are franchised by the Japanese food industries with Halal certification by the Japan Halal Business Association. Such restaurants are increasing, but their population is far from enough. To solve this situation, we developed a subsystem that estimates how much Halal or Haram each food on the menu of ordinary restaurants is by calculating the cosine similarity of its ingredients, which are acquired from the recipe site *Zexii Kitchen*, to those on the Haram list. For example, from the food menu of *Saizeriya*, Pizza Margherita and Peperoncino have the similarity score 0, and they are in fact Halal. Miso Ramen and Napolitan have the similarity score 0.01 or more to be found Haram. In that way, the Muslims in Japan can find restaurants that serve probably Halal as well as certified Halal food in Japan.

1 Introduction

Muslim visitors and students are increasing who come to Japan from across the world. The people who believe in Islam are called Muslim. Muslim means 'one who submits to God'. In the teaching of Islam, the word Halal can be translated as 'lawful', whereas Haram, it is the opposite, means 'prohibited'. Drinking alcohol and eating pork are prohibited. These terms are applied to all aspects of a Muslim's life and it extends to include food, drink, behavior and practices. Concerning food, eating Halal food is the same as worshiping God. Therefore, to eat Halal food is compulsory for a Muslim.

Halal certification is rapidly expanding on the back of the rapid economic growth and population and it is expected to expand more in the future. On the other hand, Japan has not yet been advanced to the Halal market. Halal compliant restaurants and Halal certifications are much less available in Japan compared to the other countries. Muslims have to inquire by telephone or visit the restaurant for actually confirming whether the menu is Halal or not. It is very inconvenient for Muslims to travel to Japan due to these problems.

© Springer Nature Switzerland AG 2020
L. Barolli et al. (Eds.): NBiS-2019, AISC 1036, pp. 683–691, 2020.
https://doi.org/10.1007/978-3-030-29029-0_67

The number of Muslim tourists visiting Japan from Indonesia, Vietnam and Malaysia are increasing. Japan needs to expand the scope of Halal services. Therefore, we developed a web service for Muslims to find Halal compliant restaurants in Japan. This service provides the geographical information about the Halal compliant restaurants that are franchised by the Japanese food industries with Halal certification by the Japan Halal Business Associations. Such restaurants are increasing, but their number is far from enough. To solve this situation, we developed a subsystem that estimates how much Halal or Haram each food on the menu of ordinary restaurants is by calculating the cosine similarity of its ingredients, which are acquired from the recipe site *Zexii Kitchen*, to those on the Haram list.

2   Related Research

2.1   Handy Product-Recognition App

NTT DOCOMO, INC. developed a handy smartphone application that enables foreigners in Japan who have special diet restrictions, such as Muslims and vegetarians, to determine the detailed contents of Japanese-labeled food products simply by photographing the products with their smartphones [3]. This application leverages an image-recognition engine built on the AI technology to identify the product and then provides the user with a detailed list of its ingredients in English. This system employs an image recognition technique, which is a different approach from our system, in identifying the food, but employs the same approach in identifying Halal by food ingredients.

2.2   Halal Food Detector for Muslims

This is an android application which helps Muslims living in Japan find whether the food products sold on the shop shelves are Halal or not [1]. This application makes use of the barcode printed on the product label. Its user directs the smartphone's camera to the barcode. Then the built-in barcode scanner module reads the barcode to check the database storing the product names with the attribute of Halal or not. If the food is Halal, a notification pops up on the screen informing the user that the food product just scanned is Halal. The research target is food products and our target is restaurants. But the purpose of this system is for helping Muslims and it confirms that there is a need of this system in Japan.

3   Halal Certification

Islam has prohibited drinking alcohol and eating pork. Alcohol is prohibited to protect mentality. Pigs are prohibited because they contain substances that have a bad influence on the physical and mental health of the people. Pigs are totally forbidden, including their derivatives, and even meats other than pork

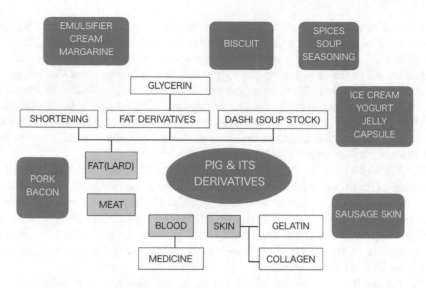

Fig. 1. Pig and its derivatives

are regarded as Haram if they are not slaughtered or processed in a manner consistent with the Islamic law. It is shown in Fig. 1 that the derivatives of the pigs are used as the ingredients in many products.

In Japan, about 150,000 Muslims live in Japan [4]. Among them, about 20,000 are Indonesians, and about 50,000 Japanese Muslims. Rest of them are from different countries. There are two types of Halal certification in Japan. Halal certification for domestic use targeting Muslim visitors in Japan and Muslims living in Japan. The other Halal certification is for exporting in Islamic areas. Halal certification ensures the proof that the product is Halal and correctly declares to Muslim consumers. Figure 2 shows the process of applying for Halal certification by *Japan Halal Association*. First of all, the association receives the inquiry, then they arrange the interview with the company and

Fig. 2. Process of Halal certification

discuss the products. After that, company applies the pre inspection. The association thoroughly examine the documents and produces the estimate. After the acceptance of estimate, the association goes for pre inspection and makes the judgment. After the approval of the judgment, the company is asked to apply for the Halal certification. The association goes for document audit and issues the report of audit. If some improvement is required, the company is requested for it. After the confirmation of improvements, the association will visit the site and will further request for improvement if any. Finally, the panel meeting approves, and the association issues the Halal certificate.

Table 1 shows the Japan's Certification Group that has received Halal Certification of food from Islamic Certification Agencies in Malaysia, Singapore, Indonesia and UAE. In Saudi Arabia to get the certification is difficult because Saudi Arabia is the most strict in the world. With Halal certification in Malaysia, Muslims can consume Halal food with relief. Japanese companies are also paying attention to the merit that these certification groups will be widely used anywhere in the Islamic world.

In this research, we collected information on Japanese food industries that obtained Halal certification from those associations to construct a web service that provides geographical information about the Halal compliant restaurants on Google Maps.

Table 1. Table of Halal certification organization

Japan Certification Group	Government Agency JAKIM (Malaysia)	Government Agency MUIS (Singapore)	Religion Agency LPPOM MUI (Indonesia)	Government Agency ESMA (UAE)
Japan Muslim Association	○	○	○ Processed food, flavor	○
NPO Japan Halal Association	○	○		
Japan Islamic Trust (Otsuka masjid)	○			○ Slaughter certificate issuance
Nippon Asia Halal Association	○	○		
Islamic Center of Japan				○ Slaughter certificate issuance
JAPAN HALAL FOUNDATION	○	○		
Islamic Cultural Center Kyushu (Fukuoka masjid)			○ Slaughter	
Muslim Professional Japan Association	○		○ Processed food , Slaughter	
Japan Halal Unit Association	○			

Fig. 3. Overview of the system

4   System Construction

This section describes how the web system has been constructed. Figure 3 shows an overview of our system. This web service provides all the information Muslims need for finding Halal compliant restaurants in Japan. In addition to the official information, they can find restaurants that serve probably Halal food as well as certified Halal food.

4.1   A Web Page for Information of Halal Compliant Restaurants

This application is a web page written in HTML5 and JavaScript. A web server is also written in JavaScript in the Node.js environment. The server sends an HTTP request to the page of *HALAL MEDIA JAPAN* for the list of certified companies [2]. From the HTTP response, our system picks out the information about the companies under which restaurants are franchised. In that way, our web service collects and provides the users with the company names, types, prefectures, city information as shown in Fig. 4. The links are directed to the restaurants plotted on Google Maps.

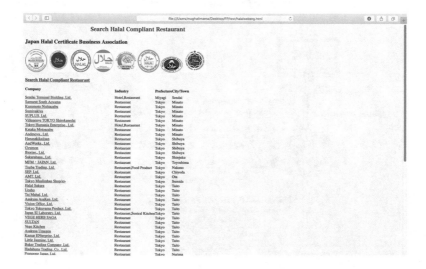

Fig. 4. Web page of search Halal compliant restaurants

4.2   Halal Estimation System by Calculating Cosine Similarity

The number of Japanese restaurants with Halal certification is increasing, but their population is far from enough. To solve this situation, we developed a subsystem that estimates how much Halal or Haram each food on the menu of ordinary restaurants is. In this system, we create a parent process to get ingredients from recipe site *Zexii Kitchen* [5] and create a child process to calculate the cosine similarity of the ingredients to the Haram list. An overview of the system is illustrated in Fig. 5. In the parent process, the server sends two HTTP requests to the recipe page of *Zexii Kitchen* to get the ingredients of the menu. Then, the server sends the data of ingredients to callback function and output as standard. In the child process, the server reads the standard input and estimates the Halal by using ingredients and Haram list. We used the cosine similarity for estimation, which is a measure of similarity between two one-hot vectors.

Let the vector $\mathbf{A} = (a_1, a_2, \cdots, a_n)$ represent the standard list of ingredients, where a_i is set 0 or 1 if the ingredient i $(i = 1, 2, \cdots, n)$ is Halal or Haram, respectively. Let $\mathbf{B} = (b_1, b_2, \cdots, b_n)$ represent the ingredients of a food in question, where b_i is set 0 or 1 if the ingredient i $(i = 1, 2, \cdots, n)$ is absent or present, respectively, in the food. Then the cosine similarity of $\mathbf{A}$ and $\mathbf{B}$ is calculated by

$$\cos\theta = \frac{\mathbf{A} \cdot \mathbf{B}}{|\mathbf{A}||\mathbf{B}|}$$

where

$$|\mathbf{A}| = \sqrt{(a_1^2 + a_2^2 + \cdots + a_n^2)}, \quad |\mathbf{B}| = \sqrt{(b_1^2 + b_2^2 + \cdots + b_n^2)},$$

and

$$\mathbf{A} \cdot \mathbf{B} = a_1 b_1 + a_2 b_2 + \cdots + a_n b_n.$$

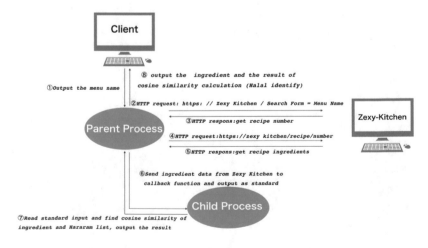

Fig. 5. System configuration of Halal estimation

Two vectors with the same orientation have the cosine similarity of '1'. Two vectors at $\theta = 90°$ have the similarity of '0', which means the vectors are completely unrelated.

Table 2 shows the results of the similarity score for foods served at several restaurants without Halal certificates. Foods having the score '0' are estimated to be Halal, whereas those having the similarity score '0.01' or more are considered Haram.

Table 2. Similarity score of foods from the menu of restaurants without Halal certificates

Restaurant	French fries	HALAL/HARAM	
Saizeriya	Margherita	HALAL	
	Anchovy pizza	HALAL	
	Chocolate cake	HALAL	
	Mushroom soup		0.056
	Clam chowder		0.077
	Seafood doria		0.04
Gusto	Tomato sauce spaghetti	HALAL	
	Margherita	HALAL	
	Fried oysters	HALAL	
	Meat sauce spaghetti		0.079
	Mackerel boiled in miso		0.15
	Keema Curry		0.099
Yumean	Tofu chige	HALAL	
	Fried oysters	HALAL	
	Tofu salad		0.05
	Cold udon(Noodles)		0.066
	Oboro tofu(Been curdled)		0.04
Bamiyan	Lettuce fried rice	HALAL	
	Mabo Tofu(Bean curd Szechwan style)		0.038
	Prawn in Chili Sauce		0.027
	Chukadon(Chinese thick sauce over rice)		0.064
Subway	Egg sandwich	HALAL	
	Tuna sandwich	HALAL	
	Teriyaki Chicken sandwich		0.034
COCO'S	Margherita	HALAL	
	Chicken steak		0.100
	Taco rice		0.096
	Seafood paella		0.048
Denny's	French toast	HALAL	
	Shrimp pilaf	HALAL	
	Carbonara		0.041
	Meat sauce pasta		0.074
	Cut steak and beef stew		0.041
Kurazushi	Warabimochi	HALAL	
	Cheese cake	HALAL	
	Miso ramen		0.15
	Tantanmen (Szechuan noodles)		0.049
	Chawanmushi(non-sweet egg custard)		0.099
	Carbonara		0.041
Hoshino Coffee shop	Souffle pancake	HALAL	
	Roll cake	HALAL	
	Bacon egg		0.051
	Crab Croquette		0.041
	Bacon sand		0.058

In the results for *Saizeriya*, the Halal foods such as Margherita, Anchovy pizza and Chocolate cake were correctly decided to be Halal. On the other hand, because of meat and gelatin contained in Mushroom soup, Clam chowder and Seafood doria, these were correctly estimated as Haram.

In the results for *Gusto*, Mackerel boiled in miso was correctly estimated as Haram after we included 'sake' in addition to 'Osake' (Japanese alcoholic drink) in the Haram list. We should include all the possible orthographical variants of Haram ingredients.

In the results for *Yumean*, Tofu salad was correctly estimated as Haram. In the *Zexii Kitchen*, 'Ponzu (soy sauce)' is used as an ingredient of Tofu salad. Generally, Ponzu contains alcohol, which means that it is correct that the Tofu salad is Haram. However, there are many Halal seasoning. Our system can not recognize whether the seasoning is Halal or not. Therefore, we need to recognize the ingredient of seasoning in our future works.

4.3   Finding Halal Compliant Restaurants on Google Maps

For finding Halal compliant restaurant geographically, we developed a web service to show the restaurants plotted on Google Maps as shown in Fig. 6. We used the JavaScript API to customize the map and display it on the web page.

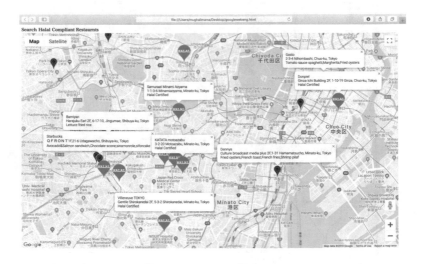

Fig. 6. Screen shot of the page for finding Halal compliant restaurants

We provide two overlays, Marker and Info Window, to show the location and information of the restaurants. The system sets the green marker on the Halal compliant restaurants, which are franchised by the Japanese food industries with Halal certification. The system also marks the probably Halal compliant restaurants, which have been found to serve Halal foods by Halal estimation system, by red pins. When a user clicks on the green marker, the system shows the name, address and information of Halal certification. If the user clicks on a red pin, the system shows the name, address and information about the availability of Halal menu.

5 Conclusions

This paper proposed a web service for finding Halal compliant restaurants in order to help the visitors from the Muslim culture in Japan find the restaurants that serve Halal food. In our web service, a user can visually find the Halal compliant restaurants on Google Maps and get the information of the restaurants. We also developed a subsystem that estimates how much Halal or Haram each food on the menu of ordinary restaurants is by calculating the cosine similarity of the ingredients to the Haram list. In the Halal estimation system, most of the Halal foods were detected correctly. But we were unable to grasp the ingredients of seasonings and the orthographical variants of ingredients, which becomes our next task.

References

1. Emad, K., Ichimura, S.: Halal food detector for Muslims, IPSJ-SIG Technical report, Vol.2015-GN-94 No. 20 (2015)
2. Halal Media Japan: Halal certified company. https://www.halalmedia.jp/. Accessed 5 May 2019
3. NTT docomo Press Release: Development of a "Food Judgment System" to show on the app the foods that Muslims and vegetarians can purchase (2018). https://www.nttdocomo.co.jp/info/news_release/2018/09/26_01.html. Accessed 5 May 2019
4. Tanaka, A.: Expanding Halal market and current situation (2014). http://www.maff.go.jp/tohoku/kihon/yusyutu/kyougikai/pdf/tanaka-siryou.pdf. Accessed 5 May 2019
5. Zexy Kitchen. https://zexy-kitchen.net/. Accessed 5 May 2019

Implementation of Interactive Tutorial for IslayPub by Hooking User Events

Daisuke Tanaka[1], Masaki Kohana[2], Michitoshi Niibori[3], Yasuhiro Ohtaki[1], Shusuke Okamoto[4], and Masaru Kamada[1(✉)]

[1] Ibaraki University, Hitachi, Ibaraki 316-8511, Japan
masaru.kamada.snoopy@vc.ibaraki.ac.jp
[2] Chuo University, Ichigaya, Tokyo 162-8473, Japan
[3] Learning-i, Ltd., c/o Ibaraki University, Hitachi, Ibaraki 316-8511, Japan
[4] Seikei University, Musashino, Tokyo 180-8633, Japan

Abstract. We present a mechanism to overwrap the existing web-based graphical programming environment *IslayPub* with another JavaScript program that interactively tutors the user where to click and how to operate IslayPub along the tutorial scenario. The original JavaScript program of IslayPub has only to be modified to include event listeners to hook the user events within the opening ⟨body⟩ tag and to include a tutorial program at the end. The tutorial program instructs where to click by an arrow near the target component in the IslayPub screen and what to do in a dialog box in each step of the scenario and hooks the user event to check if it complies with the scenario. If the user event implies the expected operation in the step, the tutorial program allows the event to propagate toward the IslayPub program to react to the event and lets the user proceed to the next step. Otherwise, the tutorial program abandons the event and prompts the user to do the right thing. In that way, the user is guided to learn how to operate IslayPub along the tutorial scenario.

1 Introduction

It has been getting common for the commercial video games to guide novice users interactively step by step in the tutorial mode where the operation is instructed by graphical symbols such as arrows and icons superimposed on the game screen [1]. In the tutorial mode, new states for the sake of presenting instructions and checking user actions must be inserted into every state transition of the game program in the regular mode. It is not usually possible to switch between the regular and tutorial modes unless the entire software design has been made to accommodate the extra states for the tutorial mode.

This study aims at a mechanism that allows for a tutorial mode added onto a web application program that has already been finished. The idea is to implement a tutorial mode by an additional tutorial program that overwraps the original web application program by exploiting the following two features of HTML and JavaScript:

- User events can be hooked by placing event listeners at the top level node ⟨body⟩ in the HTML DOM tree.
- An additional JavaScript program can be activated on load, handle the hooked user events, get and manipulate the attributes of HTML elements under the DOM tree.

In this work, the additional program for tutorial is activated on load to overwrap the original JavaScript program, shows a tutorial instruction superimposed on the target HTML element, hooks every user event, propagates the event to the target HTML element for the original web application program to react to the event only if the event matches the instruction and has taken place at the position of the target HTML element. That mechanism is applied to a web application IslayPub3.0 [2] that offers a graphical programming environment for children.

2 Requirements for the Original Web Applications

The original web application should meet the following requirements:

- The whole application is written only in HTML5 and JavaScript.
- It deterministically reacts to the user events so that the tutorial scenario always goes along a single thread without any branches.
- Every element has its id or name in order for the additional tutorial program to access its attributes.
- No event listeners are registered inside the opening ⟨body⟩ tag where the additional tutorial program will set event listeners to hook user events.

Those requirements may be satisfied by the ordinary web applications such as web-based editors including IslayPub as their example.

3 Tutorial Program

The tutorial program is written in JavaScript, and its file named *tutorial.js* will be inserted by ⟨script type="text/javascript" src="tutorial.js"⟩⟨/script⟩ at the end of the original web application file. The opening ⟨body⟩ tag of the original web application file will be rewritten as ⟨body onload="initializeTutorial()"⟩ to start the function to initialize the tutorial program. Those two additions are the only modifications to be made on the original web application program.

3.1   Function to Initialize the Tutorial

On load of the original web application file, the function *initializeTutorial()* is started. It first adds event listeners for hooking the user events to the body of the original web application by document.body.addEventListener(*event, handling function*, capture = true) in order to hook the *event* and call the *handling function* that takes care of the *event* in the tutorial program instead of the originally responsible function in the original web application. The events are click, mousedown, mouseup, change, mousewheel, and keydown in the case of Islay-Pub. The mousemove event is not hooked so that the original web application can react to the user operation while the user is dragging an element. For the drag & drop operation, it is checked by the mousedown and mouseup events if the selected element is the right one and if the drop position is the right place, respectively.

3.2   Functions to Manage the Tutorial

Table 1 shows an example of the tutorial scenario prepared for instructing the first experience of the web application IslayPub. We use a progress counter that indicates the current step to be instructed. Its initial value is 0 to indicate Step 0.

Table 1. Example tutorial scenario

Step	Icon/Position	Instructions	Event/position expected
0	Right-arrow for Left-click at $(x : 200\text{px}, y : 325\text{px})$	Let's start the first project! First, left-click where the arrow points to.	Mousedown at $(x : 200 \sim 350, y : 250 \sim 400)$
1	Right-arrow for Left-click at $(x : 550\text{px}, y : 375\text{px})$	Good! Next, left-click where the new arrow points to.	Mousedown at $(x : 550 \sim 700, y : 300 \sim 450)$
2	Up-arrow for Left-click at the element id: transition_button	Great! Next, left-click the button named "transition".	Mousedown at the element id: transition_button
3	Right-arrow for Left-click at the element id: state0	OK! Next, left-click "state0".	Mousedown at the element id: state0
4	Right-arrow for Left-click at the element id: state1	OK! Next, left-click "state1".	Mousedown at the element id: state1
5	Up-arrow for Left-click at the element id: move_button	Great job! You made a transition! Now, we are gonna move the states. Left-click the button named "move".	Mousedown at the element id: move_button

Step 0 is the very first step for the user to create a new state at the place where the mouse pointer is clicked in the work space. The instruction is displayed on the screen as shown in Fig. 1. The thick orange arrow points to the place to left-click at (x : 200px, y : 325px) as specified in Table 1. The instruction texts in Table 1 are displayed in the dialogue boxes at the bottom of the screen. The expected event is mousedown within the rectangular region (x : 200 ~ 350px, y : 250 ~ 400px) specified in Table 1.

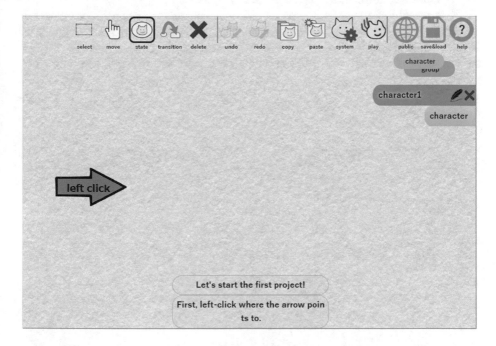

Fig. 1. Step 0: Create a state by clicking a point.

Suppose that the event was not the expected mousedown but something else such as keydown. The event listener set in the body captures the keydown event set in the body and calls the function islayTutorialKeydown(). This function abandons the event by stopPropagation() not to let it propagate to the body since the event was not the expected one specified in Table 1. Then the function lets the dialogue boxes pop up again to prompt the user to take the right action.

Suppose that the user event was the expected mousedown. Then the event listener set for mousedown in the body calls the function islayTutorialMouse-down() of the tutorial program. This function gets the coordinates where the mousedown event took place and checks them against the expected rectangular region (x : 200 ~ 350px, y : 250 ~ 400px) specified in Table 1. If the mousedown event took place outside the expected region, the function islayTutorialMouse-down() abandons the event by stopPropagation(). Then the function repeats showing the dialogue boxes to prompt the user to take the right action.

Finally in the correct case that the event is mousedown and that the mouse-down coordinates are within the region, the function simply returns to the caller after incrementing the progress counter to indicate Step 1. Then the mousedown event automatically propagates to the body toward the originally intended can-vas element of the original web application. Its associated function reacts to the event to create a circle element that represents a state in the workspace.

That finishes Step 0 of the tutorial scenario and the tutorial lesson moves on to the following steps as shown in Figs. 2, 3, 4, 5 and 6.

It is possible to specify the expected actions and positions in terms of the element id. In Step 2 of Table 1, for example, the target to click is specified by the id of the transition button element in Fig. 3. Then the event listener set for mousedown in the body calls the function islayTutorialMousedown(), which gets the rectangular region surrounding the button by getElement-ById("transition_buttton").getBoundingClientReact() for checking the mouse-down coordinates.

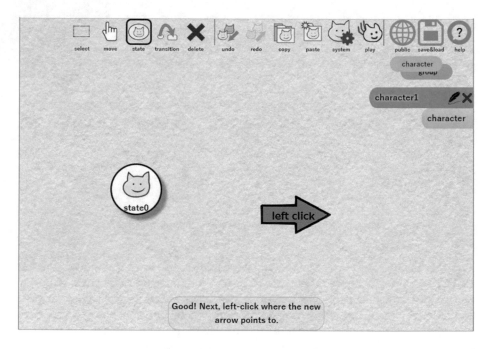

Fig. 2. Step 1: Create another state by clicking another point.

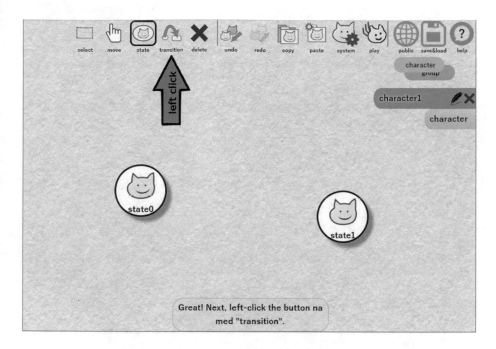

Fig. 3. Step 2: Change the mode to draw a transition by clicking the transition button.

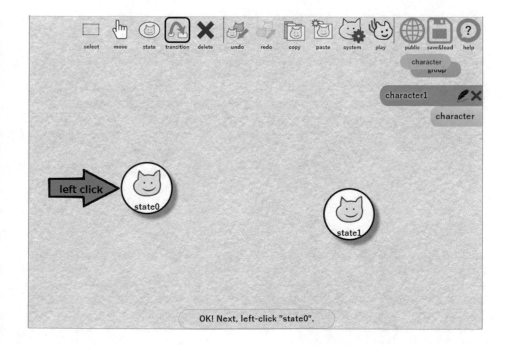

Fig. 4. Step 3: Select the source state.

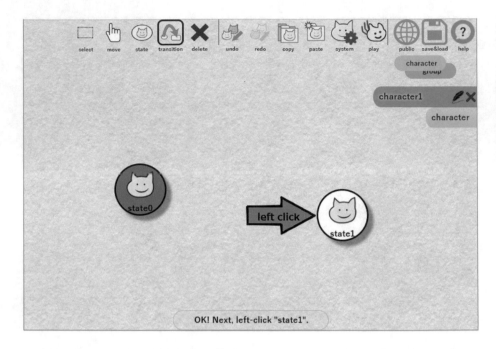

Fig. 5. Step 4: Select the destination state.

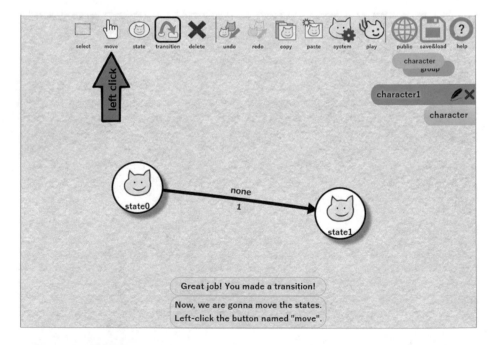

Fig. 6. Step 5: Change the mode to move elements by clicking the move button.

4   Conclusions

A mechanism is proposed and applied to the graphical programming environment *IslayPub* where the existing web-based application program is overwrapped by another JavaScript program that interactively tutors the user where to click and how to operate along a tutorial scenario. The original JavaScript of the existing web-based application program has only to be modified to load an initializer function within the opening ⟨body⟩ tag that sets event listeners for the sake of hooking the user events and to include the tutorial program at the end of the web application program.

The next step is to improve the precision in checking the coordinates where the events took place against the expected ones. Currently the shape of coordinate regions is limited to the rectangles. That causes problems when the target elements are non-rectangular. The tutorial program shall clone the target element in the original web application to make a shadow element for checking if it is clicked on behalf of the original target element at better precisions. A further step will be a tool for the instructors to create the tutorial scenario interactively and graphically by recording their model operations.

References

1. School of Game Design: How to make a good video game tutorial. https://schoolofgamedesign.com/project/good-video-game-tutorial/. Accessed 29 May 2019
2. Suzuki, K., Niibori, M., Rashed, A.S., Okamoto, S., Kamada, M.: Development of IslayPub3.0 — educational programming environment based on state-transition diagrams. In: The 4th International Workshop on Web Service and Social Media (WSSM 2015), Proceedings of the 18th International Conference on Network-Based Information Systems, (NBiS 2015), Taipei, pp. 702–705 (2015)

A Smart Lock Control System with Home Situation

Katsumi Ohkura[1] and Masaki Kohana[2(✉)]

[1] Ibaraki University, Hitachi, Ibaraki, Japan
kuraringo.no2@gmail.com
[2] Chuo University, Shinjuku, Tokyo, Japan
kohana@tamacc.chuo-u.ac.jp

Abstract. This paper proposes a way to control a smart lock using a home situation. A user can lock/unlock a smart lock using an application. And the application can unlock the smart lock with the GPS information. However, the smart locks have some problems. The operating the application is troublesome. And the GPS range is too wide to control the smart locks. In this study, we propose a way to control the smart locks using Bluetooth and the home situation. Our proposed system observes the home situation and controls the smart locks automatically using the Bluetooth and the home situation.

1 Introduction

Internet of Things (IoT) gets much attention. To leverage IoT, various networking services and various IoT devices appears. One of these devices is a smart lock. The smart lock replaces an existing traditional lock. If a door has the smart lock, a user can unlock the door using an application for smartphone. The user does not need any physical keys.

However, the user needs to control the application. It is not so much convenient rather than the physical keys. There are some features to unlock the door without the controlling application. One of the ways to unlock is knocking the door. A sensor detects the knock, and the smart lock unlocks the door. However, detection is not stable. Another way to unlock is a GPS feature. A smartphone sensing the location of the user and sends a notification to the smart lock when the user approaches the home. However, the range of GPS is wide. If the user goes to the neighborhood, the smartphone does not decide the user goes outside of the home. Therefore, a method to detect the going out and returning home accurately.

This paper proposes a system to control the smart lock. This system uses two aspects of the features. One is the Bluetooth sensing system that detects the returning home. Another is the observing home situation, such as the open/close the door. Combining the Bluetooth and the home situation, the system can detect the behavior of the user accurately.

This paper consists of the following sections. Section 2 introduces some studies related to our work. Section 3 describes the system construction. Section 4

© Springer Nature Switzerland AG 2020
L. Barolli et al. (Eds.): NBiS-2019, AISC 1036, pp. 700–704, 2020.
https://doi.org/10.1007/978-3-030-29029-0_69

describes the observation of the home situation and the web page. Section 5 describes how to control the smart lock. Section 6 shows the experimental result, and Sect. 7 concludes this paper.

2   Literature Survey

This section introduces studies related to this paper.

Pavithra et al. proposed a system of home automation [8]. This system collects information from several sensors and detects the fire. If the system detects the fire, it reports to the fire department.

Kajioka et al. proposed an attendance system using the BLE [6]. This system checks the attendance using BLE beacon. The using BLE beacon can confirm the user exists in the classroom or not.

Kikawa et al. evaluated a method to detect that a user leaves the PC desk using the Bluetooth [7]. Their experimental results show that the probability of the false positive becomes less than five percents.

Kohana et al. proposed a system that provides attendance information about teachers [9]. This system limits the range of information using GPS and network IP address. If a user joins the network of the organization such as a university, the system shows the information about the teacher. On the other hand, if a user is on a campus of a university, the system also shows the information. As a result, the user can access the information via the mobile network from a campus.

3   System Overview

This section describes system construction of this study. Figure 1 shows the system construction. This system uses two small computers and a smart lock device.

The Web server receives information from other devices and provides the home situation information on a web page. The server controls the smart lock device based on the information from the other devices. Tessel is a small computer that has several sensor modules [1]. Tessel can run a JavaScript program. This system uses the BLE sensor, the environment sensor, the climate sensor, and camera.

The micro:bit is also a small computer that provided by BBC [2]. The micro:bit can run a JavaScript or Python program. Moreover, it provides a development environment on a web page. This system uses the BLE sensor and the accelerometer.

SESAME is a smart lock device provided by CANDY HOUSE [3]. SESAME provides an application for smartphones. A user can control the SESME using the application. Moreover, SESAME provides Web APIs [4]. The proposed system controls the SESAME using Web APIs.

Slack is a communication tool, and it provides Web APIs [5]. The proposed system sends a notification to the user via Slack when an unusual situation occurs.

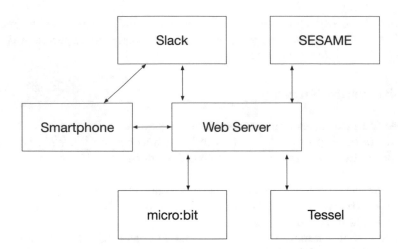

Fig. 1. System construction

The system decides the "absence" or "staying at home" using the accelerometer on the micro:bit. The micro:bit detects the open/close the door. If the distance between the smartphone and the home is longer than the threshold after moving the door, the user might be the outside of the home. If the distance does not change, the user stays at home. On the other hand, if the micro:bit detects the movement of the door while the home situation is the "absence," it is an unusual situation.

4   Home Situation Observation

The system observes the home situation using Tessel. The user confirms the situation on a web page. The Tessel senses the sound, the lightness, the temperature, and the humidity. The web page also provides information about staying at home or absence. If the user sees the web page, and the situation is staying at home, it is an unusual situation. When an unusual situation occurs, the system sends a message to the user using Slack.

5   Smart Lock Control

The smart lock control feature detects a device that has Bluetooth feature such as a smartphone. The controller unlocks the smart lock when the situation satisfies the following conditions.

- The system detects the Bluetooth device.
- The home situation is "absence."
- The distance between the device and the home is less than a threshold (4.5 m in this study).

The system calculates the distance between the smartphone and the home using Friis Transmission Equation.

$$d = \frac{10^{((TxPower - RSSI)/(n \times 10))}}{100} \tag{1}$$

The value of d is the distance. The $TxPower$ is the transmit signal strength, while the $RSSI$ is the received signal strength indicator. The value of n is a constant value that is 2.0 in an ideal space. The system decides the user is in the outside of the home when the distance is more than 4.5 m. If the distance is less than 4.5 m, the system decides the user is in the home and change the home situation from the "absence" to "staying home." The system unlocks the door when the system detects the smartphone, and the distance is less than the threshold.

The system can handle an unexpected situation. If the user turns off the device or the Bluetooth feature, the system cannot detect the device. In this situation, the system decides that the opening the door is illegal. After that, if the user turns on the device or the Bluetooth feature, the system decides that the user returns the home. Therefore, the system unlocks the door. The system avoids this situation using the distance to decide the user location.

6   Experimental Result

This section shows the experimental result. This study measured the distance information because the distance is essential in decide that the user is

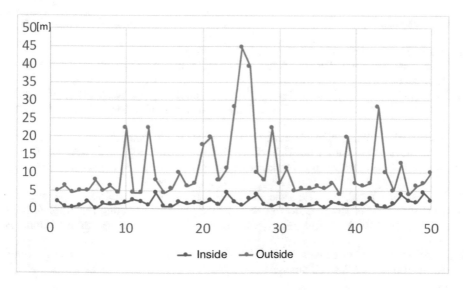

Fig. 2. Distance experiment

inside/outside of the home. This study decides the threshold based on the experimental result.

The computer that scans the Bluetooth device exists in the home. There are some shields, such as a door and a wall. The experiment measures the distance between the smartphone and the scanning device at the inside or the outside of the home 50 times, respectively.

Figure 2 shows the experimental result. The y-axis indicates the distance in meter, while the x-axis indicates the number of times. For the inside of the home, the shortest distance is 0.05 m, while the longest distance is 4.47 m. Therefore, the threshold of the system is 4.5 m.

On the other hand, most cases of the outside of the home have a longer distance than that of the inside. However, in some cases, the distance is less than 4.5. The shortest distance of the outside is 3.98 m. Therefore, the system needs to decrease the number of false positive, which is the future works.

7   Conclusion

This paper proposes a smart lock control system using the home situation. The proposed system using several sensors to observe the situation in the home, such as the temperature, the humidity, and the "absence" or "staying at home." The system also uses the Bluetooth to detect that the user returns home. The system unlocks the door if the home situation is the "absence" and the distance between the Bluetooth device of the user and the home is less than 4.5 m. Combining the home situation and the distance, the user can unlock the door without any operation of the application or any physical keys.

References

1. Tessel 2. https://tessel.io. Accessed 31 May 2019
2. Micro:bit Educational Foundation — micro:bit. https://microbit.org. Accessed 31 May 2019
3. SESAME, CANDY HOUSE Inc. https://jp.candyhouse.co. Accessed 31 May 2019
4. Sesame API, CANDY HOUSE Inc. https://docs.candyhouse.co. Accessed 31 May 2019
5. Slack API — Slack. https://api.slack.com. Accessed 31 May 2019
6. Kajioka, S., Yamamoto, D., Uchiya, T., Saito, S., Matsuo, H., Takumi, I.: Operation of localization-based attendance check system using BLE beacons, IPSJ SIG Technical report, vol. 2016-IOT-35, no. 12, September 2016
7. Kikawa, M., Yoshikawa, T., Ookubo, S., Takeshita, A., Takahashi, O.: A proposal and evaluation of the method to detect leaving one's desk using the RSSI of bluetooth, IPSJ SIG Technical report, vol. 2009-MBL-48, no. 13, pp. 95–102, January 2009
8. Pavithra, D., Balakrishnan, R.: IoT based monitoring and control system for home automation. In: Global Conference on Communication Technologies, pp. 169–173 (2015)
9. Kohana, M., Okamoto, S.: Access control for a confirming attendance system. Int. J. Space-Based Situated Comput. 6(2), 121–128 (2016)

Card Price Prediction of Trading Cards Using Machine Learning Methods

Hiroki Sakaji[1]([✉]), Akio Kobayashi[2], Masaki Kohana[3], Yasunao Takano[4], and Kiyoshi Izumi[1]

[1] The University of Tokyo, 7-3-1 Hongo, Bunkyo-ku, Tokyo 113-8654, Japan
{sakaji,izumi}@sys.t.u-tokyo.ac.jp
[2] RIKEN AIP Center, Nihonbashi 1-chome Mitsui Building, 15th floor,
1-4-1 Nihonbashi, Chuo-ku, Tokyo 103-0027, Japan
akio.kobayashi@riken.jp
[3] Chuo University, 1-18 Ichigata-Tamachi, Shinjuku-ku, Tokyo 162-8478, Japan
kohana@tamacc.chuo-u.ac.jp
[4] Kitasato University, 1-15-1 Kitasato, Minami-ku, Sagamihara-shi, Kanagawa
252-0373, Japan
tyasunao@kitasato-u.ac.jp

Abstract. In this paper, we try to predict the card prices of the trading card game using their information. The trading card game market is growing by the increasing popularity of the board game or the digital card game in the e-sports in recent years. The trading card game is a kind of card game which two or more people plays a card with some text or symbols those characteristics expresses a ruling or interaction to the other card. This interaction of cards may work effectively in the game, prices of those card pairs will be increased with the popularity of its combination. We have a hypothesis that card text is useful for prediction of card prices from the importance of card combinations. Therefore, in this research, we focus on not only the basic card information but also card text. Moreover, we use several machine learning method for prediction of card prices, and we analyze which machine learning method is an effect.

1 Introduction

The trading card game market is growing by the increasing popularity of the board game or the digital card game in the e-sports in recent years. The trading card game is a kind of card game which two or more people plays a card with some text or symbols those characteristics expresses a ruling or interaction to the other card. For example, Magic: The Gathering (MTG) released in 1993 by Wizards of the Coast is one of the famous trading card game. MTG has approximately twenty million players as of 2015.

In MTG cards, "Black Lotus" is a well-known card, and is the most expensive card. The card was traded 166,100 U.S. dollar on eBay in 2019. Many MTG cards are expensively traded on Internet sites such as eBay. There are two reasons for expensive cards. One of these is that the cards have antique values. Another one

© Springer Nature Switzerland AG 2020
L. Barolli et al. (Eds.): NBiS-2019, AISC 1036, pp. 705–714, 2020.
https://doi.org/10.1007/978-3-030-29029-0_70

is that the cards are valuable for card game; which means that using that cards are more likely to win than using other cards.

Moreover, there are two types of valuable cards: one that is strong alone and one that is strong in combination with other cards. Many valuable cards are expensive from the combination. Therefore, since the combinations depend on card contents, the card text is useful to decide card prices. From the above reasons, we considered that we could predict card prices using card contents.

In this research, we propose a method for predicting card prices using machine learning methods. In recently, machine learning methods, including deep learning, show high performance on any tasks. We employ some machine learning methods for predicting card prices and evaluate which method is the effect of this task. Additionally, we extract feature weights from learned machine learning models and analyze which features are important to predict card prices.

2 Data of Cards

In this research, we treat card information and card prices. The card information is determined by Wizards of the Coast. On the other hand, the card prices are determined in the transaction. Usually, there are card transaction prices in the trading site such as eBay. Therefore, we gathered the latest card prices from websites.

2.1 Information of Cards

To predict card price from card data, we adopt many part of the card as features of our machine learning inputs. Figure 1 shows the structure of the card. The card consists parts of below structure.

Mana cost Mana cost is the cost to play a card. Players need to prepare some resources (called Mana source in the rule of MTG) to playing cards. Basically, cards with strong effect needs more cost than weak effect.

Card type Cards has some types which shows card is a single-used or able to use every turn.

Subtype Cards has subtypes which often referred from card text.

Card text Most price related part of the card. Which shows how card works and how effective. Card text is written in natural language and consists by some sentences.

Ability Some card has ability in the card text. It is a keyword for shorten a card text and identify and encourage to understand the card effect.

Power/Toughness If the card type is a "Creature," it has a number pairs at right down side of the card. If those are large numbers, it means simply strong to beat an opponent. Basically, Mana costs increase when Power/Toughness are large number.

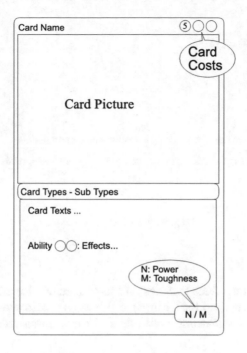

Fig. 1. Structure of the card.

Rarity Rarity of the card. MTG has a 4 levels of rarity that are "Mythic Rare," "Rare," "Uncommon," "Common" in order of rarity.
Card Title The name of the card.
Card Picture It shows the image of the thing which card means.

The Mana cost has a number which can be paid any resource and some kind of symbols which can be paid only with the appropriate resource. The symbol of number is a generic cost which can pay any kind of resource. The Wizards of the coast provides a API to get these card information[1]. The card information provided by the API as Fig. 2.

At the API response, "name" is a "Card title," "cmc" means a "Converted Mana Cost" which is a total amount of cost to play this card that summing costs in "manaCost" field. In the value of "manaCost" field, "2" means a generic cost and "G" is called "colored mana symbol" which means a cost which can pay appropriate resource. The "text" means a "Card text." In this example, "Changeling" and "Protection" are keywords of "Ability".

---

[1] https://docs.magicthegathering.io/.

```
{
    "name":"Chameleon Colossus",
    "manaCost":"{2}{G}{G}"
    "cmc":4,
    "types":[
        "Creature"
    ],
    "subtypes":[
        "Shapeshifter"
    ],
    "rarity":"Rare",
    "text":"Changeling This card is every creature type.
Protection from black
{2}{G}{G}: Chameleon Colossus gets +X/+X until end of turn, where X is its power.",
    "power":"4",
    "toughness":"4",
    ...
```

Fig. 2. The API response

3   Methodology

In this section, we introduce features and used machine learning methods. First, we show what data is used for features and how to create the features. Then, we describe used machine learning methods and them parameters.

3.1   Features

From the Fig. 1, we decide to use each part of the cards as features of our methods without "Card Title" and "Card Image." We adopt the uni-gram (single word) of the card text because we want to know the effectiveness of the words to determine the card prices. Moreover, we assume that if we adopt the combination of words will introduce a combination explosion. Moreover, we add a generic cost and "colord mana symbols" to our features. Finally, we adopt features in Table 1.

Table 1. Adopted features for experiments

Feature label	Part of the card
allcost	Converted mana cost ("cmc" in the API response)
cost	Generic cost in the "manaCost"
colorcost	colored mana symbols in the "manaCost"
Type: xxxx	Card type
SubType: xxxx	Subtype of the card
Rare	Rarity of the card
Ability: xxxx	Keyword of ability in the card text
Word: xxxx	Uni-gram extracted from card text
Power	"Power" of the "Creature" type card
Toughness	"Toughness" of the "Creature" type card

3.2   Machine Learning Methods

We employ following 4 machine learning methods for predicting card prices.

- Linear Regression (LR)
- Random Forest Regressor (RFR)
- Support Vector Regression (SVR)
- Multi–Layer Perceptron Regressor (MLPR)

These methods are provided by scikit–learn[2]. We did grid search for searching the best parameters of each machine learning method. Table 2 shows searched parameters and best parameter of SVR. Table 3 shows searched parameters and best parameter of RFR. Table 4 shows searched parameters and best parameter of MLPR. Here, we used linear kernel as kernel of SVR. Because the linear kernel of SVR can extract the feature's coefficient from the learned model. In this research, we focus on not only predicting card prices but also which features are the effect. Therefore, the linear kernel was selected.

Table 2. The parameters of SVR

Parameters	C: 1, 10, 100
Best parameters	1

Table 3. The parameters of RFR

Parameters	The number of trees: 5, 10, 20, 30, 50
Best parameters	50

Table 4. The parameters of MLPR

Parameters	Hidden layer sizes: (50), (100), (50, 50), (100, 50), (100, 100)
Best parameters	(100, 100)

From the results of parameter selection, we experiment card price prediction using acquired parameters. Additionally, the parameters not selected for the search used the default. We did grid search in addition to the search for other parameters, but since it did not work well, we searched for the above parameters in this research.

---

[2] https://scikit-learn.org/stable/.

4 Experiment

In this section, we describe an experiment of card prediction. For the experiment, we use cards of two formats "Modern" and "Standard." There are several formats for playing games in MTG, and "Modern" and "Standard" are included in the formats. Available cards are different depending on the formats. In MTG, rarity is determined for each card, and there are "Common," "Uncommon," "Rare," and "Mythic Rare." In this research, we target "Rare" and "Mythic Rare" cards for card prediction experiment. We use cards of "Modern" as training data and cards of "Standard" as test data (EX1). Moreover, we perform the another experiment that uses cards of "Modern" excluding cards of "Standard" as training data (EX2). Additionally, to perform stemming and lemmatization, spaCy[3] is employed.

Table 5 shows results of machine learning methods. Table 6 shows coefficients of LR. Table 7 shows coefficients of SVR. Table 8 shows feature weights of RFR. Table 9 shows 1st layer weights of MLPR. Tables 6, 7, 8 and 9 show the weights when the experiment was EX1.

Table 5. Experiment results

	$R^2_{train}(EX1)$	$R^2_{test}(EX1)$	RMSE(EX1)	$R^2_{train}(EX2)$	$R^2_{test}(EX2)$	RMSE(EX2)
LR	0.623	−0.338	2.591	0.636	N/A	N/A
SVR	0.122	0.626	1.369	0.124	0.322	1.844
RFR	0.860	0.659	1.308	0.871	−0.486	2.730
MLPR	0.984	0.911	0.668	0.986	−0.615	2.846

Table 6. Coefficients of LR

+Feature	+Score	−Feature	−Score
SubType: Vehicle	1712343965918.104004	SubType: Nissa	−623935535717.686279
SubType: Siren	1152922421234.881104	SubType: Ral	−623971443287.413452
Ability: Dethrone	1101705623700.050781	SubType: Homunculus	−645279027490.468384
Ability: First Strike	835551864330.235962	SubType: Chandra	−699963898224.478638
Word: shade	752901770496.617188	Ability: Hero's Reward	−708231006199.925659
Ability: Horsemanship	730412983974.350098	SubType: Shade	−752901770495.298462
Ability: Cipher	672762008447.895020	Ability: Tempting Offer	−802603371523.855469
Ability: Gotcha!	655221674389.863525	Rare	−877399826722.916870
Word: fblthp	645279027498.411011	SubType: Sorin	−1008978707049.888550
Ability: Spectacle	632784900651.179443	Word: crew	−1383696140557.461670

---

[3] https://spacy.io/.

Table 7. Coefficients of SVR

+Feature	+Score	−Feature	−Score
Word: watery	3.000000	Word: inspire	−1.209778
Word: mountains	3.000000	Word: else	−1.209808
Word: b}{b}.	2.750818	Word: juggernaut	−1.264546
SubType: Scarecrow	2.593727	Word: 200	−1.270101
Type: Land	2.544081	Word: precombat	−1.362631
Type: Planeswalker	2.139428	Word: shell	−1.389525
Word: example	2.080129	Word: g}.	−1.413683
Word: liege	2.000000	Word: b}.	−1.452901
Word: p	2.000000	Word: c}.	−1.546689
Word: mutavault	2.000000	Word: sanctum	−1.824294

Table 8. Feature weights of RFR

Feature	Score	Feature	Score
allcost	0.072161	colorcost	0.012382
Word: onto	0.042872	power	0.010269
Word: void	0.026556	Word: upkeep	0.009899
Word: canopy	0.026228	Word: life	0.009548
cost	0.021094	SubType: Lhurgoyf	0.009514
Word: soul	0.019511	Word: a	0.008077
Word: verdant	0.017614	Word: u}.	0.007638
Word: add	0.014537	Word: chalice	0.007585
Word: permanent	0.012747	Word: battlefield	0.007580
SubType: Wizard	0.012502	Word: for	0.007506

Table 9. 1st layer weights of MLPR

+Feature	+Score	−Feature	−Score
Word: verdant	0.138728	Word: nontoken	−0.062205
Word: marsh	0.096701	Word: random	−0.062350
Word: canopy	0.074166	Word: multicolore	−0.064453
Word: void	0.063751	SubType: Insect	−0.065072
Word: island	0.059858	Word: choice	−0.068444
Word: among	0.049176	Word: w}{u}{b}{r}{g	−0.070325
Word: scarecrow	0.047986	Word: ally	−0.073121
Word: horizons	0.047964	SubType: Aura	−0.076007
Word: type	0.045957	Word: inspire	−0.078242
Word: every	0.045565	Word: bloom	−0.083934

5   Discussion

From the result of the EX1, LR method to classify a test set (cards in "Standard" format) was the worst despite to the experiment result with training set (cards in "Modern" format) was not such a low score. The reason for this result can presume from the coefficient features of the LR shows in Table 6. In the Table, the score of features shows how easy to determine the card price. Most of the positive features doesn't exist in the test set. Moreover, some "Ability" type features doesn't exist in the training sets too. Because there are too many "Ability" types in the MTG cards and we did not separate the "Ability" list for the data set ("Standard" or "Modern" format). Thus, the LR algorithm found those abilities do not have any cards in the data set, so those "Ability" features to determine the card price easily as "0," and those scores became very high. By the score of R^2 (Table 5) and features (Table 6), LR was over-fitted to training set. LR in EX2 could not predict almost of the prices of cards; therefore, test results of LR in Table 5 were N/A.

In the SVR result, it shows effective features in the positive and negative support vector. In positive results, some features like as "watery" or "mutavault" extracted from a text of single card each other. If the "Word" does not a comes from a single card, "Word" type feature in this result included very few (5 or less) amount of cards. When to detect the support vector, SVR looked for a specific vector which clear to judge as positive or negative.

Thus, SVR found some specific word feature which included in a single card which has a very high price (or low price). Otherwise than a "Word" type features, the score of "SubType: Scarecrow," "Type: Land" and "Type: Planeswalker" were high. "Scarecrow" comes from only two cards and those are little expensive in the price data. The "Land" is the card type which needs for every deck to play a game. The "Land" has a more considerable demand than other card types and mostly its price becomes high compared to other types in the same rarity. "Planeswalker" types only included in "Mythic Rare" rarity which is most high and eight times rare than next rarity "Rare." Those "Planeswalker" type has strong enough effect in a single card compared to, and their prices are expensive in most cases because of its effect and rarity. In contrast to the positive result, negative results doesn't include non "Word" type result. Maybe the types other than "Land," "Planeswalker" and "Scarecrow" has a various prices and it was difficult to judges as specific types has low price or it was simply too many cards in that type. In the result of EX2, SVR outperformed other machine learning methods. It supposes to card "types" was not regarded in other machine learning methods and this kind of cards are expensive in the test set.

Same to the LR result, RFR results of EX1 show how easy to determine a card price by the feature. From the result, playing cost of the card was the most useful feature. Probably, same to the SVR, most of the "Land" type cards has "0" cost, and some other "0" cost cards are expensive by the similar reason to the "Land". Thus, it considered that "all cost" was strongly be affected by the "0" cost. Some "Word" features considered as it comes from "Land" cards such as "Word: canopy" or "Word: verdant." Other features in the result included in many cards such as "Word: onto," "Word: soul," "Word add" and "Word: permanent" because those features are middle nodes in the tree and they classify cards into rough groups.

By the result, MLPR was the best and excellent score (over 0.91) at the test. It shows that most of the card price can predict from its features shown on the card text itself. More precisely, from the Table 9, words from card texts are efficient to detect expensive cards. Same to the RFR result, specific words in the Land type cards have a high score.

However, some words that mean the card refers to other cards such as "Word: among" or "Word: type" has a high score too. On the card text, "among" often shows as "... from among them". This expression is the kind of effect that a player can choose some other cards. "type" means to refer card types or subtypes which are same to the features in the experiment. Those results show that card price becomes expensive by the combination expressed in the text introduces a high price.

For the future work, to predict each card price precisely, it is necessary to detect and separate the strong combination effect and weak combination effect that effect to the card price.

6   Related Work

With regards price prediction using text, Bollen et al. showed that tweet moods are useful for forecasting the Dow Jones Industrial Average [1]. In their research, they used self-organizing fuzzy neural network for forecasting. As a result, they could predict rise and fall with an accuracy of more than 80%. Schumaker et al. proposed a machine learning approach for predicting stock prices using financial news article analysis [9]. Yamamoto et al. proposed a method to predict crypto-asset price using Twitter [10].

Concerning text mining for predicting prices, Koppel et al. proposed a method for classifying news stories of a company according to its apparent impact on the performance of the company's stock [5]. Low et al. proposed a semantic expectation-based knowledge extraction methodology (SEKE) for extracting causal relationships [6] that used WordNet [2] as a thesaurus for extracting terms representing movement concepts. Ito et al. proposed a neural network model for visualizing online financial textual data [3,4]. Additionally, their neural network model could acquire word sentiment and its category. Milea et al. predicted the MSCI euro index (upwards, downwards, or constant) based on fuzzy grammar fragments extracted from the report published by the European Central Bank [7]. Sakaji et al. proposed a method to automatically extract basis expressions that indicate economic trends from newspaper articles by using a statistical method [8].

Their research predicts indicators stock prices and crypto-asset prices, or analyzes sentiments and causal relations; it does not predict trading card prices.

7 Conclusion

We experimented to predict card prices using several machine learning methods. Moreover, we extracted feature weights from learned machine learning models and analyzed which features were important to predict card prices. As a result, MLPR outperformed other machine learning methods and features concerning "Land cards" were effect for card price prediction in EX1. Additionally, some feature weights in the 1st layer of MLPR suggest existences of card combinations having impact for card prices. On the other hand, in EX2, SVR outperformed other machine learning methods. As future work, we will try to develop a method for analyzing the combination of cards.

References

1. Bollen, J., Mao, H., Zeng, X.: Twitter mood predicts the stock market. J. Comput. Sci. **2**(1), 1–8 (2011)
2. Fellbaum, C.: WordNet: An Electronic Lexical Database. The MIT Press, Cambridge (1998)
3. Ito, T., Sakaji, H., Izumi, K., Tsubouchi, K., Yamashita, T.: GINN: gradient interpretable neural networks for visualizing financial texts. Int. J. Data Sci. Anal. (2018)
4. Ito, T., Sakaji, H., Tsubouchi, K., Izumi, K., Yamashita, T.: Text-visualizing neural network model: understanding online financial textual data. In: Pacific-Asia Conference on Knowledge Discovery and Data Mining (PAKDD) (2018)
5. Koppel, M., Shtrimberg, I.: Good news or bad news? Let the market decide, pp. 297–301. Springer, Dordrecht (2006)
6. Low, B.T., Chan, K., Choi, L.L., Chin, M.Y., Lay, S.L.: Semantic expectation-based causation knowledge extraction: a study on Hong Kong stock movement analysis. In: Pacific-Asia Conference on Knowledge Discovery and Data Mining (PAKDD), pp. 114–123 (2001)
7. Milea, V., Sharef, N.M., Almeida, R.J., Kaymak, U., Frasincar, F.: Prediction of the MSCI EURO index based on fuzzy grammar fragments extracted from European central bank statements. In: 2010 International Conference of Soft Computing and Pattern Recognition, pp. 231–236 (2010)
8. Sakaji, H., Sakai, H., Masuyama, S.: Automatic extraction of basis expressions that indicate economic trends. In: Pacific-Asia Conference on Knowledge Discovery and Data Mining (PAKDD), pp. 977–984 (2008)
9. Schumaker, R.P., Chen, H.: Textual analysis of stock market prediction using breaking financial news: the AZFin text system. ACM Trans. Inf. Syst. **27**(2), 12:1–12:19 (2009)
10. Yamamoto, H., Sakaji, H., Matsushima, H., Yamashita, Y., Osawa, K., Izumi, K., Shimada, T.: Forecasting crypto-asset price using influencer tweets. In: International Conference on Advanced Information Networking and Applications, pp. 940–951. Springer (2019)

Kawaii in Tweets: What Emotions Does the Word Describe in Social Media?

Jun Iio[✉]

Faculty of Global Informatics, Chuo University,
1-18 Ichigaya-Tamachi, Shinjuku-ku, Tokyo 162-8478, Japan
iiojun@tamacc.chuo-u.ac.jp

Abstract. The Japanese word kawaii is a particular word that represents a kind of emotion. It is similar to "pretty" or "cute" in English. However, there is some difference between kawaii and these English words. This study reveals how the word kawaii is used in message exchanges over social networking services (SNS) and in what context it is used. The experiment compared two sets of tweets: one set with the word kawaii and the other without it. Also, it compared several sets of words within five predetermined emotional categories. The experiment revealed that there was a marked difference between the two groups, especially in the context of expressing the emotions of "enjoy, " "happy," and "like."

1 Introduction

In Japan, the word kawaii is often used by all generations. Young girls, especially, tend to use it. While it is relatively difficult to directly translate the word, its meaning is similar to "pretty" or "cute" in English. However, there is some difference between kawaii and these English words. The Japanese word is used in various contexts, and its different meanings depend on them.

There have been many studies on the concept of kawaii.

Nittono *et al.* [1] considered the kawaii-related expressions as favorite because they evoked positive feelings. They believed that kawaii had more power. Therefore, they conducted some experiments to confirm the effect of viewing cute pictures on subsequent task performance. According to their experiments, the performance indexes of the subjects who had viewed cute images increased. On the contrary, the performance indexes of those who had viewed less attractive pictures did not increase. They concluded that the cuteness-triggered positive emotion gave the subjects a narrowed attentional focus and was tied to approach motivation. He also proposed a two-layer model of kawaii based on psychological and behavioral science research to represent the word as an emotion as well as a social value [2].

Laohakangvalvit *et al.* [3] considered that kawaii is a positive adjective that carries positive meanings as "cute" or "lovable." They tried to disclose the relationship between the word and the human motion based on biological signals by adopting the eye-tracking system. Their results clarified the relationship between

© Springer Nature Switzerland AG 2020
L. Barolli et al. (Eds.): NBiS-2019, AISC 1036, pp. 715–721, 2020.
https://doi.org/10.1007/978-3-030-29029-0_71

kawaii-related feelings and eye movements represented in the simple two new indexes. Their study group also tried to design industrial products around the kawaii concept [4,5].

Sugiyama [6] examined the function of emoji in the context of kawaii-related feelings. Her investigation revealed that emoji has two functions in the mobile communication space. It allows Japanese teens to manage the communication climate as well as construct and express themselves. In essence, it means kawaii is used not only to represent "pretty" or "cute" but also as a buffer in mobile interactions.

Several works of literature characterizing Japanese pop culture mention that kawaii plays an important part in the idea of Cool Japan [7–10]. Consequently, the word kawaii is unavoidable when we have to study Japanese pop culture and communication, especially among younger generations.

Communication using Twitter is very widespread in Japan, as well. Blank [11] concluded that using Twitter is suitable for social science, even though there is a digital divide among British Twitter users. There is a similar divide in Japan, but Japanese Twitter users are younger than users of other social network services (SNS). Therefore, to investigate how kawaii is used in SNS, it seemed imperative to deeply mine the data of the tweets.

To clarify how the word kawaii is used in social media and in what context, this study conducted an in-depth investigation on a bunch of tweets from Japanese Twitter users.

2   Trending Topic-Map Graph

Since January 1, 2019, our study team has operated a system for collecting Twitter's trending topics. The primary aim of the system named "TWtrends" is to visualize the structure of trending topics and enable the analysts to grab the trends at a glance [12].

Figure 1 illustrates the screenshot of TWtrends that shows Twitter's trending topics collected on January 4, 2019. On the upper half of the screen, there are approximately 300 trending topics collected by Twitter's trends application programming interface (API). Each phrase button is clickable. When a user clicks a link, the page presents a co-occurrence network graph based on a set of tweets related to the trending topic (Fig. 2).

The bottom half of Fig. 1 represents a topic-map network graph. It is calculated based on the similarity between each trending topic. Each topic k has a vector of words frequency v_k within the whole word-space. The word-space is a hyper-dimensional space because all the words that appear in the co-occurrence network graphs for all the trending topics collected on that day are gathered in one place.

The clusters in the topic-map network are calculated by cosine similarity— that is, the similarity between two nodes k_1 and K_2 is defined by the cosine value of the angle between the two vectors of v_{k1} and v_{k2}. If the value exceeds a threshold, then the two topics are connected to each other with an edge. This time, I set the threshold value to 0.75.

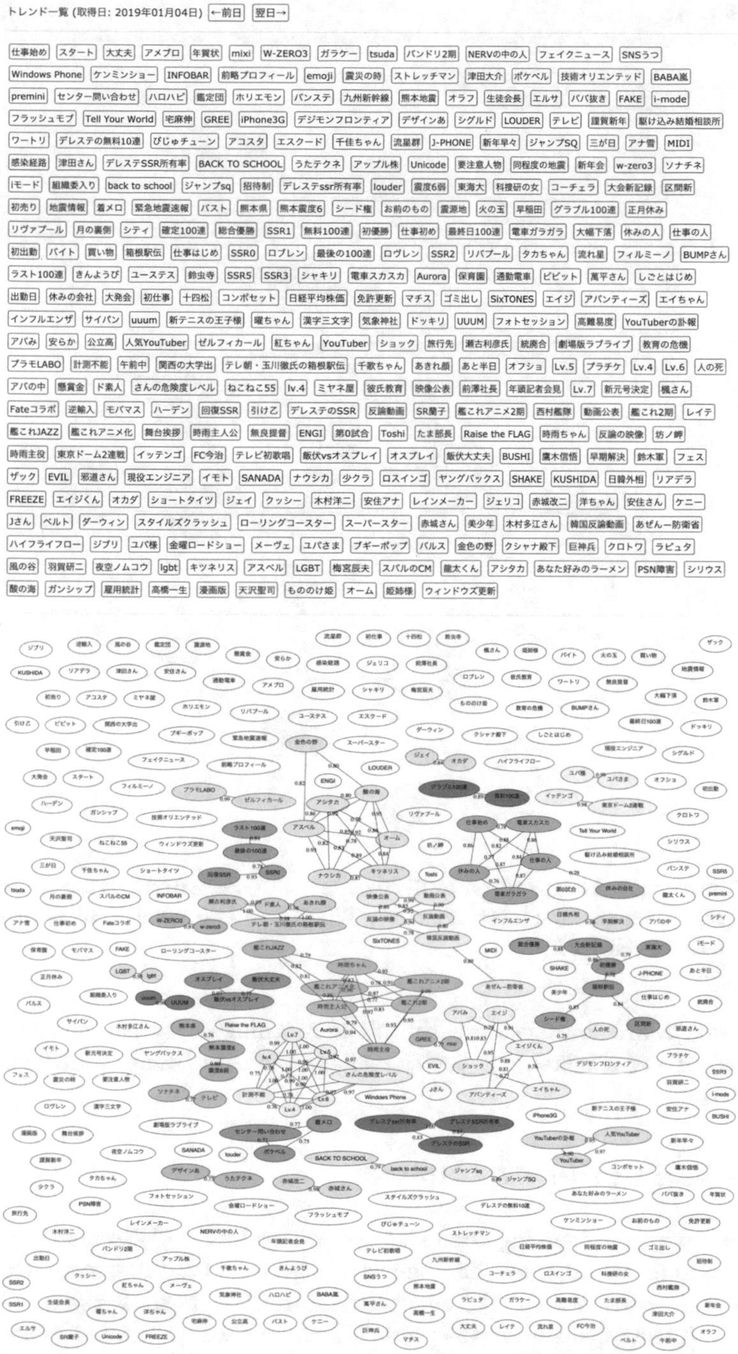

Fig. 1. The screenshot of the system "TWtrends": The upper half of this screen displays the list of trending topics. The lower half shows the topic-map network graph that consists of several clusters of nodes regarding similar issues.

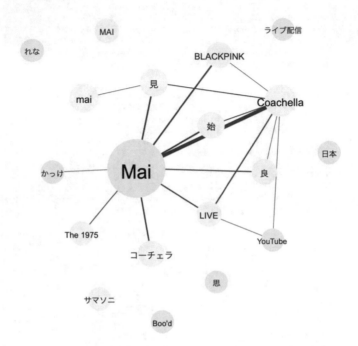

Fig. 2. The screenshot of the system "TWtrends": This page displays the co-occurrence network graph created from a set of tweets. These tweets were also collected by Twitter's standard search API.

Topic-maps are constructed early in the morning every day because it must be charted as per the trending topics collected within twenty-four hours of the previous day.

3   Method of the Experiment

Based on Twitter's trending topics collecting system "TWtrends," an experiment was conducted to reveal how the word kawaii is used in tweets. During the experiment, Twitter search API collected several tweets to investigate the context in which the word kawaii was used. The tweets were collected in accordance with the trending topics stored in the TWtrends' database in January 2019. Due to the restriction of the API search, a maximum of a hundred tweets were picked for every trending topic. After collecting the tweets, I separated them into two groups: one with the word kawaii, and the other without it.

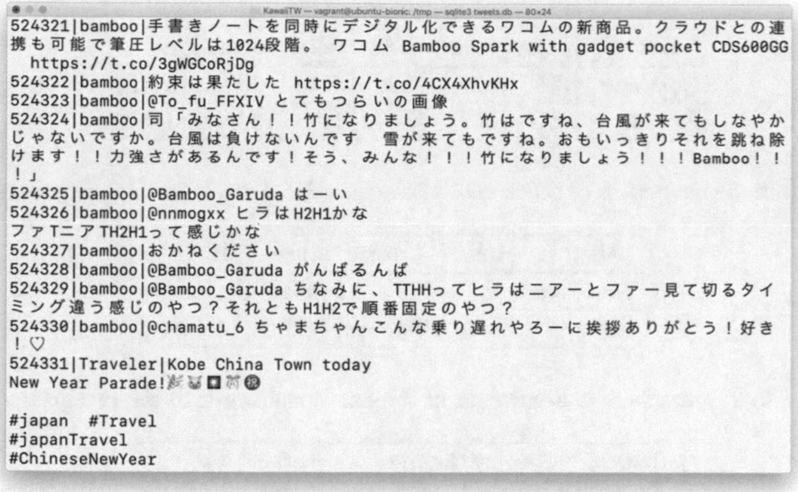

Fig. 3. The screenshot of the database which stores the collected tweets.

Subsequently, some emotional words were counted up. Uetani and Hishida [13] proposed that there are several word-groups to represent the five emotional categories: "enjoy," "happy," "like," "relief," and "surprise." Keeping in line with their classification, these emotional words were counted and the tendency of word frequency was compared between the two groups.

Figure 3 shows the screenshot indicating the content of the database. Here you can see the id number, trending topic, and the tweet stored in the database, respectively.

4   Results and Discussion

From the database implemented in the TWtrends application, I collected the tweets on the trending topics every day of January 2019. As a result, 524,430 tweets were collected and stored in the database. After that, the collected tweets are divided into two groups. The control group consisted of general words, while the other group hosted tweets with the word kawaii. Naturally, the number of tweets in the control group was larger than the other group. Hence, to equalize the number of tweets in the two sets, some tweets in the control group were decimated. The control group was labeled as NotKawaii and the last group was labeled as Kawaii.

Table 1 illustrates the number of words counted in each emotional category within the two groups, NotKawaii and Kawaii. This table can be divided into two tables, Tables 2 and 3.

Table 1. The numbers of occurrence of the emotional words within two groups.

Label	enjoy	happy	like	relief	surprise	all
NotKawaii	399	529	823	157	115	141698
Kawaii	590	867	1632	147	143	145030

Table 2. The numbers of occurrence of the emotional words in the two groups (I).

Label	relief	surprise	other	all
NotKawaii	157	115	141426	141698
Kawaii	147	143	144740	145030

Table 3. The numbers of occurrence of the emotional words in the two groups (II).

Label	enjoy	happy	like	other	all
NotKawaii	399	529	823	139947	141698
Kawaii	590	867	1632	141941	145030

Pearson's χ^2 test was applied to these two results. For the case shown in Table 2, $\chi^2 = 3.0261$, the degree of freedom is 6, and the $p = 0.8056$. These results cannot explain the small differences between the two distributions of the word frequency for "relief" and "surprise."

On the other hand, for the case shown in Table 2, $\chi^2 = 360.75$, the degree of freedom is 8, and $p < 2.2 \times 10^{-16}$. It means that there is a significant difference between the two distributions of the word frequency for "enjoy," "happy," and "like."

These results imply that the use of kawaii is notably observed in the representations of the emotions such as "enjoy," "happy," and "like." Therefore, at least in the context of Twitter interactions, the study disclosed that the word kawaii was not only used to represent "cute" or "pretty," but also used to express particular emotions such as "enjoy," "happy," and "like."

5 Conclusions

In a contemporary cyber-world, the word kawaii has complicated directions for use. It is notably the case when the younger generations use the word. To clarify how the word kawaii is used in message exchanges over SNS and in what contexts it is used, I conducted a simple experiment, comparing the two sets of tweets: one with the word kawaii and the other without it. The comparison of the two groups indicated that there were some differences between them. Especially in the context of expressing the emotions of "enjoy," 'happy,' and "like," the comparison showed a significant difference between Kawaii and NotKawaii.

References

1. Nittono, H., Fukushima, M., Yano, A., Moriya, H.: The power of Kawaii: viewing cute images promotes a careful behavior and narrows attentional focus. PLoS ONE **7**(9), e46362 (2012). https://doi.org/10.1371/journal.pone.0046362
2. Nittono, H.: The two-layer model of "kawaii": a behavioural science framework for understanding kawaii and cuteness. East Asian J. Popul. Cult. **2**(1), 79–95 (2016)
3. Laohakangvalvit, T., Iida, I., Charoenpit, S., Ohkura, M.: A study of Kawaii feeling using eye tracking. Int. J. Affect. Eng. **16**(3), 183–189 (2017). https://doi.org/10.5057/ijae.IJAE-D-16-00016
4. Laohakangvalvit, T., Achalakul, T., Ohkura, M.: A proposal of model of kawaii feelings for spoon designs. In: Kurosu, M. (ed.) Human-Computer Interaction. User Interface Design, Development and Multimodality, HCI 2017. Lecture Notes in Computer Science, vol. 10271. Springer, Cham (2017)
5. Laohakangvalvit, T., Achalakul, T., Ohkura, M.: Evaluation of attributes of cosmetic bottles using model of kawaii feelings and eye movements. In: Bagnara, S., Tartaglia, R., Albolino, S., Alexander, T., Fujita Y. (eds.) Proceedings of the 20th Congress of the International Ergonomics Association (IEA 2018) – Volume VII: Ergonomics in Design, Design for All, Activity Theories for Work Analysis and Design, Affective Design. Advances in Intelligent Systems and Computing, vol. 824, pp. 2108–2118. Springer (2018). https://doi.org/10.1007/978-3-319-96071-5_220
6. Sugiyama, S.: *Kawaii meiru* and *Mayonaka neko*: mobile emoji for relationship maintenance and aesthetic expression among Japanese teens. First Monday **20**(10) (2015)
7. Schules, D.: Kawaii Japan: defining JRPGs through the Cultural Media Mix. Kinephanos J. Media Stud. Pop Cult. **5**, 53–76 (2015). Geemu and media mix: theoretical approaches to Japanese video games
8. Lewis, B.: "Cool Japan" and the Commodification of Cute: Selling Japanese National Identity and International Image (2015). https://www.academia.edu/10476256/. Accessed 1 June 2019
9. Pellitteri, M.: Kawaii aesthetics from Japan to Europe: theory of the Japanese "Cute" and transcultural adoption of its styles in Italian and French comics production and commodified culture goods. Arts **7**(24) (2018). https://doi.org/10.3390/arts7030024
10. Jackson, K.: Cool Japan: case studies from Japan's cultural and creative industries. Asia Pac. Bus. Rev. **24**(5), 740–744 (2018). https://doi.org/10.1080/13602381.2018.1485895
11. Blank, G.: The digital divide among twitter users and its implications for social research. Soc. Sci. Comput. Rev. **35**(6), 679–697 (2016). https://doi.org/10.1177/0894439316671698
12. Iio, J., Stevie, P., Aoki, Y., Nakamura, E., Kim, S., Lee, T.: Visualization of Twitter Trends using a co-occurrence network. In: The 12th IEEE Pacific Visualization Symposium (PacificVis2019) Poster Proceedings, Bangkok, Thailand, pp. 321–322 (2019)
13. Uetani, R., Hishida, T.: A study of an emotional words dictionary for a review analysis system, In: Proceedings of Multimedia, Distributed, Cooperative, and Mobile Symposium 2016 (DICOMO 2016), pp. 456–460 (2016). (in Japanese)

Development of Goals Achievement Sharing System Based on Approach Goals in Positive Psychology

Yoshihiro Kawano[(✉)]

Department of Informatics, Tokyo University of Information Sciences,
Chiba, Japan
ykawano@rsch.tuis.ac.jp

Abstract. Recently, GNH (Gross National Happiness) has been emphasized as a policy goal of the national and local governments. In positive psychology, people feel the happiness in the process towards the goal achievement. Additionally, the goal is classified as approach and avoidance goal. And to achieve the goal, determination of approach goal is effective. In this study, we developed a goals achievement sharing system based on approach goals in positive psychology to investigate the change in the goal achievement rate and happiness scale in order to share the type of goal.

1 Introduction

Recently, GNH (Gross National Happiness) has been emphasized as a policy goal of the national and local governments [1]. The GNH is a concept that aims to "national happiness" in consideration of the traditional society and culture and environment than economic growth. In France, The Commission on the Measurement of Economic Performance and Social Progress (CMEPSP) was launched. This commission evaluates quality of life in order to advance economic and social and is studying the sustainable development theme [2]. In Japan, the measurement of social indicators, including the level of happiness in the local government has been carried out [3]. Thus, people pursuits happiness, and happiness is a mainly popular research theme in developed countries.

In this study, we adopt a definition of "happiness" in the positive psychology to separate from culture, religion, and philosophy. Positive psychology is the study founded by Martin E.P. Seligman who is American Psychological Association chairman [4]. Positive psychology focuses on the state in which the person is directed toward certain to be the original, is a psychology to attempt to scientifically verify and experiment with various elements. In positive psychology, happiness is defined as "a state of mind" that varies depending on whether the world who live their own and how to feel. In positive psychology, people feel the happiness in the process towards the goal achievement. Additionally, the goal is classified as approach and avoidance goal. Approach goal is a goal to pursue their dreams and ideals. On the other hand, the avoidance goal is a goal to avoid undesirable results so as not to feel stress. And to achieve the goal, determination of approach goal is effective. In other studies, change of

© Springer Nature Switzerland AG 2020
L. Barolli et al. (Eds.): NBiS-2019, AISC 1036, pp. 722–730, 2020.
https://doi.org/10.1007/978-3-030-29029-0_72

goal achievement rate and approach/avoidance goal by the subject experiment was investigated [5]. Because it was a questionnaire survey by paper media, the subjects were not shared approach/avoidance goal with others.

In this study, we develop a goals achievement sharing system based on approach goals in positive psychology to investigate the change in the goal achievement rate and happiness scale in order to share the type of goal. This system is based on positive psychology and designed to achieve goal for user by sharing the approach goal and the avoidance goal. On this system, we investigate change of goal achievement rate for each user. Concretely, we investigate the number of posts and the change of happiness score, grit score, and goal achievement rate before and after the experiment.

2 Proposed System

2.1 Purpose of System

The purpose of the system is to share the approach goal and the avoidance goal with other users. And in this research, we investigate the change of goal achievement rate of each user. The system was designed like a social media that the approach goal and the avoidance goal of other users are presented to all users. And we estimate the relationship of the amount and the type of the goal, the goal achievement rate, and the level of happiness.

2.2 Functions

Functions of this system are stated below.

- Function 1. Login
  This function had been developed using Ruby library. Shown as Fig. 1, the user logs in to this system by using Twitter account. Post of goals, achievement, and others are restricted until the user finished registration their profile. The profile has name, user's photo, and goals (Fig. 2).
- Function 2. Post goals
  Figure 3 is the screen shot the function for posting goals. The user posts the goals about oneself with tagging the type of goal, that is the approach goal and the avoidance goal. For example, "I will finish report until tomorrow." The goals are presented to other users.
- Function 3. Post actual achievements
  Figure 4 is the screen shot of posting actual achievements. In this function, the user reports output, for example product, picture, movie, music, software, artistic creature, blog post, and article. The achievement has the title, description, and image. This achievement and the original goal are presented to the other users.
- Function 4. My page
  Figure 5 is the screen shot of my page. In this page, the user can check one's goals and the ratio of the approach goals and the avoidance goals briefly.

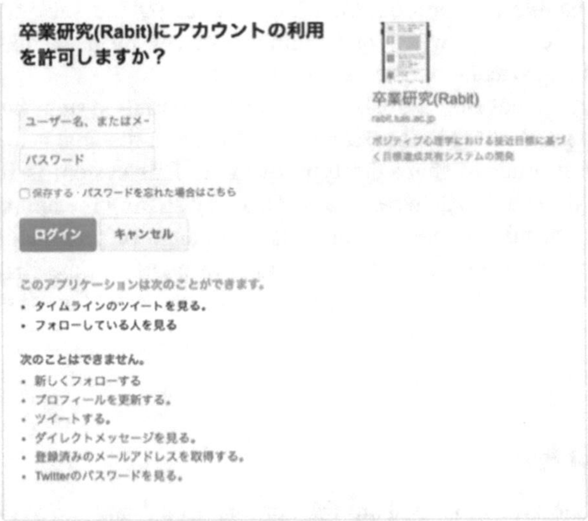

Fig. 1. Screen shot of login screen.

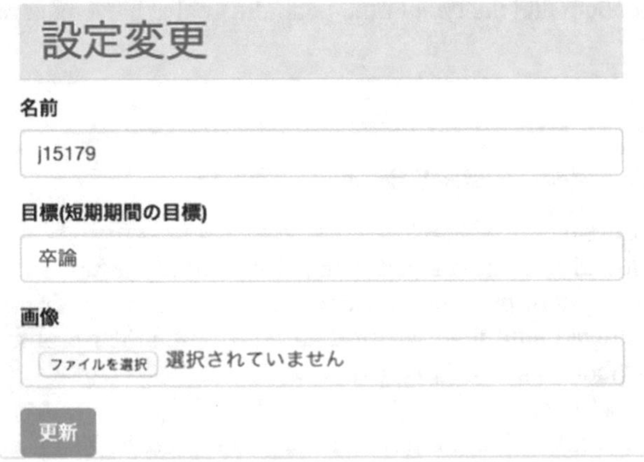

Fig. 2. Screen shot of profile registration screen.

– Function 5. Analysis of trend

Figure 6 is the screen shot of analysis of trend. In this function, the amount number of goals of all users is presented. On the other hand, the number of goals of one user also can be presented. With this function the user can analyze the trend of all goals and one's tendency.

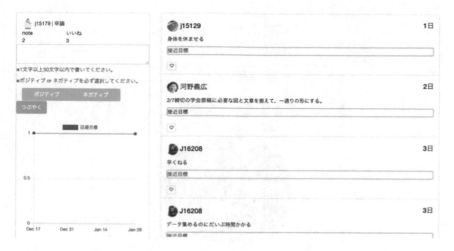

Fig. 3. Screen shot of posting goals.

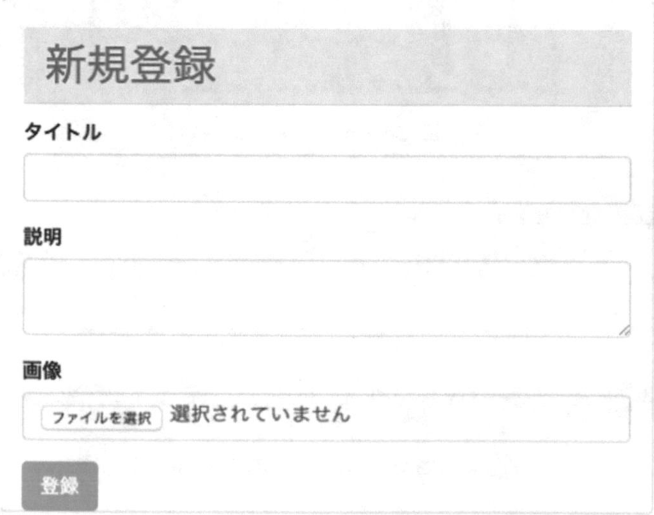

新規登録

タイトル

説明

画像

ファイルを選択　選択されていません

登録

Fig. 4. Screen shot of posting actual achievement.

3   Subject Experiment

3.1   Purpose of the Experiment

The purpose of our research is to investigate the change of goal achievement rate of each user with sharing the approach goal and the avoidance goal with other users. By developing the system sharing the approach goal and the avoidance goal of other users, we estimate the relationship of the amount and the type of the goal, the achievement rate, and the level of happiness.

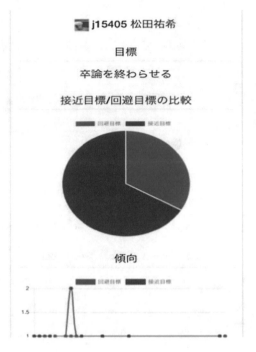

Fig. 5. Screen shot of my page.

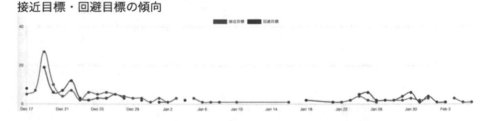

Fig. 6. Screen shot of analysis of trend.

3.2 Overview

The experiment period and evaluation items are as below. Users are divided to two groups, in order to equalize the average of "happiness scale" which Sonja Lyubomirsky suggested. For these groups, the experiment period is divided three period. That are the period when only the approach goal is presented, the period when only the avoidance goal is presented, the period when the approach goal and the avoidance goal are presented to users. During the experiment, on every week, we monitored "grit score [6]" which Angela Duckworth suggested and "happiness scale [5]". From this result, we estimated correlation between the number of posting goals and grit score, and between user and happiness scale. The end of the all experiment, we had a question-naire to estimate the goal achievement rate.

- Subject: 22 university students (3rd and 4th grade)
- Period of the experiment: December 17, 2018 – January 10, 2019
- Process: Users used the system, and they recorded the happiness scale and grit score on every week.

The questions of questionnaire are as below:

1. Could you achieve your goal?
2. How often did you post your goals?
3. What is the reason you did not post.
4. Was your post influenced by the others?
5. By which goals were you influenced, the approach goal or the avoidance goal?
6. By who were you influenced?

3.3   Result and Discussions

Table 1 is the system log. From this table, we confirmed that 20% of 22 users used the system continually, after the registration to the system. The number of "Good" was 11% of all posts. This score was less than we supposed. We consider the reason of this result is that users could not view all of posts which they marked "Good."

Table 1. System log.

The number of	Score
All users	22
Users accessed to the system	7
Posts	188
Posts marked "Good"	22
Posts of actual achievement	7

Figure 7 shows the ratio of the approach goal and the avoidance goal which users posted. From this figure, we found that the post of the avoidance goal had been marked "Good" more than the approach goal. We considered that users had interest to the avoidance goal more than the approach goal.

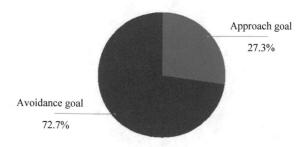

Approach goal
27.3%

Avoidance goal
72.7%

Fig. 7. The ratio of goals which was marked "Good".

– Assessment 1. By sharing the approach goal with others, could user achieve one's goal?

Figure 8 shows the ratio of the posts which are the approach goal and the avoidance goal from the start to the end of the experiment. In this figure, the result is shown for each period, that is the period 1 when only the approach goal is presented, the period 2 when only the avoidance goal is presented, the period 3 when the approach goal and the avoidance goal are presented to users. The ratio of the approach goal increased from the period 1 to the period 3. From the result we could consider that the user was influenced by other user's posts during the period 1, and posted the approach goal during next week that is the period 2.

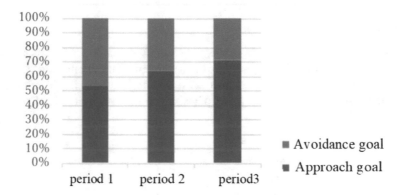

Fig. 8. The ratio of the approach goal and the avoidance goal.

Figure 9 is the result of the questionnaire after the period of experiment. From this figure, we confirmed that 57.1% of users could have achieved their goals. Additionally, many users answered that they were influenced the post of the approach goal (Fig. 10). Hence, we could consider that the user could have achieved their goals by sharing the approach goal.

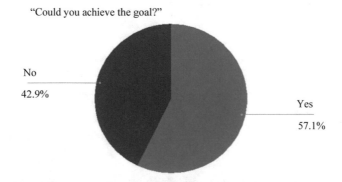

Fig. 9. The ratio of the answer to "Could you achieve your goal?"

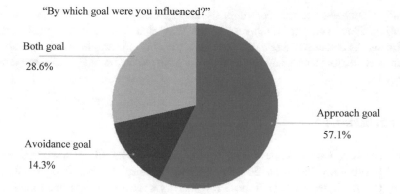

"By which goal were you influenced?"

Both goal

28.6%

Avoidance goal

14.3%

Approach goal

57.1%

Fig. 10. The ratio of the answer to "By which goal were you influenced?"

– Assessment 2. Correlation between the number of posts, happiness scale, and grit score

Table 2 shows correlation between the number of posts and happiness scale, and between the number of posts and grit score. We used results of users who posted more than 10 posts to estimate the correlation. The average of the number of posts was 22.43, and the correlation with happiness scale was 0.9. The number of posts and happiness scale are strongly correlated.

Table 2. Correlation between the number of posts, happiness scale, and grit score.

	The number of posts	Grit score	Happiness scale
The number of posts		0.90	0.88
Grit score	0.90		
Happiness scale	0.88		

On the other hand, the correlation between the number of posts and grit score is 0.8. This means strong correlation. Hence, we could confirm that the user who posts more has strong tendency to get score high happiness scale and grit score.

From the result of the experiment, there are correlation between the number of posts and happiness scale, and between the number of posts and grit score. The average number of posts was 22.4, and two subjects had posted more than the average number. 5 subjects had scored over the average of grit score. In the questionnaire after the experiment, many users answered they could have achieved their goals. However, posts of the actual achievement were only 7. We considered the reason of this was the user had changed their goals during the experiment. We estimated correlation between the number of posts and happiness scale, and there was strong correlation. However, the number of users was low, and we need additional experiment. We also estimated correlation between the number of posts and grit score, and there was also strong correlation. However, the number of users was also low, and we need additional experiment. As discussed above, we could not get enough sample to estimate about the happiness scale, grit score, and the number of posts, therefore we need more continual experiment.

4 Conclusions

In this research, Goals Achievement Sharing System based on approach goals in positive psychology has been developed. This system was based on positive psychology and designed to achieve goal for user by sharing the approach goal and the avoidance goal. On this system, we investigated change of goal achievement rate for each user. Concretely, we investigated the number of posts and the change of happiness score, grit score, and goal achievement rate before and after the experiment.

22 subjects participated in the experiment, however only 7 subjects posted their goals more than 10. This was lower than we expected, therefore we needed to consider more about structure for encouraging users to post their goals, at UI design and introduction of the system for subjects. Especially, we needed to interview subjects who did not use the system.

In the questionnaire, 4 subjects answered that they could have achieved their goals. And 7 subjects answered that they had been influenced by the other subject's post when they posted their own goals. The many of goals that was influenced by the other post were the approach goal. Being several subjects influenced the others, we can consider that the type of goals changed affected by goals shared in the system.

The correlation between the number of post and grit score, and happiness scale was strong. However, the number of effectual posts was low, and we could not confirm significant difference.

In future works, we need to make effectual instructions for user, and design user-friendly UI, for example application for smartphone. Additionally, in order to build multiple experiment environment, we need to develop API server to provide functions of this system.

References

1. Harata, S.: Happiness-Controlled Society, Shizuoka University, Departmental Bulletin Paper, vol. 20, pp. 86–90 (2015). (in Japanese)
2. Yaguchi, K.: "Sustainable development" and sustainability indices. National Diet Library, Japan, Reference **60**(4), 3–27 (2010). (in Japanese)
3. Ishida, A., Ichikawa, K.: Efforts of Local Governments on Social Indicators, ESRI Research Note No. 30, Economic and Social Research Institute (2017). (in Japanese)
4. Japan Positive Psychology Association: What is Positive Psychology. https://www.jppanetwork.org/what-is-positivepsychology. Accessed 31 May 2019. (in Japanese)
5. Lyubomirsky, S.: The How of Happiness: A Scientific Approach to Getting the Life You Want. Penguin Press, New York (2008)
6. Duckworth, A.: Grit: The Power of Passion and Perseverance. Scribner, New York (2016)

A Real-Time Programming Battle Web Application by Using WebRTC

Ryoya Fukutani[1]([✉]), Shusuke Okamoto[1], Shinji Sakamoto[1],
and Masaki Kohana[2]

[1] Seikei University, Tokyo, Japan
dm196209@cc.seikei.ac.jp, okam@st.seikei.ac.jp, shinji.sakamoto@ieee.org
[2] Chuo University, Tokyo, Japan
kohana@tamacc.chuo-u.ac.jp

Abstract. We are developing a web system for learning Python programming. The multiple users can join the system simultaneously through programming battles. This system consists of two parts which are web browsers and a web server. The browser side program written in JavaScript uses an idea from Robocode, which is a open sourced programming learning tool. A user of Robocode plays a battle by writing robot programs in Java. On the other hand, the user of our system writes a program in Python, and the program is converted to a JavaScript of robot program. The conversion is done on the web sever as soon as the user changes his or her code in real-time. In our previous work, we used WebSocket to implement a client-server model that is one-on-one fight. However, since the CPU usage ratio of both browsers and server were high, it was difficult to increase the number of players. Therefore, in this implementation, we use WebRTC to build an application on communicating among browsers that supports more than two players efficiently. We will compare CPU usage ratio on WebSocket model over WebRTC model. Moreover, we will consider which method is suitable for our system.

1 Introduction

We are developing a web system for learning Python programming by using WebRTC. This system uses an idea from Robocode and the user can learn Python programming by playing robot battles. It is important for IT skills learners to learn programming. The web system we are developing has a real-time and interactive property. We focus on multiplayer programming battle games on web browsers.

Robocode is a programming game, where the goal is to develop a program of a robot battle tank to defeat other tanks in Java or .NET [1]. A user writes a robot program in advance and runs the simulator to let the multiple robots battle. It was originally started by Mathew A. Nelson, and then IBM promoted it as a fun way to get started with learning programs in Java. It is also used as a machine learning subject such as automatic generation of a program by genetic programming [2].

© Springer Nature Switzerland AG 2020
L. Barolli et al. (Eds.): NBiS-2019, AISC 1036, pp. 731–737, 2020.
https://doi.org/10.1007/978-3-030-29029-0_73

Figure 1 is a screenshot of Robocode. The name of robots and remaining energy levels are listed on the right side pane, and battle simulation is displayed on the left side window. The user selects robots and makes them fight in the battle field. At the beginning of a battle, all robot have the same energy level. They shoot a bullet by using that energy. When a bullet hits a robot, the energy level is decreased. When it reaches zero, the robot is defeated and is removed from the field.

Our system uses an idea of Robocode, however it is little different. The user struggle with programming quiz called as a stage mission in addition to robot battles. And not only just watching a battle simulation but also rewriting own program among the battle. When the user rewrites the program, the behavior of the robot is changed instantly and the stage or battle continues without restarting.

In order to implement our system, it is important to have a quick reloading of robot code when changing. Furthermore, it is also significant that the game users see the same result of the winner and the small difference of drawing contents among their browsers. We use WebRTC to build an application on communicating among browsers under these conditions. We will compare CPU usage ratio on WebSocket model over WebRTC model. Moreover, we will consider which method is suitable for our system.

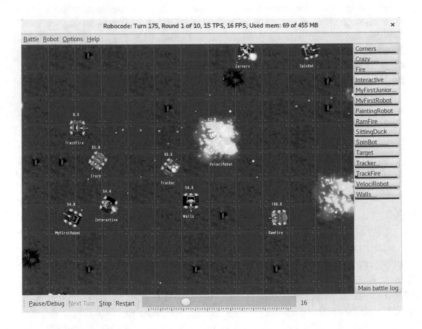

Fig. 1. A battle screen of Robocode

2   Literature Survey

This section introduces some studies related to our system such as web technologies and applications to learn programming.

In our previous work, we implemented a web application by using WebSocket to communicate between a web server and browsers. The technologies we used was based on an implementation of Robocode by Lee [3], where the run-time code is written by LiveScript, HTML and JavaScript. Each robot on the battle field is controlled by Web Worker. The user writes a robot program in LiveScript and translates it into JavaScript by themself since the system has no user interface for translating code. We implemented our application based on this run-time code and added functionalities of translating a user code on the server and sharing battle screen image for two users.

Kohana et al. proposed an information sharing method for a web-based virtual world by WebRTC [4]. They build a ring-form topology with the web browsers to communicate for the information of a virtual world that is not divided into blocks. Each web browser can get only the necessary information to update own status although they share a huge area of virtual world.

CodeCombat is a personal web programming game [5]. It is used for learning programming, where the user writes and rewrites their program to complete game stages. For example, a stage consists of a corridor, a treasure and an avatar. The user writes a program to move their avatar to treasure and presses RUN-button to see the result. Several programming languages such as Python, JavaScript, CoffeeScript and so on can be used to learn.

3   Key Technologies

This section describes four key features to build our system.

WebRTC (Web Real-Time Communication) [6] is a real-time communication protocol among web browsers and mobile applications. It can be used to build a peer-to-peer network of browsers. WebRTC enables us to build a video chat application, file sharing application and so on without any server communications. In our system, we use WebRTC to communicate robot information such as locations, angles and energy levels among browsers. The browsers can advance simulation steps without any server communications.

Web Worker is a way to run multithreads on a web browser [7]. It is a JavaScript functionality. Using Web Worker, it can run calculations in the background without disturbing user interfaces. We use a Web Worker to calculate a robot behavior.

WebSocket is a way to communicate each other between clients and a server [8]. It is a simultaneous two-way communication protocol over TCP connection. The protocol enables interaction between a web browser and a web server with lower overhead. We use it to transfer a robot program when a user rewrites their program.

Transcrypt is a one of programming languages for web browsers. It has the same syntax of Python programming language so that it can be used as a compiler from Python to JavaScript to run a code on a web browser. It also offers seamless access to any JavaScript library. The source code is opened at GitHub [9].

4   System Overview

This section describes an overview of our system. It consists of a server and two or more web browsers as clients. The role of server is deciding a connection peer for a newly connected browser and compiling robot codes. One of the browsers executes a robot simulation and shares the simulation status among other browsers.

Figure 2 is a screenshot of a web browser using our system. The canvas on the left side is a battle field and the text on the right side is a program of the robot. There are two robots, one is operated by the user of this browser and the other is a stage enemy that is prepared by the system or controlled by the other user. As the user changes their robot code, the behavior of the robot is changed immediately.

Figure 3 is an example image of a mission stage for two users. There are two tanks (tankA and tankB) at the start field and they try to get to the goal field. There is a movable gate between the start and the goal. To complete this stage, users need to cooperate with their programs. For example, while tankA pushes the gate to open from the right side, tankB moves to the left side. After that, while tankB pushes the gate to open from the left side, tankA can move to the left side and the mission is completed.

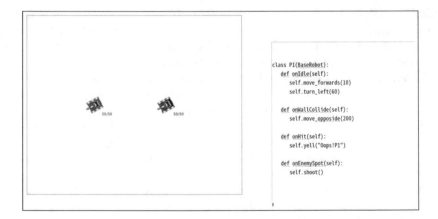

Fig. 2. Screenshot of robot battle

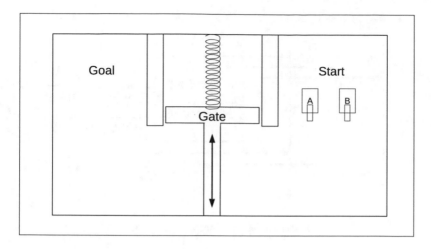

Fig. 3. An image of mission stage

The system flow goes as follows; when a new user joins the system, the browser connects the server to get a HTML file and then connects the peer-to-peer web browser network by using WebRTC. The server determines which browser should connect with which browser. Figure 4 shows an example of our system network that has a server and a tree-form topology with six browsers. The root node of our network called super browser advances a simulation step of the robot battle and sends the result of simulation. When the server receives a robot program written in Python that is modified by a user, it translates the program into a JavaScript code and sends it to the super browser.

Figure 5 shows a time chart of our system. The super browser regularly advances a simulation step, and sends the result to browser1 and browser2. And then the browser1 relays it to browser3 and browser4 and so on. Each received browser and the super browser show a robot battle scene with the result. A user watching the battle scene may want to decide to modified their program timely

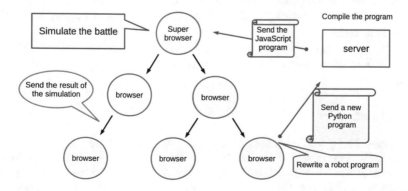

Fig. 4. Network of our system

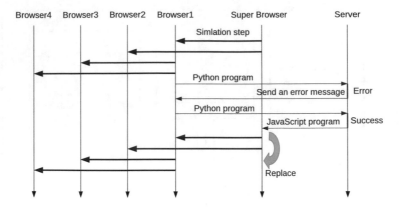

Fig. 5. A time chart of our system

since their robot is about to fail. When a user rewrites one line of their program and pushes the Enter key, the browser sends the whole program to the server. If the server fails to translate the program due to errors, it sends an error message back to the browser. On the other hand, if the server translates it successfully, it sends the robot program of JavaScript to the super browser. Once the super browser receives a program, it keeps the program in a queue and it replaces old programs with the received programs just before advancing a new simulation step. In this manner, all user can rewrite their program any time they want and enjoy a real-time robot battle.

5   Conclusion

We are developing a web system for learning Python programming. This system uses an idea from Robocode and the user can learn Python programming by playing robot battles. Using WebRTC, we have built an application on communicating among browsers which connect in a tree-form topology. A user of our system is not only just watching a battle simulation but also rewriting own program timely among the battle. When the user rewrites the program, the behavior of the robot is changed instantly and the stage or battle continues without restarting.

We will compare CPU usage ratio on WebSocket model over WebRTC model. Moreover, we will measure the transfer time from the super browser to each browser.

References

1. Robocode. http://robocode.sourceforge.net/. Accessed 9 May 2019
2. Shichel, Y., Ziserman, E,. Sipper, M.: GP-Robocode: using genetic programming to evolve Robocode players. In: Proceedings of the 8th European Conference on Genetic Programming (EuroGP 2005), pp. 143–154 (2005). https://doi.org/10. 1007/978-3-540-31989-4_13

3. LiveScript (JavaScript) implementation of Robocode. https://github.com/youchenlee/robocode-js/. Accessed 9 July 2018
4. Kohana, M., Okamoto, S.: A data sharing method using WebRTC for Web-based virtual world. In: Proceedings of the 6th International Conference on Emerging Internet, Data and Web Technologies (EIDWT 2018), pp. 880–888 (2018)
5. CodeComba. https://codecombat.com/. Accessed 9 May 2019
6. WebRTC. https://webrtc.org/. Accessed 9 May 2019
7. Web Worker. https://html.spec.whatwg.org/multipage/workers.html#worker. Accessed 23 July 2018
8. WebSocket. https://tools.ietf.org/html/rfc6455. Accessed 23 July 2018
9. Transcrypt. https://github.com/JdeH/Transcrypt. Accessed 16 May 2019

Author Index

A

Aburada, Kentaro, 281, 305, 560, 570
Ajmar, Andrea, 223
Al-Hadhrami, Nasser, 212
Al-Hadhrami, Yahya, 212
Aoki, Satoshi, 669

B

Barolli, Admir, 604
Barolli, Leonard, 27, 37, 595, 604
Beránek, Jakub, 235
Boyer, Benoît, 292
Bylykbashi, Kevin, 37

C

Cha, ByungRae, 317
Chang, Ching-Lung, 487, 526
Chang, Chuan-Yu, 487, 526, 545
Chang, Chun-Cheng, 487
Chang, Chun-Hsiang, 86
Chen, Ching-Ju, 545
Chu, Chao-Ting, 508, 517

D

D'Amico, Carmine, 223
Dang, Quang Hieu, 325
Dechen, Tenzin, 413
Dubrulle, Paul, 223
Duolikun, Dilawaer, 15, 120
Duong, Tan Nghia, 325

E

Enokido, Tomoya, 3, 15, 120, 423, 447

F

Fan, Chih-Peng, 178
Fan, Yu-Cheng, 86
Fuji, Ryusei, 305
Fujisaki, Kiyotaka, 620
Fukutani, Ryoya, 731
Funabiki, Nobuo, 247

G

Gima, Kosuke, 447
Golasowski, Martin, 235
Gotoh, Yusuke, 650, 658
Goubier, Thierry, 223
Grita, Susanna, 223

H

Habuchi, Hiromasa, 341, 350
Haraguchi, Kazuki, 669
Haramaki, Toshiyuki, 62
Hasebe, Takayuki, 270
Hayashibara, Naohiro, 167
He, Jheng-Jhan, 178
Higashinaka, Naoki, 630
Hinatsu, Shun, 292
Hirayama, Daiki, 258
Ho, Chian C., 508, 517
Hsug, Yi-Chieh, 493
Huang, Pingguo, 96
Huang, Yao-Pao, 502
Huang, Yueh-Min, 545
Hussain, Farookh Khadeer, 108, 212

I

Iio, Jun, 715
Ikeda, Makoto, 37, 595

© Springer Nature Switzerland AG 2020
L. Barolli et al. (Eds.): NBiS-2019, AISC 1036, pp. 739–741, 2020.
https://doi.org/10.1007/978-3-030-29029-0

Imada, Shotaro, 154
Imama, 683
Ishibashi, Yutaka, 96
Ishida, Tomoyuki, 341, 350
Ishii, Hazuki, 423
Izumi, Kiyoshi, 705

K
Kagawa, Tsuneo, 62, 142
Kaiya, Kohei, 189
Kamada, Masaru, 683, 692
Kammabut, Kamolchanok, 475
Kanai, Atsushi, 413
Kanzaki, Akimitsu, 650
Kao, Wen-Chun, 247
Katayama, Tetsuro, 305, 570
Kattiyanet, Aunnop, 475
Kaveeta, Vivatchai, 465, 475
Kawano, Yoshihiro, 722
Ke, Junxing, 434
Khan, Shuraia, 108
Khwanngern, Krit, 465, 475
Kim, JongWon, 317
Kitamura, Tatsuya, 669
Kitanosono, Iku, 62
Kiyohara, Ryozo, 387
Kobayashi, Akio, 705
Kohana, Masaki, 683, 692, 700, 705, 731
Koyama, Akio, 189
Kuribayashi, Minoru, 247

L
Lee, Hao-Ting, 526
Li, Hao, 434
Li, Yen-Ju, 86
Lin, Chien-Chou, 487
Lin, Chih-Yang, 502
Lin, Min-Hui, 502
Lin, Wen-Shan, 508, 517
Louise, Stephane, 223
Lu, Yangzhicheng, 341

M
Maeda, Hiroshi, 630
Martinovič, Jan, 223, 235
Martinovič, Tomáš, 223
Martkamjan, Somboon, 465
Matsumoto, Shinpei, 247
Matsuo, Keita, 37
Mazume, Shinya, 154
Mentré, David, 292
Miyachi, Hideo, 369
Miyakwa, Akihiro, 341
Murakami, Takumi, 49

N
Nadamoto, Akiyo, 669
Nagao, Takashi, 641
Nagashima, Kaoru, 396
Nagata, Akira, 396
Nagatomo, Makoto, 281, 560
Nakai, Ryo, 350
Nakamura, Katsuichi, 396
Nakamura, Shigenari, 3, 15, 423, 447
Nakashima, Makoto, 154
Natwichai, Juggapong, 465, 475
Ng, Joseph K., 434
Nguyen, Duc Minh, 325
Niibori, Michitoshi, 692
Nishino, Hiroaki, 62

O
Ochi, Akihito, 630
Ogi, Tetsuro, 359
Ogiela, Lidia, 132, 137
Ogiela, Marek R., 132
Ogiela, Urszula, 137
Ohara, Seiji, 27, 604
Ohkura, Katsumi, 700
Ohtaki, Yasuhiro, 692
Oikawa, Takanori, 557
Okada, Yoshihiro, 258
Okamoto, Shusuke, 27, 692, 731
Okazaki, Naonobu, 281, 305, 560, 570
Okubo, Takao, 270
Oma, Ryuji, 15, 423, 447
Osborn, Wendy, 73

P
Park, Mirang, 281, 305, 560, 570
Park, Sun, 317
Pham, Hung Manh, 325

Q
Qafzezi, Ermioni, 37
Qian, Qin, 96

R
Rakchittapoke, Arakin, 465
Rakowsky, Natalja, 223
Rapant, Lukáš, 235
Ruedeeniraman, Natwadee, 595

S
Saito, Masashi, 387
Sakaji, Hiroki, 705
Sakamoto, Shinji, 27, 604, 731
Sakuraba, Akira, 377
Sato, Hiroyuki, 413

Sato, Keizo, 154
Savio, Paolo, 223
Schorlemmer, Danijel, 223
Scionti, Alberto, 223
Shi, Wei, 258
Shibata, Yoshitaka, 341, 350, 377
Shimizu, Koichi, 292
Shin, Byeong-Chun, 317
Shiratori, Norio, 305
Sitthikamtiub, Watcharaporn, 465
Sitthikham, Suriya, 465, 475
Slaninová, Kateřina, 235
Someya, Kazuki, 387
Song, Xing-Xiu, 581, 586
Sueda, Tomoyuki, 167
Sugawara, Shinji, 49
Sugisaki, Ryoa, 537
Šurkovský, Martin, 235
Suzuki, Hiroyuki, 189
Szturcová, Daniela, 235

T
Taenaka, Yuzo, 396
Takahashi, Yuuki, 413
Takano, Yasunao, 705
Takatsuka, Kayoko, 570
Takeishi, Ayaka, 413
Takenaka, Masahiko, 557
Takizawa, Makoto, 3, 15, 120, 137, 423, 447
Takumi, Ichi, 537
Tamura, Hitomi, 396
Tanabe, Nasanori, 641
Tanaka, Daisuke, 692
Taniguchi, Hideo, 641
Tanimoto, Shigeaki, 413
Tateiwa, Yuichiro, 96
Terzo, Olivier, 223
Than, Viet Duc, 325

Tiangtae, Narathip, 475
Tran, Trong Hiep, 325
Tsukamoto, Kazuya, 396

U
Uchida, Noriki, 350
Uchiya, Takahiro, 537
Ueda, Kazunori, 406
Ueda, Takeshi, 292
Umetani, Tomohiro, 669
Unno, Yuki, 557
Urueta, Steven H., 359
Usuzaki, Shotaro, 305, 570

V
Vuong, Tuan Anh, 325

W
Wangyal, Sonam, 413
Watanabe, Kazuki, 281, 560
Wint, Su Sandy, 247
Wu, Chien-Chung, 493
Wu, Jian-Shiun, 545

Y
Yajima, Jun, 270
Yamaba, Hisaaki, 305, 570
Yamakami, Toshihiko, 200
Yamamoto, Noriyasu, 459
Yeh, Chia-Hung, 502
Yelamandala, Chitra Meghala, 86
Yokoyama, Kazutoshi, 641
Yoshigai, Yuki, 620
Yoshihisa, Tomoki, 650

Z
Zhang, Liyang, 189

Printed in the United States
By Bookmasters